VOLUME ONE HUNDRED AND FIFTEEN

PROGRESS IN MOLECULAR BIOLOGY AND TRANSLATIONAL SCIENCE

Oligomerization and Allosteric Modulation in G-Protein Coupled Receptors

VOLUME ONE HUNDRED AND FIFTEEN

PROGRESS IN MOLECULAR BIOLOGY AND TRANSLATIONAL SCIENCE

Oligomerization and Allosteric Modulation in G-Protein Coupled Receptors

Edited by

TERRY KENAKIN

Professor, Department of Pharmacology,
University of North Carolina School of Medicine,
North Carolina, USA

ELSEVIER

AMSTERDAM • BOSTON • HEIDELBERG • LONDON
NEW YORK • OXFORD • PARIS • SAN DIEGO
SAN FRANCISCO • SINGAPORE • SYDNEY • TOKYO
Academic Press is an imprint of Elsevier

Academic Press is an imprint of Elsevier
The Boulevard, Langford Lane, Kidlington, Oxford, OX51GB, UK
32, Jamestown Road, London NW1 7BY, UK
Radarweg 29, PO Box 211, 1000 AE Amsterdam, The Netherlands
225 Wyman Street, Waltham, MA 02451, USA
525 B Street, Suite 1900, San Diego, CA 92101-4495, USA

First edition 2013

Library of Congress Cataloging-in-Publication Data
A catalog record for this book is available from the Library of Congress

British Library Cataloguing in Publication Data
A catalogue record for this book is available from the British Library

ISBN: 978-0-12-394587-7
ISSN: 1877-1173

Printed and bound by CPI Group (UK) Ltd, Croydon, CR0 4YY
Transferred to Digital Printing, 2013

CONTENTS

CONTRIBUTORS

Yves Audigier
Cancer Research Center of Toulouse, INSERM U1037—Université Paul Sabatier Toulouse III, Toulouse, France

Carline Chaves-Almagro
Cancer Research Center of Toulouse, INSERM U1037—Université Paul Sabatier Toulouse III, Toulouse, France

Peter Chidiac
Department of Physiology and Pharmacology, Western University, London, Ontario, Canada

Wendy Cladman
Department of Physiology and Pharmacology, Western University, London, Ontario, Canada

Van A. Doze
Department of Pharmacology, Physiology and Therapeutics, School of Medicine and Health Sciences, University of North Dakota, Grand Forks, North Dakota, USA

Frederick J. Ehlert
Department of Pharmacology, School of Medicine, University of California, Irvine, California, USA

Jesús Giraldo
Laboratory of Systems Pharmacology and Bioinformatics, Institut de Neurociències and Unitat de Bioestadística, Universitat Autònoma de Barcelona, Bellaterra, Spain

Karen J. Gregory
Vanderbilt Center for Neuroscience Drug Discovery; Department of Pharmacology, Vanderbilt University Medical Center, Nashville, Tennessee, USA, and Drug Discovery Biology, Monash Institute of Pharmaceutical Sciences, Monash University, Parkville, Victoria, Australia

David A. Hall
Fibrosis DPU, GlaxoSmithKline, Stevenage, Herts, United Kingdom

Tracy M. Handel
Skaggs School of Pharmacy and Pharmaceutical Science, University of California, San Diego, La Jolla, California, USA

Jeff Jerman
MRC Technology, Centre for Therapeutics Discovery, London, United Kingdom

Michael T. Klein
Department of Pharmacology, Vanderbilt Center for Neuroscience Drug Discovery, Vanderbilt University Medical Center, Nashville, Tennessee, USA

Evi Kostenis
Molecular-, Cellular-, and Pharmacobiology Section, Institute of Pharmaceutical Biology, University of Bonn, Bonn, Germany

Julio Martin
GlaxoSmithKline, Centro de Investigacion Basica, Parque Tecnologico de Madrid, Tres Cantos, Spain

Bernard Masri
Cancer Research Center of Toulouse, INSERM U1037—Université Paul Sabatier Toulouse III, Toulouse, France

Colleen M. Niswender
Vanderbilt Center for Neuroscience Drug Discovery, and Department of Pharmacology, Vanderbilt University Medical Center, Nashville, Tennessee, USA

Meredith J. Noetzel
Vanderbilt Center for Neuroscience Drug Discovery, and Department of Pharmacology, Vanderbilt University Medical Center, Nashville, Tennessee, USA

Dianne M. Perez
Department of Molecular Cardiology, Lerner Research Institute, The Cleveland Clinic Foundation, Cleveland, Ohio, USA

François-Xavier Picault
Cancer Research Center of Toulouse, INSERM U1037—Université Paul Sabatier Toulouse III, Toulouse, France

Magalie Rocheville
GlaxoSmithKline, Medicines Research Centre, Stevenage, United Kingdom

Bryan Stephens
Skaggs School of Pharmacy and Pharmaceutical Science, University of California, San Diego, La Jolla, California, USA

Hubert H.M. Van Tol
Department of Psychiatry, University of Toronto; Department of Pharmacology, University of Toronto, and Laboratory of Molecular Neurobiology, Centre for Addiction and Mental Health, Toronto, Ontario, Canada

Paige N. Vinson
Vanderbilt Institute of Chemical Biology, Vanderbilt University School of Medicine, Nashville, TN 37232, USA

Peishen Zhao*
Department of Physiology and Pharmacology, Western University, London, Ontario, Canada

*Current affiliation: Monash Institute of Pharmaceutical Sciences, Parkville, VIC, Australia

PREFACE

Seven transmembrane receptors (7TMRs) are nature's prototypical allosteric protein designed to read chemical information brought to the cell through the extracellular space and transmit this information to the cell cytosol to affect cellular behavior. As 7TMRs can adopt numerous conformations and as the make-up of these collections of conformations can be altered by ligand binding, there is a tremendous potential for chemical intervention into 7TMR behavior, much of which can be immensely useful for therapy. This volume offers the latest in thought on how receptors function and also, importantly, how ligands affect this function. This latter idea helps us determine how drug discovery efforts can be better designed to change receptor behavior for therapeutic gain.

Beginning with the general topic of allosteric drug discovery, Chapter 1, by Michael Klein, Paige Vinson, and Colleen Niswender provides an overview of the concepts related to the discovery of three types of allosteric drug: positive and negative allosteric modulators (PAMs and NAMs) as well as direct allosteric agonists. The advantages of these chemical drug types and the important issues involved with screening for these molecules and characterizing their activity are outlined. This theme continues with an in-depth discussion of metabotropic glutamate receptors, a prototypical allosteric receptor target highly relevant to central nervous system diseases in Chapter 2, by Karen Gregory, Meredith Noetzel, and Colleen Niswender. The application of allosteric molecules to the eight subtype receptors in this class forms a model for the therapeutic exploitation of allosteric mechanisms for diseases of the central nervous system. Finally, a link between allosteric therapeutics and *in vitro* phenotypic testing of allosteric molecules is given by Magalie Rochville, Julio Martin, Jeffrey Jerman, and Evi Kostenis in their discussion of the application of label-free assays (Chapter 3). The enormous power of this new technology is reviewed, as well as the unique "privileged niche" label-free technology has in providing enormously textured drug information from natural (and in some cases, directly therapeutic) systems is discussed.

From the discussion of allosteric drug discovery, the volume moves on to distinctly therapeutic aspects of allosteric mechanisms in Chapter 4, by Yves Audigier, François-Xavier Picault, Carline Chaves-Almagro, and Bernard Masri on G protein-coupled receptors in cancer. This is a new area of interest that begs to be exploited therapeutically as it involves a number of allosteric mechanisms employed by receptors in their control of cellular signaling. Similarly, the application of allosteric mechanisms such as oligomerization is reviewed in the new area of application of stem cells to drug discovery by Van Doze and Dianne Perez (Chapter 5). The regulation of cell signaling by 7TMRs through allosteric mechanisms is a new and exciting therapeutic strategy in the use of stem cells in therapy.

The volume then focuses on the receptor itself with the chapter by David Hall on the application of receptor theory to allosteric modulation of receptors (Chapter 6). This chapter illustrates the surprising versatility of a parsimonious two-state receptor model in describing allosteric function as opposed to a multistate model. In addition, the complex impact of receptor constitutive activity on allostery is discussed for the first time in published literature. The theme of receptor allosterism at the level of the receptor is continued by Frederick Ehlert in his discussion of receptor regulation of ligand-gated ion channels (Chapter 7). The theme of two-state versus multistate allosterism is progressed in this chapter through analysis of $GABA_A$ receptor gating constants and muscarinic M_2 receptors; the analysis shows surprising flexibility of a two-state model. A third chapter on allosteric receptor mechanisms, this time with special reference to membrane-directed receptor oligomerization, is provided by Jesús Giraldo (Chapter 8). In this chapter, three models of allosterism, namely, the two-state model, the metabotropic glutamate receptor model, and an asymmetric/symmetric three-state model, are used to describe receptor homo- and heterodimerization. This analysis reveals the potential complexity of intrareceptor association but also furnishes a solid theoretical basis for prediction and quantitative analysis. The theme of oligomerization is continued with Chapter 9, by Bryan Stephens and Tracy Handel, with a discussion of a prototypical receptor system for oligomerization and allostery, namely, the chemokine receptor. This chapter discusses how chemokine receptor dimers can demonstrate negative or positive allostery as well as provide completely new hetero-oligomer-specific signaling complexes that change not only the magnitude of agonist efficacy but also the quality of that efficacy. Finally, Chapter 10, by Peishen Zhao, Wendy Cladman, Hubert Van Tol, and Peter Chidiac, brings allosteric signaling into kinetic focus through a discussion of the importance of regulators of G protein signaling (RGS proteins) and G protein signaling modifier proteins (GPSM proteins).

In the early 1900s, receptors were thought to be "switches" on the cell membrane (as described by John Newport Langley (1852–1926)). With the ability to observe complex protein and cellular behaviors has come the knowledge that receptors are much more than switches and instead can be likened to protein microprocessors capable of receiving and processing many different types of signal and channeling those to many other different kinds of cytosolic receiver. The mechanism for this transmission is allosterism and through an understanding of this molecular process will come a better understanding of the actions of drugs on 7TMRs. The time-honored process of comparing experimental data to increasingly sophisticated models continues to move pharmacology forward; this volume presents the state-of-the-art models used in this process and how they are applied.

Terry Kenakin
Chapel Hill, NC 2012

CHAPTER ONE

Approaches for Probing Allosteric Interactions at 7 Transmembrane Spanning Receptors

Michael T. Klein[1,*], Paige N. Vinson[1,†], Colleen M. Niswender[*]
[*]Department of Pharmacology, Vanderbilt Center for Neuroscience Drug Discovery, Vanderbilt University Medical Center, Nashville, Tennessee, USA
[†]Vanderbilt Institute of Chemical Biology, Vanderbilt University School of Medicine, Nashville, TN 37232, USA
[1]These authors contributed equally to this work.

Contents

Progress in Molecular Biology and Translational Science, Volume 115
ISSN 1877-1173
http://dx.doi.org/10.1016/B978-0-12-394587-7.00001-4

Abstract

In recent years, allosteric modulation of 7 transmembrane spanning receptors (7TMRs) has become a highly productive and exciting field of receptor pharmacology and drug discovery efforts. Positive and negative allosteric modulators (PAMs and NAMs, respectively) present a number of pharmacological and therapeutic advantages over conventional orthosteric ligands, including improved receptor-subtype selectivity, a lower propensity to induce receptor desensitization, the preservation of endogenous temporal and spatial activation of receptors, greater chemical flexibility for optimization of drug metabolism and pharmacokinetic parameters, and saturability of effect at target receptors, thus improving safety concerns and risk of overdose. Additionally, the relatively new concept of allosteric modulator-mediated receptor signal bias opens up a number of intriguing possibilities for PAMs, NAMs, and allosteric agonists, including the potential to selectively activate therapeutically beneficial signaling cascades, which could yield a superior tissue selectivity and side effect profile of allosteric modulators. However, there are a number of considerations and caveats that must be addressed when screening for and characterizing the properties of 7TMR allosteric modulators. Mode of pharmacology, methodology used to monitor receptor activity, detection of appropriate downstream analytes, selection of orthosteric probe, and assay time-course must all be considered when implementing any high-throughput screening campaign or when characterizing the properties of active compounds. Yet compared to conventional agonist/antagonist drug discovery programs, these elements of assay design are often a great deal more complicated when working with 7TMRs allosteric modulators. Moreover, for classical pharmacological methodologies and analyses, like radioligand binding and the assessment of compound affinity, the properties of allosteric modulators yield data that are more nuanced than orthosteric ligand–receptor interactions. In this review, we discuss the current methodologies being used to identify and characterize allosteric modulators, lending insight into the approaches that have been most successful in accurately and robustly identifying hit compounds. New label-free technologies capable of detecting phenotypic cellular changes in response to receptor activation are powerful tools well suited for assessing subtle or potentially masked cellular responses to allosteric modulation of 7TMRs. Allosteric modulator-induced receptor signal bias and the assay systems available to probe the various downstream signaling outcomes of receptor activation are also discussed.

1. INTRODUCTION

7 Transmembrane spanning receptors (7TMRs) are proteins that represent the majority of drug targets to date.[1–4] These membrane proteins communicate information received via extracellular signals, such as

hormones, sensory stimuli, and neurotransmitters, to intracellular components by regulating signal transduction via heterotrimeric G-proteins and G-protein independent pathways (e.g., β-arrestins).[5,6] 7TMR proteins share a similar topology in that they possess an extracellular amino-terminus, seven transmembrane helices connected via three extracellular and three intracellular loops, and an intracellular carboxy-terminal tail.[5,6] Based on sequence homology and functional roles, 7TMRs are commonly divided into six main families,[7] A through F, with Families A–C being the most commonly studied. These receptors are distinguished by variation in their amino acid sequences and binding sites for endogenous ligands. For example, Family A 7TMRs contain a short N-terminus and generally interact with ligands via the 7 transmembrane spanning region. In contrast, Family B receptors are characterized by a longer extracellular domain, and ligands for these receptors bind within this region as well as the extracellular loop areas.[8] Finally, Family C receptors interact with ligands via a large N-terminal region, termed the Venus flytrap domain, which is a bilobed structure that "closes" when agonists bind.[9]

Endogenous ligands interact with their receptor partner via a site that is commonly termed the "orthosteric" site. The orthosteric ligands for Family A and C are generally small, and characterized by ligands such as neurotransmitters, hormones, and ions. In contrast, the orthosteric ligands for Family B 7TMRs are peptides, such as glucagon-like peptide (GLP-1). Within the 7TMR superfamily, there are smaller groups of related proteins that interact with the same endogenous ligand. Examples of these subfamilies include muscarinic and serotonergic receptors for Family A, the corticotropin and GLP receptors for Family B, and the taste and metabotropic glutamate receptors for Family C.[7] In many cases of highly conserved orthosteric binding sites, it has been very difficult to develop ligands that are specific for a given receptor. For example, in the case of the five muscarinic receptors, this lack of subtype selectivity has made it extremely difficult to probe the therapeutic potential of activating the predominantly CNS-expressed M_1, M_4, and M_5 receptors due to simultaneous activation of peripherally expressed M_2 and M_3 subtypes, resulting in severe side effects.

Currently, the majority of FDA-approved drugs that act at 7TMRs bind at the orthosteric site and regulate receptor function by directly stimulating a receptor response (agonist), blocking constitutive activity of the receptor (inverse agonism), or competing with binding of the native agonist (competitive antagonism). Orthosteric pharmacology is expected for most 7TMR marketed drugs, as the majority of these compounds were identified by employing radioligand binding assays targeting the orthosteric site. While there are certainly many therapeutically relevant 7TMR ligands that interact via the orthosteric site (e.g., haloperidol and other D_2 dopamine receptor antagonists for the treatment of psychosis or β-adrenergic receptor agonists and antagonists for the control of heart rate and blood pressure), the lack of receptor selectivity inherent to many orthosteric ligands has, for some time, thwarted exploration of the therapeutic potential of many of the 7TMRs.

Roadblocks in orthosteric ligand development have prompted the approach of exploring compounds that interact with 7TMRs at "allosteric," or "other" binding sites on the receptor to modulate receptor activity.[10,11] Binding of an allosteric modulator to a topographically distinct site induces conformational changes of the protein that can modify receptor activity in, most simply, a positive, negative, or neutral direction. Mechanistically, modulators can affect affinity of the orthosteric ligand, the efficacy of the orthosteric ligand, both, or bind to the receptor and not alter orthosteric agonist activity.[11] A modulator that changes the affinity of an orthosteric agonist does so by changing the association or dissociation rate (or both) of the orthosteric ligand. For efficacy modulation, the conformational change in the 7TMR upon allosteric ligand binding leads to a change in signaling capacity and thereby either facilitates or inhibits receptor coupling to downstream effectors. The functional potency of an allosteric ligand is a combination of its affinity for its allosteric site as well as the degree of cooperativity exerted with the orthosteric ligand. It is possible for a modulator to possess very weak affinity for its own binding site but induce strong functional responses due to high cooperativity; in contrast, some compounds may possess high affinity but exert only weak cooperativity with the orthosteric binding site.

Allosteric modulators of 7TMRs fall into several categories. Some compounds directly activate a receptor via an allosteric site; these compounds are termed allosteric agonists. Positive allosteric modulators (PAMs) and negative allosteric modulators (NAMs) are compounds that, respectively, increase or decrease the affinity or efficacy of an orthosteric ligand. In addition to PAMs and NAMs, there are also silent allosteric modulators (SAMs or neutral allosteric ligands) that block the activity of PAMs and NAMs but do not affect the response to the orthosteric agonist. SAMs have no effect on orthosteric ligand responses, but they can block the activity of PAMs and NAMs and are quite useful tool compounds to probe receptor function (discussed in more detail below). It is also possible for allosteric compounds to act as "partial antagonists," a mode of pharmacology that is impossible for orthosteric antagonists, as orthosteric ligands block agonist activity by a competitive binding interaction. "Partial" antagonists, in contrast, are compounds that can fully occupy their allosteric binding site but induce very weak, yet saturable, negative cooperativity with the orthosteric ligand.

In addition to improvements in target selectivity, allosteric modulators of 7TMRs and other proteins have a number of potential advantages in terms of drug development. Many compounds that interact via orthosteric sites are structural mimics of the endogenous ligand and are often small, polar amino acids, peptides, or more complex ligands that are not easily optimized in terms of human drug metabolism and pharmacokinetic (DMPK) parameters. In contrast, allosteric ligands are often small molecules with structures that are much more amenable to chemical optimization for drug candidates. Additionally, since the effects of an allosteric modulator are saturable, once allosteric sites are occupied, no additional response is observed. This ceiling effect greatly improves the potential safety of an allosteric modulator and has been used with clinical success in manipulation of other protein classes, such as the γ-aminobutyric acid A ($GABA_A$) chloride channel.[12] In this case, direct activation of the $GABA_A$ channel results in severe toxicity or death; however, the use of benzodiazepines, which potentiate the activity of GABA at these channels, results in a safe and effective clinical profile. Additionally, as orthosteric agonists may chronically activate a receptor, true PAMs require the presence of the endogenous ligand for activity and therefore maintain temporal and spatial activity of the endogenous agonist. It should be noted that a potential disadvantage of PAMs is the requirement for agonist; in neurodegenerative diseases such as Alzheimer's disease, for example, endogenous levels of acetylcholine must remain high enough for potentiation to occur. In this situation, a compound with allosteric agonist, or ago-PAM, activity may actually be preferred; the final profile of a modulator may differ depending upon the disease being targeted.

2. GENERAL CONSIDERATIONS FOR ASSAYS DESIGNED TO IDENTIFY AND CHARACTERIZE ALLOSTERIC MODULATORS

Due to potential advantages in selectivity, DMPK properties, saturable efficacy, and chemical tractability, recent years have seen an explosion in the development of allosteric modulators for 7TMRs. This is, in large part, due to the development of functional assays that permit identification of ligands that modulate a receptor without regard to binding site (although there are radioligands for some allosteric sites, the majority of *de novo* "screening" approaches rely on methods that do not require knowledge of the binding site, thereby broadening the search for compounds that modulate agonist activity via diverse mechanisms).

The majority of currently used functional methods for detecting the activity of allosteric modulators fall into two broad categories, kinetic assays and endpoint assays. Many of the considerations needed to move from a focus on orthosteric to allosteric ligands are similar for these two assay types. For example, one issue of critical importance is the choice of agonist used to profile the activity of allosteric compounds. While it is often assumed that compounds that compete with orthosteric ligands are binding to the exact same site on the receptor protein, and are, therefore, also orthosteric, this idea is somewhat naïve. Different "orthosteric" ligands may contact distinct amino acids or engender specific receptor conformations that affect the ability of an allosteric ligand to cooperatively modulate that agonist. There are now clear examples of allosteric modulators producing drastically different levels of cooperativity in the presence of distinct orthosteric agonists. This phenomenon has been termed "probe dependence"[13] and has implications for both the functional characterization of allosteric modulators and the determination of allosteric ligand structure–activity relationships (SARs). In extreme cases, probe dependence can result in a complete alteration in the mode of pharmacology or a loss of selectivity of an allosteric ligand. This has been shown convincingly using compounds with the M_4 muscarinic receptor (Refs. 14,15 and Fig. 1.1). As shown in Fig. 1.1, the M_4 muscarinic receptor PAM VU0152100 shows varying degrees of positive cooperativity with various orthosteric agonists, ranging from almost no potentiation of the agonist xanomeline, intermediate levels of potentiation for the endogenous ligand, acetylcholine, and strong potentiation of the agonist oxotremorine. Additionally, as previously shown for the M_4 PAM LY2033298,[14,15] the use of oxotremorine or xanomeline as agonists unmasks PAM LY2033298 activity of VU0152100 at M_2 muscarinic receptors. These results suggest that codosing or coapplication of these PAMs with oxotremorine or xanomeline would result in potential off-target activity at M_2 receptors, confounding data interpretation. An additional point regarding probe dependence is that there may be differences in probe dependence between species homologues of a given receptor; Suratman *et al.* showed that LY2033298 exhibits probe-dependent effects between rodent and human M_4 receptors.[16] Finally, there are many 7TMRs that have more than one endogenous orthosteric agonist. For example, the GLP-1 receptor responds

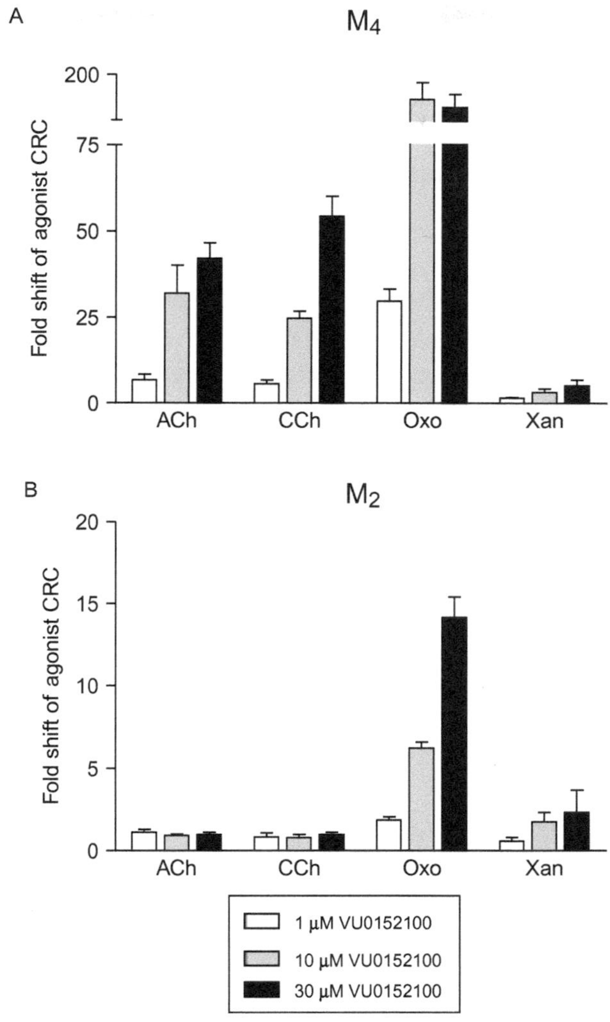

Figure 1.1 Probe dependence revealed in the presence of the M_4 muscarinic acetylcholine receptor PAM, VU0152100, and various orthosteric agonists. (A) Varying concentrations of VU0152100 (1 μM (white bars), 10 μM (gray bars), and 30 μM (black bars)) were applied 2.5 min prior to the addition of increasing concentrations of the orthosteric agonists acetylcholine (ACh), carbachol (CCh), oxotremorine (Oxo), or xanomeline (Xan) and calcium mobilization was measured in cells expressing the human M_4 receptor and the chimeric G-protein G_{qi5}. The potentiation activity of VU0152100 is clearly different in the presence of agonists other than the endogenous agonist, ACh, ranging from robust potentiation in the presence of Oxo versus almost no potentiation in the presence of Xan. (B) The selectivity profile of VU0152100 is altered in the presence of different orthosteric agonists, with the Oxo+VU0152100 and Xan+VU0152100 combinations showing potentiation at the M_2 receptor in calcium mobilization assay using cells coexpressing the human M_2 receptor and G_{qi5}. Data represent three independent experiments performed in triplicate (C.M. Niswender, previously unpublished data). Similar findings were previously reported in Refs. 14,15.

to at least two endogenous peptides, GLP-1(7–36) and oxyntomodulin[17] as well as other endogenous ligands, such as flavonoids.[18] It has recently been shown that a novel allosteric ligand potentiates the response to oxyntomodulin and not GLP-1(7–36).[17] These observations highlight the requirement for careful consideration in the choice of orthosteric ligands to assess the effects of allosteric modulators. For screening purposes, the agonist that is considered to be the primary endogenous orthosteric ligand should be the most likely choice if it is readily available and chemically stable during the assay employed for compound discovery and characterization. "Alternative" agonists should then be employed when more deeply characterizing an allosteric modulator of interest, particularly if coapplication (e.g., in electrophysiology experiments) or codosing (for *in vivo* assessment) of a distinct agonist/modulator combination is planned.

An additional point is warranted regarding differential signaling that may be induced by allosteric modulators. As receptor conformations may be unique when a 7TMR interacts with its endogenous agonist in the absence versus the presence of an allosteric modulator, the presence of the modulator may change the efficacy or types of signaling cascades that are engaged when the modulator is present. Additionally, the modulator may induce effects on its own that are unique from those generated when the receptor is engaged by its endogenous agonist. This concept has been termed functional selectivity, biased agonism, ligand-directed trafficking of stimulus, or stimulus bias.[19–23]

Historically, the study of 7TMR pharmacology has been dominated by radioligand binding and the measurement of a single downstream second messenger (i.e., the stimulation or inhibition of cyclic adenosine monophosphate (cAMP) accumulation for G_s and $G_{i/o}$ receptors, respectively, and phosphatidylinositol (PI) hydrolysis for G_q-coupled receptors).[6] With the discovery of the many other members of the heterotrimeric G-protein family, the identification of arrestin-mediated 7TMR signaling pathways, and a greater appreciation for the various cellular responses 7TMRs can mediate, the identification of stimulus bias has come to the forefront of 7TMR drug discovery.[10,24] The theoretical framework that supports the development of biased ligands is that certain signaling cascades may mediate greater therapeutic benefits than other signaling pathways or that some signaling pathways might be deleterious and cause unwanted side effects. This is clearly exemplified by the action of carvedilol and its superior properties as a therapeutic for the treatment of congestive heart failure.[25] Additionally, receptors transcribed from the same gene may be endogenously coupled to different signaling pathways in different cell types of the body, and the development of ligands that stimulate one pathway preferentially over another can provide for greater tissue specificity of drug effects.

Allosteric modulators have clearly been shown to bias the signaling of 7TMRs.[21,26,27] Biased PAMs or NAMs might be advantageous therapeutics if they respectively potentiate or inhibit the response of the endogenous agonist for one signaling cascade versus another. A particularly intriguing concept is the identification of allosteric modulators that may switch receptor signaling to pathways that the endogenous agonist normally does not signal through (e.g., switching the signaling of a G_s-coupled receptor to $G_{i/o}$). The range of responses that allosteric modulators could affect is immense and opens up 7TMR pharmacology to even greater diversity. Thus, it is prudent that the in-depth characterization of allosteric modulators be performed in more than one assay system and that these assays sample a diverse range of potential receptor responses, even those that are not normally mediated by the receptor of interest. Figure 1.2 demonstrates a situation where a hypothetical compound (compound "B") appears to be inactive in a typical assay for G_s-coupled receptors (i.e., the accumulation of cAMP) but shows robust potentiation of β-arrestin recruitment. In this situation, if only cAMP was measured, an experimenter who finds both compounds "A" and "B" to be active *in vivo* may erroneously conclude that the effect is not specific to the receptor of interest. Yet, with knowledge of other signaling pathways, concluding that the effect is nonspecific would be premature; moreover, these data would suggest a role for β-arrestin in mediating the effects observed *in vivo*, which may significantly impact drug discovery efforts going forward. With greater understanding of a new compound's properties *in vitro*, the more accurately *in vivo* pharmacodynamic properties can be predicted and the more rapidly compounds can be optimized *in vitro* and advanced through drug development programs.

In addition to the interaction of orthosteric and allosteric ligands in mediating preference for certain signaling pathways, there are also contributions of cellular proteins (G-proteins, β-arrestins, kinases, etc.) that can induce signal pathway-dependent modulation. These signaling proteins can also be considered "allosteric" regulators of orthosteric agonist function. This phenomenon has been termed "context dependence," indicating that the cellular environment can impact receptor, and compound-induced, pharmacology. This case has been demonstrated for the metabotropic glutamate receptor 7 antagonist MMPIP, which exhibits strikingly different efficacies and potencies when the receptor is expressed in distinct cell

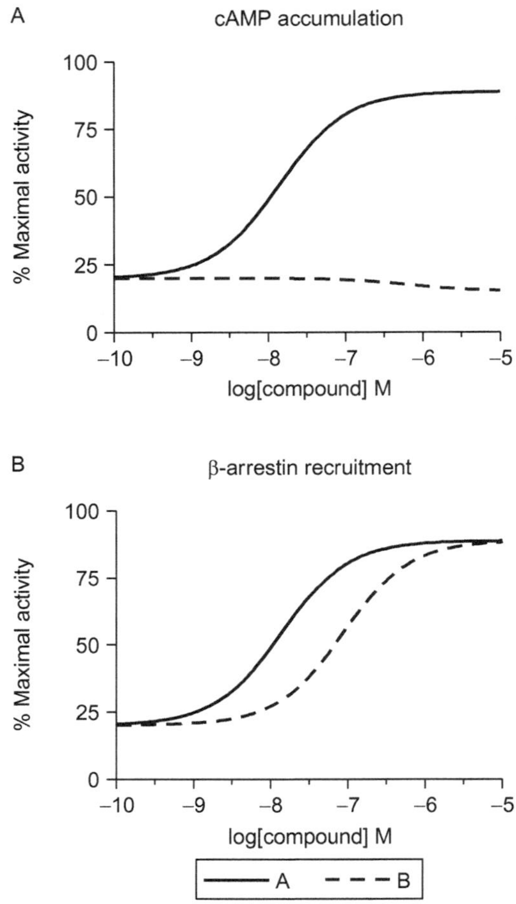

Figure 1.2 Allosteric ligands can show a range of efficacies as modulators of agonist activity. Displayed are CRCs for hypothetical compounds "A" and "B" that act at a G_s-coupled receptor. (A) One of the standard assays for measuring activity of G_s-coupled receptors is the accumulation of cAMP, and for these compounds, A appears to be a robust PAM at this receptor while compound B appears to be essentially inactive. However, in a β-arrestin recruitment assay (B), both compounds robustly potentiate the agonist's recruitment of β-arrestin to the receptor, indicating that compound B is an example of a β-arrestin-biased PAM. Having both datasets in hand allows for a more complete understanding of the pharmacology of individual compounds.

backgrounds, and this appears to translate to native tissues.[28] Additionally, the level of receptor expression can dictate the *in vitro* pharmacological profile of a compound, with compounds appearing as ago-PAMs when receptor expression is high and pure PAMs when receptor expression is decreased.[29] Therefore, employment of assays in a variety of cell backgrounds and with differential levels of receptor expression may be helpful in translating the *in vitro* profiles of allosteric modulators to the results observed in native tissues or *in vivo* models.

In addition to issues regarding the choice of agonist and the recognition of the potential for functionally selective signal transduction and pharmacological profiles, experimental design to assess the activity of allosteric modulators is also influenced by the goals of each particular experiment. For example, some studies are designed to screen large numbers of compounds for which pharmacology is, yet, unknown. Other experiments are required to provide a more quantitative aspect of allosteric interaction at the receptor. In general, experiments designed to detect PAM activity are typically designed such that the observed response to a low concentration of orthosteric agonist (e.g., a concentration eliciting a response that is 10–20% of the maximal agonist response, EC_{10} or EC_{20}) is enhanced. In contrast, for NAMs, the dampening effect on a submaximal concentration of orthosteric agonist (e.g., EC_{80} or EC_{90}) is measured. By definition, true neutral or "silent" allosteric modulators do not affect the response to the endogenous ligand; therefore, experiments designed to detect their activity must include the presence of a PAM or NAM that interacts at the same site on the receptor. These neutral modulators serve as highly useful tools when characterizing PAMs and NAMs at the site of interaction on the receptor, discussed in detail below, and may serve as valuable controls when investigating the effect of target engagement during native tissue or *in vivo* animal experiments.

3. GENERAL WORKFLOW USED IN IDENTIFYING AND CHARACTERIZING ALLOSTERIC MODULATORS

3.1. High-throughput screening

Most HTS campaigns designed to search for allosteric modulators are focused on the identification of PAMs, NAMs, or both. For HTS purposes, the design of the functional assay typically assesses compound activity at a single concentration (e.g., 10 μM), although some users of 1536-well technology may screen using multiple concentrations of a ligand to quickly understand if the effects are concentration dependent. As the hallmark of a strong HTS campaign is low variability and a robust signal window from day-to-day, signal windows (e.g., EC_{20} vs. EC_{max} response for PAMs or vehicle vs. EC_{80} response for NAMs) must be validated over multiple days during assay development. When searching for both PAMs and NAMs in a single HTS campaign, investigators may consider using an EC_{50} concentration of agonist; this may allow a wide enough signal window to detect both PAMs and NAMs but would also need to be experimentally validated. Additional parameters to be optimized include the timing of addition of the compound relative to the agonist. For example, coaddition may prevent receptor desensitization that can occur during the experiment. In contrast, ligands with slow association rates require more time to occupy their binding site and so some consideration should be given to allowing the receptor to incubate with allosteric compounds prior to addition of agonist. If reference compounds are available, these parameters can be more easily evaluated; often, however, such decisions are made by taking a conservative approach (i.e., a short preincubation period may be a good compromise).

Common screening platforms used for the search for allosteric ligands of 7TMRs are those that allow for multiple additions of compounds and agonists to a single well, enabling the detection of multiple modes of pharmacology within one experiment. While not always practical, many HTS-based assays for 7TMRs rely on kinetic plate reader technology (e.g., fluorometric imaging plate reader (FLIPR, Molecular Devices) or functional drug screening systems (FDSS, Hamamatsu)). It is possible to use these types of systems to perform multiadd assays, where compound can be added first, followed by a low (EC_{20}) and then a high (EC_{80}) concentration of orthosteric agonist.[30] Figure 1.3 demonstrates an example of a triple-add primary screening experimental design. In this experiment, glutamate responses in metabotropic glutamate receptor 5 ($mGlu_5$) expressing cells were assessed using a calcium mobilization assay (assay details and considerations are discussed in more detail in Section 5.2). A baseline read was taken for 3 s, followed by compound addition. An EC_{20} concentration of agonist was added at 144 s, and then an EC_{80} concentration at 241 s (solid line, panels A–C). In panel A, the vehicle response is compared to the response observed in the presence of the PAM VU0360172 (dotted line), a pure PAM. This compound elicits no effect alone (compound addition) but potentiates the EC_{20} glutamate response. In panel B, cells expressing a much higher density of $mGlu_5$ ($B_{max} = 25.2$ vs. 7.8 fmol/10^5 cells[29]) were employed for compound profiling; in this situation, VU0360172 induces a response alone in addition to potentiation of the EC_{20} response and would be categorized as an "ago-PAM." In panel C, the $mGlu_5$ NAM VU0405395 induces reductions in both the EC_{20} and EC_{80} responses.

It should be noted that different receptors may require distinct differences in the timing of addition of ligands when such a strategy is employed.

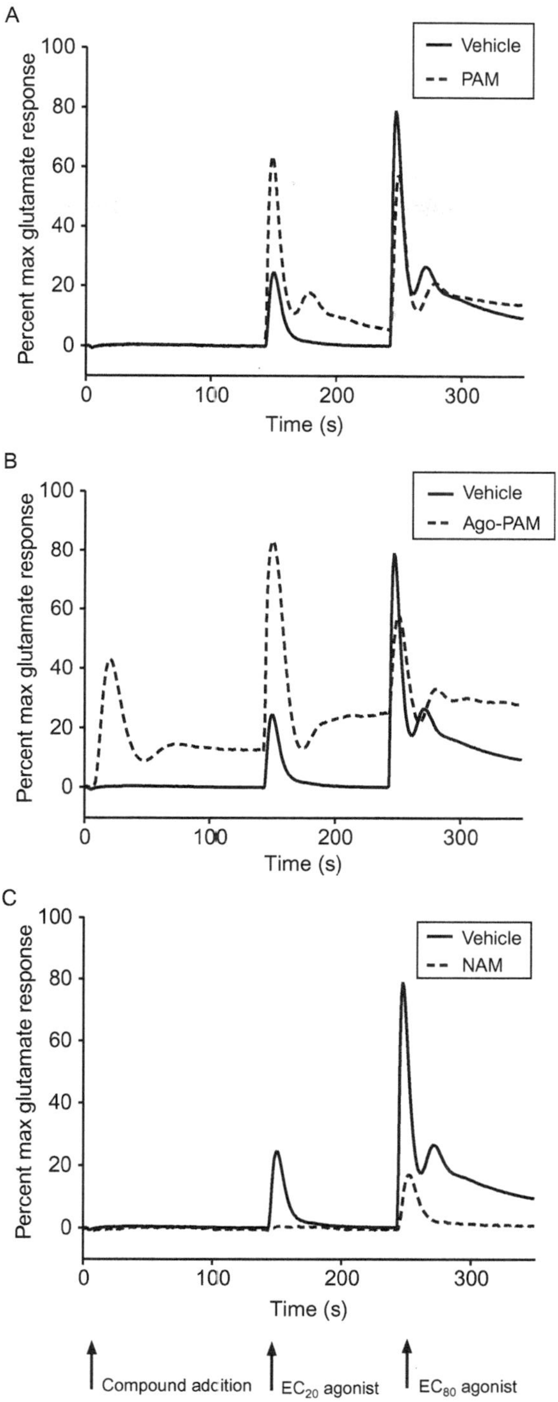

Figure 1.3 A triple-addition protocol allows for the detection of multiple modes of pharmacology in a single experiment. Calcium assay triple-add traces are shown for an $mGlu_5$ PAM (A), Ago-PAM (B), and NAM (C). The protocol consists of the addition

Addition of vehicle, generally DMSO, instead of test compound for a sample of wells is required for proper comparison and statistical calculations required for "hit" determination. In addition, a set of wells should be set aside for the addition of a maximally effective agonist concentration, allowing for response normalization. Criteria for hit selection may vary, but reliability is dependent on the signal window and variability of data for the high and low responses; screening statistics are generally performed either within or across plates for a given run, with acceptable signal windows generating Z' values > 0.5 with this parameter represented by Eq. 1.1)[31]

$$Z' = 1 - \frac{3\,\mathrm{SD\,of\,control\,(a)} + 3\,\mathrm{SD\,of\,control\,(b)}}{|\mathrm{mean\,of\,control\,(a)} - \mathrm{mean\,of\,control\,(b)}|} > 0.5 \qquad [1.1]$$

where control (a) and control (b) are the two boundaries within which the experimental response is being monitored, for example, the EC_{20} and EC_{max} responses for a PAM assay. These types of statistics provide confidence in the selection of hits for follow-up experiments.

3.2. Potency determinations

Primary screening follow-up or characterization during more advanced hit optimization usually requires the generation of CRCs for (1) confirmation of a concentration-dependent response and (2) the determination of *in vitro* potency and efficacy of the compound of interest. For allosteric modulators, these studies are generally performed by adding increasing concentrations of the allosteric ligand in the presence of a constant amount of agonist (e.g., either a low, EC_{20}, or high, EC_{80}, concentration of agonist for PAMs and NAMs, respectively). It is important to note that the parameters of the allosteric modulator response that are observed can be dramatically

of test compound followed by EC_{20} and EC_{80} concentrations of orthosteric agonist to detect agonist, PAM, and NAM activity, respectively. Data shown are expressed as percent of the response of a maximally effective glutamate concentration. (A) The $mGlu_5$ PAM VU0360172 was added to cells expressing a low density of $mGlu_5$ receptors and traces were compared to those of vehicle (vehicle, solid line; VU0360172, dotted line). (B) VU0360172 was added to cells expressing a high level of $mGlu_5$, invoking an Ago-PAM response (note response observed in the "compound addition" window). (C) The $mGlu_5$ NAM, VU0405395, was added to the same cell line as that in panel (A) and results in inhibition of both the EC_{20} and EC_{80} peaks. Each data set is from a single representative experiment (P.N. Vinson, previously unpublished data).

influenced by the agonist concentration (Fig. 1.4); therefore, it is important to maintain a tight tolerance on the range of responses observed with agonist alone. For example, one may place a range of 10–30% for a nominal EC_{20} addition of agonist and require that the agonist response be maintained within this range for appropriate comparison across datasets, especially for SAR interpretation. As can be seen in Fig. 1.4A, CRCs generated for a PAM of the metabotropic glutamate receptor 4 ($mGlu_4$) PAM, VU0400195, in the presence of increasing concentrations of the orthosteric agonist glutamate demonstrate dramatic differences in potency depending upon the glutamate concentration. For example, in the presence of a glutamate concentration that elicits no apparent effect alone (330 nM, −6.5, open triangles), VU0400195 weakly potentiates the receptor with a potency of 5 μM. Raising the glutamate concentration to 10 μM (−5, inverted triangles), a concentration that elicits a response that is approximately 80% of the glutamate maximal response, results in a leftward shift of the VU0400195 potency to approximately 120 nM, an almost 60-fold change.

Another issue that should be noted relates to the assay system chosen and the ability of the orthosteric agonist to induce a maximal response alone. This "%max" response is often tracked as it is a measure of compound efficacy, although it should be noted that the true efficacy of a compound may be underestimated due to a ceiling effect. Depending upon the assay, the orthosteric agonist may be able to elicit a response that is equivalent to the maximal response possible within that system. However, an orthosteric agonist may elicit a response that is significantly less than the maximal response capable of being achieved within that system or assay. An example of this is shown in Fig. 1.4. When $mGlu_4$ PAM activity is assessed in one assay, the response achieved in the presence of increasing concentrations of PAM is well above the "max" response elicited by glutamate alone (panels A and C, dotted line). In contrast, when assessed using another assay (panels B and D), the maximal response observed in the presence of PAM is the same as the maximal response elicited by glutamate.

Potency determinations are often used to support chemical optimization of allosteric modulators and in interpreting SAR. One interesting observation for multiple classes of allosteric ligands is the ability of ligands to "switch" modes of pharmacology with rather simple chemical modifications. For example, subtle changes to an $mGlu_5$ PAM scaffold that includes the compound DFB generates compounds ranging from NAM, PAM, to SAM activity.[32] Within another $mGlu_5$ scaffold, a "partial antagonist" was initially

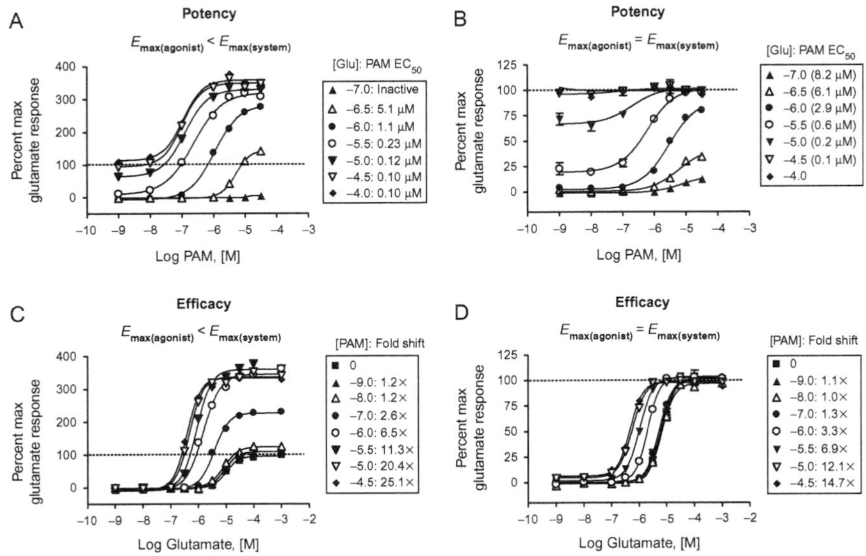

Figure 1.4 Profiles of allosteric modulators in various assays and coupling scenarios. (A and B) Potency data for the $mGlu_4$ PAM VU0400195 in the presence of different concentrations of the orthosteric agonist glutamate are shown for two assay systems. In a G_{qi5}-mediated calcium assay (A), a maximal concentration of glutamate does not induce the maximal effect possible within the system ($E_{max(agonist)} < E_{max(system)}$), shown as a dotted line for the maximal glutamate alone response. In a different assay system, $mGlu_4$-mediated activation of G protein inwardly rectifying potassium channels (B), glutamate induces a maximal response alone that cannot be further potentiated by PAMs ($E_{max(agonist)} = E_{max(system)}$); again, dotted line indicated the maximal glutamate response. Data for both assay systems reveal that there is a substantial leftward shift in the potency of the PAM at various glutamate concentrations. In the case of VU0400195, glutamate concentrations that induce no apparent effects alone can be potentiated by this PAM. For example, addition of 330 nM glutamate does not induce an apparent response in either calcium or GIRK assays, yet a concentration-dependent effect is observed in the presence of VU0400195 with a potency of 5–6 μM (white triangles). The presence of increasing concentrations of glutamate results in a dramatic 60-fold left shift in VU0400195 potency (e.g., addition of VU0400195 + 10 μM glutamate shifts the potency of VU0400195 to 0.1 μM, closed inverted triangles). Therefore, determination of PAM potencies can be dramatically affected by the concentration of orthosteric agonist used in the assay. (C and D). These efficacy experiments again show the ability of a PAM to increase the maximal response in one assay (C), but not in another (D). In both assays, PAM activity can also be observed as a leftward shift of the agonist concentration–response. Data are representative of three independent experiments performed in triplicate (C.M. Niswender, previously unpublished data).

identified via an HTS program.[33] This compound maximally inhibited 71% of the glutamate response with an $mGlu_5$ $IC_{50}=486$ nM. Introduction of a 3′-methyl group transformed this partial antagonist into a very potent full NAM ($mGlu_5$ $IC_{50}=7.5$ nM). When this methyl group was moved just one position, the resulting compound was a PAM with an EC_{50} of 3.3 μM; further optimization within this series of PAMs eventually resulted in the development of an $mGlu_5$ PAM with good *in vivo* activity. These results highlight the advantages of testing for multiple modes of pharmacology when supporting an allosteric modulator chemistry optimization effort.

3.3. Efficacy determinations

As stated in Section 3.2, a measure of a modulator's efficacy may be obtained from a CRC of the modulator in the presence of a constant concentration of agonist. However, this approach is limiting in that it is but a snapshot of the "strength" of an allosteric modulator. By measuring the shift of a CRC curve of the agonist in the presence of either a maximally effective concentration of modulator or a range of modulator concentrations (Fig. 1.4), one can measure the change in the EC_{50} of the agonist and express the change between the EC_{50} in the absence of modulator and presence of modulator. By expressing these two values as a ratio (EC_{50} in the absence of modulator/ EC_{50} in the presence of modulator), the resulting "fold shift" gives a quantitative expression of the modulator's efficacy with PAMs giving a fold shift of greater than 1 and NAMs a fold shift significantly less than 1. It should be noted that fold-shift ratios are normally distributed on a logarithmic scale, and interpretation of a PAM's or NAM's efficacy should take this into account. This approach can be applied using most functional assays where an effect of allosteric modulation is observed on the agonist response. Due to the noncompetitive nature between allosteric modulators and the orthosteric agonist, the agonist curves in the presence of NAMs can exhibit a decrease in the maximum response and/or a limited rightward shift (Fig. 1.5A), whereas competitive antagonists will shift the CRC in an parallel rightward fashion (Fig. 1.5B). These types of experiments can assist in confirming that a compound is acting via an allosteric mechanism of action and does not compete with the orthosteric site. Additionally, in certain assays, this type of strategy may allow for the detection of activity of the compound on its own, suggestive of allosteric agonism in that assay system (Fig. 1.5C).

If a SAM exists and has been demonstrated to interact at a specific site on the receptor (e.g., through radioligand binding competition assays), such a

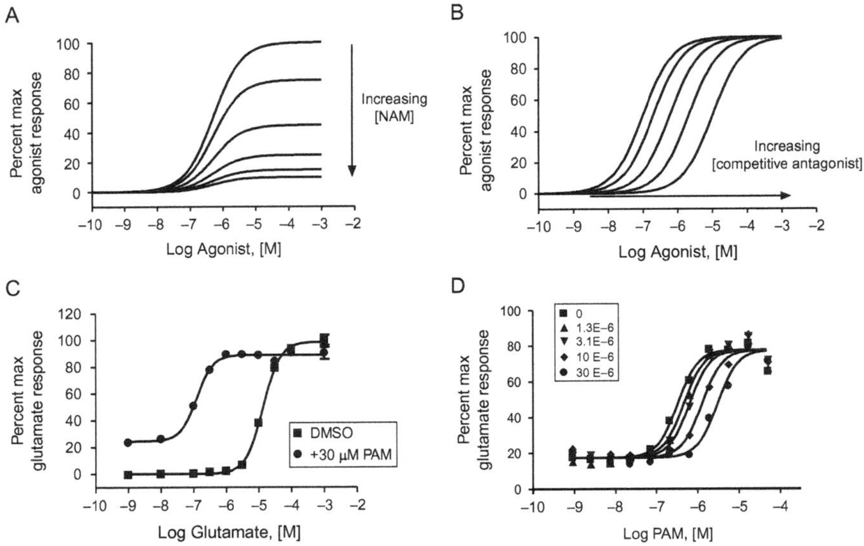

Figure 1.5 Examples of allosteric modulator profiles and use of SAMs as tools to determine ligand binding sites. (A) Simulation of the right-shift and depression of the E_{max} that is observed when increasing concentrations of a high efficacy NAM are applied with an agonist CRC. Note that the effect is saturable. (B) In comparison, this panel shows a simulation of the behavior of an antagonist that is competitive with the agonist. Note that the effect of the antagonist is surmountable as demonstrated by the return to maximal response by the agonist. (C) Example of a baseline effect observed in a fold-shift assay by a PAM that imparts a response independent of orthosteric agonist (DMSO control CRC curve, squares; curve in the presence of an ago-PAM, circles; note the increase in the response in the presence of low concentration of agonist. C.M. Niswender, previously unpublished data). (D) The effect of increasing concentrations of a neutral allosteric ligand (5-MPEP) on a PAM (VU0404211) CRC in a Ca^{+2} assay. 5-MPEP is known to interact at the MPEP site on $mGlu_5$ and may be used as a probe to determine if a PAM is competitive at the same site. Note the features of competitive behavior of the data that can be fit to the Gaddum/Schild equation to test this hypothesis (P.N. Vinson, previously unpublished data).

compound may be used as a probe in a functional assay to determine whether or not another modulator is interacting at the same site in a competitive manner. An example of this is the SAM 5-MPEP.[34] 5-MPEP induces no effect on the response to glutamate alone but completely displaces a radioligand at a site on mGlu5 known to interact with many allosteric ligands (the so-called "MPEP" site). Shown in Fig. 1.5D is an experiment in which increasing concentrations of 5-MPEP were preapplied to cells expressing $mGlu_5$. Approximately 2 min later, a CRC of PAM in the presence of an EC_{20} concentration of glutamate was added. The parallel rightward shift of the CRC indicates that

5-MPEP interacts competitively with the PAM being profiled, suggesting that this PAM interacts with $mGlu_5$ via the MPEP binding site. Had the PAM in question interacted at an alternate site, the curves would not have shifted in a parallel fashion, but would rather have decreased in maximal response as the concentration of 5-MPEP was increased (e.g., see Ref. 35).

4. DATA ANALYSIS: GENERAL FEATURES FOR ALLOSTERIC INTERACTIONS

While detailed methods are provided in other portions of this book, a brief discussion on analysis of functional data is warranted. The simplest way to examine experimental data with allosteric modulators is to simply fit CRC data or fold shift to a common four-parameter logistic equation:

$$Y = \text{Bottom} + \frac{\text{Top} - \text{Bottom}}{1 + 10^{(\log \text{EC}_{50} - X) \times \text{HillSlope}}} \quad [1.2]$$

where Y is the measured response, X is the logarithm of ligand concentration, Bottom and Top are the lower and upper bounds of a sigmoidal curve, respectively, EC_{50} is the concentration of ligand where 50% of maximal response is observed, and HillSlope indicates the Hill coefficient used to describe the steepness of the sigmoidal curve.

However, there are additional pieces of information, such as an estimate of the affinity of the modulator, which cannot be accommodated by this simple equation. While most mass action-based molecular models of allosteric modulation contain too many parameters to be fitted to real data, there are now modified methods that are useful in fitting experimental data using allosteric modulators and provide some quantitative aspects to the interactions of allosteric ligands with their targets, such as an estimate of modulator affinity. This can be done using an "operational" model of allosterism, which can be used to describe cooperativity and allosteric agonist effects.[13] The following equation describes the effects of an allosteric modulator on the signaling of an orthosteric agonist:

$$E = \frac{E_m(\tau_A[A](K_B + \alpha\beta[B]) + \tau_B[B]K_A)^n}{([A]K_B + K_AK_B + [B]K_A + \alpha[A][B])^n + (\tau_A[A](K_B + \alpha\beta[B]) + \tau_B[B]K_A)^n} \quad [1.3]$$

where E is the effect (i.e., measured response); [A] and [B] are the concentrations of orthosteric and allosteric ligand, respectively; K_A and K_B are the equilibrium dissociation constants of the orthosteric and allosteric ligands, respectively; α describes a cooperativity factor relating to the allosteric effect of each ligand on the other ligand's binding affinity; β, which ranges from zero to infinity, quantifies the extent to which the allosteric modulator modifies the efficacy of the orthosteric agonist at a given signal pathway; τ_A and τ_B relate to the ability of the orthosteric and allosteric ligands, respectively, to promote receptor activation (i.e., direct agonism) and are dictated by the intrinsic efficacy of each ligand, the efficiency of stimulus–response coupling, and the total density of receptors; E_m is the maximal possible system response (described as E_{max} in Fig. 1.3): and n is the slope factor linking occupancy to response.

This model allows an investigator to fit experimental data, and there are three key parameters that can be routinely derived from the model's application to full concentration–response and curve–shift relationships. These include the K_B of the allosteric modulator, which is an estimate of the affinity of the allosteric ligand; a composite cooperativity parameter, $\alpha\beta$, which provides information on the overall allosteric effect on the orthosteric agonist in the chosen functional assay (and, depending on the bias of the modulator for orthosteric agonist affinity vs. efficacy, can be close to the "fold-shift" value); and the modulator efficacy parameter, τ_B, which provides information on the ability of the allosteric ligand to promote agonism in the absence of orthosteric ligand. The estimate of affinity is particularly important, as many 7TMRs do not have radioligands that correspond to an allosteric site. Affinity estimates can be particularly helpful in determining which compounds might be candidates for radiolabeling or for potential development into positron emission tomography (PET) tracers, tools that are critical for relating occupancy of a target to *in vivo* efficacy and can help guide clinical development.

5. KINETIC ASSAYS TO MEASURE ALLOSTERIC INTERACTIONS AT 7TMRs

5.1. Fluorescence-based second-messenger assays

Identifying a general method for measuring specific target engagement is a highly desirable feature of an assay used in pharmacological research and drug discovery efforts. The introduction of probes that fluoresce upon binding to ionic species has met that need for many 7TMRs by measuring intracellular ion changes occurring through specific signaling pathways upon ligand binding to the receptor. This process occurs within seconds, enabling the tracking of changes in intracellular ion levels kinetically. Typically, one point corresponding to the maximum response is used to characterize the effect of a compound, but the kinetic readout allows for a more detailed assessment of the dynamics of the cellular response; multiple compound additions per well (e.g., a compound's agonist and PAM or NAM activity can be determined in a single well) can also be employed. These assays have become critical for HTS, lead optimization, and basic science investigation of 7TMR pharmacology. The application of these methods to allosteric modulators is described below and refers to plate-based formats and fluorescent imaging techniques using instruments such as the Hamamatsu FDSS or Molecular Devices FLIPR. However, lower end fluorescent readers also may be used for lower throughput and resolution.

5.2. Calcium mobilization, general principles

In general, monitoring Ca^{2+} mobilization (aka "calcium flux" assay) exploits the stimulation of phospholipase C by 7TMRs that couple to the G_q family of heterotrimeric G-proteins and the subsequent inositol triphosphate (IP_3)-stimulated release of Ca^{2+} from the endoplasmic reticulum into the cytoplasm. The binding of calcium to a preloaded fluorescent dye provides a means of monitoring the change in Ca^{2+} concentration in response to the cellular event. Fortunately, the development of chimeric G-proteins, such as G_{qi5}, and the identification of promiscuous G-proteins, such as $G_{\alpha 15}$ and $G_{\alpha 16}$, that can redirect the signaling of non-G_q-coupled receptors to the phospholipase C (PLC) pathway have enabled the application of this method beyond receptors that naturally couple to G_q.[36–38] It should also be noted that, depending upon how the assay is performed, 7TMR-mediated activation of cell surface calcium channels, often via the $\beta\gamma$ subunits of the heterotrimeric G-protein, can also contribute to the calcium signal induced via 7TMR activation. If a compound does possess functional selectivity, it is possible that some modulators may induce peculiarities in the calcium trace that may be indicative of differential modulation of these pathways and can be monitored for an early read on potentially functionally selective signaling events. Recently, Blatterman *et al.* have reported a compound that functionally uncouples the $G_{\beta\gamma}$ pathway from that of $G\alpha$[39]; therefore, the ability to monitor distinct parts of the G-protein signaling cascade could be useful in compound profiling.

Other interesting and, perhaps, informative features may exist within the raw data of a calcium trace, which may be the result of applied modulator. Some of these features include calcium oscillations, effect on a second agonist addition (i.e., in a "triple-add" design (Fig. 1.3), comparing the pharmacology of the EC_{80} peak to determine if there is a reduction, enhancement, or retention of the response at the control level), and calcium response at a plateau phase of the trace. Although the allosteric CRC curve is traditionally generated from the kinetic trace data by calculating the peak response, one could also extract and plot responses corresponding to a different point on the trace or simply qualitatively describe such features. Knowing the physiological relevance is the challenge; yet, with current data extraction, analysis, and storage techniques, it is possible to track these different features for *post hoc* comparison to a downstream assay either *in vitro* or *in vivo*.

5.3. Thallium flux, general principles

The influx of ions into the cell through ion channels may also be monitored using fluorescent probes. The discovery that thallium may be used as a surrogate for ions,[40] such as potassium, allows for the ability to use this ion to monitor the activity of potassium channels such as G-protein-coupled inwardly rectifying potassium channels (GIRKs), human ether-a-go-go, Kv7.4 (KCNQ4), K(ATP) channels, and renal outer medullary potassium channels.[40–51] As many $G_{i/o}$-coupled receptors modulate potassium channels, predominantly via the $G_{\beta\gamma}$ subunits of the heterotrimeric G-protein, this technique has been used as another kinetic assay to examine the activity of allosteric modulators targeting this class of receptors.[43] For cells expressing both $G_{i/o}$-coupled receptors and GIRKs, the release of the $G_{\beta\gamma}$ subunits upon 7TMR stimulation activates GIRK heteromers. By bathing the cells in buffer containing thallium and preloading the cells with a thallium-sensitive dye (e.g., BTC, AM, FluoZin™-2, or dyes contained within propriety kits such as FluxOR®, Invitrogen), a readout of receptor activity can be achieved. The kinetics of this ion flux is characteristically different from those of calcium mobilization (Fig. 1.6), and the conditions of the experiment are such that repeated additions of agonist may not be well tolerated, making addition of multiple boluses of thallium problematic. In this case, compound can be added in non-thallium-containing buffer and then the agonist can be coapplied with thallium. While data can be analyzed using stand-alone time points as in calcium mobilization experiments, we have found that measuring the slope within the first few seconds after the thallium +/− agonist addition results in a robust relationship from which to generate CRC data.[43] One additional note regarding thallium: because the compound is a heavy metal, handling and disposal must be performed according to local safety guidelines.

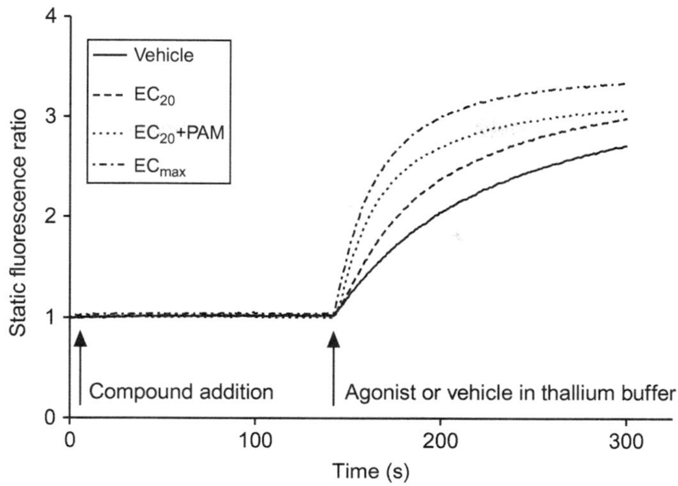

Figure 1.6 Characteristic traces observed from a thallium flux experiment. Shown are data collected during a thallium flux assay in hM_4 muscarinic receptor/GIRK-HEK cell line (P.N. Vinson, previously unpublished data). The M_4 PAM, VU0152100, was added to cells that had been incubated with the dye FluoZin TM-2 AM followed by an EC_{20} concentration of acetylcholine in buffer containing thallium. The response is measured as the slope of the linear portion of the curve after thallium addition. The vehicle response can then be subtracted, followed by normalization to the baseline-corrected EC_{max} of agonist, to normalize across experiments.

5.4. Fluorescent probe methodology

The process of performing a plate-based fluorescent probe kinetic imaging assay should be optimized for each cell line and agonist and detailed methods may be found in primary publications using this approach as well as reviews.[17,43,52–54] In general, the process begins with plating cells expressing the receptor of interest (recombinant or natively expressed) in black-welled, clear-bottomed plates (96, 384, 1536; poly-D-lysine-coated or noncoated as appropriate for adherent cells) at a density allowing for optimal response approximately 24 h later. It is important to know if other, natively expressed, proteins could possibly respond to the agonist used in the assay, resulting in confounding experimental results. One approach to test for off-target effects is to perform a parallel assay in the parental cell line that lacks the receptor of interest (if using a recombinant source) or to use a specific antagonist, if it exists. For the assay itself, the cell medium is typically removed and replaced with a buffer solution such as Hanks balanced salt solution buffered with HEPES. This may be performed manually or by using an automated plate washer. An optimized concentration of dye in buffer is added and incubated for a period of time (usually on a scale of approximately 1 h) to allow penetration of the dye into the cytosolic space. Note that many dyes require a surfactant such as pluronic acid to improve their distribution in an aqueous solution. Most dyes are designed to be activated by ester cleavage by naturally occurring esterases to produce a molecule with high affinity for the ion of interest. An additional consideration to maximize the dye concentration within the cell is to include the anion exchange inhibitor, probenecid, to the buffer used throughout the experiment. The dye-loading step is also amenable to manual or automated (e.g., bulk liquid dispenser) methods. To reduce background fluorescence, the buffer may be exchanged after dye incubation, although there are dye products where this step may not be necessary. We have observed that some assays benefit from the cells having a "settling time" of approximately 10 min after the second buffer exchange, resulting in a more stable response. The experimental readout on the fluorescence instrument begins by the acquisition of baseline followed by the collection of the readout resulting from compound and subsequent agonist additions. Each of these steps should be considered when introducing and optimizing a new assay of this type; any manipulation that occurs could have a detrimental effect on the cells' response and optimal dye concentration and incubation time and temperature may be different across cell lines. Many of these types of kinetic assays can be performed on lower end fluorescence plate readers with liquid-handling capabilities (e.g., Molecular Devices FlexStations), which lack the advantage of simultaneously adding compound and recording data from all wells on an assay plate, but come at only a fraction of the cost of an FDSS or FLIPR assay system. It should be noted, however, that because these fluorescence plate readers cannot read all wells simultaneously, different wells on the plate will be read at distinct time points after the addition of dye or compound and a drift in the assay signal may be observed that should be taken into account when analyzing data.

5.5. Alternatives to fluorescence technology

An alternative to permeable fluorescent dyes to detect changes in intracellular Ca^{2+} is the photoprotein aequorin (AEQ).[55] This protein, which originates from jellyfish, binds to calcium, resulting in oxidation of its cofactor, coelenterazine, thereby producing CO_2 and light emission, the latter of which is detected through luminescence. Aequorin is coexpressed with the receptor of interest (along with, if necessary, a PLC-coupling G-protein), with stable expression in the mitochondria providing a better signal compared to expression in the cytoplasm.[56] The assay design consists of adding coelenterazine followed by the test compound(s). Although this platform is gaining popularity for 7TMR Ca^{2+} signaling detection,[57–59] it has been applied only in limited cases for allosteric modulators.[60] The main advantage of this assay is the avoidance of fluorescence signal from test compounds, which can significantly confound data analysis where the fluorescence signal is expected to directly correlate with receptor activity, as is done for dye-based Ca^{2+} assays. It remains to be seen if this approach will be amenable to designs such the triple-add assay due to the AEQ assay typically applying cofactor-loaded cells to a plate of compound (agonist, antagonist, or modulator). Kits for AEQ luminescence assays include the transient transfection BacMam Aequorin Cellular Assay® kit from Invitrogen and stable cell lines produced by PerkinElmer under the product name AequoScreen®.

6. LABEL-FREE TECHNOLOGY

Label-free technology, so named because it does not require isolation of a specific signaling pathway and requires no biochemical techniques, such as the use of dyes or enzymatic reactions for measurement, is becoming increasingly popular for the study of allosteric modulator pharmacology. As these techniques are "unbiased" in terms of the binding site of a compound and in the signaling pathway induced, they represent a way to cast a wide net in terms of the pharmacology elicited by a modulator. Rather than reflecting effects of receptor activation on a single signaling pathway, these assays provide a measure of effects on a composite of different signaling responses to a 7TMR ligand, although certain pathways will dominate the kinetic trace that is measured.

While discussed in detail in another chapter in this volume, there are two main principles currently driving label-free techniques monitoring 7TMR activity in whole cells. The first of these is the cellular impedance technique, based on Ohm's law of voltage (V) = current (I) × resistance (Z). In this method, cells are layered onto a plate and a current is applied. The current can flow either between the cells or through them, resulting in a certain level of resistance, or impedance. When drugs are applied, the cells change shape due to intracellular cytoskeletal dynamics, resulting in a change in the impedance value from baseline. Examples of this technology are marketed in 96- and 384-well format as the CellKey® system from MSD Analytical Technologies and as the xCELLignce® from Roche in partnership with ACEA Biosciences, Inc. Impedance technology can detect signals from various 7TMRs with distinct "signatures" indicative of G_s versus G_q versus $G_{i/o}$-coupled receptors.[61–65]

An alternate technique to impedance technology involves dynamic mass distribution, or DMR; several readers are now available in 96-, 384-, and 1536-well format marketed as the Epic® (Corning, Incorporated) and the EnSpire® (PerkinElmer), which is based on the Epic® technology. This method relies on plating cells containing the receptor of interest onto a plate in which individual wells contain a resonance waveguide grating. Initial baseline readings are taken by illuminating the plate with a broadband light source and measuring the wavelength of the refracted light (Fig. 1.7A).

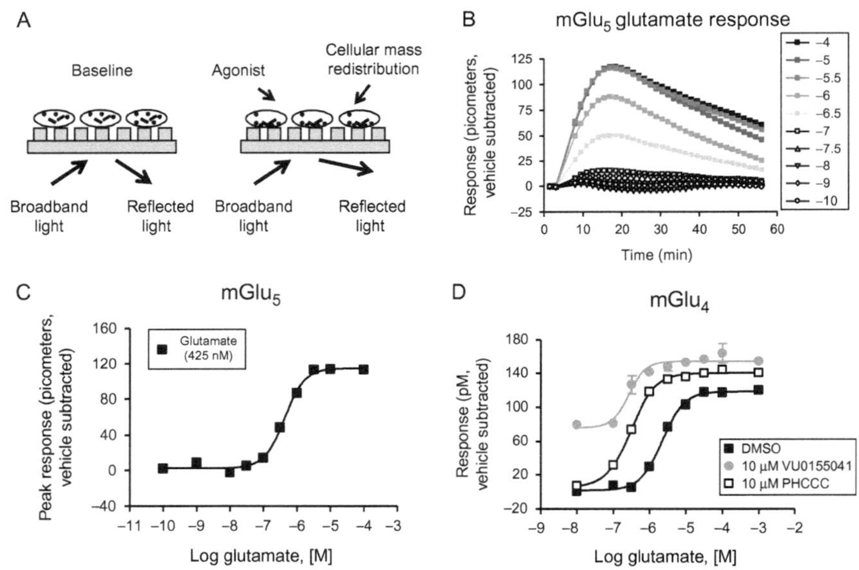

Figure 1.7 Principle of Epic® technology and examples of glutamate receptor pharmacology assessed using this method. (A) Dynamic mass redistribution principle. Cells are plated onto plates containing a resonance waveguide grating. Broadband light is shone onto the bottom of the place and then reflected. After agonist treatment, mass distribution of intracellular components occurs, resulting in a shift in the wavelength of the reflected light measured in picometers. (B) Example Epic® traces observed after the application of the agonist glutamate to cells expressing $mGlu_5$ after subtraction of vehicle responses. (C) The peak responses from panel (B) are plotted as a glutamate CRC. (D) Responses of glutamate alone and glutamate in the presence of two mGlu4 PAMs. In this case, PHCCC acts as a pure PAM, with no agonist activity as noted in other *in vitro* assays. In contrast, VU0155041 shows agonist activity in other *in vitro* settings (panel (C) of Fig. 1.5 shows data in the absence and presence of VU0155041 in a thallium flux assay); note the raised baseline in the Epic® data shown here. (C.M. Niswender, previously unpublished data).

After cell stimulation, mass redistribution of intracellular components occurs, resulting in a shift in the refracted wavelength. Responses are measured as a picometer wavelength shift from the initial recorded wavelength and can occur in either a positive or negative direction. These types of assays are proving useful for the study of 7TMR pharmacology.[28,39,65–82]

Using Corning Epic® technology, we have generated preliminary data profiling the signatures of some of the metabotropic glutamate receptors, plus and minus allosteric modulators. Shown in Fig. 1.7B are traces generated using the Epic® with HEK cells expressing $mGlu_5$. These traces have been normalized to the response that occurs in the presence of vehicle (a downward deflection of the traces). The maximum response induced in each of these traces can then be converted to a CRC (Fig. 1.7C), which results in a calculated potency for glutamate at $mGlu_5$ of approximately 425 nM, well within the range observed with this cell line in our laboratories in calcium mobilization assays.[29,34,53] In Fig. 1.7D is shown an experiment with cells expressing the related $mGlu_4$ receptor in which receptor activity has been measured in the presence and absence of the allosteric modulators PHCCC and VU0155041.[83–85] These compounds differ in that PHCCC functions as a "pure" potentiator, with no agonist activity observed in multiple *in vitro* assays and in slice physiology, whereas VU0155041 exhibits ago-PAM activity *in vitro*.[83–85] As can be seen in Fig. 1.7D, the agonist activity of VU0155041 is readily observed using Epic® technology (note the raised baseline similar to that observed in thallium flux assays, Fig. 1.5C) while PHCCC behaves as a pure PAM.

There are a number of advantages to the use of label-free technology. For example, no knowledge of the binding site is required, and functional readouts can be quite diverse, which may be useful for examining the activity of alternate signaling pathways as well as the pharmacology of "orphan" receptors for which signaling pathways and ligands may not yet have been identified. Both types of assays are capable of monitoring signaling induced by endogenously expressed, rather than overexpressed, receptors within cells. This allows for the ability to measure receptor activity in a cell background in which the receptor would normally be expressed, as well as assessing receptor modulation at potentially more physiological levels of expression. In terms of assessing functional selectivity, label-free technology adds an additional tool to enable an early read on compound profiles and potential differences in signal transduction. When performed in kinetic mode, a wealth of information is present in the traces, and "slices" of receptor activity can be taken at different time points, which may reflect distinct signaling events. The ability to read a plate and then return after some time has passed and take additional measurements may also allow for monitoring of very late receptor-mediated events, such as internalization or effects on new protein synthesis.

While label-free techniques have many advantages, there are a few points to consider. While the signals emanating from a receptor may represent a treasure trove of information regarding signaling, it can also be quite complex to tease out the contributions of various signaling events without systematic, detailed follow-up experiments. When used as a platform for HTS, it should be noted that the number of hits may escalate exponentially, as the systems are able of detecting any event that causes impedance or DMR changes, regardless of the specificity for the receptor being interrogated; carefully planned subsequent experiments (untransfected parental cells, alternate cell lines expressing the receptor of interest) are then required to understand which hits are desirable for further study. An additional factor is cost. Aside from the cost of the reader itself, plates containing the appropriate reagents to measure receptor-mediated signals are required and may be cost limiting for some investigators. Nevertheless, as label-free technology continues to evolve, it is anticipated to bring new opportunities for the study of receptor pharmacology, including that induced by allosteric modulators.

7. ENDPOINT ASSAYS

There are a number of compelling reasons for investigators to choose an endpoint assay for allosteric compound screening and characterization. The endpoint assays available for 7TMRs are numerous and diverse, and in many cases, these assays may be the only viable options for studying multiple signaling pathways for a particular receptor with reasonable throughput.[86] Many endpoint assays can be multiplexed with each other so long as the readout for one signaling pathway does not interfere with the readout of another (e.g., multiplexed luminescent and fluorescent readouts,[87] fluorescent readouts with nonoverlapping spectral windows). Additionally, many endpoint assays can be run using cells or tissue subsequent to a kinetic assay (e.g., kinetic readout of receptor-induced Ca^{2+} mobilization followed by the detection of IP_3 levels). These multiplexing options allow for multiple signaling pathways to be studied and can help in the assessment of allosteric modulator-induced signal bias. Some signaling pathways may be more sensitive to putative agonist or inverse agonist properties of allosteric compounds and with these multiplexing options, a compound's agonist or inverse agonist effects may be more readily uncovered.[16] Moreover, running kinetic assays for 7TMRs may not be feasible for all laboratories as they might require hardware that is not widely available (e.g., robotic and liquid-handling systems, FDSS or FLIPR systems, plate readers equipped with evanescent waveguide sensors), while many endpoint assays are designed to produce robust readouts detectable on fairly common equipment (e.g., luminometers, spectrophotometers) and can be hand-pipetted if necessary.

There are, however, important caveats that must be considered when working with allosteric modulators in 7TMR endpoint assays. These assays typically provide lower throughput for screening allosteric modulators than kinetic readouts, and end point assays are not as flexible as kinetic assays in terms of assessing multiple properties of a compound in a single assay run (e.g., the Ca^{2+} mobilization triple-add protocol discussed above that allows for the detection of agonist, PAM, and NAM/antagonist activity in a single well). A significant drawback of endpoint assays is the inability to rapidly adjust agonist concentrations during the assay run in order to achieve optimal agonist concentration for PAM and NAM assessments (i.e., agonist EC_{20} and EC_{80}, respectively). This can be particularly troublesome if the receptor being studied demonstrates a significant day-to-day variability in agonist potency or if the agonist CRC has a large Hill coefficient. One solution to this problem is to run several concentrations of agonist against test compound CRCs to hedge against missing the optimal EC levels (e.g., running EC_{10}, EC_{50}, and EC_{90} concentrations of agonist provides two concentrations, EC_{10} and EC_{50}, for the assessment of PAM activity and two concentrations, EC_{50} and EC_{90}, for that assessment of NAM activity). It may be advantageous to run a full matrix of agonist and allosteric modulator concentrations, which will yield both modulator potencies and agonist fold-shift data needed to assess the cooperative coefficients of allosteric compounds (see Eq. 1.3). Assays designed in this manner will provide a robust platform for measuring the properties of an allosteric modulator but will also limit throughput.

Novel compounds may exhibit unexpected effects *in vivo* or lack activity *in vivo*, thus it is important to study signaling pathways *in vitro* that are relevant to the *in vivo* situation to gain a better understanding of allosteric drug action. The high degree of biological relevance that many 7TMR endpoint assays offer makes these assays important components of allosteric drug discovery efforts. The measurement of cAMP and PI hydrolysis are classic endpoint assays used for the detection of 7TMR activity.[88,89] These techniques were established several decades ago and were used to tease apart the receptor–G-protein–enzyme (i.e., adenylate cyclase, phospholipase) system.[90,91] Measurement of cAMP accumulation and PI hydrolysis continues to be important endpoint assays for 7TMR drug discovery/tool development, but more recently, other endpoint assays have gained ground in this arena, including extracellular regulated kinase (ERK) phosphorylation[92–94] and arrestin recruitment.[95–97] The assay systems discussed below for the detection of cAMP accumulation, PI hydrolysis, ERK1/2 phosphorylation, arrestin recruitment, and transcriptional regulation will focus particularly on those methods that are (1) nonradioactive, thus requiring fewer safety precautions compared to older methodologies that rely upon radiochemicals, (2) applicable to HTS and robotic automation, (3) significantly less expensive than traditional techniques like radioligand binding, Western blot analysis, and microscopy, particularly when scaled to 384- and 1536-well plates, and (4) detectable on relatively common laboratory plate readers.

7.1. cAMP accumulation

Traditional methods of cAMP quantification like ion-exchange chromatography[88] and [^{3}H]cAMP protein kinase A (PKA) competition binding assays[98] require the use of radiochemicals, are laborious and low throughput, and come with a prohibitively high cost per data point for the screening of thousands of compounds in typical HTS campaigns or during follow-up chemical optimization and SAR support. Advances in molecular biology and the proliferation of high sensitivity plate readers have allowed for the development of nonradioactive, miniaturized cAMP assays capable of detecting sub-nanomolar levels of cAMP produced by intact cells, membrane homogenates, and tissue preparations. Two examples of such assay kits are the LANCE® cAMP/LANCE® ultra cAMP assays system (PerkinElmer) and the HitHunter® cAMP assay system (DiscoveRx). Several other assay systems are commercially available (e.g., CatchPoint® cyclic-AMP, Molecular Devices; HTRF® cAMP kits, Cisbio; AlphaScreen® cAMP Assay, PerkinElmer; and various cAMP enzyme-linked immunosorbent assay (ELISA) kits). In this review, the LANCE® cAMP and HitHunter® cAMP assay systems will serve as examples of affordable assay kits whose applications are not limited to proprietary or stably transfected cell lines.

The assay principle of the LANCE® cAMP kits is based on the disruption of a fluorescence resonance energy transfer (FRET) complex formed between a fluorescent europium (Eu) chelator linked to a biotinylated-cAMP molecule though a streptavidin–biotin interaction (or direct covalent linking of the Eu chelator with cAMP in the case of the LANCE® ultra version of this kit) and a fluorescently labeled anti-cAMP antibody. In the absence of unlabeled cAMP, the FRET complex readily forms and maximal FRET is achieved, but in the presence of cAMP, the FRET complex is disrupted by unlabeled cAMP competition with biotinylated-cAMP at the anti-cAMP antibody binding site, and FRET is significantly reduced. Therefore, the concentration of cAMP produced by adenylate cyclase is proportional to FRET output along a sigmoidal cAMP competition curve. The concept of the assay is similar to the classic PKA [^{3}H]cAMP competition binding methodology, in that increased cAMP production yields a reduction in the overall assay signal and the cAMP standard curve conforms to a sigmoidal competition curve. Due to the nonlinear nature of the cAMP standard curve, this assay system requires that FRET values be converted to cAMP levels before data are analyzed or normalized and requires that the investigator take care not to saturate the anti-cAMP antibody by using excessive amounts of cells or membranes, as this would yield inaccurate data. The LANCE® kit's time-resolved FRET (TR-FRET) readout is resistant to interference by autofluorescence of cells, membranes, test compounds, and the assay plate. This readout is achievable with most filter-based fluorescence plate readers with the appropriate filter sets and xenon flash lamp light source. The LANCE® cAMP kits are not well suited for HTS in cell monolayers; although cAMP levels can be assessed in cell monolayers, for use in HTS applications, adherent cells must be lifted and suspended.

Aside from the assay detection technology, the HitHunter® cAMP system differs from LANCE® cAMP in that increasing cAMP levels yield an increase in the assay signal, though the standard curve is still a sigmoidal one. In this assay, a cAMP tracer is covalently linked to a fragment of the beta-galactosidase enzyme (the enzyme donor, ED); ED-cAMP is captured by an anti-cAMP antibody and is not available to complement with the remaining fraction of beta-galactosidase (the enzyme acceptor, EA); thus, when the beta-galactoside substrate is added, no enzyme activity is present and chemiluminescence is not produced. In the presence of increasing levels of cAMP, the unlabeled cAMP competes with the ED-cAMP at the anti-cAMP antibody, allowing for increased levels of free ED-cAMP capable of complementing the EA fragment and robust chemiluminescence is produced. Like LANCE® cAMP and other competitive assays, the standard curve of this assay is sigmoidal and requires the conversion of raw luminescence units to units of cAMP for the reasons stated above. The potential to saturate the system with unlabeled cAMP is also present; thus, the optimization of cell number or membrane concentration is required. An advantage of the HitHunter® system is that adenylate cyclase activity can be easily measured in cell monolayers. Another advantage of this system is that the luminescent assay readout can be detected on most luminescence-capable plate readers, which are more prevalent and less expensive than the time-resolved fluorescence plate readers required for the LANCE® system. However, the HitHunter® system has lower sensitivity than the LANCE® cAMP kits.

With the LANCE® cAMP assay system, particularly low levels of cAMP can be quantified as the result of its higher affinity anti-cAMP antibody, accommodating the needs of primary cell culture or slow growing cell lines where measurement of subnanomolar levels of cAMP is required.[99] Because allosteric modulators may display different properties based on receptor density,[100] the ability to use cells from primary culture with endogenous levels of receptor is a significant advantage of these assay systems in allosteric drug development. The sensitivity offered by these assay systems is essential to investigating the effects of allosteric modulators, where the measurement of small changes in agonist potency is often necessary. Both assay systems are capable of measuring cAMP produced by homogenized tissue and membrane preparations, requiring only that an ATP regeneration system is present in the assay medium (e.g., Ref. 101).

7.2. PI hydrolysis

PI hydrolysis, like cAMP production, is a well-studied 7TMR signaling pathway, and, in recent years, several assay systems that are applicable to allosteric modulator HTS efforts have become commercially available to measure PI hydrolysis products. The general principle of this assay is that stimulation of G_q-coupled receptors activates PLC, which in turn hydrolyzes phosphatidylinositol 4,5-bisphosphate (PIP_2) to produce inositol 1,4,5-triphosphate (IP_3) and diacyl glycerol. IP_3 diffuses to the endoplasmic reticulum to activate IP_3-sensitive Ca^{2+} channels, releasing Ca^{2+} from the endoplasmic reticulum and increasing cytosolic Ca^{2+} levels. Thus, the measurement of IP_3 levels is a highly relevant biological endpoint for studying 7TMR activity; however, the half-life of IP_3 is very short compared to its downstream metabolite inositol 1-phosphate (IP_1), which accumulates in cells in the presence of lithium (usually LiCl).[102] Because of the relative ease of measuring IP_1 and the biological significance of IP_3, accumulation of both molecules is commonly used to assess G_q-mediated activity.

The only currently available plate-based assay for measuring IP_3 is the HitHunter® Inositol 1,4,5 Triphosphate assay system (DiscoveRx), which provides a reasonable degree of assay simplicity fit for automation of HTS campaigns and higher sensitivity than most other IP_3 detection methods. In this assay system, an IP_3-binding protein is used to slow the rotation of a fluorescent IP_3 tracer molecule; in this state, plane-polarized light is minimally disrupted and a high florescence polarization (FP) signal is detected. In the presence of high levels of IP_3, the unlabeled IP_3 competes for binding at the IP_3-binding protein, increasing the amount of unbound fluorescent IP_3 tracer molecules free to rotate more rapidly, reducing the FP signal. This assay system is in effect very similar to the LANCE® cAMP readout: increasing IP_3 correlates with a reduction in the assay signal along a nonlinear standard curve, thus necessitating the conversion of FP values to units of IP_3 amounts before data can be analyzed. The principle advantage of this assay system is its miniaturized format compared to traditional radioactive means of detecting IP_3. It should be noted that fluorescent compounds may interfere with FP signals in homogenous assay systems using this method for detection.[103]

The IP-One HTRF® assay system (Cisbio) is an example of a microtiter plate assay system for the detection of IP_1. Because IP_1 levels remain relatively stable over extended periods when in the presence of lithium, the measurement of IP_1 can be more readily assessed with greater sensitivity than IP_3 and lends itself better to HTS applications, particularly for allosteric drug discovery campaigns where assay sensitivity is at a premium. The IP-One HTRF® system provides a miniaturized detection method of IP_1 accumulation and is built on TR-FRET technology, which can be detected on any plate reader capable of detecting LANCE® and other HTRF® assays. The principle behind this assay system is again one of competition binding: IP_1 labeled with a FRET acceptor competes with cellularly produced IP_1 for binding to an anti-IP_1 antibody labeled with a FRET donor molecule. In the presence of low cellular IP_1, FRET acceptor-labeled IP_1 readily interacts with the anti-IP_1 antibody, producing a robust FRET signal; high levels of cellular IP_1 inhibit the formation of the FRET complex. This, like the LANCE® cAMP assay, produces a TR-FRET signal that inversely correlates with IP_1 levels along a nonlinear standard curve, necessitating the conversion of TR-FRET values to units of IP_1.

Both the HitHunter® IP_3 and IP-One HTRF® assay systems can be applied to native cells and primary cell culture; however, neither assay system is well suited for membrane preparations or tissue homogenates.

Considering the greater sensitivity of the IP-One HTRF® assay and the robust nature of TR-FRET readouts, which do not suffer significantly from autofluorescence and fluorescent test compounds,[104] the IP-One HTRF® assay system is better suited for allosteric drug HTS efforts, while IP_3 detection systems are best used for verifying the activity of small groups of compounds in follow-up assays.

7.3. ERK1/2 phosphorylation

Receptor stimulation of mitogen-activated protein kinase (MAPK) pathways has in recent years become appreciated as major signaling pathways for 7TMRs.[105] Various receptors coupled to G_q, $G_{i/o}$, and G_s have been found to stimulate the phosphorylation of ERKs (e.g., ERK1/2) either through G-protein-mediated mechanisms or through arrestin signaling.[94,95,105,106] With mounting evidence for the biological relevance of 7TMR-mediated ERK1/2 phosphorylation and the availability of plate-based assays, measurement of pERK1/2 is becoming a significant endpoint assay for drug discovery efforts, especially for allosteric modulators, which have been shown to exhibit pERK responses that significantly differ from classical assay systems.[15]

Traditionally, Western blot analysis has been used to assess levels of ERK1/2 phosphorylation (e.g., Ref. 107); however, this method is time-consuming, expensive, very low throughput, and difficult to accurately quantify. The most widely used microtiter plate-based assay system for measuring pERK accumulation is the AlphaScreen® SureFire® pERK1/2 (Thr202/Tyr204) assay system (PerkinElmer); this method addresses many of the issues that make Western blot analyses unsuitable for detailed characterization of novel compounds. This system was developed from the luminescent oxygen channeling assay (LOCI),[108,109] and, much like an ELISA, the principle behind the pERK1/2 AlphaScreen® system is one of immunocapture by an antibody with affinity for one epitope of the ERK molecule and immunocapture by a second phospho-specific antibody. The first antibody is tagged with an AlphaScreen® donor bead that, upon light excitation, produces a short-lived reactive oxygen species ($^1O_2$) that diffuse less than 200 nm before settling to a ground state[110]; if the phospho-specific antibody (tagged with an AlphaScreen® acceptor bead) is nearby, i.e. is bound to the same pERK molecule as the first antibody, the reactive oxygen can interact with the acceptor bead before it returns to a ground state and a chemiluminescent reaction is produced. Low levels of pERK will allow for the formation of very few of these complexes and little luminescence will be detectable; higher levels of pERK will allow for many of these complexes to form and the luminescent signal will increase. However, excessive levels of pERK will cause a reduction in luminescence, as few of the donor bead-conjugated antibodies will interact with the same ERK molecules as the acceptor bead-conjugated antibodies, producing a bimodal standard curve.[111] Fortunately, levels of pERK produced in most cells are low enough that this "hook effect" is not commonly observed, though one should still be aware of this caveat. The nature of this assay system and the low levels of pERK produced by most cells allow for a direct, essentially linear correlation between pERK1/2 accumulation and AlphaScreen® signal, thus no conversion of raw Alpha values to units of pERK is necessary. Fetal bovine serum contains a plethora of mitogens that stimulate a robust pERK response and can be used as a positive control for pERK accumulation and a control that experimental values can be normalized to for day-to-day comparisons. Just as with Western blot analysis of pERK levels, various cell types can be used in this assay, including primary cultured cells (e.g., Ref. 112). Being a plate-based assay system, AlphaScreen® pERK1/2 is better suited for the characterization of allosteric modulator potencies; assessing the effects of PAMs or NAMs on agonist EC_{20} and EC_{80} concentrations is not easily accommodated by Western blot analysis.

Several caveats accompany this assay system. Care must be taken not to mechanically disrupt cells during the compound treatment phase, as this will typically induce MAPK activity and confound results. Unlike cAMP and PI hydrolysis assays, no convenient inhibitors of the enzymes that dephosphorylate pERK exist, and depending on the cell background, maximal pERK production may occur within as little as 5 min with a precipitous drop-off afterward.[111] It is advisable to establish agonist time-courses to identify the optimal assay window for a particular receptor and cell type. pERK assays often cost more to run than other commonly used 7TMR assay systems; however, the higher cost may be warranted since detection of pERK levels relays highly relevant biological data that often sheds light on critical and unique properties of allosteric modulators that would otherwise go unobserved in other assays.

7.4. Arrestin recruitment

Many 7TMRs have the ability to signal through more than just the canonical G-protein-mediated pathways. Arrestins, which were originally identified as molecules that uncouple 7TMRs from their cognate G-proteins and were thus implicated in receptor desensitization,[113,114] have now been shown to mediate a variety of signaling cascades upon recruitment to receptors.[95,97] Activation of MAPKs has been highlighted as a major arrestin signaling pathway; however, no high-throughput assay systems are available that can clearly isolate arrestin-mediated signaling cascades from G-protein-mediated signaling cascades. Thus, several groups have devised methods to detect arrestin recruitment to various 7TMRs (e.g., Refs. 115–117); these methods, however, tend not to provide the throughput and large signal window required for HTS campaigns. Better suited to HTS needs are the two most common commercially available arrestin recruitment assays: the Tango™ GPCR Assay System (Life Technologies) and the PathHunter® β-Arrestin GPCR Assay system (DiscoveRx). Unfortunately, both systems rely upon proprietary cell lines expressing tagged 7TMRs and tagged arrestin molecules, which limits the flexibility of these assays.

The Tango™ GPCR Assay System is a gene reporter based proximity assay for arrestin recruitment to 7TMRs.[118] The receptor of interest is tagged with a viral transcription factor linked to the receptor via a peptide sequence containing a protease cleavage site. This receptor construct is expressed in a cell line co-expressing arrestin molecules tagged with a protease specific for the protease site of the receptor construct. Upon recruitment of arrestin, the protease site is cleaved and the viral transcription factor translocates to the nucleus and promotes the transcription of a beta-lactamase reporter gene. The activity of beta-lactamase is then measured by its ability to cleave a FRET substrate molecule, reducing the FRET signal. This assay, though highly engineered, does provide a robust readout of arrestin recruitment specific to the receptor of interest without cross-reactivity with other endogenously expressed receptors. Though this assay system can be automated and applied to HTS platforms, when planning multiple experiments, investigators should take into account the long incubation times required between compound addition and the detection phase (at least several hours). An advantage of this assay system over others is that it allows for imaging of cellular responses to the FRET substrate[118]; however, this is typically not a quality required for HTS efforts.

A more streamlined proximity assay for arrestin recruitment is the PathHunter® β-Arrestin GPCR Assay system (DiscoveRx). In this system, the receptor of interest is tagged with a portion of the beta-galactosidase enzyme and expressed in cells that also express β-arrestin2 tagged with the remaining portion of the enzyme. Upon recruitment to the tagged 7TMR, the enzyme fragments complement and with the addition of the beta-galactoside substrate, a chemiluminescent reaction is catalyzed that is readily detectable on most luminescence-equipped plate readers. Due to its luminescent readout, the signal-to-noise ratio of this assay is very high and provides Z' values suitable for most HTS programs.

Neither the PathHunter® nor Tango™ assay systems can be applied to primary cell culture or native (i.e., untagged) receptors, limiting the use of these assays to *in vitro* screening of compounds. A significant strength of these and other arrestin recruitment assays is that many 7TMRs interact with arrestin,[118] and the measurement of arrestin recruitment may be used as a primary measurement of receptor activity when the G-protein coupling of a receptor is unknown or not easily measured.[118] Additionally, arrestin recruitment is highly proximal to ligand binding and receptor conformational changes, thus little receptor reserve is typically detected in these assay systems, helping to simplify the interpretation of results. The investigation of allosteric modulators' abilities to affect arrestin recruitment to 7TMRs is still in its early stages and will undoubtedly yield compelling and crucial data for future allosteric drug development efforts.[21]

7.5. Transcriptional regulation

Many of the signaling cascades induced by 7TMRs ultimately impact transcription of a wide array of genes. Activation of 7TMRs has been linked to transcription regulated by CREB, STATs, SRF, NFkB, NFAT, Elk-1, c-Jun, and c-Fos. Of these, induction of c-Fos has become an important preclinical marker of antipsychotic efficacy of novel compounds. The exact mechanisms that relate c-Fos induction to antipsychotic action are not fully understood; it is known that c-Fos binds to c-Jun to form the AP-1 transcription factor, which induces transcription of numerous genes involved in cell proliferation, differentiation, cell damage, and apoptosis.[119] Drugs known to produce robust antipsychotic activity in humans (e.g., haloperidol, clozapine, olanzapine, risperidone) show a direct correlation of antipsychotic efficacy with their ability to induce c-Fos in striatal and prefrontal cortical regions of the brain.[120] *In situ* hybridization has been frequently used in the past to assess c-Fos expression, and this technique will continue to be vital to the investigation of psychoactive drugs, as the specific brain regions where c-Fos is induced is critical to behavioral outcomes. However, *in situ* hybridization is not amenable to the characterization of numerous novel compounds; to screen compounds for the ability of induce c-Fos, more efficient cell-based assays have been developed. One of the most commonly used plate-based, nonradioactive assay systems for determining c-Fos activation is the c-Fos EZ-TFA Transcription Factor Assay (Millipore).

The principle behind this assay is similar to an ELISA. A double-stranded oligonucleotide, containing the DNA-binding consensus sequence of the TPA response element (5′-TGAG/CTCA-3′), is used as a probe that binds c-Fos/c-Jun dimers present in nuclear extracts prepared from cells treated with drug or control cells. This oligonucleotide probe is biotinylated and, when transferred to a streptavidin-coated micro plate, the oligonucleotide-c-Fos/c-Jun complex adheres to the plate surface via a high-affinity biotin–streptavidin interaction. After a wash step, antibodies specific for c-Fos and c-Jun are added, and with successive washing and addition of a horse radish peroxidase secondary antibody, activated c-Fos levels can be determined. The assay readout is colorimetric and can be detected on a variety of common and inexpensive plate readers. A caveat of this assay system is the relatively laborious preparation of experimental samples, which includes the purification of nuclear extracts. This limits the application of this assay and others like it to follow-up experiments and later in-depth analysis of a novel compound's properties. These tradeoffs may be warranted if the induction of c-Fos is a critical marker for the development of a new drug. Most published investigations of c-Fos, and the induction of other transcription factors by 7TMRs for that matter, have not employed 7TMRs allosteric modulators. Considering the potential of allosteric compounds to induce these transcription factors, this area of research will be important for understanding the impact of allosteric compounds *in vivo* and in the clinic.

CREB (cAMP response element-binding) is another well-studied transcription factor that is clearly linked to the activity of 7TMRs. CREB can be activated through several pathways that are linked to 7TMRs. Cellular cAMP, a second messenger of G_s-coupled receptors and inhibited by $G_{i/o}$ receptors, mediates the activation of CREB by stimulating PKA, which then translocates to the nucleus and phosphorylates CREB. CREB can also be activated via phosphorylation by Ca^{2+}/calmodulin-dependent protein kinases, and can thus be linked to G_q-coupled receptor activity. Additionally, CREB can be activated through G-protein-induced MAPK signaling cascades.[121] Once phosphorylated, CREB binds the cAMP response element (CRE), driving the expression of hundreds of genes (e.g., tyrosine hydroxylase, c-Fos, brain-derived neurotrophic factor) and has been most commonly associated with the processes involved in learning and memory[122] but has also been implicated in depression, anxiety, and addiction.[123] Because of these observations, CREB activation is seen as an important outcome of 7TMR stimulation, and several assay systems have been developed to detect phosphorylated CREB. The AlphaScreen® SureFire® CREB (p-Ser133) Assay (PerkinElmer) is one such kit and is based on the same assay principle at the AlphaScreen® SureFire® pERK kit discussed above and comes with the same benefits and caveats of this assay system. There are also numerous ELISAs available for the detection of phosphorylated CREB, including those that use specific antibodies to capture phosphorylated CREB and those that use oligonucleotides containing the CRE-binding region to capture activated CREB. Both the ELISA methods and AlphaScreen kit present rather expensive approaches for detecting CREB activation in a plate-based format. Alternatively, gene reporter assays present a method for detecting CREB activity that is more amenable to HTS screening. Cells can be transfected with a vector that encodes an enzyme whose activity can be easily measured (e.g., luciferase, GFP) and is under the control of CRE. Compounds that activate CREB would thus result in an increase in the reporter gene's expression. This method is limited to recombinant cell lines, and as with any gene reporter assay, cells must undergo a prolonged incubation (several hours to overnight) before the activity can be measured.

8. RADIOLIGAND BINDING ASSAYS FOR ALLOSTERIC INTERACTIONS

Due to the reliance on binding assays during HTS campaigns before current functional assays were available, most small molecules discovered were competitive with the orthosteric radioligand or, at least, inhibited its binding to a measureable degree. In the case of allosteric (or putatively allosteric) ligands, binding assays are applied to more thoroughly characterize a compound's interaction with the receptor to better understand its affinity, site of interaction, and influence on endogenous ligand binding. This information may then be combined with functional data to more fully understand the ligand–target dynamic. The equilibrium dissociation constant of a compound can be derived from the law of mass action and describes the strength of association between the ligand and its compliment target (e.g., a small molecule and a 7TMR). This is expressed as

$$K_d = \frac{k_{off}}{k_{on}} = \frac{[\text{ligand}]\cdot[\text{receptor}]}{[\text{ligand}\cdot\text{receptor}]} \quad [1.4]$$

where K_d is the concentration of ligand that occupies half of the target at equilibrium, and k_{off} and k_{on} are dissociation rate (reciprocal time units) and association rate (reciprocal concentration-time units), respectively.

It is upon this relation that further equilibrium binding principles are built. This also assumes that the ligand(s) involved in the experiment does not bind at more than one target on the receptor. This basic equation holds true, given certain caveats, regardless of the site where the ligand binds.[124,125] However, as discussed below, more complex models are needed when describing systems where more than one site and/or ligand are being considered, as is often the case in studies involving allosteric modulators and orthosteric radioligands.

This section will focus on describing approaches used to determine binding characteristics for an allosteric compound at the target of interest using radiochemical techniques. Practical considerations and applications will be covered with references to other sources for detailed methods including the derivation of quantitative models of allosteric modulation that may be applied to binding data.

8.1. General methodology and initial considerations

8.1.1 Receptor source

Because 7TMRs are transmembrane proteins spanning between the extracellular and cytosolic space, cellular membranes must be isolated by established methods.[126] Alternatively, whole cell fractions may be utilized when there is concern of disrupting particular protein–protein interactions that may be necessary for proper receptor conformation; however, this type of preparation may produce higher variability and potentially different results from more purified membrane preparations due to the presence of different effectors (pharmacological and physiochemical).[127] As with any assay, consideration should be taken for the system used as the source of the receptor. If possible, choosing a system that is highly enriched in the target receptor, such as in inducible cell lines capable of generating high expression levels or in lines in which the receptor is highly constitutively expressed, will increase both yield and specific binding.

8.1.2 Choice of radioligand

The experiments designed for allosteric binding studies are dictated, to a large degree, by the availability of radioligand(s) for the target of interest. As the orthosteric sites of 7TMRs have been studied for significantly longer than the allosteric sites, radioligands for the former are more readily available. High-affinity radioligands for allosteric sites do exist and these have proven to be invaluable tools for probing PAM and NAM interactions.[128–132] The utility of any radioligand is limited by its affinity, such that weakly bound ($K_d > 10$ nM) radioligands will often not provide a strong enough interaction with the binding site to remain bound during post-binding assay steps that involve washing the membranes or separating bound versus free ligand. The development of allosteric radioligands is an active area for drug discovery as more allosteric molecules move through the discovery pipeline.[133] This is critical for both *in vitro* characterization assays and applications of ligands as PET tracers in preclinical and clinical studies. An example of such development is the application of [^{3}H]-methoxy-PEPy to probe one of the allosteric sites of mGlu$_5$. [^{3}H]-methoxy-PEPy is based on the biaryl acetylene scaffold of MPEP, which is a NAM at mGlu$_5$. Several useful tools have been generated from this chemical series, including the neutral allosteric modulator 5-MPEP (Fig. 1.5D) and the PET ligand [^{18}F]-FPEB.[134] As explained below, the results from a binding experiment should not be considered as stand-alone evidence of allosteric (or lack of) interaction but should be interpreted in conjunction with other assays.

8.1.3 Experimental conditions

Conditions such as temperature, membrane protein amount needed for optimal signal, radioligand K_d at the receptor being used, and minimum time to reach steady state should be known before beginning experiments with the allosteric modulator of interest. These (with the exception of radioligand K_d) can be determined through straightforward matrix experiments to ensure that the assay design is robust and resources are used wisely throughout the study. When combining binding results with those from functional studies, one should consider aligning the conditions of the two whenever possible, for example, buffer conditions, receptor source, and presence or absence of any component that may affect the compound–target interaction such as the presence of agonist. Note that the presence of the allosteric modulator itself may affect time to reach equilibrium; therefore, it should be considered that this parameter may have been affected when an experiment generates complex binding curves that do not fit to the allosteric ternary complex model (ATCM).[135,136]

9. COMPETITIVE OR NOT COMPETITIVE

Generally, binding experiments are performed to determine the affinity of a compound at its target. However, since the compound of interest is not standardly the radiolabel, one of the most common questions that binding assays are designed to address is to determine if the compound being studied is competitive with another ligand known to interact at a specific site of the receptor. If a compound appears to completely displace a radiolabel, the next calculation that is performed is to determine the affinity, or K_i value. Competition binding assays to study allosteric compounds are performed in the same way as those in which an orthosteric compound is the compound of interest; some differences exist, however, in the data analysis and interpretation.

Briefly, the experiment is performed under steady-state conditions with a known concentration of radioligand (typically around its K_d but not so low as to result in ligand depletion) and varying concentrations of the competing compound. The radioligand, unlabeled compound being studied, and membrane preparation (at a known, constant protein concentration) are combined and should be solvent-matched if the compound is dissolved in a nonaqueous solvent. A second, unlabeled compound at a concentration known to completely displace the radioligand from the target is included in a condition to determine nonspecific binding. In addition, the binding of the radioligand in the absence of competing compound is included to determine maximal (100%) binding. The experiment is mixed at a constant temperature until the predetermined time to reach steady state. Note that this time to equilibrium may differ in the presence of an allosteric modulator compared to without, and this should be considered during the experimental design phase.[136] The contents of the wells are then transferred to a filter plate and rapidly washed to remove unbound radiolabel; this last step is performed quickly using cold buffer to minimize dissociation of the bound radioligand. Once the filter plate(s) is dried (overnight is usually sufficient), scintillant is added and the plate is read using a microplate scintillation counter. The amount of radioligand present in the experiment should be determined empirically for each experiment by taking a count measurement of the radioligand-working solution added to the experiment and converting these counts using standard conversion factors and the instrument efficiency for the isotope used (for an example of this calculation, see http://graphpad.com/curvefit/radioactivity_theory.htm). Scintillation proximity assay beads may also be employed as a homogenous approach to perform a binding experiment, although this is not applied as commonly as the filter plate approach.

9.1. Allosteric radioligand competition

The availability of an allosteric radioligand typically depends on the site being well characterized such that there has been time for a high-affinity compound to be developed for that site. The choice of test compounds for these experiments will come from those that exhibit features of allosterism in upstream assays or are of a similar structural class as other known allosteric ligands for that target. In these cases, relatively straightforward analyses of the data using the law of mass action and standard curve fitting of the percent radioligand bound versus log[ligand] may be applied in which the log(IC_{50}) of the ligand is determined using nonlinear regression of a three-parameter equation as shown below. These experiments are treated the same as orthosteric radioligand–orthosteric compound competition experiments. The more complex situation where the radioligand and compound are not expected to bind to the same site is addressed in Section 9.3.

$$\%\,\text{Bound} = \text{Bottom} + \frac{\text{Top} - \text{Bottom}}{1 + 10^{(\log[\text{ligand}] - \log \text{IC}_{50})}} \quad [1.5]$$

In the above equation, "log[ligand]" values are the log of the molar concentration of the competing ligand (allosteric modulator) added to the experiment and the "% Bound" values are the corresponding values calculated from the experiment and are usually expressed as % Specific Binding using the total and nonspecific binding conditions. By knowing the concentration of the radioligand in the experiment and the K_d of the radioligand under the same experimental conditions, if behavior consistent with a competitive mode of action (full displacement) is observed, one may calculate the K_i (which is the more physiologically relevant parameter compared to the IC_{50}) of the compound of interest at the target site, which may be determined by fitting experimental data in a similar manner as above using the following equation

$$\%\,\text{Bound} = \frac{\text{Bottom} + (\text{Top} - \text{Bottom})}{1 + 10^{\log[\text{ligand}] - \log[10^{\log K_i} \times (1 + ([\text{radioligand}]/K_d))]}} \quad [1.6]$$

where [radioligand] is its molar concentration and K_d is the radioligand's equilibrium binding constant at the target. However, caution should be applied when concluding the competitive nature of a compound for a radioligand at a specific site. Confirmation of full competition (i.e., that the compound being studied and the radioligand have fully overlapping sites of interaction on the target) should be determined using an orthogonal assay that also measures the competition of one ligand versus the other, for example, saturation isotherms of radioligand at different competing ligand concentrations (with no change in B_{max} indicating a competitive interaction), lack of effect on rate of radioligand dissociation (see Section 9.3), or a functional assay using a neutral ligand that is known to bind to the same site as the radioligand (described above in Section 3.3 and Fig. 1.5D). The challenge with performing a functional assay for this kind of confirmation is assuring that it is being performed under steady-state conditions. For experimental results that do not show full displacement of the radioligand, the interpretation of this type of binding data may be qualitative in nature although there are approaches that use models specifically describing allosteric interactions such as ATCM, which will determine the affinity of the allosteric modulator independent of site of interaction of the radioligand as discussed in Section 9.2. Lastly, studies utilizing site-directed mutagenesis of the receptor are a common approach for shedding light on the location of allosteric modulator binding as well the effect on signaling.[15,35,137–143] It should be emphasized that knowing the nature of the interaction of the compound (in these examples, allosteric modulator) with the radioligand is not a purely academic pursuit. For drug discovery purposes, this knowledge is directly applied to the development of PET tracers that are used for confirming target engagement and determining receptor occupancy needed for therapeutic effect during preclinical and clinical studies. Although translation across species cannot be guaranteed, the *in vitro* characterization of a ligand is an important first step before moving into resource-intensive studies in whole animal and humans.

9.2. Orthosteric radioligand competition

Many orthosteric 7TMR radioligands are available as tools for competition binding studies. In these studies, increasing concentrations of allosteric modulator are incubated with a single concentration of radioligand to give qualitative confirmation that the modulator is indeed interacting at a site topographically separate from the orthosteric site due to its lack of full displacement of the radioligand (except in the case of a NAM with high cooperativity, which may displace 100%). If there is a directional effect (enhancement or decrement) on radiolabel binding, the mode of action of the allosteric modulator for the ligand being used as the radiolabel may be ascertained; an increase above "100% specific binding" denotes a PAM that acts through enhancing the affinity of the radioligand at the orthosteric site, while a decrease indicates that the modulator is a NAM (Fig. 1.8). A constraint of this interpretation is that this determined mode of action cannot be assumed to be the same across radioligands (an example of probe dependence discussed earlier). In addition, if the modulator has no effect on radioligand affinity, then there will be no change observed for the binding, lending no information to its role as a PAM or NAM. For quantitative analysis of allosteric modulator

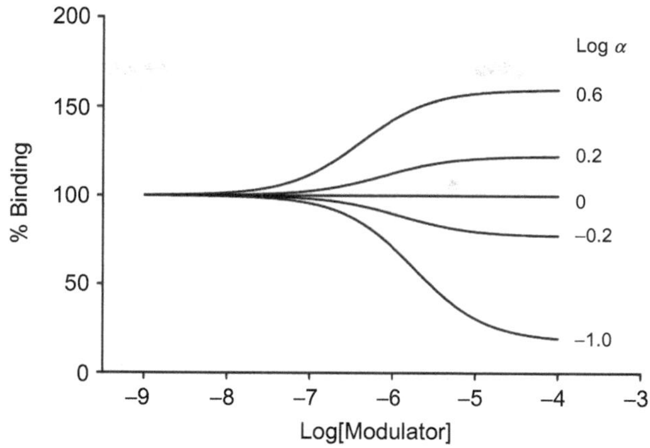

Figure 1.8 Simulation of the effect of different values of the allosteric modulator affinity parameter, α, on the binding of orthosteric radioligand using the ATCM. For the simulation, radioligand concentration = radioligand $K_D = 1$ nM and modulator affinity (K_B) = 10^{-6} M. Curves were generated with GraphPad Prism 5 using the preloaded allosteric modulator titration equation.

effect on orthosteric radioligand binding, the ATCM, which accounts for both allosteric and orthosteric binding and the effect of the former on the latter, may be applied.[144,145] Fitting the data to this model will allow determination of the affinity of the allosteric modulator and its cooperativity factor, α, for the radioligand, which is a measure of the degree that an allosteric modulator affects the binding of the orthosteric ligand ($0 < \alpha < 1$ for NAMs; $\alpha > 1$ for PAMs).[4] In the case where the modulator does not alter the affinity ($\alpha = 1$), no change in binding will be observed. Care should be taken not to interpret this as there being no allosteric modulation (especially if evidence has been found in previous functional assays); but, instead, that there is no allosteric modulation on the binding of the orthosteric compound being used as the probe in the experiment. The equation derived from the ATCM that may be used for fitting the experimental data can be found in graphing and fitting software such as GraphPad Prism 5 and is shown in a modified form below:

$$\%\text{Bound} = \%\,\text{Bound}_0 \times \left(\frac{[\text{radioligand}] + K_D}{[\text{radioligand}] + K_{\text{app}}}\right) \quad [1.7]$$

where

$$K_{\text{app}} = K_D \times \left(\frac{1 + \frac{[\text{ligand}]}{K_B}}{1 + \alpha\frac{[\text{ligand}]}{K_B}}\right) \quad [1.8]$$

In this equation, % $Bound_0$ is the amount of radioligand bound with no ligand (allosteric modulator) present; note that this is not termed "Top" as before since there may be some enhancement of binding under PAM application. K_{app} is the apparent K_D and is made up of the K_D of the radioligand modified by a factor consisting of modulator concentration, [ligand], the affinity cooperativity factor, α, and the dissociation constant of the modulator, K_B. Figure 1.8 shows examples of the effect of PAMs and NAMs with different α values on orthosteric radioligand binding using a simulation of the above equation. Note the effect not only on the change in amount bound but also on the left shift in the curve that occurs upon larger absolute values of log α.

One may also determine the effect of the allosteric modulator on an unlabeled orthosteric ligand by performing a standard competition binding assay using a single radiolabel concentration and titrating with cold orthosteric ligand of interest in the presence of increasing concentrations of allosteric modulator. This will show what, if any, effect the modulator has on the affinity of the unlabeled orthosteric ligand based on the shift imparted on the curve. This is useful when one is interested in the effect on an orthosteric ligand that is not labeled or is not suitable as a radioligand.[146] This effect may be expressed as a fold-increase on ligand affinity by the allosteric modulator for the purposes of confirmation of effect and relative comparisons across orthosteric and allosteric ligands. Similar to the application of the ATCM above, the model may be extended to account for more than one orthosteric ligand and the data sets fit globally to determine the parameters of allosteric modulator affinity and α values (modulator for cold ligand and modulator for radioligand) as detailed in Langmead (2011).[145]

9.3. Dissociation kinetic assays

Another primary means of using binding as a way to detect or confirm allosterism at a receptor is to observe the effect on dissociation kinetics of an orthosteric radioligand. When an allosteric compound binds to its target, the effect on the receptor can be detected by a change in the rate of association or dissociation of a bound orthosteric ligand if the conformational change impacts the orthosteric ligand's affinity. However, because a competitive compound would also have an effect on association due solely to its presence at the orthosteric binding site, this design should not be used as the first pass in a kinetic assay. Rather, a change in the k_{off} of a radioligand once it has reached equilibrium may be assigned unambiguously to engagement of the test compound at an allosteric site. The design and caveats of this assay are described elsewhere in detail.[136] Briefly, to test the presence and/or degree of allosterism of a test compound in relation to the site of the radioligand, a single concentration of radioligand is allowed to come to equilibrium at the receptor of interest (using a predetermined time). The test compound, along with a cold orthosteric compound at a high enough concentration to block radioligand reassociation, is added at different time points (including a zero-concentration compound condition) before experiment termination (i.e., harvesting). The remaining bound radioligand is measured and plotted as a function of time. Alternatively, one may use the "infinite dilution" design that may not be practical in some cases due to the volumes involved, but avoids any effect that the modulator may have on the unlabeled ligand used to prevent reassociation of the radioligand.[147] Equations describing monoexponential decay kinetics and the effect of one or more allosteric modulator concentrations on radioligand off rate may be combined (by substituting the right side of the latter for k_{offobs} in the monoexponential decay equation) and the data for one or more modulator concentrations fit simultaneously to solve for k_{off}, K_{offB}, and K_{b}/α (note that K_{b} and α cannot be resolved from each other using this approach).

$$\text{Bound}_t = \text{Bound}_{t=0} \cdot e^{-k_{\text{offobs}} \cdot t} \qquad [1.9]$$

$$k_{\text{offobs}} = \frac{\frac{k_{\text{offB}} \cdot [\text{B}] \cdot \alpha}{K_{\text{B}}} + k_{\text{off}}}{1 + \frac{[\text{B}] \cdot \alpha}{K_{\text{B}}}} \qquad [1.10]$$

In the above equations, Bound_t is an expression of the amount of radioligand bound at a specific time, $\text{Bound}_{t=0}$ is the amount bound at equilibrium (no allosteric ligand), k_{off}, k_{offB}, and k_{offobs} are the radioligand dissociation rate constant at the receptor in the absence of the modulator, for the receptor–modulator complex, and observed experimental rate constant, respectively. K_{B} is the modulator equilibrium dissociation constant, [B] is the modulator concentration, and α is the affinity cooperativity factor discussed earlier. As [B] approaches zero, $k_{\text{offobs}} = k_{\text{off}}$ as represented in the control curve that is generated. Likewise, an increase in the modulator concentration, affinity (i.e., lower K_{b}), and/or α (i.e., greater cooperativity) values brings the observed rate closer to the dissociation rate from the modulator-occupied receptor.

For a better understanding of the behavior that may be observed in this assay, consider the following scenarios. In one case, an allosteric modulator may not affect affinity of the orthosteric ligand, in which case $k_{\text{offB}} = k_{\text{off}} = k_{\text{offobs}}$, whereby no effect is seen and therefore a negative result is not proof of the ligand being orthosteric. This is once more an example of the necessity to perform more than one type of assay for determining allosterism. In a second example, consider two allosteric compounds that impart an identical k_{offB} to the orthosteric compound but have differing K_b/α ratios (Fig. 1.9). The compound with the lower K_b/α (greater affinity and/or greater affinity cooperativity) will have a larger effect on k_{offobs} at the same concentration. Although the use of association kinetics is not the best approach as an initial investigation of a putative allosteric compound, once the allosteric nature is corroborated by two or more assays, one may measure the effect on the orthosteric rate of association as a further characterization of the compound.[139]

This same principle can be applied using other means to detect the dissociating ligand. For example, the agonist ABA-X-BY630 was used as a fluorescent probe at human A1 and A3 adenosine receptors and was shown to be allosteric as observed by an increase in the dissociation kinetics upon addition of orthosteric agonists and antagonists.[148] The experimental design was different as described for binding, but, the same effect was observed to determine the allosteric nature of the compound of interest. Another unconventional study using dissociation kinetics to confirm allosterism was performed by Gomes *et al.*[149] to support the interaction between sites on the δ and μ opioid subunits of the corresponding heteromer.

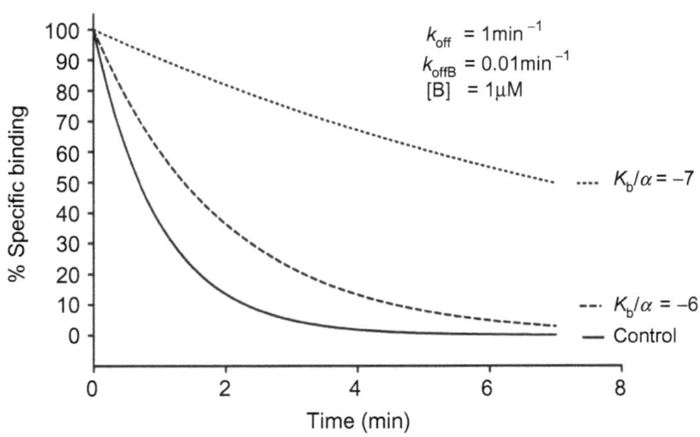

Figure 1.9 Simulation of the effect of the K_b/α parameter of an allosteric modulator on the observed dissociation rate, k_{offobs}, of an orthosteric radioligand. At a submaximally effective concentration and same k_{offB}, the modulator that has the greater absolute affinity:cooperativity ratio will impart a larger effect on k_{offobs}.

10. CONCLUSIONS AND FUTURE DIRECTIONS

As the allosteric modulator field has emerged, *in vitro* pharmacological approaches to detect and quantify their activity have rapidly developed. These techniques are now aiding in the translation of allosteric modulator pharmacology into clinical development. One area in which this is beginning to emerge for allosteric ligands of 7TMRs is in the development of PET ligands to detect target engagement.[134,150–158] The availability of a PET ligand that is truly competitive with a clinical candidate offers confidence that the relevant receptor site has been occupied; additionally, these findings can help relate occupancy to efficacy and aid in setting appropriate doses in the clinic. It should be noted that careful molecular pharmacological profiling of a test compound with a radioligand/PET ligand should be undertaken to ensure that the relationship between the two compounds is truly competitive. To this end, PET ligands unique to a given allosteric scaffold may be required for true competition, often necessitating new PET tracer development as different scaffolds progress. Additionally, because of the cooperativity, it may be much easier and straightforward to develop a NAM as a PET tracer when compared to a PAM. For NAMs, the affinity and potency of a compound are often closely aligned, whereas PAMs can show high degrees of positive cooperativity. Molecular pharmacology techniques (e.g., progressive fold-shift experiments) can aid in the determination of affinity estimates; compounds that saturate in their fold-shift values at low concentrations would be predicted to have higher affinities than those that continue to shift. This method can be used as one technique to aid in prioritization of compounds for labeling as tracers. Additional experiments with tools such as SAMs can further aid in understanding if an interaction between two compounds is truly competitive in nature.

Which assays are the best to use to find and characterize allosteric modulators? This question, of course, depends on the needs of the individual investigator. Label-free techniques certainly have the potential to cast a wide net and are being embraced for their ability to capture multiple, potentially unknown, signaling events. However, the cost and need for extensive secondary assays to eliminate nonspecific compounds and convolute signals captured in the kinetic trace information may not make them an ideal choice for some investigators if testing large numbers of compounds is required. Other notes from our own experience come from the ability of many allosteric modulators to "mode switch" during a chemical optimization program. To support a chemistry effort, the utility to screen multiple modes of pharmacology in a first pass experiment, such as that shown in Fig. 1.3 with multiple compound and agonist additions, is quite helpful. Additionally, experiments designed to definitively categorize a compound as allosteric, for example, by determining that a compound cannot compete for binding of an orthosteric radioligand, can be quite important when coupled with functional assays in terms of understanding compound profiles.

Functionally selective effects bring with them their own set of requirements. Certainly, label-free technology can act as an initial window into compound profiles, highlighting compounds that may engage distinct signaling events. The ability to measure multiple, discrete signaling events can be quite useful in "binning" compounds and then attempting to correlate *in vitro* profiles with *in vivo* efficacy. The emergence of the importance of G-protein independent events, such as β-arrestin signaling, suggests that assays to monitor these activities should be examined for a receptor of interest and to further profile allosteric modulator pharmacology.

A common question posed in small molecule allosteric modulators programs is: which parameter of an allosteric modulator is most important for *in vivo* efficacy? Unfortunately, the answer is not clear and may vary by 7TMR, allosteric binding site, allosteric ligand chemotype and therapeutic indication. In a chemical optimization program, potency is generally an important optimization parameter, but it should be noted again that efficacy (e.g., leftward fold shift of an agonist CRC) is balanced with affinity. A compound that shifts a curve 20-fold at 1 μM and then 70-fold at 10 μM would not be predicted to have high affinity; additionally, depending on pharmacokinetic parameters, it is not clear what levels of such a drug could even be achieved (i.e., is a fold shift of 70 relevant if the highest concentration of drug available is 1 μM?). A compound that shifts 10-fold at 1 and 10 μM has saturated in its effect, suggesting that it has completely occupied its binding site at 1 μM. This compound could be a better choice if higher affinity is required.

In conclusion, the number and variety of assays designed to detect and characterize allosteric modulators continue to develop at a rapid pace. The potential ability to correlate *in vitro* pharmacological and functionally selective effects with relevant *in vivo* efficacy or side effect profiles can be described as a new frontier of critical importance as basic science findings are translated into new tools and drug candidates. It is anticipated that new techniques, as well as evolution of existing methods and practices, will continue to be developed and refined by pharmacological community.

ACKNOWLEDGMENTS

The authors would like to acknowledge funding from the NIH (NS078262, NS31373 MH84659, MH093366), The Molecular Libraries Probe Centers Network, Seaside Therapeutics, Janssen Pharmaceuticals, Inc., and the International Rett Syndrome Foundation.

REFERENCES

1. Overington JP, Al-Lazikani B, Hopkins AL. How many drug targets are there? *Nat Rev Drug Discov* 2006;**5**:993–6.
2. Conn PJ, Christopoulos A, Lindsley CW. Allosteric modulators of GPCRs: a novel approach for the treatment of CNS disorders. *Nat Rev Drug Discov* 2009;**8**:41–54.
3. Bridges TM, Lindsley CW. G-protein-coupled receptors: from classical modes of modulation to allosteric mechanisms. *ACS Chem Biol* 2008;**3**:530–41.
4. Christopoulos A, Kenakin T. G protein-coupled receptor allosterism and complexing. *Pharmacol Rev* 2002;**54**:323–74.
5. Pierce KL, Premont RT, Lefkowitz RJ. Seven-transmembrane receptors. *Nat Rev Mol Cell Biol* 2002;**3**:639–50.
6. Lefkowitz RJ. Seven transmembrane receptors: something old, something new. *Acta Physiol (Oxf)* 2007;**190**:9–19.
7. Kolakowski Jr LF. GCRDb: a G-protein-coupled receptor database. *Receptors Channels* 1994;**2**:1–7.
8. Harmar AJ. Family-B G-protein-coupled receptors. *Genome Biol* 2001;**2**: REVIEWS3013.
9. Niswender CM, Conn PJ. Metabotropic glutamate receptors: physiology, pharmacology, and disease. *Annu Rev Pharmacol Toxicol* 2010;**50**:295–322.
10. Kenakin T, Miller LJ. Seven transmembrane receptors as shapeshifting proteins: the impact of allosteric modulation and functional selectivity on new drug discovery. *Pharmacol Rev* 2010;**62**:265–304.
11. Melancon BJ, Hopkins CR, Wood MR, Emmitte KA, Niswender CM, Christopoulos A, et al. Allosteric modulation of seven transmembrane spanning receptors: theory, practice, and opportunities for central nervous system drug discovery. *J Med Chem* 2012;**55**:1445–64.
12. Mohler H, Fritschy JM, Rudolph U. A new benzodiazepine pharmacology. *J Pharmacol Exp Ther* 2002;**300**:2–8.
13. Leach K, Sexton PM, Christopoulos A. Allosteric GPCR modulators: taking advantage of permissive receptor pharmacology. *Trends Pharmacol Sci* 2007;**28**:382–9.
14. Leach K, Loiacono RE, Felder CC, McKinzie DL, Mogg A, Shaw DB, et al. Molecular mechanisms of action and in vivo validation of an M4 muscarinic acetylcholine receptor allosteric modulator with potential antipsychotic properties. *Neuropsychopharmacology* 2010;**35**:855–69.
15. Valant C, Felder CC, Sexton PM, Christopoulos A. Probe dependence in the allosteric modulation of a G protein-coupled receptor: implications for detection and validation of allosteric ligand effects. *Mol Pharmacol* 2012;**81**:41–52.
16. Suratman S, Leach K, Sexton P, Felder C, Loiacono R, Christopoulos A. Impact of species variability and 'probe-dependence' on the detection and in vivo validation of allosteric modulation at the M4 muscarinic acetylcholine receptor. *Br J Pharmacol* 2011;**162**:1659–70.
17. Koole C, Wootten D, Simms J, Valant C, Sridhar R, Woodman OL, et al. Allosteric ligands of the glucagon-like peptide 1 receptor (GLP-1R) differentially modulate endogenous and exogenous peptide responses in a pathway-selective manner: implications for drug screening. *Mol Pharmacol* 2010;**78**:456–65.

18. Wootten D, Simms J, Koole C, Woodman OL, Summers RJ, Christopoulos A, et al. Modulation of the glucagon-like peptide-1 receptor signaling by naturally occurring and synthetic flavonoids. *J Pharmacol Exp Ther* 2011;**336**:540–50.
19. Mathiesen JM, Christopoulos A, Ulven T, Royer JF, Campillo M, Heinemann A, et al. On the mechanism of interaction of potent surmountable and insurmountable antagonists with the prostaglandin D2 receptor CRTH2. *Mol Pharmacol* 2006;**69**:1441–53.
20. Urban JD, Clarke WP, von Zastrow M, Nichols DE, Kobilka B, Weinstein H, et al. Functional selectivity and classical concepts of quantitative pharmacology. *J Pharmacol Exp Ther* 2007;**320**:1–13.
21. Digby GJ, Conn PJ, Lindsley CW. Orthosteric- and allosteric-induced ligand-directed trafficking at GPCRs. *Curr Opin Drug Discov Dev* 2010;**13**:587–94.
22. Violin JD, Lefkowitz RJ. Beta-arrestin-biased ligands at seven-transmembrane receptors. *Trends Pharmacol Sci* 2007;**28**:416–22.
23. Kenakin TP. Biased signalling and allosteric machines: new vistas and challenges for drug discovery. *Br J Pharmacol* 2012;**165**:1659–69.
24. Kenakin T. A holistic view of GPCR signaling. *Nat Biotechnol* 2010;**28**:928–9.
25. Wisler JW, DeWire SM, Whalen EJ, Violin JD, Drake MT, Ahn S, et al. A unique mechanism of beta-blocker action: carvedilol stimulates beta-arrestin signaling. *Proc Natl Acad Sci USA* 2007;**104**:16657–62.
26. Marlo JE, Niswender CM, Days EL, Bridges TM, Xiang Y, Rodriguez AL, et al. Discovery and characterization of novel allosteric potentiators of M1 muscarinic receptors reveals multiple modes of activity. *Mol Pharmacol* 2009;**75**:577–88.
27. Kenakin T. G protein coupled receptors as allosteric proteins and the role of allosteric modulators. *J Recept Signal Transduct Res* 2010;**30**:313–21.
28. Niswender CM, Johnson KA, Miller NR, Ayala JE, Luo Q, Williams R, et al. Context-dependent pharmacology exhibited by negative allosteric modulators of metabotropic glutamate receptor 7. *Mol Pharmacol* 2010;**77**:459–68.
29. Noetzel MJ, Rook JM, Vinson PN, Cho HP, Days E, Zhou Y, et al. Functional impact of allosteric agonist activity of selective positive allosteric modulators of metabotropic glutamate receptor subtype 5 in regulating central nervous system function. *Mol Pharmacol* 2012;**81**:120–33.
30. Weaver CD. Triple-addition assay protocols for detecting and characterizing modulators of seven-transmembrane receptors. *Current Protocols in Chemical Biology* 2011;**3**:119–40.
31. Zhang J-H, Chung TDY, Oldenburg KR. A simple statistical parameter for use in evaluation and validation of high throughput screening assays. *J Biomol Screen* 1999;**4**:67–73.
32. O'Brien JA, Lemaire W, Chen TB, Chang RS, Jacobson MA, Ha SN, et al. A family of highly selective allosteric modulators of the metabotropic glutamate receptor subtype 5. *Mol Pharmacol* 2003;**64**:731–40.
33. Sharma S, Rodriguez AL, Conn PJ, Lindsley CW. Synthesis and SAR of a mGluR5 allosteric partial antagonist lead: unexpected modulation of pharmacology with slight structural modifications to a 5-(phenylethynyl)pyrimidine scaffold. *Bioorg Med Chem Lett* 2008;**18**:4098–101.
34. Rodriguez AL, Nong Y, Sekaran NK, Alagille D, Tamagnan GD, Conn PJ. A close structural analog of 2-methyl-6-(phenylethynyl)-pyridine acts as a neutral allosteric site ligand on metabotropic glutamate receptor subtype 5 and blocks the effects of multiple allosteric modulators. *Mol Pharmacol* 2005;**68**:1793–802.
35. Hammond AS, Rodriguez AL, Townsend SD, Niswender CM, Gregory KJ, Lindsley CW, et al. Discovery of a novel chemical class of mGlu(5) allosteric ligands with distinct modes of pharmacology. *ACS Chem Neurosci* 2010;**1**:702–16.

36. Coward P, Chan SDH, Wada HG, Humphries GM, Conklin BR. Chimeric G proteins allow a high-throughput signaling assay of Gi-coupled receptors. *Anal Biochem* 1999;**270**:242–8.
37. Kostenis E, Waelbroeck M, Milligan G. Techniques: promiscuous Gα proteins in basic research and drug discovery. *Trends Pharmacol Sci* 2005;**26**:595–602.
38. Yokoyama T, Kato N, Yamada N. Development of a high-throughput bioassay to screen melatonin receptor agonists using human melatonin receptor expressing CHO cells. *Neurosci Lett* 2003;**344**:45–8.
39. Blattermann S, Peters L, Ottersbach PA, Bock A, Konya V, Weaver CD, et al. A biased ligand for OXE-R uncouples Galpha and Gbetagamma signaling within a heterotrimer. *Nat Chem Biol* 2012;**8**:631–8.
40. Weaver CD, Harden D, Dworetzky SI, Robertson B, Knox RJ. A thallium-sensitive, fluorescence-based assay for detecting and characterizing potassium channel modulators in mammalian cells. *J Biomol Screen* 2004;**9**:671–7.
41. Bhave G, Chauder BA, Liu W, Dawson ES, Kadakia R, Nguyen TT, et al. Development of a selective small-molecule inhibitor of Kir1.1, the renal outer medullary potassium channel. *Mol Pharmacol* 2011;**79**:42–50.
42. Lewis LM, Bhave G, Chauder BA, Banerjee S, Lornsen KA, Redha R, et al. High-throughput screening reveals a small-molecule inhibitor of the renal outer medullary potassium channel and Kir7.1. *Mol Pharmacol* 2009;**76**:1094–103.
43. Niswender CM, Johnson KA, Luo Q, Ayala JE, Kim C, Conn PJ, et al. A novel assay of Gi/o-linked G protein coupled receptor coupling to potassium channels provides new insights into the pharmacology of the group III metabotropic glutamate receptors. *Mol Pharmacol* 2008;**73**:1213–24.
44. Gentles RG, Grant-Young K, Hu S, Huang Y, Poss MA, Andres C, et al. Initial SAR studies on apamin-displacing 2-aminothiazole blockers of calcium-activated small conductance potassium channels. *Bioorg Med Chem Lett* 2008;**18**:5316–9.
45. Gentles RG, Hu S, Huang Y, Grant-Young K, Poss MA, Andres C, et al. Preliminary SAR studies on non-apamin-displacing 4-(aminomethylaryl)pyrrazolopyrimidine K(Ca) channel blockers. *Bioorg Med Chem Lett* 2008;**18**:5694–7.
46. Huang XP, Mangano T, Hufeisen S, Setola V, Roth BL. Identification of human Ether-a-go-go related gene modulators by three screening platforms in an academic drug-discovery setting. *Assay Drug Dev Technol* 2010;**8**:727–42.
47. Li Q, Rottlander M, Xu M, Christoffersen CT, Frederiksen K, Wang MW, et al. Identification of novel KCNQ4 openers by a high-throughput fluorescence-based thallium flux assay. *Anal Biochem* 2011;**418**:66–72.
48. Schmalhofer WA, Swensen AM, Thomas BS, Felix JP, Haedo RJ, Solly K, et al. A pharmacologically validated, high-capacity, functional thallium flux assay for the human Ether-a-go-go related gene potassium channel. *Assay Drug Dev Technol* 2010;**8**:714–26.
49. Titus SA, Beacham D, Shahane SA, Southall N, Xia M, Huang R, et al. A new homogeneous high-throughput screening assay for profiling compound activity on the human ether-a-go-go-related gene channel. *Anal Biochem* 2009;**394**:30–8.
50. Wojtovich AP, Williams DM, Karcz MK, Lopes CM, Gray DA, Nehrke KW, et al. A novel mitochondrial K(ATP) channel assay. *Circ Res* 2010;**106**:1190–6.
51. Zou B, Yu H, Babcock JJ, Chanda P, Bader JS, McManus OB, et al. Profiling diverse compounds by flux- and electrophysiology-based primary screens for inhibition of human ether-a-go-go related gene potassium channels. *Assay Drug Dev Technol* 2010;**8**:743–54.
52. Emkey R, Rankl NB. Screening G protein-coupled receptors: measurement of intracellular calcium using the fluorometric imaging plate reader. *Methods Mol Biol* 2009;**565**:145–58.

53. Rodriguez AL, Grier MD, Jones CK, Herman EJ, Kane AS, Smith RL, et al. Discovery of novel allosteric modulators of metabotropic glutamate receptor subtype 5 reveals chemical and functional diversity and in vivo activity in rat behavioral models of anxiolytic and antipsychotic activity. *Mol Pharmacol* 2010;**78**:1105–23.
54. Langmead CJ, Fry VA, Forbes IT, Branch CL, Christopoulos A, Wood MD, et al. Probing the molecular mechanism of interaction between 4-n-butyl-1-[4-(2-methylphenyl)-4-oxo-1-butyl]-piperidine (AC-42) and the muscarinic M(1) receptor: direct pharmacological evidence that AC-42 is an allosteric agonist. *Mol Pharmacol* 2006;**69**:236–46.
55. Rizzuto R, Pozzan T. Microdomains of intracellular Ca2+: molecular determinants and functional consequences. *Physiol Rev* 2006;**86**:369–408.
56. Stables J, Green A, Marshall F, Fraser N, Knight E, Sautel M, et al. A bioluminescent assay for agonist activity at potentially any G-protein-coupled receptor. *Anal Biochem* 1997;**252**:115–26.
57. Poul EL, Hisada S, Mizuguchi Y, Dupriez VJ, Burgeon E, Detheux M. Adaptation of aequorin functional assay to high throughput screening. *J Biomol Screen* 2002;**7**:57–65.
58. Gilchrist MA, Cacace A, Harden DG. Characterization of the 5-HT2b receptor in evaluation of aequorin detection of calcium mobilization for miniaturized GPCR high-throughput screening. *J Biomol Screen* 2008;**13**:486–93.
59. Jones B, Holskin B, Meyer S, Ung T, Dupriez V, Flores SY, et al. Aequorin functional assay for characterization of G-protein-coupled receptors: implementation with cryopreserved transiently transfected cells. *Anal Biochem* 2010;**400**:184–9.
60. Silvano E, Millan MJ, la Cour CM, Han Y, Duan L, Griffin SA, et al. The tetrahydroisoquinoline derivative SB269,652 is an allosteric antagonist at dopamine D3 and D2 receptors. *Mol Pharmacol* 2010;**78**:925–34.
61. Peters MF, Knappenberger KS, Wilkins D, Sygowski LA, Lazor LA, Liu J, et al. Evaluation of cellular dielectric spectroscopy, a whole-cell, label-free technology for drug discovery on Gi-coupled GPCRs. *J Biomol Screen* 2007;**12**:312–9.
62. Peters MF, Vaillancourt F, Heroux M, Valiquette M, Scott CW. Comparing label-free biosensors for pharmacological screening with cell-based functional assays. *Assay Drug Dev Technol* 2010;**8**:219–27.
63. McGuinness RP, Proctor JM, Gallant DL, van Staden CJ, Ly JT, Tang FL, et al. Enhanced selectivity screening of GPCR ligands using a label-free cell based assay technology. *Comb Chem High Throughput Screen* 2009;**12**:812–23.
64. Verdonk E, Johnson K, McGuinness R, Leung G, Chen YW, Tang HR, et al. Cellular dielectric spectroscopy: a label-free comprehensive platform for functional evaluation of endogenous receptors. *Assay Drug Dev Technol* 2006;**4**:609–19.
65. Shiau AK, Massari ME, Ozbal CC. Back to basics: label-free technologies for small molecule screening. *Comb Chem High Throughput Screen* 2008;**11**:231–7.
66. Coffman F, Hamid R, Cohen MC, Garippa R, Cohen S. Biophysical profiling of tumor cell lines. *Anal Cell Pathol (Amst)* 2011;**34**:225–34.
67. Dodgson K, Gedge L, Murray DC, Coldwell M. A 100K well screen for a muscarinic receptor using the Epic label-free system—a reflection on the benefits of the label-free approach to screening seven-transmembrane receptors. *J Recept Signal Transduct Res* 2009;**29**:163–72.
68. Li G, Lai F, Fang Y. Modulating cell-cell communication with a high-throughput label-free cell assay. *J Lab Autom* 2012;**17**:6–15.
69. Schmidt J, Liebscher K, Merten N, Grundmann M, Mielenz M, Sauerwein H, et al. Conjugated linoleic acids mediate insulin release through islet G protein-coupled receptor FFA1/GPR40. *J Biol Chem* 2011;**286**:11890–4.

70. Lee PH, Gao A, van Staden C, Ly J, Salon J, Xu A, et al. Evaluation of dynamic mass redistribution technology for pharmacological studies of recombinant and endogenously expressed g protein-coupled receptors. *Drug Dev Technol* 2008;**6**:83–94.
71. Fang Y, Ferrie A, Fontaine N, Ki Yuen P. Optical biosensors for monitoring dynamic mass redistribution in living cells mediated by epidermal growth factor receptor activation. *Conf Proc IEEE Eng Med Biol Soc* 2005;**1**:666–9.
72. Fang Y, Ferrie AM. Optical biosensor differentiates signaling of endogenous PAR1 and PAR2 in A431 cells. *BMC Cell Biol* 2007;**8**:24.
73. Fang Y, Ferrie AM. Label-free optical biosensor for ligand-directed functional selectivity acting on beta(2) adrenoceptor in living cells. *FEBS Lett* 2008;**582**:558–64.
74. Fang Y, Ferrie AM, Fontaine NH, Mauro J, Balakrishnan J. Resonant waveguide grating biosensor for living cell sensing. *Biophys J* 2006;**91**:1925–40.
75. Fang Y, Ferrie AM, Tran E. Resonant waveguide grating biosensor for whole-cell GPCR assays. *Methods Mol Biol* 2009;**552**:239–52.
76. Fang Y, Li G, Ferrie AM. Non-invasive optical biosensor for assaying endogenous G protein-coupled receptors in adherent cells. *J Pharmacol Toxicol Methods* 2007;**55**: 314–22.
77. Verrier F, An S, Ferrie AM, Sun H, Kyoung M, Deng H, et al. GPCRs regulate the assembly of a multienzyme complex for purine biosynthesis. *Nat Chem Biol* 2011;**7**: 909–15.
78. Deng H, Hu H, He M, Hu J, Niu W, Ferrie AM, et al. Discovery of 2-(4-methylfuran-2(5H)-ylidene)malononitrile and thieno[3,2-b]thiophene-2-carboxylic acid derivatives as G protein-coupled receptor 35 (GPR35) agonists. *J Med Chem* 2011;**54**:7385–96.
79. Goral V, Jin Y, Sun H, Ferrie AM, Wu Q, Fang Y. Agonist-directed desensitization of the beta2-adrenergic receptor. *PLoS One* 2011;**6**:e19282.
80. Ferrie AM, Sun H, Fang Y. Label-free integrative pharmacology on-target of drugs at the beta(2)-adrenergic receptor. *Sci Rep* 2011;**1**:33.
81. Ferrie AM, Wu Q, Fang Y. Resonant waveguide grating imager for live cell sensing. *Appl Phys Lett* 2010;**97**:223704.
82. Codd EE, Mabus JR, Murray BS, Zhang SP, Flores CM. Dynamic mass redistribution as a means to measure and differentiate signaling via opioid and cannabinoid receptors. *Assay Drug Dev Technol* 2011;**9**:362–72.
83. Maj M, Bruno V, Dragic Z, Yamamoto R, Battaglia G, Inderbitzin W, et al. (-)-PHCCC, a positive allosteric modulator of mGluR4: characterization, mechanism of action, and neuroprotection. *Neuropharmacology* 2003;**45**:895–906.
84. Marino MJ, Williams Jr DL, O'Brien JA, Valenti O, McDonald TP, Clements MK, et al. Allosteric modulation of group III metabotropic glutamate receptor 4: a potential approach to Parkinson's disease treatment. *Proc Natl Acad Sci USA* 2003;**100**:13668–73.
85. Niswender CM, Johnson KA, Weaver CD, Jones CK, Xiang Z, Luo Q, et al. Discovery, characterization, and antiparkinsonian effect of novel positive allosteric modulators of metabotropic glutamate receptor 4. *Mol Pharmacol* 2008;**74**:1345–58.
86. Zhang R, Xie X. Tools for GPCR drug discovery. *Acta Pharmacol Sin* 2012;**33**:372–84.
87. Henriksen U, Fog J, Loechel F, Praestegaard M. Profiling of multiple signal pathway activities by multiplexing antibody and GFP-based translocation assays. *Comb Chem High Throughput Screen* 2008;**11**:537–44.
88. Salomon Y, Londos C, Rodbell M. A highly sensitive adenylate cyclase assay. *Anal Biochem* 1974;**58**:541–8.
89. Agranoff BW, Murthy P, Seguin EB. Thrombin-induced phosphodiesteratic cleavage of phosphatidylinositol bisphosphate in human platelets. *J Biol Chem* 1983;**258**:2076–8.
90. Birnbaumer L, Pohl SL, Michiel H, Krans MJ, Rodbell M. The actions of hormones on the adenyl cyclase system. *Adv Biochem Psychopharmacol* 1970;**3**:185–208.

91. Gilman AG. Guanine nucleotide-binding regulatory proteins and dual control of adenylate cyclase. *J Clin Invest* 1984;**73**:1–4.
92. McDonald PH, Chow CW, Miller WE, Laporte SA, Field ME, Lin FT, et al. Beta-arrestin 2: a receptor-regulated MAPK scaffold for the activation of JNK3. *Science* 2000;**290**:1574–7.
93. Osmond RI, Crouch MF, Dupriez VJ. An emerging role for kinase screening in GPCR drug discovery. *Curr Opin Mol Ther* 2010;**12**:305–15.
94. Werry TD, Christopoulos A, Sexton PM. Mechanisms of ERK1/2 regulation by seven-transmembrane-domain receptors. *Curr Pharm Des* 2006;**12**:1683–702.
95. Lefkowitz RJ, Shenoy SK. Transduction of receptor signals by beta-arrestins. *Science* 2005;**308**:512–7.
96. Lefkowitz RJ, Inglese J, Koch WJ, Pitcher J, Attramadal H, Caron MG. G-protein-coupled receptors: regulatory role of receptor kinases and arrestin proteins. *Cold Spring Harb Symp Quant Biol* 1992;**57**:127–33.
97. Pierce KL, Lefkowitz RJ. Classical and new roles of beta-arrestins in the regulation of G-protein-coupled receptors. *Nat Rev Neurosci* 2001;**2**:727–33.
98. Gilman AG. A protein binding assay for adenosine 3′:5′-cyclic monophosphate. *Proc Natl Acad Sci USA* 1970;**67**:305–12.
99. Smith C, Toohey N, Knight JA, Klein MT, Teitler M. Risperidone-induced inactivation and clozapine-induced reactivation of rat cortical astrocyte 5-hydroxytryptamine(7) receptors: evidence for in situ G protein-coupled receptor homodimer protomer cross-talk. *Mol Pharmacol* 2011;**79**:318–25.
100. Digby GJ, Noetzel MJ, Bubser M, Utley TJ, Walker AG, Byun NE, et al. Novel allosteric agonists of M1 muscarinic acetylcholine receptors induce brain region-specific responses that correspond with behavioral effects in animal models. *J Neurosci* 2012;**32**:8532–44.
101. Klein M, Teitler M. Distribution of 5-ht(1E) receptors in the mammalian brain and cerebral vasculature: an immunohistochemical and pharmacological study. *Br J Pharmacol* 2012;**166**:1290–302.
102. Naccarato WF, Ray RE, Wells WW. Biosynthesis of myo-inositol in rat mammary gland. Isolation and properties of the enzymes. *Arch Biochem Biophys* 1974;**164**: 194–201.
103. Turek-Etienne TC, Small EC, Soh SC, Xin TA, Gaitonde PV, Barrabee EB, et al. Evaluation of fluorescent compound interference in 4 fluorescence polarization assays: 2 kinases, 1 protease, and 1 phosphatase. *J Biomol Screen* 2003;**8**:176–84.
104. Hemmilä I, Webb S. Time-resolved fluorometry: an overview of the labels and core technologies for drug screening applications. *Drug Discov Today* 1997;**2**:373–81.
105. Hawes BE, van Biesen T, Koch WJ, Luttrell LM, Lefkowitz RJ. Distinct pathways of Gi- and Gq-mediated mitogen-activated protein kinase activation. *J Biol Chem* 1995;**270**:17148–53.
106. Fukuhara S, Marinissen MJ, Chiariello M, Gutkind JS. Signaling from G protein-coupled receptors to ERK5/Big MAPK 1 involves Galpha q and Galpha 12/13 families of heterotrimeric G proteins. Evidence for the existence of a novel Ras AND Rho-independent pathway. *J Biol Chem* 2000;**275**:21730–6.
107. Kramer RM, Roberts EF, Hyslop PA, Utterback BG, Hui KY, Jakubowski JA. Differential activation of cytosolic phospholipase A2 (cPLA2) by thrombin and thrombin receptor agonist peptide in human platelets. Evidence for activation of cPLA2 independent of the mitogen-activated protein kinases ERK1/2. *J Biol Chem* 1995;**270**: 14816–23.
108. Ullman EF, Kirakossian H, Switchenko AC, Ishkanian J, Ericson M, Wartchow CA, et al. Luminescent oxygen channeling assay (LOCI): sensitive, broadly applicable homogeneous immunoassay method. *Clin Chem* 1996;**42**:1518–26.

109. Ullman EF, Kirakossian H, Singh S, Wu ZP, Irvin BR, Pease JS, et al. Luminescent oxygen channeling immunoassay: measurement of particle binding kinetics by chemiluminescence. *Proc Natl Acad Sci USA* 1994;**91**:5426–30.
110. Peppard J, Glickman F, He Y, Hu SI, Doughty J, Goldberg R. Development of a high-throughput screening assay for inhibitors of aggrecan cleavage using luminescent oxygen channeling (AlphaScreen). *J Biomol Screen* 2003;**8**:149–56.
111. Osmond RI, Sheehan A, Borowicz R, Barnett E, Harvey G, Turner C, et al. GPCR screening via ERK 1/2: a novel platform for screening G protein-coupled receptors. *J Biomol Screen* 2005;**10**:730–7.
112. Sava A, Formaggio E, Carignani C, Andreetta F, Bettini E, Griffante C. NMDA-induced ERK signalling is mediated by NR2B subunit in rat cortical neurons and switches from positive to negative depending on stage of development. *Neuropharmacology* 2012;**62**:925–32.
113. Wilden U, Hall SW, Kuhn H. Phosphodiesterase activation by photoexcited rhodopsin is quenched when rhodopsin is phosphorylated and binds the intrinsic 48-kDa protein of rod outer segments. *Proc Natl Acad Sci USA* 1986;**83**:1174–8.
114. Benovic JL, Kuhn H, Weyand I, Codina J, Caron MG, Lefkowitz RJ. Functional desensitization of the isolated beta-adrenergic receptor by the beta-adrenergic receptor kinase: potential role of an analog of the retinal protein arrestin (48-kDa protein). *Proc Natl Acad Sci USA* 1987;**84**:8879–82.
115. Oakley RH, Hudson CC, Cruickshank RD, Meyers DM, Payne Jr RE, Rhem SM, et al. The cellular distribution of fluorescently labeled arrestins provides a robust, sensitive, and universal assay for screening G protein-coupled receptors. *Assay Drug Dev Technol* 2002;**1**:21–30.
116. Namkung Y, Dipace C, Javitch JA, Sibley DR. G protein-coupled receptor kinase-mediated phosphorylation regulates post-endocytic trafficking of the D2 dopamine receptor. *J Biol Chem* 2009;**284**:15038–51.
117. Gurevich VV, Richardson RM, Kim CM, Hosey MM, Benovic JL. Binding of wild type and chimeric arrestins to the m2 muscarinic cholinergic receptor. *J Biol Chem* 1993;**268**:16879–82.
118. Hanson BJ, Wetter J, Bercher MR, Kopp L, Fuerstenau-Sharp M, Vedvik KL, et al. A homogeneous fluorescent live-cell assay for measuring 7-transmembrane receptor activity and agonist functional selectivity through beta-arrestin recruitment. *J Biomol Screen* 2009;**14**:798–810.
119. Hess J, Angel P, Schorpp-Kistner M. AP-1 subunits: quarrel and harmony among siblings. *J Cell Sci* 2004;**117**:5965–73.
120. Butini S, Gemma S, Campiani G, Franceschini S, Trotta F, Borriello M, et al. Discovery of a new class of potential multifunctional atypical antipsychotic agents targeting dopamine D3 and serotonin 5-HT1A and 5-HT2A receptors: design, synthesis, and effects on behavior. *J Med Chem* 2009;**52**:151–69.
121. Altarejos JY, Montminy M. CREB and the CRTC co-activators: sensors for hormonal and metabolic signals. *Nat Rev Mol Cell Biol* 2011;**12**:141–51.
122. Saura CA, Valero J. The role of CREB signaling in Alzheimer's disease and other cognitive disorders. *Rev Neurosci* 2011;**22**:153–69.
123. Carlezon Jr WA, Duman RS, Nestler EJ. The many faces of CREB. *Trends Neurosci* 2005;**28**:436–45.
124. Kenakin TP. *Pharmacologic analysis of drug-receptor interaction*. 3rd ed. Philadelphia: Lippincott-Raven Publishers; 1997.
125. Limbird LE. *Cell surface receptors: a short course on theory & methods*. 3rd ed. New York: Springer Science + Business Media; 2005.
126. McKinney M, Raddatz R. Practical aspects of radioligand binding. *Current Protocols in Pharmacol*. 2006;**33**:1.3.1–1.3.42.

127. Kenakin T. The classification of seven transmembrane receptors in recombinant expression systems. *Pharmacol Rev* 1996;**48**:413–63.
128. Bernat V, Heinrich MR, Baumeister P, Buschauer A, Tschammer N. Synthesis and application of the first radioligand targeting the allosteric binding pocket of chemokine receptor CXCR3. *ChemMedChem* 2012;**7**:1481–9.
129. Cosford NDP, Roppe J, Tehrani L, Schweiger EJ, Seiders TJ, Chaudary A, et al. [H-3]-methoxymethyl-MTEP and [H-3]-methoxy-PEPy: potent and selective radioligands for the metabotropic glutamate subtype 5 (mGlu5) receptor. *Bioorg Med Chem Lett* 2003;**13**:351–4.
130. Heitman LH, Oosterom J, Bonger KM, Timmers CM, Wiegerinck PHG, IJzerman AP. [3H]Org 43553, the first low-molecular-weight agonistic and allosteric radioligand for the human luteinizing hormone receptor. *Mol Pharmacol* 2008;**73**:518–24.
131. Shin N, Solomon K, Zhou N, Wang KH, Garlapati V, Thomas B, et al. Identification and characterization of INCB9471, an allosteric noncompetitive small-molecule antagonist of C-C chemokine receptor 5 with potent inhibitory activity against monocyte migration and HIV-1 infection. *J Pharmacol Exp Ther* 2011;**338**:228–39.
132. Tränkle C, Mies-Klomfass E, Cid MHB, Holzgrabe U, Mohr K. Identification of a [3H]ligand for the common allosteric site of muscarinic acetylcholine M2 receptors. *Mol Pharmacol* 1998;**54**:139–45.
133. Jeffrey Conn P, Christopoulos A, Lindsley CW. Allosteric modulators of GPCRs: a novel approach for the treatment of CNS disorders. *Nat Rev Drug Discov* 2009;**8**:41–54.
134. Wang JQ, Tueckmantel W, Zhu A, Pellegrino D, Brownell AL. Synthesis and preliminary biological evaluation of 3-[(18)F]fluoro-5-(2-pyridinylethynyl)benzonitrile as a PET radiotracer for imaging metabotropic glutamate receptor subtype 5. *Synapse* 2007;**61**:951–61.
135. Avlani V, May LT, Sexton PM, Christopoulos A. Application of a kinetic model to the apparently complex behavior of negative and positive allosteric modulators of muscarinic acetylcholine receptors. *J Pharmacol Exp Ther* 2004;**308**:1062–72.
136. Leach K, Sexton PM, Christopoulos A. Quantification of allosteric interactions at G protein-coupled receptors using radioligand binding assays. *Curr Protoc Pharmacol* 2011; Chapter 1, Unit 1.22.
137. de Kruijf P, Lim HD, Roumen L, Renjaän VA, Zhao J, Webb ML, et al. Identification of a novel allosteric binding site in the CXCR2 chemokine receptor. *Mol Pharmacol* 2011;**80**:1108–18.
138. Gregory KJ, Dong EN, Meiler J, Conn PJ. Allosteric modulation of metabotropic glutamate receptors: structural insights and therapeutic potential. *Neuropharmacology* 2011;**60**:66–81.
139. Valant C, Gregory KJ, Hall NE, Scammells PJ, Lew MJ, Sexton PM, et al. A novel mechanism of G protein-coupled receptor functional selectivity. Muscarinic partial agonist McN-A-343 as a bitopic orthosteric/allosteric ligand. *J Biol Chem* 2008;**283**:29312–21.
140. Zhong H, Hansen KB, Boyle NJ, Han K, Muske G, Huang X, et al. An allosteric binding site at the human serotonin transporter mediates the inhibition of escitalopram by R-citalopram: kinetic binding studies with the ALI/VFL–SI/TT mutant. *Neurosci Lett* 2009;**462**:207–12.
141. Leach K, Simms J, Sexton PM, Christopoulos A. Structure-function studies of muscarinic acetylcholine receptors. *Handb Exp Pharmacol* 2012;**208**:29–48.
142. Gregory KJ, Hall NE, Tobin AB, Sexton PM, Christopoulos A. Identification of orthosteric and allosteric site mutations in M2 muscarinic acetylcholine receptors that contribute to ligand-selective signaling bias. *J Biol Chem* 2010;**285**:7459–74.

143. Koole C, Wootten D, Simms J, Savage EE, Miller LJ, Christopoulos A, et al. Second extracellular loop of human glucagon-like peptide-1 receptor (GLP-1R) differentially regulates orthosteric but not allosteric agonist binding and function. *J Biol Chem* 2012;**287**:3659–73.
144. Gregory KJ, Sexton PM, Christopoulos A. Overview of receptor allosterism. *Curr Protoc Pharmacol* 2010; Chapter 1, Unit 1.21.
145. Langmead CJ. Determining allosteric modulator mechanism of action: integration of radioligand binding and functional assay data. *Methods Mol Biol* 2011;**746**:195–209.
146. Nawaratne V, Leach K, Felder CC, Sexton PM, Christopoulos A. Structural determinants of allosteric agonism and modulation at the M4 muscarinic acetylcholine receptor: identification of ligand-specific and global activation mechanisms. *J Biol Chem* 2010;**285**:19012–21.
147. Christopoulos A, Lanzafame A, Ziegler A, Mitchelson F. Kinetic studies of co-operativity at atrial muscarinic M2 receptors with an "infinite dilution" procedure. *Biochem Pharmacol* 1997;**53**:795–800.
148. May LT, Bridge LJ, Stoddart LA, Briddon SJ, Hill SJ. Allosteric interactions across native adenosine-A3 receptor homodimers: quantification using single-cell ligand-binding kinetics. *FASEB J* 2011;**25**:3465–76.
149. Gomes I, IJzerman AP, Ye K, Maillet EL, Devi LA. G protein-coupled receptor heteromerization: a role in allosteric modulation of ligand binding. *Mol Pharmacol* 2011;**79**:1044–52.
150. Sephton SM, Dennler P, Leutwiler DS, Mu L, Schibli R, Kramer SD, et al. Development of [(18)F]-PSS223 as a PET tracer for imaging of metabotropic glutamate receptor subtype 5 (mGluR5). *Chimia (Acrau)* 2012;**66**:201–4.
151. Wanger-Baumann CA, Mu L, Honer M, Belli S, Alf MF, Schubiger PA, et al. In vitro and in vivo evaluation of [18F]-FDEGPECO as a PET tracer for imaging the metabotropic glutamate receptor subtype 5 (mGluR5). *Neuroimage* 2011;**56**:984–91.
152. Baumann CA, Mu L, Wertli N, Kramer SD, Honer M, Schubiger PA, et al. Syntheses and pharmacological characterization of novel thiazole derivatives as potential mGluR5 PET ligands. *Bioorg Med Chem* 2010;**18**:6044–54.
153. Baumann CA, Mu L, Johannsen S, Honer M, Schubiger PA, Ametamey SM. Structure-activity relationships of fluorinated (E)-3-((6-methylpyridin-2-yl)ethynyl) cyclohex-2-enone-O-methyloxime (ABP688) derivatives and the discovery of a high affinity analogue as a potential candidate for imaging metabotropic glutamate recepors subtype 5 (mGluR5) with positron emission tomography (PET). *J Med Chem* 2010;**53**:4009–17.
154. Ametamey SM, Treyer V, Streffer J, Wyss MT, Schmidt M, Blagoev M, et al. Human PET studies of metabotropic glutamate receptor subtype 5 with 11C-ABP688. *J Nucl Med* 2007;**48**:247–52.
155. Treyer V, Streffer J, Ametamey SM, Bettio A, Blauenstein P, Schmidt M, et al. Radiation dosimetry and biodistribution of 11C-ABP688 measured in healthy volunteers. *Eur J Nucl Med Mol Imaging* 2008;**35**:766–70.
156. Treyer V, Streffer J, Wyss MT, Bettio A, Ametamey SM, Fischer U, et al. Evaluation of the metabotropic glutamate receptor subtype 5 using PET and 11C-ABP688: assessment of methods. *J Nucl Med* 2007;**48**:1207–15.
157. Nordquist RE, Steckler T, Wettstein JG, Mackie C, Spooren W. Metabotropic glutamate receptor modulation, translational methods, and biomarkers: relationships with anxiety. *Psychopharmacology (Berl)* 2008;**199**:389–402.
158. Hostetler ED, Eng W, Joshi AD, Sanabria-Bohorquez S, Kawamoto H, Ito S, et al. Synthesis, characterization, and monkey PET studies of [(1)(8)F]MK-1312, a PET tracer for quantification of mGluR1 receptor occupancy by MK-5435. *Synapse* 2011;**65**:125–35.

CHAPTER TWO

Pharmacology of Metabotropic Glutamate Receptor Allosteric Modulators: Structural Basis and Therapeutic Potential for CNS Disorders

Karen J. Gregory[*,†,‡], **Meredith J. Noetzel**[*,†], **Colleen M. Niswender**[*,†]

[*]Vanderbilt Center for Neuroscience Drug Discovery, Vanderbilt University Medical Center, Nashville, Tennessee, USA
[†]Department of Pharmacology, Vanderbilt University Medical Center, Nashville, Tennessee, USA
[‡]Drug Discovery Biology, Monash Institute of Pharmaceutical Sciences, Monash University, Parkville, Victoria, Australia

Contents

Progress in Molecular Biology and Translational Science, Volume 115
ISSN 1877-1173
http://dx.doi.org/10.1016/B978-0-12-394587-7.00002-6

Abstract

The metabotropic glutamate receptors (mGlus) mediate a neuromodulatory role throughout the brain for the major excitatory neurotransmitter, glutamate. Seven of the eight mGlu subtypes are expressed within the CNS and are attractive targets for a variety of psychiatric and neurological disorders including anxiety, depression, schizophrenia, Parkinson's disease, and Fragile X syndrome. Allosteric modulation of these class C 7-transmembrane spanning receptors represents a novel approach to facilitate development of mGlu subtype-selective probes and therapeutics. Allosteric modulators that interact with sites topographically distinct from the endogenous ligand-binding site offer a number of advantages over their competitive counterparts. In particular for CNS therapeutics, allosteric modulators have the potential to maintain the spatial and temporal aspects of endogenous neurotransmission. The past 15 years have seen the discovery of numerous subtype-selective allosteric modulators for the majority of the mGlu family members, including positive, negative, and neutral allosteric modulators, with a number of mGlu allosteric modulators now in clinical trials.

ABBREVIATIONS

2,2,2,TEMPS 2,2,2-trifluoroethyl [3-(1-methyl-butoxy)-phenyl]-pyridine-3-ylmethyl-sulphonamide

5MPEP 5-methyl-6-(phenylethynyl)-pyridine

6-OHDA 6-hydroxydopamine

7TMR 7-transmembrane spanning receptor

7TMs seven transmembrane-spanning domains

ACPT-1 (1*S*,3*R*,4*S*)-1-aminocyclopentane-1,3,4-tricarboxylic acid

AD Alzheimer's disease

ADX47273 *S*-(4-fluoro-phenyl)-(3-[3-(4-fluoro-phenyl)-[1,2,4]oxadiazol-5-yl]-piperidin-1-yl)-methanone

ADX88178 5-methyl-*N*-(4-methylpyrimidin-2-yl)-4-(1H-pyrazol-4-yl)thiazol-2-amine

AFQ056 (3a*S*,5*S*,7a*R*)-methyl 5-hydroxy-5-(*m*-tolylethynyl)octahydro-1H-indole-1-carboxylate

AMN082 *N*,*N*′-bis(diphenylmethyl)-1,2-ethanediamine

ATCM allosteric ternary complex model
Aβ β-amyloid
BINA biphenyl-indanone A
CBiPES *N*-[4′-cyano-biphenyl-3-yl]-*N*-(3-pyridinylmethyl)-ethanesulfonamide hydrochloride
CDPPB 3-cyano-*N*-(1,3-diphenyl-1H-pyrazol-5-yl)benzamide
CFMMC 3-cyclohexyl-5-fluoro-6-methyl-7-(2-morpholin-4-ylethoxy)-4H-chromen-4-one
CPCCOEt 7-(hydroxyimino)cyclopropa[*b*]chromen-1a-carboxylate ethyl ester
CPPHA *N*-(4-chloro-2-[(1,3-dioxo-1,3-dihydro-2H-isoindol-2-yl)methyl]phenyl)-2-hydroxybenzamide
CPPZ 1-(4-(2-chloro-4-fluorophenyl)piperazin-1-yl)-2-(pyridin-4-ylmethoxy)ethanone
DCG-IV (2*S*,2′*R*,3′*R*)-2-(2′3′-dicarboxycyclopropyl)glycine
DCPG (*S*)-3,4-dicarboxyphenylglycine
DFB [(3-fluorophenyl)methylene]hydrazone-3-fluorobenzaldehyde
EAE experimental autoimmune encephalomyelitis
EM-TBPC 1-ethyl-2-methyl-6-oxo-4-(1,2,4,5-tetrahydro-benzo[*d*]azepin-3-yl)-1,6-dihydro-pyrimidine-5-carbonitrile
ERK1/2 extracellular signal-regulated kinases 1 and 2
FMRP Fragile X mental retardation protein
FTIDC 4-[1-(2-fluoropyridin-3-yl)-5-methyl-1H-1,2,3-triazol-4-yl]-*N*-isopropyl-*N*-methyl-3,6-dihydropyridine-1(2H)-carboxamide
FXS Fragile X syndrome
GABA γ-aminobutyric acid
GRN-529 (4-difluoromethoxy-3-(pyridine-2-ylethynyl)phenyl)5H-pyrrolo[3,4-b]pyridine-6(7H)-yl methanone
HYDIA (1*S*,2*R*,3*R*,5*R*,6*S*)-2-amino-3-hydroxy-bicyclo[3.1.0]hexane-2,6-dicarboxylic acid
JNJ16259685 (3,4-dihydro-2H-pyrano[2,3]b quinolin-7-yl) (cis-4-methoxycyclohexyl) methanone
L-AP4 L-2-amino-4-phosphonobutyrate
L-DOPA L-3,4-dihydroxyphenylalanine
LSP1-2111 ((2*S*)-2-amino-4-[hydroxy[hydroxy(4-hydroxy-3-methoxy-5-nitro-phenyl) methyl] phosphoryl]butanoic acid)
LSP1-3081 (3*S*)-3-[(3-amino-3-carboxypropyl(hydroxy)phosphinyl)-hydroxymethyl]-5-nitrothiophene
LSP4-2022 (2*S*)-2-amino-4-(((4-(carboxymethoxy)phenyl)(hydroxy)methyl)(hydroxy) phosphoryl)butanoic acid
LTD long-term depression
LTP long-term potentiation
Lu AF21934 (1*S*,2*S*)-N^1-(3,4-dichlorophenyl)cyclohexane-1,2-dicarboxamide
LY2140023 (1*R*,4*S*,5*S*,6*S*)-2-thiabicyclo[3.1.0]-hexane-4,6-dicarboxylic acid,4-[(2*S*)-2-amino-4-(methylthio)-1-oxobutyl]amino-,2,2-dioxide monohydrate
LY2389575 (3*S*)-1-(5-bromopyrimidin-2-yl)-*N*-(2,4-dichlorobenzyl)pyrrolidin-3-amine methanesulfonate hydrate
LY341495 (2*S*,1′*S*,2′*S*)-2-(9-xanthylmethyl)-2-(2′-carboxycyclopropylglycine)
LY354740 (1*S*,2*R*,5*R*,6*S*)-2-amino-bicyclo[3.1.0]hexane-2,6-dicarboxylic acid
LY379268 (–)-2-oxa-4-aminobicyclo[3.1.0]hexane-4,6-dicarboxylate
LY404039 (–)-(1*R*,4*S*,5*S*,6*S*)-4-amino-2-sulfonylbicyclo[3.1.0]hexane-4,6-dicarboxylic acid

LY456066 2-[4-(indan-2-ylamino)-5,6,7,8-tetrahydro-quinazolin-2-ylsulfanyl]-ethanol hydrochloride
LY487379 2,2,2-trifluoro-*N*-[4-(2-methoxyphenoxy) phenyl]-*N*-(3-pyridinylmethyl) ethanesulfonamide
LY541850 (1*S*,2*S*,4*R*,5*R*,6*S*)-2-amino-4-methylbicyclo[3.1.0]hexane2,6-dicarboxylic acid
LY566332 *N*-(4′-cyano-biphenyl-3-yl)-*N*-(3-pyridinylmethyl)-ethanesulfonamide hydrochloride
M-5MPEP 2-(2-(3-methoxyphenyl)ethynyl)-5-methylpyridine
methoxyPEPy 3-methoxy-5-(2-pyridinylethynyl)pyridine
mGlu metabotropic glutamate receptor
MGS0008 5-[2-[4-(6-fluoro-1H-indole-3-yl) piperidin-1-yl]ethyl]-4-(4-fluorophenyl) thiazole-2-carboxylic acid amide
MGS0028 (1*R*,2*S*,5*S*,6*S*)-2-amino-6-fluoro-4-oxobicyclo[3.1.0]hexane-2,6-dicarboxylic acid monohydrate
MGS0039 (1*R*,2*R*,3*R*,5*R*,6*R*)-2-amino-3-(3,4-dichlorobenzyloxy)-6-fluorobicyclo [3.1.0] hexane-2,6-dicarboxylic acid
MMPIP 6-(4-methoxyphenyl)-5-methyl-3-(4-pyridinyl)-isoxazolo [4,5-c]pyridine-4 (5H)-one hydrochloride
MPEP 2-methyl-6-(phenylethynyl)pyridine
MRLSD-650 *N*-(4-(1H-tetrazol-5-yl)phenyl)-2-((6,7-dichloro-2-cyclopentyl-2-methyl-1-oxo-2,3-dihydro-1H-inden-5-yl)oxy)acetamide
MTEP 3-[(2-methyl-1,3-thiazol-4-yl)ethynyl] pyridine
NAM negative allosteric modulator
NMDA *N*-methyl-D-aspartate
PAM positive allosteric modulator
PCP phencyclidine
PD Parkinson's disease
PET positron emission tomography
PHCCC *N*-phenyl-7-(hydroxyimino)cyclopropa[*b*] chromen-1a-carboxamide
PPI prepulse inhibition
R214127 1-(3,4-dihydro-2H-pyrano[2,3-b]quinolin-7-yl)-2-phenyl-1-ethanone
Ro01-6128 ethyl diphenylacetylcarbamate
RO4491533 4-[3-(2,6-dDimethylpyridin-4-yl)phenyl]-7-methyl-8-trifluoromethyl- 1, 3-dihydrobenzo[b][1,4]diazepin-2-one
RO4988546 5-[7-trifluoromethyl-5-(4-trifluoromethyl-phenyl)-pyrazolo[1,5-a] pyrimidin-3-ylethylnyl]-pyridine-3-sulphonic acid
RO5488608 3′-(8-methyl-4-oxo-7-trifluoromethyl-4,5-dihydro-3H-benzo[*b*][1,4] diazepin-2-yl)-biphenyl-3-sulphonic acid
Ro67-4853 butyl (9H-xanthene-9-carbonyl)carbamate
Ro67-7476 (*S*)-2-(4-fluorophenyl)-1-(toluene-4-sulfonyl)pyrrolidine
***S*-4C3H-PG** (*S*)-4-carboxy-3-hydroxyphenylglycine
SAM silent allosteric modulator
SAR structure–activity relationship
SC-CA1 Schaffer-collateral-CA1
SIB-1757 6-methyl-2-(phenylazo)-3-pyridinol
SIB-1893 2-methyl-6-(2-phenylethenyl)pyridine

THIIC *N*-(4-((2-(trifluoromethyl)-3-hydroxy-4-(isobutyryl)phenoxy)methyl)benzyl)-1-methyl-1H-imidazole-4-carboxamide
TSC tuberous sclerosis complex
VFD Venus-Flytrap domain
VU0155041 cis-2-([(3,5-dichlorophenyl)amino]carbonyl) cyclohexanecarboxylic acid
VU0360172 *N*-cyclobutyl-6-((3-fluorophenyl)ethynyl)nicotinamide
VU0364289 2-(4-(2-(benzyloxy)acetyl)piperazin-1-yl)benzonitrile
VU0364770 *N*-(3-chlorophenyl)picolinamide
VU29 4-nitro-*N*-(1,3-diphenyl-1H-pyrazol-5-yl)benzamide
VU48 4-nitro-*N*-(1-(2-bromophenyl)-3-phenyl-1H-pyrazol-5-yl)benzamide
VU71 4-nitro-*N*-(1,4-diphenyl-1H-pyazol-5-yl)benzamide
YM298198 6-amino-*N*-cyclohexyl-*N*,3-dimethylthiazolo[3,2-a]benzimidazole-2-carboxamide

1. INTRODUCTION

The 7-transmembrane spanning receptor (7TMR), G protein-coupled receptor, superfamily represents the largest class of cell-surface receptors. Typified by their extracellular N terminus, seven transmembrane-spanning domains (7TMs) and intracellular C terminus, 7TMRs translate extracellular stimuli to intracellular responses. 7TMRs are highly tractable drug targets; however, the majority of 7TMR-based drug discovery programs have failed to yield highly selective compounds. This is particularly prevalent for 7TMR targets for CNS disorders, a therapeutic area with one of the highest rates of attrition in drug discovery.[1]

The traditional approach to 7TMR drug discovery has been to either mimic or block the actions of the endogenous neurotransmitter or hormone by competitively interacting with the endogenous ligand (orthosteric)-binding site. However, orthosteric agents typically lack suitable subtype selectivity, likely a consequence of the highly conserved nature of orthosteric binding pockets within a single 7TMR family. An alternative approach that has gained significant momentum in the past two decades is to target allosteric sites that are topographically distinct from the orthosteric site. The binding of an allosteric modulator has the potential to alter the affinity and/or efficacy of an orthosteric ligand. Allosteric modulators that enhance receptor function are referred to as positive allosteric modulators (PAMs), while those that inhibit are termed negative allosteric modulators (NAMs). Arguably, the metabotropic glutamate receptor (mGlu) family is one of the most well-studied 7TMR families with respect to allosteric

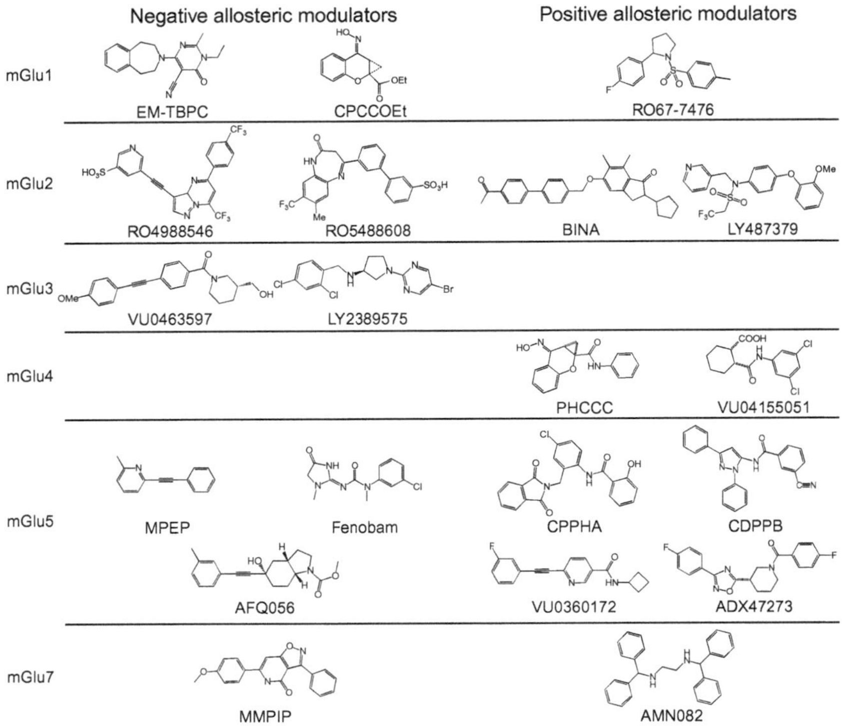

Figure 2.1 Representative positive and negative allosteric modulators across different mGlu subtypes.

modulation. Selective allosteric modulators have been discovered for most family members, including a full spectrum of allosteric modulators and multiple chemotypes for some subtypes (Fig. 2.1). Glutamatergic synapses are ubiquitously distributed throughout the brain, with members of the mGlu family involved in a broad range of functions within the CNS. Collectively, the mGlus are attractive targets for therapeutic intervention for a range of neurological and psychiatric disorders. Thus, allosteric modulation of mGlus represents a novel, exciting therapeutic avenue for the treatment of numerous CNS diseases.

2. PHYSIOLOGY AND PHARMACOLOGY OF mGlus

The mGlu family is comprised of eight subtypes, sub-classified into three major groups based on sequence homology, pharmacological properties, and coupling to second messenger pathways. Group I includes $mGlu_1$

and $mGlu_5$; group II, $mGlu_2$ and $mGlu_3$; and group III, $mGlu_4$, $mGlu_6$, $mGlu_7$, and $mGlu_8$. Group I mGlus preferentially couple to activation of the $G_{q/11}$ family of G proteins, activating phosphoinositide hydrolysis and mobilizing intracellular Ca^{++} stores as the major signaling mechanism. Group II and group III mGlus preferentially couple to $G_{i/o}$, resulting in inhibition of adenylyl cyclases and cAMP accumulation; additionally, signaling from these receptors modulates many ion channels. Individual subtypes are further identified by their pharmacological profile through use of specific agonists, and, increasingly, selective allosteric modulators.[2]

With the exception of $mGlu_{6,}$ which is localized to the retina, mGlus are ubiquitously expressed throughout the CNS and found in both neurons and glia. Activation of mGlus by the neurotransmitter glutamate modulates neuronal excitability, synaptic transmission, and other cellular functions within the CNS. Individual mGlu subtypes demonstrate differential localization and can exhibit distinct functions in specific brain regions (CNS localization for each of the mGlu subtypes is reviewed in detail in Ref. 3). Despite the potential for varied functions at individual synapses, some generalizations can be made with respect to the function of the different mGlu groups. Group I mGlus are predominantly found postsynaptically, with activation resulting in cell depolarization and increases in neuronal excitability via modulation of a variety of ion channels. Modulation by group I mGlus spans robust excitation to more subtle changes in the pattern and frequency of cell firing and responses to excitatory inputs.[4–6] Group II and III mGlus are predominantly localized on presynaptic terminals or preterminal axons and, in some cases, are also found postsynaptically. Activation of presynaptic group II and III mGlus inhibits neurotransmitter release on a variety of excitatory (glutamatergic), inhibitory (GABAergic), and neuromodulatory (monoamines, ACh, peptides) synapses. The physiological roles of mGlus have been presented in detail in multiple reviews.[4–9] All of the mGlu subtypes have now been genetically deleted in mice, providing further insights into the biology and pathophysiology of mGlus.[10] Our collective understanding of the varied roles of mGlus within the CNS suggests that selective activation or inhibition of specific mGlu subtypes could modulate neurotransmission in a therapeutically beneficial manner. Furthermore, it is postulated that targeting mGlus will allow for modulation of glutamergic neurotransmission while avoiding the adverse effects associated with directly targeting the ionotropic glutamate receptor family.

2.1. Advantages and complexities of allosteric modulation

Allosteric modulators offer a number of theoretical advantages compared with their competitive counterparts. As mentioned earlier, allosteric modulators have the potential for greater subtype selectivity when compared to orthosteric ligands. Such selectivity can be driven through interactions with novel binding sites that show greater sequence divergence among subtypes or via selective cooperativity for an individual subtype at the exclusion of others.[11] Furthermore, selectivity can be engineered by combining orthosteric and allosteric moieties within the same compound, where selectivity is driven by the allosteric moiety, yielding a bitopic (also referred to as dualsteric) ligand.[12–17] A second advantage relates to allosteric modulators that possess no intrinsic activity in their own right. Such modulators are assumed to be quiescent in the absence of an endogenous agonist and only modulate receptor function when the endogenous agonist is present. In this manner, allosteric modulators have the potential to retain spatial and temporal aspects of endogenous receptor signaling. This is of particular interest for CNS targets where "fine-tuning" neurotransmission is likely to have an improved therapeutic outcome as opposed to sustained receptor blockade or activation. However, it should be noted that this can also be a disadvantage as pure allosteric modulators are reliant upon the presence of the endogenous ligand. In this case, the endogenous agonist tone must be sufficient to permit potentiation. A further benefit of allosteric modulators is that the degree of cooperativity is saturable such that there is a "ceiling" level to their effect. This mechanism of receptor modulation may lead to an improved therapeutic window in the case of overdose. Furthermore, for NAMs with weak cooperativity, there is the potential to "dial down" receptor activity, retaining a degree of receptor activity in the presence of a NAM, which may be a more desirable effect than full inhibition of a target response. While allosteric modulators offer numerous advantages over orthosteric agents, drug discovery programs focused on the discovery and optimization of allosteric modulators do face challenges.

Small molecule drug discovery programs, whether they be for allosteric or orthosteric ligands, encounter similar obstacles during optimization concerning target validation, solubility and formulation, brain penetrance, clearance, and generation of active metabolites. Furthermore, it is becoming increasingly apparent that the consequences of 7TMR activation or inhibition are not limited to G protein coupling and activation of traditional second messenger pathways, and integration of signaling pathways affected by a modulator can be distinct and lead to different outputs in pharmacology, both *in vitro* and *in vivo*. This phenomenon has been referred to by many monikers, including "stimulus trafficking," "biased agonism," and "functional selectivity".[18–20] Indeed, orthosteric ligands for $mGlu_1$ were recently reported to display signaling bias.[21] Biased pharmacology is attributed to the stabilization of different receptor conformations that favor one signaling outcome over another. Since allosteric modulators engender unique receptor conformations to either potentiate or inhibit the response to an orthosteric ligand, it is not surprising that there is potential for biased pharmacology of allosteric modulators. A number of examples of biased modulation have been reported for allosteric modulators of the mGlus. For example, *N*-(4-chloro-2-[(1,3-dioxo-1,3-dihydro-2H-isoindol-2-yl) methyl]phenyl)-2-hydroxybenzamide (CPPHA), a group I PAM, shows a different profile for extracellular signal-regulated kinases 1 and 2 (ERK1/2) phosphorylation in rat cortical astrocytes compared with [(3-fluorophenyl) methylene]hydrazone-3-fluorobenzaldehyde (DFB), another $mGlu_5$ PAM, despite displaying similar potentiator profiles in a Ca^{++} mobilization assay.[22] The $mGlu_1$ PAMs ethyl diphenylacetylcarbamate (Ro01-6128), butyl (9H-xanthene-9-carbonyl)carbamate (Ro67-4853), and (*S*)-2-(4-fluorophenyl)-1-(toluene-4-sulfonyl)pyrrolidine (Ro67-7476) are agonists for cAMP accumulation and ERK1/2 phosphorylation, but not for Ca^{++} mobilization.[23] 6-(4-Methoxyphenyl)-5-methyl-3-(4-pyridinyl)-isoxazolo [4,5-c]pyridine-4 (5H)-one hydrochloride (MMPIP), an $mGlu_7$ NAM, shows differential inhibition of agonist-dependent receptor activation dependent upon the assay and cellular background utilized.[24] Biased pharmacology of $mGlu_4$ PAMs has also been observed when G_q-coupled receptors are co-activated.[25] The therapeutic relevance of such biased pharmacology has yet to be realized. It is conceivable that if a desirable therapeutic outcome or adverse-effect profile can be definitively attributed to the modulation of one pathway over another, then biased modulators may actually be preferred in potential therapeutics.

Increasingly, allosteric modulators are being identified that, in addition to modulating activation elicited by orthosteric agonists, act as agonists in their own right. For the mGlu family, numerous examples of these so-called allosteric agonists have been identified.[22,23,26–30] It should be noted that allosteric agonism by mGlu PAMs may be attributed to receptor overexpression in recombinant systems and may not translate to native cells and tissues.[26] However, recent efforts have identified allosteric agonists that are capable of engendering receptor activation in native systems.[22,31] Given that allosteric agonists elicit receptor activation via an alternate binding site, there is likely to be an even greater propensity for biased pharmacology, adding additional layers of complexity and scope for therapeutic avenues. Biased modulation or agonism may allow for even more refined "fine-tuning" of disease pathophysiology if selective modulation of a particular pathway can be linked to a desirable therapeutic outcome. It remains to be seen whether pure PAMs versus PAMs with agonist activity, or alternatively full versus weak NAMs, are the best approach for modulating mGlu function in the CNS. In all likelihood, distinct allosteric modulator phenotypes will be required for different therapeutic outcomes.

In addition to the prospect of biased pharmacology, allosteric interactions are also "probe dependent" such that the cooperativity manifested between allosteric and orthosteric sites is dependent upon the chemical nature of the compounds occupying the two sites; the same allosteric modulator may be positively cooperative with one ligand and negatively or neutrally cooperative with another. A number of examples of this phenomenon have been described for the mGlu family. For example, NAMs of $mGlu_2$ modulate the affinity of the radiolabeled orthosteric agonist [^{3}H]-(1*S*,2*R*,5*R*,6*S*)-2-amino-bicyclo[3.1.0] hexane-2,6-dicarboxylic acid (LY354740), but exhibit neutral cooperativity toward the binding of the antagonist [^{3}H]-(1*S*,2*R*,3*R*,5*R*,6*S*)-2-amino-3-hydroxy-bicyclo[3.1.0]hexane-2,6-dicarboxylic acid (HYDIA).[32] The $mGlu_2$ PAMs Biphenyl-indanone A (BINA) and *N*-[4′-cyano-biphenyl-3-yl)-*N*-(3-pyridinylmethyl)-ethanesulfonamide hydrochloride (CBiPES) potentiate the activity of the orthosteric agonist (1*S*,2*S*,4*R*,5*R*,6*S*)-2-amino-4-methylbicyclo[3.1.0]hexane2,6-dicarboxylic acid (LY541850) but not (−)-2-oxa-4-aminobicyclo[3.1.0]hexane-4,6-dicarboxylate (LY379268), also an orthosteric agonist.[33] The probe-dependent nature of allosteric interactions presents a challenge with respect to translating *in vitro* activity to *in vivo* efficacy, especially if a surrogate orthosteric ligand is utilized in characterization of a novel allosteric modulator.

Structure–activity relationships (SARs) for mGlu modulators are also notoriously difficult. During the discovery of CPPHA, the SAR was found to be "steep" with minimal changes to the structure resulting in a complete loss of activity.[34] Furthermore, numerous mGlu modulator chemotypes display "molecular switches" whereby a PAM or silent (or neutral) allosteric modulator (SAM) arises from a NAM scaffold or vice versa.[35] For the mGlus, molecular switching was originally observed during the discovery of the first $mGlu_5$ PAM, DFB.[36] This phenomenon continues to be a challenge for medicinal chemists, particularly with respect to targeting $mGlu_5$, with PAMs often being derived from NAM scaffolds[37–39]; SAMs from either

NAM or PAM chemotypes[39,40]; and NAMs from PAMs.[41] Furthermore, molecular switches have also been described with respect to unanticipated alterations in mGlu selectivity.[42] Despite the inherent complexity in the discovery and characterization of allosteric modulators, the potential advantages have resulted in concerted efforts globally pursuing allosteric modulators of many 7TMRs for a variety of indications.

3. ALLOSTERIC MODULATION OF mGlus

3.1. Quantification of allosteric modulation

Occupation of an allosteric site will perturb receptor conformations such that the "geography" of the orthosteric site and any other potential receptor-ligand/protein interfaces may be changed. Therefore, a 7TMR simultaneously bound by both an allosteric and an orthosteric ligand is no longer the same entity as the 7TMR bound by either ligand alone. Changes in the "geography" of the orthosteric site can be manifested as an alteration in the binding affinity and/or signaling efficacy of the orthosteric ligand, either in a positive or negative manner. This phenomenon is referred to as "cooperativity." The simplest allosteric 7TMR model, the allosteric ternary complex model (ATCM), accounts for allosteric modulation of ligand affinity only. Changes in ligand affinity are assumed to be reciprocal in nature, where the binding of an allosteric modulator either enhances or inhibits the affinity of an orthosteric ligand and vice versa. This modulation of affinity is defined by the "cooperativity factor" α, denoting the magnitude and direction of the allosteric interaction between the two conformationally linked sites.[43,44] In this model, it is assumed that the 7TMR stimulus imparted by the ternary receptor complex is no different to that of the binary complex. While this has been the case for many allosteric modulators studied to date, there is no reason why the conformational change engendered by an allosteric modulator cannot perturb 7TMR signaling efficacy in addition to, or independently of, affinity modulation. Indeed, many mGlu allosteric modulators, particularly NAMs, influence orthosteric ligand efficacy in the absence of any modulation of affinity.[29,45–48]

For the mGlu family, the lack of effect of a modulator on orthosteric agonist affinity most likely reflects the fact that the orthosteric and allosteric binding pockets are found on very distinct receptor regions (Ref. 49, see later for discussion). Furthermore, it should also be considered that an allosteric modulator may have differential effects on affinity versus efficacy.[50] The potential for allosteric modulation of affinity and efficacy has necessitated the development of more complex models to account for allosteric modulation of 7TMRs.[51,52] While such models account for the array of possible effects an allosteric interaction may have on the overall functional response to 7TMR activation, the large number of parameters prevents them from being amenable to fitting experimental biological data. An "operational model of allosterism" has been reported[53] that combines the simple ATCM with an operational model of agonism.[54] Efficacy modulation is incorporated into this model by the introduction of the parameter, β. Furthermore, the potential for allosteric modulator intrinsic efficacy, also termed allosteric agonism, is also included. Allosteric modulation is thus governed by two cooperativity parameters, α and β, that may differ for each and every set of interacting ligands. Both α and β should be constant between different assays of 7TMR function, unless pathway-specific allosteric modulation is occurring, in which case cooperativity values will differ. Quantitative analysis of allosteric modulator behaviors provides the basis for both verification of an allosteric mode of action and optimization of medicinal chemistry efforts for allosteric modulator affinity and/or cooperativity.

3.2. Verification of an allosteric mechanism of action

Cell-based functional assays are the predominant primary screening method employed for the identification of small-molecule allosteric modulators of 7TMRs. The most common approach for lead identification of an allosteric modulator is to assess modulator concentration–response curves in the presence of either a low dose (for PAMs) or sub-maximal dose (for NAMs) of agonist. Modulator potency reflects a composite of modulator affinity and cooperativity with agonist and is dependent upon the concentration of agonist present. Typically, progressive fold-shift analysis of agonist potency curves and, where possible, radioligand-binding assays are performed as second-tier assays to quantify and verify an allosteric mechanism of action. Simple competition between two ligands for a common orthosteric binding site will manifest as progressive dextral displacement of the agonist concentration–response curve in a functional assay with no change in the maximal agonist response; in binding assays, complete inhibition of orthosteric radioligand binding by increasing concentrations of competitor will be observed, irrespective of the concentration of the radiolabeled probe. For $mGlu_1$, $mGlu_2$, and $mGlu_5$, allosteric radioligands have now been developed, allowing assessment of novel modulators using simple competition binding to address allosteric modulation via engagement of a common allosteric binding site.[32,55–59] However, for an allosteric interaction either between an orthosteric and allosteric site or between two allosteric sites, the change in orthosteric agonist potency or radioligand binding can deviate dramatically from the expectations of simple competition.

The hallmark of an allosteric interaction is the saturable nature of modulation, as defined by the cooperativity between the two binding sites. As for many 7TMRs, mGlu modulators may also be assessed for their ability to perturb orthosteric radioligand binding, for example, [^{3}H]quisqualate for group I mGlus and [^{3}H]LY354740 or [^{3}H]-(2*S*,2′*R*,3′*R*)-2-(2′3′-dicarboxycyclopropyl)glycine (DCG-IV) for group II mGlus. While the majority of mGlu allosteric modulators elicit their effects exclusively via efficacy modulation, there are a number of exceptions where affinity modulation is also involved. Increased affinity of [^{3}H]quisqualate and [^{3}H]DCG-IV has been reported in the presence of $mGlu_1$ and $mGlu_2$ PAMs, respectively.[60,61] Furthermore, the orthosteric agonists LY345740 and glutamate increased binding of the $mGlu_2$ PAM radioligand [^{3}H]-2,2,2-trifluoroethyl [3-(1-methyl-butoxy)-phenyl]-pyridine-3-ylmethyl-sulphonamide (2,2,2TEMPS), while quisqualate enhanced the apparent affinity of some, but not all, $mGlu_5$ PAMs.[32,48] On the other hand, a NAM may decrease radioligand binding, as seen for several $mGlu_2$ NAMs and [^{3}H]LY354740, indicating that the radioligand occupies the receptor in the presence of NAM, albeit with significantly reduced affinity.[32]

Radioligand dissociation kinetic assays also present a useful means for validating an allosteric mode of action. The dissociation rate of a pre-equilibrated radioligand-receptor complex can be modified only if the test ligand interacts with another site on the complex, changing the receptor conformation prior to the radioligand dissociation. Interestingly, despite the success of this technique with respect to identification of allosteric modulators of class A 7TMRs, modulation of dissociation kinetics has yet to be shown for an allosteric interaction at an mGlu. Radioligand-binding assays provide the means of detecting allosteric ligands that modulate orthosteric ligand affinity and can also be used to quantify the allosteric effect in terms of the simple ATCM, providing estimates of modulator K_B and α. However, efficacy modulation cannot be quantified nor detected through the use of radioligand-binding assays. In a functional assay, the presence of an allosteric modulator can cause concentration-dependent translocation of the agonist concentration–response curve to either the right (NAM; Fig. 2.2A) or left (PAM; Fig. 2.2B) that will approach a limit as defined by the cooperativity (either α or β) between the two ligands. When the change in agonist potency by a NAM is analyzed using

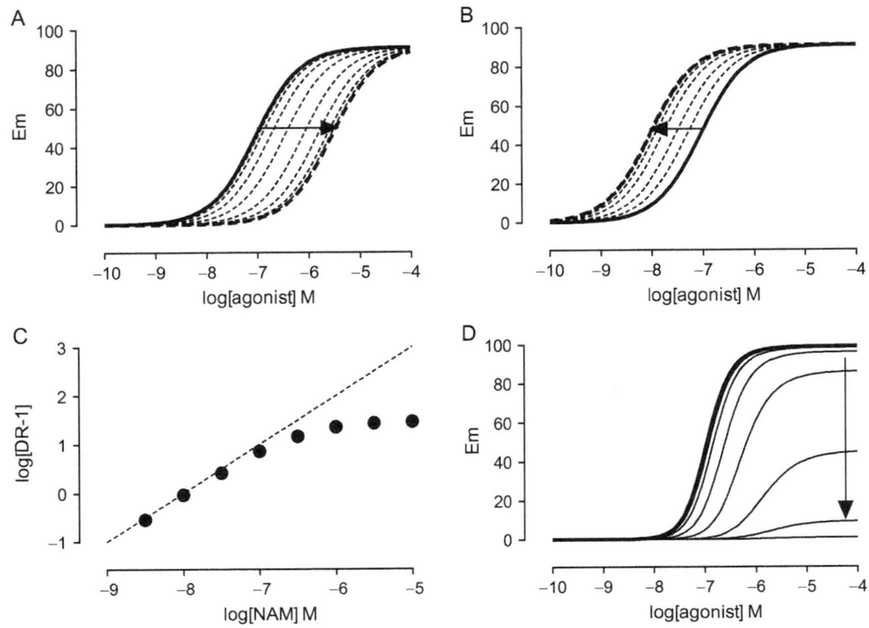

Figure 2.2 Simulations of the possible effect of allosteric modulators on agonist functional responses. (A) A negative allosteric modulator may cause a progressive rightward shift in the agonist concentration–response curve that approaches a limit as defined by the cooperativity between the two sites. (B) On the other hand, a positive allosteric modulator will cause a progressive leftward shift of the agonist concentration–response curve. (C) Schild regression analysis of the effect of the NAM in A, results in a relationship that deviates significantly from unity (dashed line). (D) Negative modulation of efficacy can also result in a depression in the maximal response to agonist, a profile that is commonly observed for mGlu NAMs.

Schild regression,[62] a curvilinear relationship may be observed (Fig. 2.2C). Furthermore, changes in the maximal agonist response may also be observed, indicative of efficacy modulation (a function of β; Fig. 2.2D). In addition to deviations in a pharmacological profile from the expectations of simple competition, chimeric and mutant receptor constructs are also routinely used to confirm an allosteric mechanism of action (further discussed later).

4. LOCATION OF ALLOSTERIC SITES

4.1. Structural features of mGlus

The mGlus belong to the class C 7TMR subfamily along with the Calcium-sensing, γ-aminobutyric acid $(GABA)_B$, pheromone and taste receptors, and are characterized by their large extracellular N-terminal domain, in addition

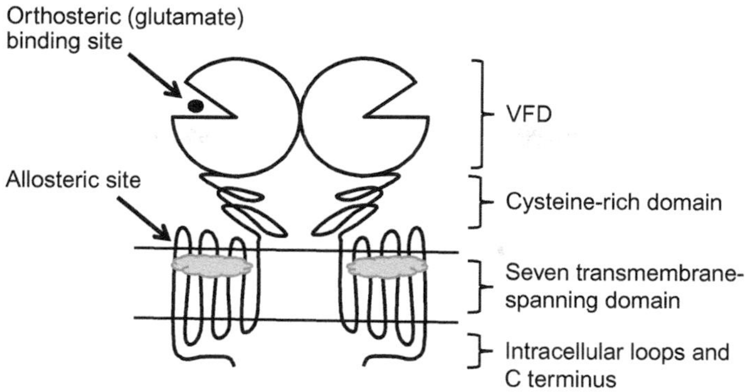

Figure 2.3 Cartoon representation of the structural features of mGlus.

to the seven transmembrane-spanning α-helical domains (7TMs) that are the hallmark of a 7TMR (Fig. 2.3). The large N-terminal domain, commonly referred to as the Venus-Flytrap domain (VFD), contains the endogenous ligand-binding site. Crystal structures of the VFD of $mGlu_1$, $mGlu_3$, and $mGlu_7$ have been solved, revealing that the VFD is made up of two lobes that form a clam shell-shaped structure, with the glutamate-binding site found between the two lobes.[63–65] The mGlus function as dimers, both homo- and heterodimers, with a disulfide bond predicted between the VFDs of individual protomers.[66] Binding of glutamate results in closing of the clam-shell-like lobes of the VFD; this conformational change is transmitted via the cysteine-rich domain to the 7TMs and subsequently to intracellular signaling partners.[63–65,67–70] The cysteine-rich domain is a unique feature of the class C 7TMR family (with the exception of $GABA_B$ receptor, which does not possess one) that links the VFD and 7TMs via a conserved disulfide bridge.[67,71] All mGlu small molecule allosteric modulators discovered to date are believed to interact with the 7TM domains (see below).

4.2. Localization of allosteric binding sites

Chimeric receptor constructs have proven very useful in localizing small molecule allosteric modulator binding pockets of the mGlus to the 7TM domains. Initially utilized to determine orthosteric agonist selectivity,[72,73] exchange of the VFD from one mGlu subtype with another, or with another class C 7TMR family member, has proven fruitful in validating an allosteric mechanism of action. For example, 7-(hydroxyimino)cyclopropa[*b*]chromen-1a-carboxylate ethyl ester (CPCCOEt) was validated as a selective $mGlu_1$ allosteric modulator through the use of chimeric constructs between $mGlu_1$ and the calcium-sensing receptor as well as with other mGlu subtypes.[46,74,75] Subsequently, this strategy has been employed as the first step to characterize numerous selective mGlu NAMs and PAMs.[45,61,76–78]

In addition to chimeric receptor constructs, another method of confirming that allosteric modulators elicit their effects exclusively through interactions with the 7TM domains is through the use of truncated receptor constructs. For example, the extracellular VFD of an mGlu can be eliminated, resulting in a "headless" mGlu that retains intact and functional 7TM domains and the C terminus.[79] Thus, headless mGlus retain intracellular signaling functionality but lack the orthosteric binding site. Importantly, headless mGlus can be positively, or negatively, regulated by ligands like any other 7TMR, such that a PAM becomes an agonist and a NAM an inverse agonist.[79–81]

While chimeric and headless receptors constructs are useful for validating an allosteric mechanism of action and gross localization of allosteric binding sites, they do not provide detailed structural information. A crystal structure of the 7TM domains for a class C 7TMR has yet to be solved; however, utilization of class A 7TMR templates in homology modeling has provided some structural insights with respect to allosteric binding pockets of mGlus. Unfortunately, there is low sequence identity between class A and C 7TMRs (<20%), preventing atomic resolution structural information from homology models of class C 7TMRs. However, homology modeling alongside experimental mutagenesis studies has proven fruitful. Early studies of allosteric binding pockets of $mGlu_1$ and $mGlu_5$ localized an allosteric binding pocket within the 7TM domains in a region analogous to the rhodopsin/biogenic amine orthosteric binding pocket.[45,55,82,83,83a] This general localization continues to be further refined to residues that govern selectivity, are functionally important, or are binding determinants through site-directed mutagenesis (see later for discussion).

4.3. Common allosteric sites within and between subtypes

The idea that allosteric modulators may share a common binding pocket across subtypes arises from the identification of modulators that lack subtype selectivity. For example, the $mGlu_5$ NAMs 2-methyl-6-(2-phenylethenyl)pyridine (SIB-1893) and 2-methyl-6-(phenylethynyl)pyridine (MPEP) are also $mGlu_4$ PAMs, potentiating the response to L-2-amino-4-phosphonobutyrate (L-AP4), a group III mGlu orthosteric agonist.[84] DFB and CPPHA ($mGlu_5$ PAMs) are also weak $mGlu_4$ NAMs,[34,36] while *N*-phenyl-7-(hydroxyimino)cyclopropa[*b*] chromen-1a-carboxamide (PHCCC), an $mGlu_4$ PAM, is also an $mGlu_1$ NAM.[85] Efforts toward understanding allosteric binding sites have been greatly enhanced by the development of radiolabeled mGlu allosteric modulators. For $mGlu_1$, three selective allosteric radioligands have been developed: [^{3}H]-1-(3,4-dihydro-2H-pyrano[2,3-b] quinolin-7-yl)-2-phenyl-1-ethanone (R214127),[56] [^{3}H]-1-ethyl-2-methyl-6-oxo-4-(1,2,4,5-tetrahydro-benzo[*d*]azepin-3-yl)-1,6-dihydro-pyrimidine-5-carbonitrile (EM-TBPC),[55] and [^{3}H]-6-amino-*N*-cyclohexyl-*N*,3-dimethylthiazolo[3,2-a]benzimidazole-2-carboxamide (YM298198).[57] In addition, multiple selective tritiated radioligands and positron emission tomography (PET) ligands have been developed for $mGlu_5$,[58,59,86–91] providing capabilities for probing allosteric small molecule binding of $mGlu_5$, both *in vitro* and *in vivo*. Very recently, the first radiolabeled PAM, [^{3}H]2,2,2-TEMPS, a $mGlu_2$ selective compound, was reported.[32] Conducting competition-binding assays with allosteric radioligands allows for direct assessment of a competitive versus allosteric binding mode for novel allosteric modulators at known allosteric sites.[29,32,47,55–57,81,92] Furthermore, the identification of neutral allosteric ligands, or SAMs that compete or allosterically interact with other allosteric modulators, has also facilitated efforts in understanding allosteric modulator binding and classifying novel modulators as either competitive or noncompetitive with a known allosteric site.[40,93,94] For further details on the use of SAMs in *in vitro* profiling of allosteric compounds, see Chapter 1.

Initial mutational sites to identify a common mGlu allosteric modulator binding pocket have been chosen based on models predicting a pocket that may resemble the site where 11-*cis*-retinal, an inverse agonist, binds in bovine rhodopsin.[95,96] Indeed, critical residues for *cis*-retinal binding in rhodopsin accurately predicted residues that are important for binding of mGlu allosteric ligands, validating the use of rhodopsin as a template for the transmembrane region of class C 7TMRs.[55,83] Residues important for allosteric modulator binding and cooperativity determinants have been characterized for $mGlu_1$, $mGlu_2$, and $mGlu_5$ (Table 2.1). The binding site for [^{3}H]EM-TBPC at $mGlu_1$ has been localized to TMs 6 and 7,[55] which is also where functionally important residues are located that are significant for the interaction of other $mGlu_1$ NAMs: 3-cyclohexyl-5-fluoro-6-methyl-7-(2-morpholin-4-ylethoxy)-4H-chromen-4-one (CFMMC), 2-[4-(indan-2-ylamino)-5,6,7,8-tetra-hydro-quinazolin-2-ylsulfanyl]-ethanol hydrochloride (LY456066), YM298198, 4-[1-(2-fluoropyridin-3-yl)-5-methyl-1H-1,2,3-triazol-4-yl]-*N*-isopropyl-*N*-methyl-3,6-dihydropyridine-1(2H)-carboxamide (FTICD), and CPCCOEt.[46,80,98,99] Similarly, binding determinants for MPEP and fenobam at $mGlu_5$ are found on TMs 3, 6, and 7, as seen in

Table 2.1 Amino acids implicated in binding and cooperativity of mGlu allosteric modulators

	mGlu$_1$						mGlu$_2$		mGlu$_5$				
	PAMs	**NAMs**					**PAMs**	**NAMs**	**PAMs**			**NAMs**	
Amino acid[a]	**CPPHA**	**Ro67-7476**	**EM-TBPC**	**CFMMC**	**LY45606, YM298198 FTICD**	**CPCCOEt**	**LY487379, MRLSD-650, BINA**	**RO4988546, RO5488608**	**CPPHA**	**VU29**	**DFB**	**MPEP**	**Fenobam**
1.42	F599								F585				
3.28								R635					
3.29								R636				R648	R648
3.36		S668						F643			P655	**P655**	**P655**
3.39		C671									S658	**S658**	**S658**
3.40			Y672									**Y659**	**Y659**
4.45							S688						
4.46							G689						
4.55				I725									
45.51			N747										
45.52								H723					
45.54			N750								N734		
5.47		L757	L757					L732[b]			L744		**L744**

5.48					A733[c]					
5.50		N760	N760[d]		N735	N735[e]				
6.44								T781	**T781**	**T781**
6.48	**W798**	W798	W798[d]			W773		W785	**W785**	**W785**
6.51	**F801**	F801	F801					F788	**F788**	**F788**
6.55	**Y805**	Y805	Y805			F780		Y792	**Y792**	**Y792**
7.32	**T815**	T815	T815	T815				M802	**M802**	
7.35				A818		V798			**S805**	
7.40							A810	A810	**A810**	**A810**

[a]The position of each residue in the mGlu 7TMD is given by the numbering system proposed by Ballesteros and Weinstein[338] and is based on the alignments with Rhodopsin and β_2-adrenoceptor described in Gregory *et al.*[49] Residues in the second extracellular loop (ECL2) are labeled "45" to indicate their location between helices 4 and 5. Amino acids are numbered based on their location within the human mGlu sequence.
[b]Mutation of this residue had no effect on RO4988546.
[c]BINA not tested at mutations of this residue.
[d]Effects only LY456066.
[e]Mutation of this residue had no effect on NAM potency of MNI compounds.[101]
In bold: residues implicated in allosteric modulator binding via use of a selective allosteric radioligand.
Refs. 30,32,45–47,55,60,61,80,83,83a,97–101.

Table 2.1.[45,83,83a,97] Subsequent studies of the $mGlu_5$ PAMs DFB, 4-nitro-*N*-(1,3-diphenyl-1H-pyrazol-5-yl)benzamide (VU29) and 3-cyano-*N*-(1,3-diphenyl-1H-pyrazol-5-yl)benzamide (CDPPB) show overlap between point mutations that caused a loss in PAM activity and those that reduce the binding and NAM activity of MPEP.[30,81,97] As further evidence for this commonality in the location of allosteric binding pockets among the mGlu receptors, substituting nonconserved residues of $mGlu_1$, known to be important for CPCCOEt and Ro67-7476 allosteric modulation, onto $mGlu_5$ results in a gain of function for these $mGlu_1$ selective modulators at $mGlu_5$.[46,61] Conversely, substitution of residues important for MPEP binding onto $mGlu_1$ results in the gain of [^{3}H] MPEP binding.[45] Point mutations that perturb modulation by $mGlu_2$ selective PAMs have been mapped to TMs 4 and 5.[60,100,101] The two residues identified in TM5 cluster near the common allosteric site of group I mGlus, while those in TM4 do not. More recently, characterization of the binding pocket of novel $mGlu_2$ NAMs, 5-[7-trifluoromethyl-5-(4-trifluoromethyl-phenyl)-pyrazolo [1,5-a]pyrimidin-3-ylethylnyl]-pyridine-3-sulphonic acid (RO4988546) and 3′-(8-methyl-4-oxo-7-trifluoromethyl-4,5-dihydro-3H-benzo[*b*][1,4] diazepin-2-yl)-biphenyl-3-sulphonic acid (RO5488608), mapped residues within TMs 3, 5, 6, and 7 that showed significant overlap with the common pocket identified for the group I mGlus.[32] Importantly, both of these NAMs fully displaced the binding of the radiolabeled PAM [^{3}H] 2,2,2-TEMPS, indicating a common, or at least overlapping, binding pocket utilized for NAMs and PAMs of $mGlu_2$, mirroring that observed for $mGlu_5$. To date, unlike the mutations in group I mGlus, the existing $mGlu_2$ data relies entirely upon functional assays; thus, it remains to be determined whether these residues are required for modulator binding or the expression of cooperativity.

4.4. Multiple allosteric sites within a receptor subtype

In addition to the concept of a common allosteric binding pocket across the mGlu family, there are multiple lines of evidence for additional allosteric sites per receptor protein. The most definitively established second allosteric site is that of CPPHA for the group I mGlus. Initially, CPPHA was found to show no displacement of the binding of [^{3}H]-3-methoxy-5-(2-pyridinylethynyl)pyridine (methoxyPEPy) to the common allosteric site at $mGlu_5$, even at concentrations several orders of magnitude higher than its potency as a PAM.[30,34] At $mGlu_1$, CPPHA does not compete with [^{3}H]R214127, a compound which is thought to bind to a site on $mGlu_1$ that is homologous with the MPEP binding site on $mGlu_5$,[45,56] despite having PAM activity at both group I mGlus.[30] More recently, CPPHA was found to not fully displace [^{3}H]MPEP binding, indicative of co-occupation of the receptor with both CPPHA and the radioligand.[48] Furthermore, 5-methyl-6-(phenylethynyl)-pyridine (5MPEP), an $mGlu_5$ SAM, noncompetitively inhibits CPPHA potentiation of glutamate, suggesting that ligands at the distinct MPEP and CPPHA sites can allosterically regulate one another.[30] Mutational studies further support the existence of a second allosteric site on the group I mGlus. Mutation of a phenylalanine in TM1 at position 585 to isoleucine in $mGlu_5$, and its equivalent at $mGlu_1$ (F599I), eliminated the potentiation of $mGlu_5$ by 1 μM CPPHA[30] without affecting the interaction of PAMs at the common allosteric site. Conversely, substitution of A810V in $mGlu_5$, a point mutation known to perturb MPEP binding, had no effect on the PAM activity of CPPHA.[30] Additionally, two $mGlu_1$ selective PAMs, 4-nitro-*N*-(1-(2-bromophenyl)-3-phenyl-1H-pyrazol-5-yl)benzamide (VU48) and 4-nitro-*N*-(1,4-diphenyl-1H-pyazol-5-yl) benzamide (VU71), act noncompetitively with [^{3}H]R214127 and are not susceptible to mutations known to affect NAMs that bind the common group I mGlu allosteric site.[47] Collectively, these data strongly support the presence of at least one other allosteric site on group I mGlus.

Multiple binding sites have also been suggested for $mGlu_4$ PAMs and $mGlu_2$ NAMs. At $mGlu_4$, PHCCC was unable to influence the concentration–response curve of cis-2-([(3,5-dichlorophenyl)amino]carbonyl) cyclohexanecarboxylic acid (VU0155041), a PAM from a different chemical class, suggesting that these two ligands do not act at a single allosteric site.[27] Additional support for two allosteric sites on $mGlu_4$ comes from data showing that PHCCC was able to displace an $mGlu_4$ PAM radioligand (4-PAM2) but was not able to displace compounds within the VU0155041 scaffold.[102] Mutational studies have yet to be performed to elucidate the binding and/or cooperativity determinants contributing to $mGlu_4$ modulation by either chemotype. At $mGlu_2$, N735D, known to perturb the allosteric modulation by PAMs,[60,100] had no effect on NAMs from the MNI series of compounds.[101] Moreover, the double point mutation of S6888L + G689V reduces $mGlu_2$ PAM activity but has no effect on the function of NAMs.[32,60,100] In the second extracellular loop of $mGlu_2$, substitution of H723V affects inhibition by NAMs but has no effect on $mGlu_2$ PAMs *N*-(4-(1H-tetrazol-5-yl)phenyl)-2-((6,7-dichloro-2-cyclopentyl-2-methyl-1-oxo-2,3-dihydro-1H-inden-5-yl) oxy)acetamide (MRLSD-650) or 2,2,2-trifluoro-*N*-[4-(2-methoxyphenoxy) phenyl]-*N*-(3-pyridinylmethyl) ethanesulfonamide (LY487379).[32,60,100]

However, as mentioned earlier, given that these $mGlu_2$ data rely upon functional measures alone, it remains to be seen whether or not these differential effects correspond to different binding sites or merely reflect the manifestation of positive versus negative cooperativity. Allosteric modulators acting at different sites could potentially engender different receptor conformations and thus regulate mGlus via different pathways. Indeed, as mentioned previously, CPPHA has been shown to differentially affect ERK1/2 phosphorylation relative to inositol phosphate accumulation in rat cortical astrocytes, while PAMs competitive for the common allosteric site have been shown to have equivalent effects in both assays.[22] Our understanding of the structure–function relationships of mGlus is currently in its infancy, requiring the development of more subtype-selective tools, particularly high-affinity radiolabels for many of the subtypes and for additional allosteric sites. Delineation of the determinants for modulator cooperativity and affinity has the potential to facilitate our understanding of many of the complexities of mGlu allosteric modulators such as steep SAR, modulator bias, and molecular switches within a scaffold, thereby aiding development of mGlu therapeutics.

5. THERAPEUTIC INDICATIONS FOR ALLOSTERIC MODULATORS

mGlus are known to be expressed widely throughout the central nervous system and the periphery. The role of each of the subtypes has been explored utilizing mouse knockouts of individual receptor subtypes and through the development of ligands acting at specific receptor subtypes. Furthermore, mGlus are thought to be therapeutic targets for a multitude of central nervous systems disorders, including disorders such as schizophrenia, cognition, drug abuse, Parkinson's disease (PD), autism spectrum disorders, anxiety, and depression. Research on these receptors as targets spans from basic science studies to human clinical trials. The rational and current therapeutics at each receptor subtype will be discussed below and are summarized in Table 2.2.

5.1. $mGlu_1$ NAMs for neuropathic pain

Neuropathic pain is a chronic condition that results in long-term hyperalgesia and allodynia.[103] $mGlu_1$ receptors are found in brain regions associated with the perception of pain such as the dorsal root ganglia, dorsal horn of the spinal cord, thalamus, amygdala, and cerebral cortex.[104] $mGlu_1$ knockout animals exhibit decreased sensitivity to pain, suggesting that

Table 2.2 Therapeutic indications for mGlu allosteric modulation and representative allosteric modulators

		Positive allosteric modulators and/or allosteric agonists		Negative allosteric modulators	
mGlu		**Therapeutic indication**	**Allosteric modulators**	**Therapeutic indication**	**Allosteric modulators**
Group I	1		CPPHA[a], Ro67-7476, VU71, Ro01-6128, Ro67-4853, and VU48	Neuropathic pain	CPCCOEt, EM-TBPC, FTICD, JNJ16259685, CFMMC, LY456066, and YM298198
	5	Cognition, schizophrenia, and TSC	CDPPB, CPPHA[a], ADX47273, VU0360172, VU29, CPPZ, VU0364289, DFB, and VU0092273	Astrocytic disorders, autism, depression, drug abuse, migraine, FXS, PD-LID[b], and AD	MPEP, MTEP, AFQ056, ADX48621, fenobam, RO4917523, and M-5MPEP
Group II	2	Anxiety, cognition, depression, drug abuse, schizophrenia, and AD	BINA, LY487379, THIIC, CBiPES, and MRLSD-650	Depression	RO4988546[a], RO5488608[a], MNI-167, and RO4491533[a]
	3	Anxiety, cognition, depression, drug abuse, schizophrenia, and AD	None reported	Depression	RO4491533[a], LY2389575, and VU0463597/ML289
Group III	4	PD, neuroinflammation, and neuroprotection	PHCCC, VU04155051, VU0364770, Lu AF21934, and ADX88178		None reported
	7	Anxiety, depression, and PD	AMN082	Anxiety and depression	MMPIP
	8	Anxiety, drug abuse, pain, and PD	None reported		None reported

[a]Non-selective between group members.
[b]Parkinson's disease levodopa-induced dyskinesia.

antagonism of $mGlu_1$ may be a therapeutic strategy for the treatment of neuropathic pain.[103] The first compound to be identified as an $mGlu_1$ NAM was CPCCOEt,[46,85] which exhibited a neuroprotective effect in cortical cultures and organotypic slices following ischemic injury.[105] Subsequently, the $mGlu_1$ selective NAM (3,4-dihydro-2H-pyrano[2,3]b quinolin-7-yl) (cis-4-methoxycyclohexyl) methanone (JNJ16259685) was shown to have efficacy in the formalin model of hyperalgesia in rat.[106] Recently, Bennett and colleagues demonstrated that two structurally related fused tricyclic $mGlu_1$ antagonists exhibited analgesic properties in a rat spinal nerve ligation assay,[107] suggesting that $mGlu_1$ antagonists may have efficacy for the treatment of neuropathic pain.

5.2. $mGlu_{2/3}$ PAMs and agonists for schizophrenia and cognition

Schizophrenia is a devastating disease that affects approximately one percent of the population worldwide. It is characterized by three symptom clusters: positive symptoms (hallucinations, delusions), negative symptoms (flattened affect, withdrawal, lack of motivation), and cognitive dysfunction (impaired memory attention, working memory deficits, and affective (anxiety/depression)).[108] Current antipsychotics primarily treat the positive symptoms of this disorder, having little or no efficacy in addressing the negative and cognitive aspects of the disease. Thus, there is a great need for the development of better therapeutics for schizophrenia. Research suggests that there is hypofunction of *N*-methyl-D-aspartate (NMDA) receptors in schizophrenia, leading to decreased downstream activation of GABAergic neurons, which in turn results in disinhibition and ultimately an overall increase in excitation.[109,110] In further support of the glutamatergic hypothesis of schizophrenia, NMDA receptor antagonists such as phencyclidine (PCP) and ketamine induce symptoms in rats and humans that mirror the positive, negative, and cognitive symptoms observed in schizophrenic patients.[111] Additionally, ligands that enhance NMDA receptor function, such as glycine co-agonists, have been shown to be efficacious in the treatment of schizophrenia.[112]

Group II mGlus are expressed in regions thought to be involved in schizophrenia, such as circuitry within the prefrontal cortex, striatum, thalamus, hippocampus, and amygdala.[113] In addition to providing a feedback mechanism to regulate glutamate release, these receptors can negatively modulate the transmission of other neurotransmitters including GABA, norepinephrine, and dopamine.[114] Two common behavioral assays often used to assess antipsychotic-like activity are PCP-induced hyperlocomotion and stereotypy and amphetamine-induced hyperlocomotion.[115–119] Group II agonists demonstrate efficacy in preclinical models of psychosis and anxiety.[109,120–123] LY354740 was the first $mGlu_{2/3}$ agonist shown to reverse PCP-induced hyperlocomotion and stereotypy[120]; the structurally related $mGlu_{2/3}$ agonists LY379268, 5-[2-[4-(6-fluoro-1H-indole-3-yl) piperidin-1-yl]ethyl]-4-(4-fluorophenyl)thiazole-2-carboxylic acid amide (MGS0008), and (1*R*,2*S*,5*S*,6*S*)-2-amino-6-fluoro-4-oxobicyclo[3.1.0] hexane-2,6-dicarboxylic acid monohydrate (MGS0028) also demonstrate efficacy in this model.[123,124] In addition, LY354740 and LY379268 also reverse amphetamine-induced hyperlocomotion.[123]

Encouragingly, group II agonists have been shown to have clinical efficacy in schizophrenic patients. (1*R*,4*S*,5*S*,6*S*)-2-thiabicyclo[3.1.0]-hexane-4,6-dicarboxylic acid,4-[(2*S*)-2-amino-4-(methylthio)-1-oxobutyl] amino-,2,2-dioxide monohydrate (LY2140023), an oral prodrug of orthosteric agonist (−)-(1*R*,4*S*,5*S*,6*S*)-4-amino-2-sulfonylbicyclo[3.1.0] hexane-4,6-dicarboxylic acid (LY404039), improves the positive and negative symptoms of patients with schizophrenia. LY2140023 exhibits similar efficacy when compared to the atypical antipsychotic olanzapine but without the side effects, such as weight gain or extrapyramidal symptoms, associated with typical and atypical antipsychotics.[125] A multicenter Phase 2 clinical trial has also been conducted in patients with schizophrenia; however, neither LY2140023 nor the active control olanzapine produced effects that differed from placebo.[126] Due to the lack of response of olanzapine, further clinical studies are warranted to determine the true efficacy of LY2140023.

Assessments of cognitive enhancing effects of $mGlu_{2/3}$ agonists have produced mixed results. For example, LY354740 improves the performance in a T-maze task following administration of PCP.[120] Additionally, LY487379 promotes cognitive flexibility and facilitates behavioral inhibition.[127] LY379268 prevents PCP-induced accuracy deficits observed in mice using a five-choice serial reaction time task.[128] In contrast, others have reported no effect or an impairment of cognitive performance following administration of $mGlu_{2/3}$ agonists.[129–131]

The development of LY354740 shows that it is possible to target the orthosteric site of the group II receptors and achieve selectivity versus the other six mGlus; however, LY354740 does not discriminate between $mGlu_2$ and $mGlu_3$. It is of note that the effects of LY404039 are absent in $mGlu_2$ knockout mice but present in $mGlu_3$ knockout mice,[132] suggesting that activation of $mGlu_2$ mediates the clinical efficacy of group II agonists.

The role of $mGlu_2$ in mediating antipsychotic effects has recently been explored with the availability of selective $mGlu_2$ PAMs, such as the compounds BINA and LY487379. BINA modulates excitatory transmission in the medial prefrontal cortex, a region thought to be involved in the pathology of schizophrenia.[119] LY487379 potentiates the inhibition of field excitatory postsynaptic potentials by the group II selective agonist, DCG-IV, at the medial perforant path-dentate gyrus synapse in the hippocampus, an effect known to be mediated by $mGlu_2$.[60] In rodent models of psychosis, both BINA and LY487379 reverse PCP and amphetamine-induced hyperlocomotion.[118,133,134] Taken together, these results suggest that selectively targeting $mGlu_2$ receptors may have efficacy in the treatment of schizophrenia, addressing both psychosis and cognitive deficits. The $mGlu_2$ PAM ADX71149, a compound resulting from a partnership between Addex and Janssen Pharmaceuticals Inc., is now undergoing Phase II clinical trials.

Psychosis has been associated with Alzheimer's disease (AD) in addition to the well-known accumulation of β-amyloid (Aβ) protein. LY379268 was used to assess the potential for the use of $mGlu_{2/3}$ agonist for the treatment of AD. LY379268 reduced neurodegeneration mediated by Aβ in cortical cell cultures.[135] Interestingly, the $mGlu_2$ PAM *N*-(4′-cyano-biphenyl-3-yl)-*N*-(3-pyridinylmethyl)-ethanesulfonamide hydrochloride (LY566332) increased the toxicity of Aβ peptides.[135] Additionally, the $mGlu_3$ NAM (3*S*)-1-(5-bromopyrimidin-2-yl)-*N*-(2,4-dichlorobenzyl) pyrrolidin-3-amine methanesulfonate hydrate (LY2389575) blocked the neuroprotective effect of LY379268,[135] suggesting that coactivation of $mGlu_{2/3}$, rather than selective activation of $mGlu_2$, may have therapeutic potential for the treatment of psychosis associated with AD. These results indicate that $mGlu_2$ selective compounds may not always be ideal for the treatment of psychosis, depending on the accompanying disease state.

5.3. $mGlu_{2/3}$ in anxiety

Anxiety disorders can be triggered by a multitude of factors including increased stress and poor coping mechanisms. Dysregulation of neurotransmitters such as glutamate, serotonin, and GABA are thought to contribute to anxiety disorders[136,137] and areas of the brain such as amygdala, cortex, thalamus, and hypothalamus have all been implicated in anxiety.[136,137] Group II mGlu selective orthosteric agonists LY354740, LY379268, and LY404039 exhibit efficacy in rodent behavioral models predictive of anxiolytic efficacy.[109,138,139] Furthermore, LY354740 demonstrates efficacy in patients in treating panic attacks and generalized anxiety disorder.[140,141] The $mGlu_2$ PAMs LY487379, CBiPES, and BINA also exhibit anxiolytic effects when assessed in rodent models of anxiety such as the fear-potentiated startle response, stress-induced hyperthermia, and elevated plus maze tasks.[116,134,140] Furthermore, the effect of LY487379 on the fear-potential startle response can be blocked by a group II/III selective antagonist, supporting the concept that this effect is mediated by group II mGlus.[116,140] *N*-(4-((2-(trifluoromethyl)-3-hydroxy-4-(isobutyryl)phenoxy)methyl)benzyl)-1-methyl-1H-imidazole-4-carboxamide (THIIC), an $mGlu_2$ PAM representing a different chemical scaffold distinct from BINA and LY487379, also reversed marble-burying, stress-induced hyperthermia, and stress-induced increases in cerebellar cGMP,[142] further suggesting that $mGlu_{2/3}$ activation is a viable target for the treatment of anxiety.

5.4. $mGlu_{2/3}$ PAMs for drug abuse

In addition to potential roles in modulating circuitry involved in psychosis, memory, and anxiety, $mGlu_{2/3}$ receptors are expressed in the ventral tegmental area and nucleus accumbens, brain regions key for the development of drug addiction. LY354740 was the first compound shown to prevent nicotine withdrawal in rats.[143] Other agonists, LY379268 and LY404039, also reduce drug-seeking behaviors in animal models of self-administration of cocaine, nicotine, and ethanol.[144–150] Furthermore, direct injection of LY379268 into the ventral tegmental area, nucleus accumbens, or amygdala attenuates drug-seeking behavior.[145,151–153] Recently, Kufahl and colleagues demonstrated that administration of LY379268 decreased cue-induced and drug-primed reinstatement of methamphetamine seeking.[154] The $mGlu_2$ PAM BINA and a BINA analog also suppress drug self-administration.[147,155] Collectively, these data demonstrate that targeting $mGlu_{2/3}$ has potential for treatment of drug abuse.

5.5. $mGlu_{2/3}$ PAMs for depression

Major depressive disorder is one the most costly disorders of the central nervous system. The incidence of depression varies throughout the world, ranging from 2% to 16% of the population depending on the region.[156,157] Currently, monoamine oxidase inhibitors and tricyclic antidepressants are used in the treatment of depression; however, these drugs are not effective in many patients, especially those with drug-resistant depression. Additionally, the onset of action for current therapeutics is slow, with the first improvements in mood taking 3–4 weeks to occur. Therefore, there is a great need for the development of better therapeutics for the treatment of depression.

The pathophysiology of depression is not well understood although recent research has suggested that the glutamatergic system may be altered in depression. Evidence for the involvement of the glutamate in major depressive disorder first came from the antidepressant effects of ketamine, an NMDA receptor antagonist. The first study was conducted in a small group of patients by Berman and colleagues[158] and followed with a double-blind crossover trial.[159] In both cases, ketamine induced a rapid and robust antidepressive effect in patients. As mGlu receptors are known to modulate the activity of NMDA receptors and $mGlu_{2/3}$ receptors are expressed in brain regions thought to be involved in depression, including prefrontal cortex, hippocampus, and amygdala,[113,160] these receptors may be candidates for therapeutics to treat depression. The $mGlu_{2/3}$ agonist LY379268 modulates the activity of NMDA receptors through postsynaptic alterations in Akt/GSK3β signaling to reverse the effects of an NMDA receptor antagonist.[161] Furthermore, LY379268 increases NMDA currents in the prefrontal cortex.[161] If depression results from increased NMDA receptor function, as suggested by NMDA antagonist studies, it is predicted that $mGlu_{2/3}$ antagonists could be beneficial for the treatment of depression.

The efficacy of atypical antipsychotics is thought to occur through an increase in extracellular dopamine in the prefrontal cortex.[162,163] Interestingly, systemic administration of LY379268 also resulted in increased levels of dopamine and its metabolites in the medial prefrontal cortex, along with increasing levels of serotonin.[114,164] These results suggest that $mGlu_{2/3}$ *agonists* could be therapeutic for treating depression. These contrasting results suggest that it is important to further understand which signaling pathways are altered in depression. It is possible that the pathology of different forms of depression may arise from changes in distinct pathways.

The preceding information indicates that there is a complex role of $mGlu_{2/3}$ in depression. The development of $mGlu_{2/3}$ compounds has allowed an initial evaluation into the role of these receptors in depression and into the preferred mode of pharmacology that should be explored for disease treatment; these studies have led to mixed results, indicating that both $mGlu_{2/3}$ agonists and antagonists can demonstrate antidepressant activity. For example, Fell *et al.*[142] demonstrated that the $mGlu_2$ PAM THIIC exerted antidepressant-like activity in three rodent models of depression. Co-administration of group II agonists along with traditional antidepressants has been shown to shorten the time between first administration and achievement of efficacy.[165,166] Somewhat paradoxically, $mGlu_{2/3}$ antagonists have also been found to be effective in treating depression. Antagonists such as (1*R*,2*R*,3*R*,5*R*,6*R*)-2-amino-3-(3,4-dichlorobenzyloxy)-6-fluorobicyclo[3.1.0]hexane-2,6-dicarboxylic acid (MGS0039), (2*S*,1′*S*,2′*S*)-2-(9-xanthylmethyl)-2-(2′-carboxycyclopropylglycine) (LY341495), and the $mGlu_{2/3}$ NAM 4-[3-(2,6-dimethylpyridin-4-yl)phenyl]-7-methyl-8-trifluoromethyl-1,3-dihydrobenzo[b][1,4]diazepin-2-one (RO4491533), exhibit antidepressant efficacy in multiple rodent models of depression, such as the mouse tail-suspension test and learned helplessness paradigm.[167–169] Clinical trials are currently underway to assess the efficacy and safety of RO4995819 (Phase II) and have just been completed for BCI-632 (a prodrug of MGS0039; Phase I) (www.clinicalTrials.gov). Clearly, further research is necessary to determine if $mGlu_{2/3}$ PAMs/agonists and/or NAMs will be efficacious in the treatment of depression. It is possible that the use of PAMs or NAMs may depend on the brain region and state of depression.[170]

5.6. $mGlu_4$ PAMs for PD and neuroprotection

PD is characterized by changes in motor behaviors including bradykinesia, tremors, rigidity, and postural instability, along with disturbances in sleep, depression, and cognitive impairments.[171] The primary pathology in PD is the loss of dopaminergic neurons in the substantia nigra pars compacta that normally project to the striatum and other nuclei in the basal ganglia. The loss of these neurons results in disruptions in the basal ganglia-thalamocortical motor circuit, ultimately causing abnormalities in motor function (for review Ref. 172). Current therapies for PD are focused on dopamine replacement with L-3,4-dihydroxyphenylalanine (L-DOPA) or dopamine agonists; however, long-term treatment with L-DOPA results in a loss of efficacy and the development of side effects, such as motor complications and psychiatric issues.[173–175] Current therapies address PD symptoms, but have no effect on the progression of the disease. Glutamate receptors are known to be expressed throughout the basal ganglia, both pre- and postsynaptically, and have been shown to modulate neurotransmission at multiple synapses, thus making them potential therapeutic targets for PD.

$mGlu_4$ is expressed in the basal ganglia, hippocampus, cerebellum, and olfactory bulbs[176] and its expression in the basal ganglia makes this receptor a potential therapeutic target for the treatment of PD. The highest level of $mGlu_4$ expression in the forebrain is in the external and internal globus pallidus, while lower levels are present in the caudate nucleus, putamen, thalamus, subthalamic nucleus, substantia nigra pars compacta, substantia nigra pars reticulata, hippocampus, and cerebral cortex.[177] Activation of group III mGlus reduces synaptic transmission at the striato-pallidal, corticostriatal, and subthalamic nucleus-substantia nigra pars reticulata synapses.[177–183] Both L-AP4 and L-serine-*O*-phosphate, selective group III agonists, reduce synaptic transmission at these synapses[178–180,184] and have antiparkinsonian effects in rodent models of PD.[179,181] Furthermore, certain doses of the $mGlu_4$ preferring orthosteric agonists (2*S*)-2-amino-4-[hydroxy[hydroxy(4-hydroxy-3-methoxy-5-nitro-phenyl)methyl]phosphoryl]butanoic acid (LSP1-2111) and (2*S*)-2-amino-4-(((4-(carboxymethoxy)phenyl)(hydroxy)methyl)(hydroxy)phosphoryl)butanoic acid (LSP4-2022) demonstrate efficacy in rodent models of PD, such as reversal of haloperidol-induced catalepsy and 6-hydroxydopamine (OHDA)-induced motor deficits.[185,186] Direct injection of the orthosteric group III agonist (3*S*)-3-[(3-amino-3-carboxypropyl (hydroxy)phosphinyl)-hydroxymethyl]-5-nitrothiophene (LSP1-3081) into the striatum improves akinesia in a rodent model of PD.[182] The agonist (1*S*,3*R*,4*S*)-1-aminocyclopentane-1,3,4-tricarboxylic acid (ACPT-1), which activates $mGlu_4$, $mGlu_6$, and $mGlu_8$, also reverses motor deficits in rodent models.[187,188] Interestingly, this compound decreases hyperalgesia in a model of neuropathic pain and also exhibits antidepressant-like effects,[189,190] suggesting that a group III mGlu activator might also be beneficial in treating the non-motor symptoms of PD. In a primate model of PD, direct injection of L-AP4 into the external globus pallidus or striatum modulated cellular activity, suggesting that $mGlu_4$ activation at multiple synapses in the basal ganglia may lead to a reduction in Parkinsonian symptoms.[177]

The first selective $mGlu_4$ PAM characterized was PHCCC.[77,191,192] This compound has no apparent agonist activity in most *in vitro* systems but enhances glutamate potency and efficacy in these assays. In slice recordings, PHCCC potentiates the effects of L-AP4 at the striato-pallidal synapse.[191] In animal models, administration of PHCCC reduces reserpine-induced akinesia[191] and exerts a neuroprotective effect in a model of dopaminergic cell death.[193] Another $mGlu_4$ PAM from a different structural class, VU0155041, demonstrates improved potency, selectivity, and aqueous solubility compared to PHCCC.[27] VU0155041 is efficacious in reversing haloperidol-induced catalepsy and reserpine-induced akinesia when administered intracerebroventricularly[27] and also demonstrates neuroprotective effects in the 6-OHDA model of PD.[194] It should be noted that VU0155041 exhibits agonist activity *in vitro* in addition to its ability to potentiate the response to glutamate; at this point it is unclear if this is a desired characteristic for disease treatment.

Numerous other $mGlu_4$ PAMs have now been developed[195–202] and many exhibit efficacy in rodent PD models. *N*-(3-chlorophenyl)picolinamide (VU0364770) is a selective and systemically active $mGlu_4$ PAM. This compound is active in multiple models of PD either when administered alone or in combination with either an adenosine A_{2A} receptor antagonist, another 7TMR with therapeutic target potential for PD, or L-DOPA.[197] (1*S*,2*S*)-*N*^{1}-(3,4-dichlorophenyl)cyclohexane-1,2-dicarboxamide (Lu AF21934), another selective and systemically active compound related to the VU0155041 series, inhibits corticostriatal transmission in brain slice electrophysiology, both alone and by potentiating the response to the $mGlu_4$ selective agonist, LSP1-2111; this compound also reverses haloperidol-induced catalepsy.[202] Similar to VU0364770, when Lu AF21934 was administered in combination with L-DOPA, it decreased akinesia. Encouragingly, Lu AF21934 also decreased the incidence of L-DOPA-induced dyskinesias.[202] Interestingly, Lu AF21934 and 5-methyl-*N*-(4-methylpyrimidin-2-yl)-4-(1H-pyrazol-4-yl) thiazol-2-amine (ADX88178) have been reported to also induce anxiolytic-like affect in rodent models, including stress-induced hypothermia, four-plate and marble-burying.[203,204] These studies demonstrate the $mGlu_4$ is a viable target for the treatment of PD at multiple synapses in the basal ganglia circuit, perhaps for treating motor and non-motor symptoms and for slowing disease progression.

5.7. $mGlu_4$ PAMs in pain and neuroinflammation

There is evidence that $mGlu_4$ PAMs may be beneficial in the treatment of chronic pain and neuroinflammation. Chronic pain is associated with affective disorder. $mGlu_4$ receptors are found in the dorsal horn of the spinal cord and amygdala, regions known to modulate pain responses.[205,206] The group III agonist L-AP4 reduces capsaicin-induced sensitization of mechanical stimulation.[207] Furthermore, L-AP4 inhibits synaptic transmission in the amygdala in a rodent model of arthritic pain.[208] The group III agonist ACPT-1 and the $mGlu_4$ selective PAM PHCCC reduce inflammatory responses to mechanically induced hyperalgesia and also reduce neuropathic pain in rodent models.[190]

Neuroinflammation is commonly observed in the brains of patients with multiple sclerosis. In rodent models, experimental autoimmune encephalomyelitis (EAE) is exacerbated when $mGlu_4$ is absent.[209] Furthermore, administration of PHCCC attenuates EAE, suggesting that $mGlu_4$ activation may play a neuroprotective role in inflammation[209] and possibly in other disorders involving inflammation. Data continue to emerge regarding the potential of $mGlu_4$ activation in a variety of disease areas, spurred in part due to the development of highly selective allosteric ligands.

5.8. $mGlu_5$ NAMs for anxiety

Currently, there are limited therapeutics for the treatment of anxiety disorder. Interestingly, $mGlu_5$ is expressed in the regions implicated in anxiety disorders.[210] An increase in NMDA receptor function has been associated with anxiety,[6] and $mGlu_5$ antagonists have been shown to reduce NMDA signaling.[211] Chojnacka-Wójcik and colleagues demonstrated that administration of the group I antagonist (*S*)-4-carboxy-3-hydroxyphenylglycine (*S*-4C3H-PG) directly into the hippocampus-induced anxiolytic properties when assessed using the conflict drinking test in rats.[212] However, Hayashi *et al.*[213] showed that this compound is also a partial agonist at group II receptors, complicating the assessment of the role of $mGlu_5$ alone. The first reported selective $mGlu_5$ NAMs were 6-methyl-2-(phenylazo)-3-pyridinol (SIB-1757) and SIB-1893.[214] Shortly thereafter, MPEP was developed, followed by 3-[(2-methyl-1,3-thiazol-4-yl)ethynyl] pyridine (MTEP); both of these compounds showed improved potency, selectivity, and brain penetrance.[215,216] The radioligands and PET ligands developed for $mGlu_5$ based on these scaffolds have proven useful for clinical development to validate target engagement and aid dose-finding.[58,59,86–91] MPEP and other structural analogs exhibit anxiolytic effects in rodent models of anxiety.[217–221] Furthermore, fenobam, a selective $mGlu_5$ NAM, demonstrates efficacy in preclinical rodent models of anxiety[222–224] and in human trials.[225] Interestingly, $mGlu_1$ selective NAMs, such as JNJ16259685, have also been shown to have efficacy in rodent models of anxiety; however, these compounds have also been associated with adverse effects such as memory impairments prohibiting them from further development.[226–229] Although $mGlu_5$-selective NAMs are indicated for treatment of anxiety, the potential effect of these compounds on memory needs to be further assessed.

5.9. $mGlu_5$ NAMs for major depression disorder

As discussed above, there is a great need for the development of drugs to treat major depressive disorder and targeting alterations in glutamatergic signaling may help alleviate depression. $mGlu_5$ is a close signaling partner with

NMDA receptors and antagonism of $mGlu_5$ has been shown to reduce NMDA activity in multiple brain regions,[230–232] suggesting that antagonizing $mGlu_5$ may be a potential treatment for depression. In fact, targeting $mGlu_5$ may avoid adverse side effects observed with direct blockade of NMDA receptors. The $mGlu_5$ NAMs MPEP, MTEP, and (4-difluoromethoxy-3-(pyridine-2-ylethynyl)phenyl)5H-pyrrolo[3,4-b]pyridine-6(7H)-yl methanone (GRN-529) produce antidepressant-like activity in rodent behavioral models of depression, such as the forced swim test and tail-suspension test.[216,217,233–235] The $mGlu_5$ NAM AZD2066 was in phase II clinical trials for major depression, but the study has been terminated. Currently, a phase II clinical trial is underway for the use of the $mGlu_5$ NAM RO4917523 as an adjunctive therapy in patients with major depressive disorder (www.clinicalTrials.gov).

5.10. $mGlu_5$ NAMs for Autism

Fragile X syndrome (FXS) is the leading cause of autism and the most common inherited form of mental retardation.[236,237] FXS is caused by a mutation in the gene encoding fragile X mental retardation protein (FMRP) that represses the translation of specific mRNAs regulating protein translation in neuronal dendrites.[236,238] Patients with FXS exhibit symptoms that include moderate to severe mental retardation, attention deficits, anxiety, developmental delays, heightened responses to sensory stimuli and an increased incidence of epilepsy.[239] $mGlu_5$ plays roles in both long-term depression (LTD) and long-term potentiation (LTP) at the Schaffer collateral-CA1 (SC-CA1) synapse in the hippocampal formation[240–245] and a balance between these two forms of synaptic plasticity is thought to be important for normal cognitive function. A loss of FMRP in mutant mice increases group I mGlu-dependent LTD at the SC-CA1 synapse in hippocampus.[246] Specifically, over-activity of $mGlu_5$ has been suggested to play a role in the cognitive impairments observed in patients with FXS and $mGlu_5$ antagonists have been proposed as a treatment for the disorder.[246–248] In behavioral assays utilizing FXS mice, administration of MPEP reduces anxiety and seizures.[249] Furthermore, crossbreeding FXS mice with $mGlu_5$ knockout mice corrects the majority of FXS phenotypes,[250] suggesting a specific interplay between $mGlu_5$ and FMRP. Acute treatment of Fragile X mice with the $mGlu_5$ NAM CTEP rescued LTD at the SC-CA1 synapse along with correcting protein synthesis and preventing audiogenic seizures.[251] Chronic treatment with CTEP rescued learning and memory deficits, elevated locomotor activity and increased spine density,[251] suggesting that $mGlu_5$ NAM treatment may be effective in addressing multiple symptoms. Importantly, these studies demonstrated that it is possible to treat FXS after the development of the phenotype, suggesting that treatment with $mGlu_5$ NAMs may be clinically efficacious in patients in which the disease is already established.[251] In clinical trials, patients with FXS receiving fenobam displayed improvements in clinical behaviors and prepulse inhibition with no adverse effects.[252] AFQ056, another $mGlu_5$ NAM, is in phase II/III clinical trials for adults with FXS and is now beginning phase I clinical trials in children (www.ClinicalTrials.gov). In addition to the treatment of FXS, there is emerging evidence to suggest that $mGlu_5$ NAMs may be beneficial in treating other forms of autism. Silverman *et al.*[253] found that administration of GRN-529, a selective $mGlu_5$ NAM, reduces repetitive behaviors and aberrant social interactions in mouse models of autism. However, the mechanism of action and the efficacy in other autism models still needs to be assessed.

In addition to the beneficial effects of $mGlu_5$ NAMs, there also may be potential for negative side effects. $mGlu_5$ has been shown to be important in cognitive processing and $mGlu_5$ NAMs such as MPEP have been shown to exacerbate psychosis induced by PCP[254] and cause cognitive impairments and psychotomimetic effects. For example, early clinical trials with fenobam reported adverse effects (psychostimulant); however, the mechanism behind these adverse effects is unclear.[225,255,256] In contrast, no psychostimulant effects or cognitive impairments were reported for fenobam in a recent clinic trial in FXS patients, or for another $mGlu_5$ NAM, ADX10059.[252,257,258] Additional work is needed to assess the potential for negative side effects in larger and more diverse patient populations.

5.11. $mGlu_5$ NAMs for drug abuse

$mGlu_5$ receptors in the nucleus accumbens may play a role in mediating drug reinforcement. Administration of MPEP reduces cocaine self-administration in mice[259] and additional studies demonstrated that administration of $mGlu_5$ antagonists/NAMs decrease the rewarding effects of other drugs of abuse such as amphetamine, morphine, and alcohol.[260–262] Furthermore, literature suggests that blocking $mGlu_5$ activation may prevent drug relapse.[263–266] Thus, $mGlu_5$ may be a viable target for treating substance abuse and relapse. It is interesting to note that $mGlu_1$ antagonists may also be a therapeutic option for substance abuse treatment[267]; these studies are still in the early stages of exploration.

5.12. $mGlu_5$ NAMs for potential treatment of astrocytic disorders

Astrocytes were previously considered to be structural and support cells for neurons; however, recent research suggests that astrocytes play more than just a supportive role in brain function. Astrocytes are known to express mGlus, specifically $mGlu_5$ and $mGlu_3$.[268,269] They are important for the uptake of glutamate from the synapse, which is then converted to glutamine and cycled back to neurons.[270,271] Additionally, astrocytes may also be involved in the clearance of GABA and glycine from the synapse.[272] The idea of a tripartite synapse has been developed that posits that the action at synapse involves the interaction of the pre- and postsynaptic neurons along with astrocytes.[273] Activation of $mGlu_5$ by agonists and PAMs increases calcium oscillations in astrocytes, whereas $mGlu_5$ NAMs decrease calcium oscillations.[48,274–277] Astrocyte dysfunction has been implicated in epileptic activity[278] and MPEP administration is neuroprotective in a mouse model of epilepsy.[279] In addition, antagonism of $mGlu_3$ activity has also been shown to have a neuroprotective effect.[280–282] Reactive astrogliosis has also been observed in diseases such as amyotrophic lateral sclerosis, PD, and AD,[283,284] suggesting that targeting $mGlu_5$ and $mGlu_3$ on astrocytes may be relevant therapeutic directions to explore.

5.13. $mGlu_5$ for treatment of PD, iatrogenic dystonias, gastroesophogeal reflux disorder, and migraine

In addition to their potential as treatments for anxiety, depression, and addiction disorders, $mGlu_5$ NAMs have also been suggested as treatment options for other disorders such as iatrogenic dystonias, chronic pain, gastroesophogeal reflux disorder, and migraine,[285] although in some cases feasibility or physiological basis needs to be further assessed. MPEP and MTEP demonstrate efficacy in rodent models of PD.[286–289] The $mGlu_5$ NAM (3a*S*,5*S*,7a*R*)-methyl 5-hydroxy-5-(*m*-tolylethynyl)octahydro-1H-indole-1-carboxylate (AFQ056) is in phase II clinical trials for L-DOPA-induced dyskinesia associated with the treatment of PD (www.ClinicalTrials.org; Ref. 290) in addition to its potential for use in FXS. Another $mGlu_5$ NAM, ADX48621, is also advancing in phase II clinical trials for the treatment of L-DOPA-induced dyskinesia (www.ClinicalTrials.org). The $mGlu_5$ NAM ADX10059 demonstrated efficacy in an acute trial for gastroesophogeal reflux disorder[257,258]; however, a separate trial investigating its efficacy in migraine was terminated due to liver toxicity (www.ClinicalTrials.org).

As addressed briefly above, there is some uncertainty regarding the development of $mGlu_5$ NAMs for clinical treatments due to the potential for target-related adverse effects. One possible way to overcome this may be the development of "partial antagonists," or antagonists with low levels of negative cooperativity with the orthosteric site. Rodriguez *et al.*[94] reported the discovery of $mGlu_5$ NAMs with weak negative cooperativity, also referred to as "partial antagonists." These compounds may have improved clinical success since they do not completely block the activity of glutamate at 100% receptor occupancy, thus allowing $mGlu_5$ receptors to maintain some level of activity. Incomplete blockade may decrease the adverse liability profile, although this is still a hypothesis that requires further testing using *in vivo* studies.

5.14. $mGlu_5$ PAMs for schizophrenia and cognition

There is evidence to suggest that changes in glutamatergic transmission through specific brain circuits occur in schizophrenia.[291,292] Specifically, NMDA receptor hypofunction has been proposed to play a role in schizophrenia (discussed further above). However, direct targeting of NMDA receptors is unlikely to be effective due to potential toxicity that can result from over-activation of these receptors.[293] Therefore, alternative ways to modulate NMDA receptor activity have been sought. One possible target is $mGlu_5$, which is a closely related signaling partner with NMDA receptors in brain regions involved in cognitive function and implicated in the pathology of schizophrenia such as the hippocampus, striatum, and prefrontal cortex.[34,231,232,294–301] In addition, $mGlu_5$ has also been shown to play a role in cognition.[302,303] Thus, it is possible that activation of $mGlu_5$ may enhance not only NMDA receptor function, leading to the improvement of positive symptoms, but also may provide a strategy to treat the cognitive impairments associated with schizophrenia, an improvement over current therapies.

The first selective $mGlu_5$ PAMs include compounds from three different chemical series: DFB, CDPPB, and CPPHA. These modulators induce a leftward shift in the glutamate response curve of approximately 10-fold for CPPHA and CDPPB and two- to four- fold for DFB, while having no impact on the affinity of glutamate.[29,34,36,304] As mentioned earlier, CDPPB, DFB, and related compounds are thought to interact competitively with the MPEP site,[36,81] whereas CPPHA interacts with the receptor at a second distinct allosteric site.[30] The development of these compounds has now provided tools to study potentiation of $mGlu_5$ activity in electrophysiological preparations studying responses such as LTP, LTD, and depolarization of subthalamic nucleus neurons.[34,36,81,245] Unfortunately, DFB and CPPHA have poor pharmacokinetic properties, making it difficult to utilize them as *in vivo* tools; however, CDPPB has been assessed in multiple animal models. Behavioral paradigms commonly used to assess the antipsychotic-like activity of compounds include amphetamine-induced hyperlocomotion, amphetamine-induced disruption of prepulse inhibition (PPI), and conditioned avoidance responding. CDPPB reverses both amphetamine-induced hyperlocomotion and amphetamine-induced disruption of PPI,[29] suggesting that $mGlu_5$ PAMs exhibit an antipsychotic profile. In addition, $mGlu_5$ PAMs may also be viable tools for enhancing cognition and attenuating the negative symptoms of schizophrenia. For example, Vales *et al.*[129] demonstrated that CDPPB blocked the cognitive deficits induced by MK-801, while having no effect on locomotor activity. Sub-chronic administration of CDPPB also improves PCP-induced cognitive deficits in mice.[305] Vardigan *et al.*[306] showed that CDPPB attenuates MK-801-induced deficits in sucrose preference, a behavioral paradigm thought to model the negative symptoms of schizophrenia. This deficit was also reversed by clinically efficacious compounds such as clozapine and D-serine.[306]

Subsequently, multiple $mGlu_5$ PAMs from a variety of different scaffolds have been developed, including *S*-(4-fluoro-phenyl)-(3-[3-(4-fluoro-phenyl)-[1,2,4]oxadiazol-5-yl]-piperidin-1-yl)-methanone (ADX47273), *N*-cyclobutyl-6-((3-fluorophenyl)ethynyl)nicotinamide (VU0360172), 2-(4-(2-(benzyloxy)acetyl)piperazin-1-yl)benzonitrile (VU0364289), and 1-(4-(2-chloro-4-fluorophenyl)piperazin-1-yl)-2-(pyridin-4-ylmethoxy) ethanone (CPPZ).[307] Many of these compounds possess improved pharmacokinetic properties compared to CDPPB and other earlier PAMs, and, despite their structural diversity (Fig. 2.1), interact with the common allosteric site. Unfortunately the SAR around CPPHA is very "steep" such that it has been difficult to develop additional analogs[308] to probe the *in vivo* efficacy of PAMs interacting with the second allosteric site on group I mGlus. A number of these MPEP-site $mGlu_5$ PAMs have been assessed in rodent behavioral paradigms.[29,309–313] VU0360172,[39] VU0364289[37] and CPPZ[314] are efficacious in the amphetamine-induced hyperlocomotion model. Furthermore, ADX47273 displays efficacy in rodent behavioral models similar to known antipsychotics and, importantly, improves cognitive functioning in impaired animals.[28,315,316] Thus, $mGlu_5$ PAMs have the potential to treat the various symptom clusters associated with schizophrenia, representing a significant improvement over the current therapeutics.

5.15. $mGlu_5$ PAMs for treatment of TSC

Tuberous sclerosis complex (TSC) is an autosomal dominant disorder that results in hamartomas in multiple organ systems. Patients with TSC are also frequently diagnosed with autism spectrum disorders, intellectual disabilities, behavioral abnormalities, and epilepsy.[317] Mutations in the either of the two TSC genes (*TSC1* or *TSC2*) leads to alterations in protein synthesis. Interestingly, FXS (discussed above) manifests with similar symptoms and is also a result of alterations in protein synthesis. Thus, it was proposed that the treatments may be similar between the two disorders. Surprisingly, while studies investigating mGlu-induced LTD at the SC-CA1 synapse were enhanced in FXS, similar studies showed that LTD was deficient in TSC models.[318,319] Furthermore, treatment with an $mGlu_5$ PAM reversed the deficit in mGlu-induced LTD in brain slices and reversed cognitive deficits in TSC mice.[318] Altogether, these results suggest that $mGlu_5$ PAMs may be a therapeutic strategy for TSC. Moreover, these data accentuate the need to understand disorders on a molecular level, as opposing synaptic abnormalities may manifest with the same or similar clinical symptoms.

5.16. $mGlu_7$ for stress and anxiety

$mGlu_7$ is expressed widely throughout the CNS, and $mGlu_7$ knockout mice show a decrease in stress responses, depression-like behaviors, and fear.[320,321] Unfortunately, there are relatively few selective tool compounds for $mGlu_7$ at this time, making it difficult to assess the therapeutic potential of $mGlu_7$ activation or inhibition. In social anxiety disorder and post-traumatic stress disorder, a normally innocuous stimulus (conditioned stimulus) induces feelings such as fear and anxiety. To reverse these behavioral effects, an extinction paradigm is used in which exposure to the conditioned stimulus is repeated in the absence of adverse effects, referred to as extinction training.[322–324] Treatment with *N,N'*-bis(diphenylmethyl)-1,2-ethanediamine (AMN082), an $mGlu_7$ allosteric agonist, decreases within-session fear extinction and facilitated extinction retention in rodents,[325] suggesting therapeutic potential for $mGlu_7$ PAMs/agonists. $mGlu_7$ is also expressed in the basal ganglia, positioning it be a potential target for the treatment of PD. AMN082 reverses haloperidol-induced catalepsy and apomorphine-induced rotation following 6-OHDA lesions in rats[128] and reverses reserpine-induced akinesia,[326] supporting its potential as a target for PD, either alone or in combination with other mGlu modulators, such as those for $mGlu_4$.

Activation of $mGlu_7$ has also been indicated as a potential therapeutic avenue in a number of other CNS-related disorders. Administration of AMN082 is efficacious in preclinical pain[327–329] and depression models,[330] while administration of the $mGlu_7$-selective NAM MMPIP alone produces cognitive deficits,[331] implicating a role for $mGlu_7$ in cognition. In addition, AMN082 reduces alcohol intake/preference and cocaine-induced drug-seeking behavior in rats; an effect blocked by co-administration of MMPIP.[332–334] It should, however, be stressed that AMN082, and metabolites thereof, have significant off-target activity, in particular as inhibitors of monoamine transporters.[335] Additionally, MMPIP exhibits differential pharmacology when the receptor is expressed in distinct cell backgrounds, a finding that may translate to *in vivo* settings.[24] For these reasons, the strictly $mGlu_7$-based effects of AMN082 and MMPIP in various models disease models can be difficult to interpret since many studies have not employed $mGlu_7$ knockout mice to validate the mechanism of compound activity. In order to further substantiate $mGlu_7$ as a therapeutic target, selective pharmacological tools are sorely needed; however, current work is promising.

5.17. $mGlu_8$ potential for the treatment of PD and anxiety

Assessing the feasibility of targeting $mGlu_8$ as a potential therapeutic target has also been difficult due to the lack of subtype-selective compounds for group III mGlus. Many of the compounds that currently exist act across all group III receptors or have selectivity for $mGlu_4$ over the other subtypes. Despite these challenges, several studies suggest that group III mGlus have potential therapeutic utility in PD. The use of $mGlu_4$-preferring agonists and $mGlu_4$ PAMs suggests that $mGlu_4$ is important in the antiparkinsonian effects observed in animal models (see discussion above); however, this does not preclude other group III receptors as potential therapeutic targets. To this end, Johnson *et al.*[336] recently reported that icv administration of the $mGlu_8$-preferring agonist (*S*)-3,4-dicarboxyphenylglycine (DCPG) reversed motor deficits in PD models in which dopamine deletion was prolonged for at least 24 h, such as reserpine-induced akinesia, haloperidol-induced catalepsy, and 6-OHDA PD paradigms.[336] Interestingly, DCPG did not reverse deficits in locomotor activity observed after acute treatment with reserpine or haloperidol.[336] Lopez *et al.*[337] also demonstrated that DCPG, when injected directly into the globus pallidus or substantia nigra pars reticulata, was not effective at reversing the motor deficits observed in rodent models of PD.[337] However, these studies were conducted with acute dopamine depletion. Therefore, more work is needed to determine if $mGlu_8$ activation is a target for PD therapeutics. There is additional evidence that $mGlu_8$ may also be a target for the treatment of anxiety. $mGlu_8$ knockout mice have been reported to have higher levels of anxiety compared to control animals in the elevated plus maze.[139] Overall, the therapeutic potential for $mGlu_8$ is still under investigation, with additional subtype-selective tools with varying degrees of cooperativity and activity needed for this receptor subtype.

6. CONCLUDING REMARKS

The mGlu allosteric modulator field has seen a number of significant advances since the discovery of the first allosteric modulator in 1996. Subtype selective, positive, negative, and neutral modulators have been identified for many of the individual family members. The studies reported to date support the notion of at least one common allosteric binding site location, at least between group I and group II mGlus, analogous to that of the biogenic amine/retinal binding pocket of class A 7TMRs. It remains to be seen whether targeting this or alternate allosteric sites will yield the best therapeutics, with more pharmacological tools required to discover and optimize modulators binding to alternate allosteric sites, and especially to group III mGlus. Inroads have been made into our understanding of the molecular basis for mGlu allosteric interactions, but further work is needed to elucidate and quantify determinants of selectivity, affinity, cooperativity, biased pharmacology, and molecular switches. As highlighted above, a number of $mGlu_5$ NAMs have entered and are currently in clinical trials for an array of CNS-related disorders. Furthermore, given the physiological and pathophysiological roles for the majority of mGlu subtypes, these 7TMRs are promising therapeutic targets and modulators for several of these receptors are currently in preclinical and clinical development for numerous CNS disorders and diseases.

ACKNOWLEDGMENTS

The authors greatly acknowledge funding support from the National Institute of Mental Health [Grant 2R01 MH062646-13]; National Institute of Neurological Disorders and Stroke (NINDS) [R01 NS-31373 and R21 NS-578262]; Molecular Libraries Probe Production Centers Network [Grant 5 u54 MH84659-03, 5 u54 MH84659-03S1]; National Research Service Award from NINDS [M. J. N.: F32 NS071746]; American Australian Association Merck Foundation Fellowship 2010 [K. J. G.]; NARSAD Maltz Investigator Fellowship 2010 [K. J. G.]; and National Health and Medical Research Council (Australia) Overseas Biomedical Postdoctoral Training Fellowship [K. J. G.] for their work on metabotropic glutamate receptors.

REFERENCES

1. Kola I, Landis J. Can the pharmaceutical industry reduce attrition rates? *Nat Rev Drug Discov* 2004;**3**:711–5.
2. Niswender CM, Conn PJ. Metabotropic glutamate receptors: physiology, pharmacology, and disease. *Annu Rev Pharmacol Toxicol* 2010;**50**:295–322.
3. Ferraguti F, Shigemoto R. Metabotropic glutamate receptors. *Cell Tissue Res* 2006;**326**:483–504.
4. Coutinho V, Knopfel T. Metabotropic glutamate receptors: electrical and chemical signaling properties. *Neuroscientist* 2002;**8**:551–61.
5. Anwyl R. Metabotropic glutamate receptors: electrophysiological properties and role in plasticity. *Brain Res Brain Res Rev* 1999;**29**:83–120.
6. Valenti O, Conn PJ, Marino MJ. Distinct physiological roles of the Gq-coupled metabotropic glutamate receptors co-expressed in the same neuronal populations. *J Cell Physiol* 2002;**191**:125–37.
7. Pinheiro PS, Mulle C. Presynaptic glutamate receptors: physiological functions and mechanisms of action. *Nat Rev Neurosci* 2008;**9**:423–36.
8. Conn PJ, Pin JP. Pharmacology and functions of metabotropic glutamate receptors. *Annu Rev Pharmacol Toxicol* 1997;**37**:205–37.
9. Bellone C, Luscher C, Mameli M. Mechanisms of synaptic depression triggered by metabotropic glutamate receptors. *Cell Mol Life Sci* 2008;**65**:2913–23.
10. Nicoletti F, Bockaert J, Collingridge GL, Conn PJ, Ferraguti F, Schoepp DD, et al. Metabotropic glutamate receptors: from the workbench to the bedside. *Neuropharmacology* 2011;**60**:1017–41.
11. Lazareno S, Dolezal V, Popham A, Birdsall NJ. Thiochrome enhances acetylcholine affinity at muscarinic M4 receptors: receptor subtype selectivity via cooperativity rather than affinity. *Mol Pharmacol* 2004;**65**:257–66.
12. Valant C, Gregory KJ, Hall NE, Scammells PJ, Lew MJ, Sexton PM, et al. A novel mechanism of G protein-coupled receptor functional selectivity. Muscarinic partial agonist McN-A-343 as a bitopic orthosteric/allosteric ligand. *J Biol Chem* 2008;**283**:29312–21.
13. Newman AH, Beuming T, Banala AK, Donthamsetti P, Pongetti K, Labounty A, et al. Molecular determinants of selectivity and efficacy at the dopamine D3 receptor. *J Med Chem* 2012;**55**:6689–99.
14. Melancon BJ, Gogliotti RD, Tarr JC, Saleh SA, Chauder BA, Lebois EP, et al. Continued optimization of the MLPCN probe ML071 into highly potent agonists of the hM1 muscarinic acetylcholine receptor. *Bioorg Med Chem Lett* 2012;**22**:3467–72.
15. Jensen PC, Thiele S, Steen A, Elder A, Kolbeck R, Ghosh S, et al. Reversed binding of a small molecule ligand in homologous chemokine receptors—differential role of extracellular loop 2. *Br J Pharmacol* 2012;**166**:258–75.
16. Daval SB, Valant C, Bonnet D, Kellenberger E, Hibert M, Galzi JL, et al. Fluorescent derivatives of AC-42 to probe bitopic orthosteric/allosteric binding mechanisms on muscarinic M1 receptors. *J Med Chem* 2012;**55**:2125–43.
17. Antony J, Kellershohn K, Mohr-Andra M, Kebig A, Prilla S, Muth M, et al. Dualsteric GPCR targeting: a novel route to binding and signaling pathway selectivity. *FASEB J* 2009;**23**:442–50.
18. Urban JD, Clarke WP, von Zastrow M, Nichols DE, Kobilka B, Weinstein H, et al. Functional selectivity and classical concepts of quantitative pharmacology. *J Pharmacol Exp Ther* 2007;**320**:1–13. http://dx.doi.org/10.1124/jpet.106.104463.
19. Kenakin T. Functional selectivity through protean and biased agonism: who steers the ship? *Mol Pharmacol* 2007;**72**:1393–401.
20. Galandrin S, Oligny-Longpre G, Bouvier M. The evasive nature of drug efficacy: implications for drug discovery. *Trends Pharmacol Sci* 2007;**28**:423–30.

21. Emery AC, Diraddo JO, Miller E, Hathaway HA, Pshenichkin S, Takoudjou GR, et al. Ligand bias at metabotropic glutamate 1a receptor: molecular determinants that distinguish beta-arrestin from G protein mediated signaling. *Mol Pharmacol* 2012;**82**:291–301.
22. Zhang Y, Rodriguez AL, Conn PJ. Allosteric potentiators of metabotropic glutamate receptor subtype 5 have differential effects on different signaling pathways in cortical astrocytes. *J Pharmacol Exp Ther* 2005;**315**:1212–9.
23. Sheffler DJ, Conn PJ. Allosteric potentiators of metabotropic glutamate receptor subtype 1a differentially modulate independent signaling pathways in baby hamster kidney cells. *Neuropharmacology* 2008;**55**:419–27.
24. Niswender CM, Johnson KA, Miller NR, Ayala JE, Luo Q, Williams R, et al. Context-dependent pharmacology exhibited by negative allosteric modulators of metabotropic glutamate receptor 7. *Mol Pharmacol* 2010;**77**:459–68.
25. Yin S, Zamorano R, Conn PJ, Niswender CM. Functional selectivity induced by mGlu(4) receptor positive allosteric modulation and concomitant activation of G(q) coupled receptors. *Neuropharmacology* 2013;**66**:122–32.
26. Noetzel MJ, Rook JM, Vinson PN, Cho HP, Days E, Zhou Y, et al. Functional impact of allosteric agonist activity of selective positive allosteric modulators of metabotropic glutamate receptor subtype 5 in regulating central nervous system function. *Mol Pharmacol* 2012;**81**:120–33.
27. Niswender CM, Johnson KA, Weaver CD, Jones CK, Xiang Z, Luo Q, et al. Discovery, characterization, and antiparkinsonian effect of novel positive allosteric modulators of metabotropic glutamate receptor 4. *Mol Pharmacol* 2008;**74**:1345–58.
28. Liu F, Grauer S, Kelley C, Navarra R, Graf R, Zhang G, et al. ADX47273 [S-(4-fluoro-phenyl)-{3-[3-(4-fluoro-phenyl)-[1,2,4]-oxadiazol-5-yl]-piper idin-1-yl}-methanone]: a novel metabotropic glutamate receptor 5-selective positive allosteric modulator with preclinical antipsychotic-like and procognitive activities. *J Pharmacol Exp Ther* 2008;**327**:827–39.
29. Kinney GG, O'Brien JA, Lemaire W, Burno M, Bickel DJ, Clements MK, et al. A novel selective positive allosteric modulator of metabotropic glutamate receptor subtype 5 has in vivo activity and antipsychotic-like effects in rat behavioral models. *J Pharmacol Exp Ther* 2005;**313**:199–206.
30. Chen Y, Goudet C, Pin JP, Conn PJ. N-{4-Chloro-2-[(1,3-dioxo-1,3-dihydro-2H-isoindol-2-yl)methyl]phenyl}-2-hy droxybenzamide (CPPHA) acts through a novel site as a positive allosteric modulator of group 1 metabotropic glutamate receptors. *Mol Pharmacol* 2008;**73**:909–18.
31. Shirey JK, Brady AE, Jones PJ, Davis AA, Bridges TM, Kennedy JP, et al. A selective allosteric potentiator of the M1 muscarinic acetylcholine receptor increases activity of medial prefrontal cortical neurons and restores impairments in reversal learning. *J Neurosci* 2009;**29**:14271–86.
32. Lundstrom L, Bissantz C, Beck J, Wettstein JG, Woltering TJ, Wichmann J, et al. Structural determinants of allosteric antagonism at metabotropic glutamate receptor 2: mechanistic studies with new potent negative allosteric modulators. *Br J Pharmacol* 2011;**164**:521–37.
33. Sanger H, Hanna L, Colvin EM, Grubisha O, Ursu D, Heinz BA, et al. Pharmacological profiling of native group II metabotropic glutamate receptors in primary cortical neuronal cultures using a FLIPR. *Neuropharmacology* 2013;**66**:264–73.
34. O'Brien JA, Lemaire W, Wittmann M, Jacobson MA, Ha SN, Wisnoski DD, et al. A novel selective allosteric modulator potentiates the activity of native metabotropic glutamate receptor subtype 5 in rat forebrain. *J Pharmacol Exp Ther* 2004;**309**:568–77.
35. Wood MR, Hopkins CR, Brogan JT, Conn PJ, Lindsley CW. "Molecular switches" on mGluR allosteric ligands that modulate modes of pharmacology. *Biochemistry* 2011;**50**:2403–10.

36. O'Brien JA, Lemaire W, Chen TB, Chang RS, Jacobson MA, Ha SN, et al. A family of highly selective allosteric modulators of the metabotropic glutamate receptor subtype 5. *Mol Pharmacol* 2003;**64**:731–40.
37. Zhou Y, Manka JT, Rodriguez AL, Weaver CD, Days EL, Vinson PN, et al. Discovery of N-aryl piperazines as selective mGluR5 potentiators with improved in vivo utility. *ACS Med Chem Lett* 2010;**1**:433–8.
38. Sharma S, Kedrowski J, Rook JM, Smith RL, Jones CK, Rodriguez AL, et al. Discovery of molecular switches that modulate modes of metabotropic glutamate receptor subtype 5 (mGlu5) pharmacology in vitro and in vivo within a series of functionalized, regioisomeric 2- and 5-(phenylethynyl)pyrimidines. *J Med Chem* 2009;**52**:4103–6.
39. Rodriguez AL, Grier MD, Jones CK, Herman EJ, Kane AS, Smith RL, et al. Discovery of novel allosteric modulators of metabotropic glutamate receptor subtype 5 reveals chemical and functional diversity and in vivo activity in rat behavioral models of anxiolytic and antipsychotic activity. *Mol Pharmacol* 2010;**78**:1105–23.
40. Hammond AS, Rodriguez AL, Townsend SD, Niswender CM, Gregory KJ, Lindsley CW, et al. Discovery of a novel chemical class of mGlu(5) allosteric ligands with distinct modes of pharmacology. *ACS Chem Neurosci* 2010;**1**:702–16.
41. Lamb JP, Engers DW, Niswender CM, Rodriguez AL, Venable DF, Conn PJ, et al. Discovery of molecular switches within the ADX-47273 mGlu5 PAM scaffold that modulate modes of pharmacology to afford potent mGlu5 NAMs, PAMs and partial antagonists. *Bioorg Med Chem Lett* 2011;**21**:2711–4.
42. Sheffler DJ, Wenthur CJ, Bruner JA, Carrington SJS, Vinson PN, Gogi KK, et al. Development of a novel, CNS-penetrant, metabotropic glutamate receptor 3 (mGlu3) NAM probe (ML289) derived from a closely related mGlu5 PAM. *Bioorg Med Chem Lett* 2012;**22**:3921–3.
43. Stockton JM, Birdsall NJ, Burgen AS, Hulme EC. Modification of the binding properties of muscarinic receptors by gallamine. *Mol Pharmacol* 1983;**23**:551–7.
44. Ehlert FJ. Estimation of the affinities of allosteric ligands using radioligand binding and pharmacological null methods. *Mol Pharmacol* 1988;**33**:187–94.
45. Pagano A, Ruegg D, Litschig S, Stoehr N, Stierlin C, Heinrich M, et al. The non-competitive antagonists 2-methyl-6-(phenylethynyl)pyridine and 7-hydroxyiminocyclopropan[b]chromen-1a-carboxylic acid ethyl ester interact with overlapping binding pockets in the transmembrane region of group I metabotropic glutamate receptors. *J Biol Chem* 2000;**275**:33750–8.
46. Litschig S, Gasparini F, Rueegg D, Stoehr N, Flor PJ, Vranesic I, et al. CPCCOEt, a noncompetitive metabotropic glutamate receptor 1 antagonist, inhibits receptor signaling without affecting glutamate binding. *Mol Pharmacol* 1999;**55**:453–61.
47. Hemstapat K, de Paulis T, Chen Y, Brady AE, Grover VK, Alagille D, et al. A novel class of positive allosteric modulators of metabotropic glutamate receptor subtype 1 interact with a site distinct from that of negative allosteric modulators. *Mol Pharmacol* 2006;**70**:616–26.
48. Bradley SJ, Langmead CJ, Watson JM, Challiss RA. Quantitative analysis reveals multiple mechanisms of allosteric modulation of the mGlu5 receptor in rat astroglia. *Mol Pharmacol* 2011;**79**:874–85.
49. Gregory KJ, Dong EN, Meiler J, Conn PJ. Allosteric modulation of metabotropic glutamate receptors: structural insights and therapeutic potential. *Neuropharmacology* 2011;**60**:66–81.
50. Price MR, Baillie GL, Thomas A, Stevenson LA, Easson M, Goodwin R, et al. Allosteric modulation of the cannabinoid CB1 receptor. *Mol Pharmacol* 2005;**68**:1484–95.
51. Parmentier ML, Prezeau L, Bockaert J, Pin JP. A model for the functioning of family 3 GPCRs. *Trends Pharmacol Sci* 2002;**23**:268–74.

52. Hall DA. Modeling the functional effects of allosteric modulators at pharmacological receptors: an extension of the two-state model of receptor activation. *Mol Pharmacol* 2000;**58**:1412–23.
53. Leach K, Sexton PM, Christopoulos A. Allosteric GPCR modulators: taking advantage of permissive receptor pharmacology. *Trends Pharmacol Sci* 2007;**28**:382–9.
54. Black JW, Leff P. Operational models of pharmacological agonism. *Proc R Soc Lond Ser B* 1983;**220**:141–62.
55. Malherbe P, Kratochwil N, Knoflach F, Zenner MT, Kew JN, Kratzeisen C, et al. Mutational analysis and molecular modeling of the allosteric binding site of a novel, selective, noncompetitive antagonist of the metabotropic glutamate 1 receptor. *J Biol Chem* 2003;**278**:8340–7.
56. Lavreysen H, Janssen C, Bischoff F, Langlois X, Leysen JE, Lesage AS. [3H]R214127: a novel high-affinity radioligand for the mGlu1 receptor reveals a common binding site shared by multiple allosteric antagonists. *Mol Pharmacol* 2003;**63**:1082–93.
57. Kohara A, Toya T, Tamura S, Watabiki T, Nagakura Y, Shitaka Y, et al. Radioligand binding properties and pharmacological characterization of 6-amino-N-cyclohexyl-N,3-dimethylthiazolo[3,2-a]benzimidazole-2-carboxamide (YM-298198), a high-affinity, selective, and noncompetitive antagonist of metabotropic glutamate receptor type 1. *J Pharmacol Exp Ther* 2005;**315**:163–9.
58. Gasparini F, Andres H, Flor PJ, Heinrich M, Inderbitzin W, Lingenhohl K, et al. [(3) H]-M-MPEP, a potent, subtype-selective radioligand for the metabotropic glutamate receptor subtype 5. *Bioorg Med Chem Lett* 2002;**12**:407–9.
59. Cosford ND, Roppe J, Tehrani L, Schweiger EJ, Seiders TJ, Chaudary A, et al. [3H]-methoxymethyl-MTEP and [3H]-methoxy-PEPy: potent and selective radioligands for the metabotropic glutamate subtype 5 (mGlu5) receptor. *Bioorg Med Chem Lett* 2003;**13**:351–4.
60. Schaffhauser H, Rowe BA, Morales S, Chavez-Noriega LE, Yin R, Jachec C, et al. Pharmacological characterization and identification of amino acids involved in the positive modulation of metabotropic glutamate receptor subtype2. *Mol Pharmacol* 2003;**64**:798–810.
61. Knoflach F, Mutel V, Jolidon S, Kew JN, Malherbe P, Vieira E, et al. Positive allosteric modulators of metabotropic glutamate 1 receptor: characterization, mechanism of action, and binding site. *Proc Natl Acad Sci USA* 2001;**98**:13402–7.
62. Arunlakshana O, Schild HO. Some quantitative uses of drug antagonists. *Br J Pharmacol Chemother* 1959;**14**:48–58.
63. Tsuchiya D, Kunishima N, Kamiya N, Jingami H, Morikawa K. Structural views of the ligand-binding cores of a metabotropic glutamate receptor complexed with an antagonist and both glutamate and Gd3+. *Proc Natl Acad Sci USA* 2002;**99**:2660–5.
64. Muto T, Tsuchiya D, Morikawa K, Jingami H. Structures of the extracellular regions of the group II/III metabotropic glutamate receptors. *Proc Natl Acad Sci USA* 2007;**104**:3759–64.
65. Kunishima N, Shimada Y, Tsuji Y, Sato T, Yamamoto M, Kumasaka T, et al. Structural basis of glutamate recognition by a dimeric metabotropic glutamate receptor. *Nature* 2000;**407**:971–7.
66. Romano C, Yang WL, O'Malley KL. Metabotropic glutamate receptor 5 is a disulfide-linked dimer. *J Biol Chem* 1996;**271**:28612–6.
67. Rondard P, Liu J, Huang S, Malhaire F, Vol C, Pinault A, et al. Coupling of agonist binding to effector domain activation in metabotropic glutamate-like receptors. *J Biol Chem* 2006;**281**:24653–61.
68. Liu X, He Q, Studholme DJ, Wu Q, Liang S, Yu L. NCD3G: a novel nine-cysteine domain in family 3 GPCRs. *Trends Biochem Sci* 2004;**29**:458–61.

69. Bessis AS, Rondard P, Gaven F, Brabet I, Triballeau N, Prezeau L, et al. Closure of the Venus flytrap module of mGlu8 receptor and the activation process: insights from mutations converting antagonists into agonists. *Proc Natl Acad Sci USA* 2002;**99**: 11097–102.
70. Bessis AS, Bertrand HO, Galvez T, De Colle C, Pin JP, Acher F. Three-dimensional model of the extracellular domain of the type 4a metabotropic glutamate receptor: new insights into the activation process. *Protein Sci* 2000;**9**:2200–9.
71. Huang S, Cao J, Jiang M, Labesse G, Liu J, Pin JP, et al. Interdomain movements in metabotropic glutamate receptor activation. *Proc Natl Acad Sci USA* 2011;**108**:15480–5.
72. Tones MA, Bendali N, Flor PJ, Knopfel T, Kuhn R. The agonist selectivity of a class III metabotropic glutamate receptor, human mGluR4a, is determined by the N-terminal extracellular domain. *Neuroreport* 1995;**7**:117–20.
73. Takahashi K, Tsuchida K, Tanabe Y, Masu M, Nakanishi S. Role of the large extracellular domain of metabotropic glutamate receptors in agonist selectivity determination. *J Biol Chem* 1993;**268**:19341–5.
74. Gasparini F, Floersheim P, Flor PJ, Heinrich M, Inderbitzin W, Ott D, et al. Discovery and characterization of non-competitive antagonists of group I metabotropic glutamate receptors. *Farmaco* 2001;**56**:95–9.
75. Brauner-Osborne H, Jensen AA, Krogsgaard-Larsen P. Interaction of CPCCOEt with a chimeric mGlu1b and calcium sensing receptor. *Neuroreport* 1999;**10**:3923–5.
76. Mitsukawa K, Yamamoto R, Ofner S, Nozulak J, Pescott O, Lukic S, et al. A selective metabotropic glutamate receptor 7 agonist: activation of receptor signaling via an allosteric site modulates stress parameters in vivo. *Proc Natl Acad Sci USA* 2005;**102**: 18712–7.
77. Maj M, Bruno V, Dragic Z, Yamamoto R, Battaglia G, Inderbitzin W, et al. (-)-PHCCC, a positive allosteric modulator of mGluR4: characterization, mechanism of action, and neuroprotection. *Neuropharmacology* 2003;**45**:895–906.
78. Carroll FY, Stolle A, Beart PM, Voerste A, Brabet I, Mauler F, et al. BAY36-7620: a potent non-competitive mGlu1 receptor antagonist with inverse agonist activity. *Mol Pharmacol* 2001;**59**:965–73.
79. Goudet C, Gaven F, Kniazeff J, Vol C, Liu J, Cohen-Gonsaud M, et al. Heptahelical domain of metabotropic glutamate receptor 5 behaves like rhodopsin-like receptors. *Proc Natl Acad Sci USA* 2004;**101**:378–83.
80. Suzuki G, Kimura T, Satow A, Kaneko N, Fukuda J, Hikichi H, et al. Pharmacological characterization of a new, orally active and potent allosteric metabotropic glutamate receptor 1 antagonist, 4-[1-(2-fluoropyridin-3-yl)-5-methyl-1H-1,2,3-triazol-4-yl]-N-isopropyl-N- methyl-3,6-dihydropyridine-1(2H)-carboxamide (FTIDC). *J Pharmacol Exp Ther* 2007;**321**:1144–53.
81. Chen Y, Nong Y, Goudet C, Hemstapat K, de Paulis T, Pin JP, et al. Interaction of novel positive allosteric modulators of metabotropic glutamate receptor 5 with the negative allosteric antagonist site is required for potentiation of receptor responses. *Mol Pharmacol* 2007;**71**:1389–98.
82. Ott D, Floersheim P, Inderbitzin W, Stoehr N, Francotte E, Lecis G, et al. Chiral resolution, pharmacological characterization, and receptor docking of the noncompetitive mGlu1 receptor antagonist (+/-)-2-hydroxyimino- 1a, 2-dihydro-1H-7-oxacyclopropa[b]naphthalene-7a-carboxylic acid ethyl ester. *J Med Chem* 2000;**43**: 4428–36.
83. Malherbe P, Kratochwil N, Zenner MT, Piussi J, Diener C, Kratzeisen C, et al. Mutational analysis and molecular modeling of the binding pocket of the metabotropic glutamate 5 receptor negative modulator 2-methyl-6-(phenylethynyl)-pyridine. *Mol Pharmacol* 2003;**64**:823–32.

83a. Malherbe P, Kratochil N, Muhlemann A, Zenner MT, Fischer C, Stahl M, et al. Comparison of the binding pockets of two chemically unrelated allosteric antagonists of the mGlu5 receptor and identification of crucial residues involved in the inverse agonism of MPEP. *J Neurochem* 2006;**98**:601–15.
84. Mathiesen JM, Svendsen N, Brauner-Osborne H, Thomsen C, Ramirez MT. Positive allosteric modulation of the human metabotropic glutamate receptor 4 (hmGluR4) by SIB-1893 and MPEP. *Br J Pharmacol* 2003;**138**:1026–30.
85. Annoura H, Fukunaga A, Uesugi M, Tatsuoka T, Horikawa Y. A novel class of antagonists for metabotropic glutamate receptors, 7-(hydroxyimino)cyclopropa[b]chromen-1a-carboxylates. *Bioorg Med Chem Lett* 1996;**6**:763–6.
86. Yu M. Recent developments of the PET imaging agents for metabotropic glutamate receptor subtype5. *Curr Top Med Chem* 2007;**7**:1800–5.
87. Treyer V, Streffer J, Wyss MT, Bettio A, Ametamey SM, Fischer U, et al. Evaluation of the metabotropic glutamate receptor subtype 5 using PET and 11C-ABP688: assessment of methods. *J Nucl Med* 2007;**48**:1207–15.
88. Baumann CA, Mu L, Wertli N, Kramer SD, Honer M, Schubiger PA, et al. Syntheses and pharmacological characterization of novel thiazole derivatives as potential mGluR5 PET ligands. *Bioorg Med Chem* 2010;**18**:6044–54.
89. Anderson JJ, Rao SP, Rowe B, Giracello DR, Holtz G, Chapman DF, et al. [3H] Methoxymethyl-3-[(2-methyl-1,3-thiazol-4-yl)ethynyl]pyridine binding to metabotropic glutamate receptor subtype 5 in rodent brain: in vitro and in vivo characterization. *J Pharmacol Exp Ther* 2002;**303**:1044–51.
90. Anderson JJ, Bradbury MJ, Giracello DR, Chapman DF, Holtz G, Roppe J, et al. In vivo receptor occupancy of mGlu5 receptor antagonists using the novel radioligand [3H]3-methoxy-5-(pyridin-2-ylethynyl)pyridine). *Eur J Pharmacol* 2003;**473**:35–40.
91. Ametamey SM, Treyer V, Streffer J, Wyss MT, Schmidt M, Blagoev M, et al. Human PET studies of metabotropic glutamate receptor subtype5 with 11C-ABP688. *J Nucl Med* 2007;**48**:247–52.
92. Lavreysen H, Wouters R, Bischoff F, Nobrega Pereira S, Langlois X, Blokland S, et al. JNJ16259685, a highly potent, selective and systemically active mGlu1 receptor antagonist. *Neuropharmacology* 2004;**47**:961–72.
93. Schann S, Mayer S, Franchet C, Frauli M, Steinberg E, Thomas M, et al. Chemical switch of a metabotropic glutamate receptor 2 silent allosteric modulator into dual metabotropic glutamate receptor 2/3 negative/positive allosteric modulators. *J Med Chem* 2010;**53**:8775–9.
94. Rodriguez AL, Nong Y, Sekaran NK, Alagille D, Tamagnan GD, Conn PJ. A close structural analog of 2-methyl-6-(phenylethynyl)-pyridine acts as a neutral allosteric site ligand on metabotropic glutamate receptor subtype 5 and blocks the effects of multiple allosteric modulators. *Mol Pharmacol* 2005;**68**:1793–802.
95. Teller DC, Okada T, Behnke CA, Palczewski K, Stenkamp RE. Advances in determination of a high-resolution three-dimensional structure of rhodopsin, a model of G-protein-coupled receptors (GPCRs). *Biochemistry* 2001;**40**:7761–72.
96. Palczewski K, Kumasaka T, Hori T, Behnke CA, Motoshima H, Fox BA, et al. Crystal structure of rhodopsin: a G protein-coupled receptor. *Science* 2000;**289**:739–45.
97. Muhlemann A, Ward NA, Kratochwil N, Diener C, Fischer C, Stucki A, et al. Determination of key amino acids implicated in the actions of allosteric modulation by 3,3′-difluorobenzaldazine on rat mGlu5 receptors. *Eur J Pharmacol* 2006;**529**:95–104.
98. Surin A, Pshenichkin S, Grajkowska E, Surina E, Wroblewski JT. Cyclothiazide selectively inhibits mGluR1 receptors interacting with a common allosteric site for non-competitive antagonists. *Neuropharmacology* 2007;**52**:744–54.
99. Fukuda J, Suzuki G, Kimura T, Nagatomi Y, Ito S, Kawamoto H, et al. Identification of a novel transmembrane domain involved in the negative modulation of mGluR1 using a

newly discovered allosteric mGluR1 antagonist, 3-cyclohexyl-5-fluoro-6-methyl-7-(2-morpholin-4-ylethoxy)-4H-chromen-4-one. *Neuropharmacology* 2009;**57**:438–45.
100. Rowe BA, Schaffhauser H, Morales S, Lubbers LS, Bonnefous C, Kamenecka TM, et al. Transposition of three amino acids transforms the human metabotropic glutamate receptor (mGluR)-3-positive allosteric modulation site to mGluR2, and additional characterization of the mGluR2-positive allosteric modulation site. *J Pharmacol Exp Ther* 2008;**326**:240–51.
101. Hemstapat K, Da Costa H, Nong Y, Brady AE, Luo Q, Niswender CM, et al. A novel family of potent negative allosteric modulators of group II metabotropic glutamate receptors. *J Pharmacol Exp Ther* 2007;**322**:254–64.
102. Hess J, Clements MK, McDonald TP, Liverton N, McCauley JA, Butcher JW, et al. Identification and characterization of radioligands that bind to an allosteric modulator site on the mGlur4 receptor, In: *Society for neuroscience annual meeting,* San Diego, CA; 2010.
103. Schkeryantz JM, Kingston AE, Johnson MP. Prospects for metabotropic glutamate 1 receptor antagonists in the treatment of neuropathic pain. *J Med Chem* 2007;**50**:2563–8.
104. Martin LJ, Blackstone CD, Huganir RL, Price DL. Cellular localization of a metabotropic glutamate receptor in rat brain. *Neuron* 1992;**9**:259–70.
105. Meli E, Picca R, Attucci S, Cozzi A, Peruginelli F, Moroni F, et al. Activation of mGlu1 but not mGlu5 metabotropic glutamate receptors contributes to postischemic neuronal injury in vitro and in vivo. *Pharmacol Biochem Behav* 2002;**73**:439–46.
106. Mabire D, Coupa S, Adelinet C, Poncelet A, Simonnet Y, Venet M, et al. Synthesis, structure-activity relationship, and receptor pharmacology of a new series of quinoline derivatives acting as selective, noncompetitive mGlu1 antagonists. *J Med Chem* 2005;**48**:2134–53.
107. Bennett CE, Burnett DA, Greenlee WJ, Knutson CE, Korakas P, Li C, et al. Fused tricyclic mGluR1 antagonists for the treatment of neuropathic pain. *Bioorg Med Chem Lett* 2012;**22**:1575–8.
108. Kim DH, Maneen MJ, Stahl SM. Building a better antipsychotic: receptor targets for the treatment of multiple symptom dimensions of schizophrenia. *Neurotherapeutics* 2009;**6**:78–85.
109. Swanson CJ, Bures M, Johnson MP, Linden AM, Monn JA, Schoepp DD. Metabotropic glutamate receptors as novel targets for anxiety and stress disorders. *Nat Rev Drug Discov* 2005;**4**:131–44.
110. Conn PJ, Lindsley CW, Jones CK. Activation of metabotropic glutamate receptors as a novel approach for the treatment of schizophrenia. *Trends Pharmacol Sci* 2009;**30**:25–31.
111. Gaspar PA, Bustamante ML, Silva H, Aboitiz F. Molecular mechanisms underlying glutamatergic dysfunction in schizophrenia: therapeutic implications. *J Neurochem* 2009;**111**:891–900.
112. Lindsley CW, Zhao Z, Leister WH, O'Brien J, Lemaire W, Williams Jr DL, et al. Design, synthesis, and in vivo efficacy of glycine transporter-1 (GlyT1) inhibitors derived from a series of [4-phenyl-1-(propylsulfonyl)piperidin-4-yl]methyl benzamides. *ChemMedChem* 2006;**1**:807–11.
113. Wright RA, Johnson BG, Zhang C, Salhoff C, Kingston AE, Calligaro DO, et al. CNS distribution of metabotropic glutamate 2 and 3 receptors: transgenic mice and [(3)H] LY459477 autoradiography. *Neuropharmacology* 2013;**66**:89–98.
114. Cartmell J, Schoepp DD. Regulation of neurotransmitter release by metabotropic glutamate receptors. *J Neurochem* 2000;**75**:889–907.
115. Pinkerton AB, Cube RV, Hutchinson JH, James JK, Gardner MF, Rowe BA, et al. Allosteric potentiators of the metabotropic glutamate receptor 2 (mGlu2). Part 3: identification and biological activity of indanone containing mGlu2 receptor potentiators. *Bioorg Med Chem Lett* 2005;**15**:1565–71.

116. Johnson MP, Barda D, Britton TC, Emkey R, Hornback WJ, Jagdmann GE, et al. Metabotropic glutamate 2 receptor potentiators: receptor modulation, frequency-dependent synaptic activity, and efficacy in preclinical anxiety and psychosis model(s). *Psychopharmacology (Berl)* 2005;**179**:271–83.
117. Govek SP, Bonnefous C, Hutchinson JH, Kamenecka T, McQuiston J, Pracitto R, et al. Benzazoles as allosteric potentiators of metabotropic glutamate receptor 2 (mGluR2): efficacy in an animal model for schizophrenia. *Bioorg Med Chem Lett* 2005;**15**:4068–72.
118. Galici R, Echemendia NG, Rodriguez AL, Conn PJ. A selective allosteric potentiator of metabotropic glutamate (mGlu) 2 receptors has effects similar to an orthosteric mGlu2/3 receptor agonist in mouse models predictive of antipsychotic activity. *J Pharmacol Exp Ther* 2005;**315**:1181–7.
119. Benneyworth MA, Xiang Z, Smith RL, Garcia EE, Conn PJ, Sanders-Bush E. A selective positive allosteric modulator of metabotropic glutamate receptor subtype 2 blocks a hallucinogenic drug model of psychosis. *Mol Pharmacol* 2007;**72**:477–84.
120. Moghaddam B, Adams BW. Reversal of phencyclidine effects by a group II metabotropic glutamate receptor agonist in rats. *Science* 1998;**281**:1349–52.
121. Lorrain DS, Schaffhauser H, Campbell UC, Baccei CS, Correa LD, Rowe B, et al. Group II mGlu receptor activation suppresses norepinephrine release in the ventral hippocampus and locomotor responses to acute ketamine challenge. *Neuropsychopharmacology* 2003;**28**:1622–32.
122. Conn PJ, Tamminga C, Schoepp DD, Lindsley C. Schizophrenia: moving beyond monoamine antagonists. *Mol Interv* 2008;**8**:99–107.
123. Cartmell J, Monn JA, Schoepp DD. The metabotropic glutamate 2/3 receptor agonists LY354740 and LY379268 selectively attenuate phencyclidine versus d-amphetamine motor behaviors in rats. *J Pharmacol Exp Ther* 1999;**291**:161–70.
124. Nakazato A, Kumagai T, Sakagami K, Yoshikawa R, Suzuki Y, Chaki S, et al. Synthesis, SARs, and pharmacological characterization of 2-amino-3 or 6-fluorobicyclo [3.1.0]hexane-2,6-dicarboxylic acid derivatives as potent, selective, and orally active group II metabotropic glutamate receptor agonists. *J Med Chem* 2000;**43**:4893–909.
125. Patil ST, Zhang L, Martenyi F, Lowe SL, Jackson KA, Andreev BV, et al. Activation of mGlu2/3 receptors as a new approach to treat schizophrenia: a randomized Phase 2 clinical trial. *Nat Med* 2007;**13**:1102–7.
126. Kinon BJ, Zhang L, Millen BA, Osuntokun OO, Williams JE, Kollack-Walker S, et al. A multicenter, inpatient, phase 2, double-blind, placebo-controlled dose-ranging study of LY2140023 monohydrate in patients with DSM-IV schizophrenia. *J Clin Psychopharmacol* 2011;**31**:349–55.
127. Nikiforuk A, Popik P, Drescher KU, van Gaalen M, Relo AL, Mezler M, et al. Effects of a positive allosteric modulator of group II metabotropic glutamate receptors, LY487379, on cognitive flexibility and impulsive-like responding in rats. *J Pharmacol Exp Ther* 2010;**335**:665–73.
128. Greco B, Lopez S, van der Putten H, Flor PJ, Amalric M. Metabotropic glutamate 7 receptor subtype modulates motor symptoms in rodent models of Parkinson's disease. *J Pharmacol Exp Ther* 2010;**332**:1064–71.
129. Vales K, Svoboda J, Benkovicova K, Bubenikova-Valesova V, Stuchlik A. The difference in effect of mGlu2/3 and mGlu5 receptor agonists on cognitive impairment induced by MK-801. *Eur J Pharmacol* 2010;**639**:91–8.
130. Spinelli S, Ballard T, Gatti-McArthur S, Richards GJ, Kapps M, Woltering T, et al. Effects of the mGluR2/3 agonist LY354740 on computerized tasks of attention and working memory in marmoset monkeys. *Psychopharmacology (Berl)* 2005;**179**:292–302.
131. Aultman JM, Moghaddam B. Distinct contributions of glutamate and dopamine receptors to temporal aspects of rodent working memory using a clinically relevant task. *Psychopharmacology (Berl)* 2001;**153**:353–64.

132. Fell MJ, Svensson KA, Johnson BG, Schoepp DD. Evidence for the role of metabotropic glutamate (mGlu)2 not mGlu3 receptors in the preclinical antipsychotic pharmacology of the mGlu2/3 receptor agonist (-)-(1R,4S,5S,6S)-4-amino-2-sulfonylbicyclo[3.1.0]hexane-4,6-dicarboxylic acid (LY404039). *J Pharmacol Exp Ther* 2008;**326**:209–17.
133. Hackler EA, Byun NE, Jones CK, Williams JM, Baheza R, Sengupta S, et al. Selective potentiation of the metabotropic glutamate receptor subtype 2 blocks phencyclidine-induced hyperlocomotion and brain activation. *Neuroscience* 2010;**168**:209–18.
134. Galici R, Jones CK, Hemstapat K, Nong Y, Echemendia NG, Williams LC, et al. Biphenyl-indanone A, a positive allosteric modulator of the metabotropic glutamate receptor subtype 2, has antipsychotic- and anxiolytic-like effects in mice. *J Pharmacol Exp Ther* 2006;**318**:173–85.
135. Caraci F, Molinaro G, Battaglia G, Giuffrida ML, Riozzi B, Traficante A, et al. Targeting group II metabotropic glutamate (mGlu) receptors for the treatment of psychosis associated with Alzheimer's disease: selective activation of mGlu2 receptors amplifies beta-amyloid toxicity in cultured neurons, whereas dual activation of mGlu2 and mGlu3 receptors is neuroprotective. *Mol Pharmacol* 2011;**79**:618–26.
136. Shin LM, Liberzon I. The neurocircuitry of fear, stress, and anxiety disorders. *Neuropsychopharmacology* 2010;**35**:169–91.
137. Etkin A. Functional neuroanatomy of anxiety: a neural circuit perspective. *Curr Top Behav Neurosci* 2010;**2**:251–77.
138. Rorick-Kehn LM, Johnson BG, Knitowski KM, Salhoff CR, Witkin JM, Perry KW, et al. In vivo pharmacological characterization of the structurally novel, potent, selective mGlu2/3 receptor agonist LY404039 in animal models of psychiatric disorders. *Psychopharmacology (Berl)* 2007;**193**:121–36.
139. Linden AM, Johnson BG, Peters SC, Shannon HE, Tian M, Wang Y, et al. Increased anxiety-related behavior in mice deficient for metabotropic glutamate 8 (mGlu8) receptor. *Neuropharmacology* 2002;**43**:251–9.
140. Schoepp DD, Wright RA, Levine LR, Gaydos B, Potter WZ. LY354740, an mGlu2/3 receptor agonist as a novel approach to treat anxiety/stress. *Stress* 2003;**6**:189–97.
141. Dunayevich E, Erickson J, Levine L, Landbloom R, Schoepp DD, Tollefson GD. Efficacy and tolerability of an mGlu2/3 agonist in the treatment of generalized anxiety disorder. *Neuropsychopharmacology* 2008;**33**:1603–10.
142. Fell MJ, Witkin JM, Falcone JF, Katner JS, Perry KW, Hart J, et al. N-(4-((2-(trifluoromethyl)-3-hydroxy-4-(isobutyryl)phenoxy)methyl)benzyl)-1-methy l-1H-imidazole-4-carboxamide (THIIC), a novel metabotropic glutamate 2 potentiator with potential anxiolytic/antidepressant properties: in vivo profiling suggests a link between behavioral and central nervous system neurochemical changes. *J Pharmacol Exp Ther* 2011;**336**:165–77.
143. Helton DR, Tizzano JP, Monn JA, Schoepp DD, Kallman MJ. LY354740: a metabotropic glutamate receptor agonist which ameliorates symptoms of nicotine withdrawal in rats. *Neuropharmacology* 1997;**36**:1511–6.
144. Sidhpura N, Weiss F, Martin-Fardon R. Effects of the mGlu2/3 agonist LY379268 and the mGlu5 antagonist MTEP on ethanol seeking and reinforcement are differentially altered in rats with a history of ethanol dependence. *Biol Psychiatry* 2010;**67**:804–11.
145. Peters J, Kalivas PW. The group II metabotropic glutamate receptor agonist, LY379268, inhibits both cocaine- and food-seeking behavior in rats. *Psychopharmacology (Berl)* 2006;**186**:143–9.
146. Liechti ME, Markou A. Interactive effects of the mGlu5 receptor antagonist MPEP and the mGlu2/3 receptor antagonist LY341495 on nicotine self-administration and reward deficits associated with nicotine withdrawal in rats. *Eur J Pharmacol* 2007;**554**:164–74.
147. Jin X, Semenova S, Yang L, Ardecky R, Sheffler DJ, Dahl R, et al. The mGluR2 positive allosteric modulator BINA decreases cocaine self-administration and cue-induced

cocaine-seeking and counteracts cocaine-induced enhancement of brain reward function in rats. *Neuropsychopharmacology* 2010;**35**:2021–36.
148. Hao Y, Martin-Fardon R, Weiss F. Behavioral and functional evidence of metabotropic glutamate receptor 2/3 and metabotropic glutamate receptor 5 dysregulation in cocaine-escalated rats: factor in the transition to dependence. *Biol Psychiatry* 2010;**68**:240–8.
149. Backstrom P, Hyytia P. Suppression of alcohol self-administration and cue-induced reinstatement of alcohol seeking by the mGlu2/3 receptor agonist LY379268 and the mGlu8 receptor agonist (S)-3,4-DCPG. *Eur J Pharmacol* 2005;**528**:110–8.
150. Adewale AS, Platt DM, Spealman RD. Pharmacological stimulation of group II metabotropic glutamate receptors reduces cocaine self-administration and cocaine-induced reinstatement of drug seeking in squirrel monkeys. *J Pharmacol Exp Ther* 2006;**318**:922–31.
151. Lu L, Uejima JL, Gray SM, Bossert JM, Shaham Y. Systemic and central amygdala injections of the mGluR(2/3) agonist LY379268 attenuate the expression of incubation of cocaine craving. *Biol Psychiatry* 2007;**61**:591–8.
152. Liechti ME, Lhuillier L, Kaupmann K, Markou A. Metabotropic glutamate 2/3 receptors in the ventral tegmental area and the nucleus accumbens shell are involved in behaviors relating to nicotine dependence. *J Neurosci* 2007;**27**:9077–85.
153. Bossert JM, Gray SM, Lu L, Shaham Y. Activation of group II metabotropic glutamate receptors in the nucleus accumbens shell attenuates context-induced relapse to heroin seeking. *Neuropsychopharmacology* 2006;**31**:2197–209.
154. Kufahl PR, Watterson LR, Nemirovsky NE, Hood LE, Villa A, Halstengard C, et al. Attenuation of methamphetamine seeking by the mGluR2/3 agonist LY379268 in rats with histories of restricted and escalated self-administration. *Neuropharmacology* 2013;**66**:290–301.
155. Dhanya RP, Sidique S, Sheffler DJ, Nickols HH, Herath A, Yang L, et al. Design and synthesis of an orally active metabotropic glutamate receptor subtype-2 (mGluR2) positive allosteric modulator (PAM) that decreases cocaine self-administration in rats. *J Med Chem* 2011;**54**:342–53.
156. Wittchen HU, Jacobi F, Rehm J, Gustavsson A, Svensson M, Jonsson B, et al. The size and burden of mental disorders and other disorders of the brain in Europe 2010. *Eur Neuropsychopharmacol* 2011;**21**:655–79.
157. Kessler RC, Berglund P, Demler O, Jin R, Koretz D, Merikangas KR, et al. The epidemiology of major depressive disorder: results from the National Comorbidity Survey Replication (NCS-R). *JAMA* 2003;**289**:3095–105.
158. Berman RM, Cappiello A, Anand A, Oren DA, Heninger GR, Charney DS, et al. Antidepressant effects of ketamine in depressed patients. *Biol Psychiatry* 2000;**47**:351–4.
159. Zarate Jr CA, Singh JB, Carlson PJ, Brutsche NE, Ameli R, Luckenbaugh DA, et al. A randomized trial of an N-methyl-D-aspartate antagonist in treatment-resistant major depression. *Arch Gen Psychiatry* 2006;**63**:856–64.
160. Drevets WC. Functional anatomical abnormalities in limbic and prefrontal cortical structures in major depression. *Prog Brain Res* 2000;**126**:413–31.
161. Xi D, Li YC, Snyder MA, Gao RY, Adelman AE, Zhang W, et al. Group II metabotropic glutamate receptor agonist ameliorates MK801-induced dysfunction of NMDA receptors via the Akt/GSK-3beta pathway in adult rat prefrontal cortex. *Neuropsychopharmacology* 2011;**36**:1260–74.
162. Youngren KD, Moghaddam B, Bunney BS, Roth RH. Preferential activation of dopamine overflow in prefrontal cortex produced by chronic clozapine treatment. *Neurosci Lett* 1994;**165**:41–4.
163. Moghaddam B. Preferential activation of cortical dopamine neurotransmission by clozapine: functional significance. *J Clin Psychiatry* 1994;**55**(Suppl. B):27–9.

164. Cartmell J, Perry KW, Salhoff CR, Monn JA, Schoepp DD. Acute increases in monoamine release in the rat prefrontal cortex by the mGlu2/3 agonist LY379268 are similar in profile to risperidone, not locally mediated, and can be elicited in the presence of uptake blockade. *Neuropharmacology* 2001;**40**:847–55.
165. Matrisciano F, Zusso M, Panaccione I, Turriziani B, Caruso A, Iacovelli L, et al. Synergism between fluoxetine and the mGlu2/3 receptor agonist, LY379268, in an in vitro model for antidepressant drug-induced neurogenesis. *Neuropharmacology* 2008;**54**:428–37.
166. Matrisciano F, Scaccianoce S, Del Bianco P, Panaccione I, Canudas AM, Battaglia G, et al. Metabotropic glutamate receptors and neuroadaptation to antidepressants: imipramine-induced down-regulation of beta-adrenergic receptors in mice treated with metabotropic glutamate 2/3 receptor ligands. *J Neurochem* 2005;**93**:1345–52.
167. Yoshimizu T, Shimazaki T, Ito A, Chaki S. An mGluR2/3 antagonist, MGS0039, exerts antidepressant and anxiolytic effects in behavioral models in rats. *Psychopharmacology (Berl)* 2006;**186**:587–93.
168. Chaki S, Yoshikawa R, Hirota S, Shimazaki T, Maeda M, Kawashima N, et al. MGS0039: a potent and selective group II metabotropic glutamate receptor antagonist with antidepressant-like activity. *Neuropharmacology* 2004;**46**:457–67.
169. Campo B, Kalinichev M, Lambeng N, El Yacoubi M, Royer-Urios I, Schneider M, et al. Characterization of an mGluR2/3 negative allosteric modulator in rodent models of depression. *J Neurogenet* 2011;**25**:152–66.
170. Chaki S, Ago Y, Palucha-Paniewiera A, Matrisciano F, Pilc A. mGlu2/3 and mGlu5 receptors: potential targets for novel antidepressants. *Neuropharmacology* 2013;**66**:40–52.
171. Chaudhuri KR, Healy DG, Schapira AH. Non-motor symptoms of Parkinson's disease: diagnosis and management. *Lancet Neurol* 2006;**5**:235–45.
172. Johnson KA, Conn PJ, Niswender CM. Glutamate receptors as therapeutic targets for Parkinson's disease. *CNS Neurol Disord Drug Targets* 2009;**8**:475–91.
173. Poewe WH, Lees AJ, Stern GM. Treatment of motor fluctuations in Parkinson's disease with an oral sustained-release preparation of L-dopa: clinical and pharmacokinetic observations. *Clin Neuropharmacol* 1986;**9**:430–9.
174. Poewe WH, Lees AJ, Stern GM. Low-dose L-dopa therapy in Parkinson's disease: a 6-year follow-up study. *Neurology* 1986;**36**:1528–30.
175. Chen JJ, Swope DM. Pharmacotherapy for Parkinson's disease. *Pharmacotherapy* 2007;**27**:161S–173S.
176. Lavreysen H, Dautzenberg FM. Therapeutic potential of group III metabotropic glutamate receptors. *Curr Med Chem* 2008;**15**:671–84.
177. Bogenpohl J, Galvan A, Hu X, Wichmann T, Smith Y. Metabotropic glutamate receptor 4 in the basal ganglia of parkinsonian monkeys: ultrastructural localization and electrophysiological effects of activation in the striatopallidal complex. *Neuropharmacology* 2013;**66**:242–52.
178. Wittmann M, Marino MJ, Bradley SR, Conn PJ. Activation of group III mGluRs inhibits GABAergic and glutamatergic transmission in the substantia nigra pars reticulata. *J Neurophysiol* 2001;**85**:1960–8.
179. Valenti O, Marino MJ, Wittmann M, Lis E, DiLella AG, Kinney GG, et al. *J Neurosci* 2003;**23**:7218–26.
180. Matsui T, Kita H. Activation of group III metabotropic glutamate receptors presynaptically reduces both GABAergic and glutamatergic transmission in the rat globus pallidus. *Neuroscience* 2003;**122**:727–37.
181. MacInnes N, Messenger MJ, Duty S. Activation of group III metabotropic glutamate receptors in selected regions of the basal ganglia alleviates akinesia in the reserpine-treated rat. *Br J Pharmacol* 2004;**141**:15–22.

182. Cuomo D, Martella G, Barabino E, Platania P, Vita D, Madeo G, et al. Metabotropic glutamate receptor subtype 4 selectively modulates both glutamate and GABA transmission in the striatum: implications for Parkinson's disease treatment. *J Neurochem* 2009;**109**:1096–105.
183. Conn PJ, Battaglia G, Marino MJ, Nicoletti F. Metabotropic glutamate receptors in the basal ganglia motor circuit. *Nat Rev Neurosci* 2005;**6**:787–98.
184. Macinnes N, Duty S. Group III metabotropic glutamate receptors act as heteroreceptors modulating evoked GABA release in the globus pallidus in vivo. *Eur J Pharmacol* 2008;**580**:95–9.
185. Goudet C, Vilar B, Courtiol T, Deltheil T, Bessiron T, Brabet I, et al. A novel selective metabotropic glutamate receptor 4 agonist reveals new possibilities for developing subtype selective ligands with therapeutic potential. *FASEB J* 2012;**26**:1682–93.
186. Beurrier C, Lopez S, Revy D, Selvam C, Goudet C, Lherondel M, et al. Electrophysiological and behavioral evidence that modulation of metabotropic glutamate receptor 4 with a new agonist reverses experimental parkinsonism. *FASEB J* 2009;**23**:3619–28.
187. Lopez S, Turle-Lorenzo N, Johnston TH, Brotchie JM, Schann S, Neuville P, et al. Functional interaction between adenosine A2A and group III metabotropic glutamate receptors to reduce parkinsonian symptoms in rats. *Neuropharmacology* 2008;**55**:483–90.
188. Lopez S, Jouve L, Turle-Lorenzo N, Kerkerian-Legoff L, Salin P, Amalric M. Antiparkinsonian action of a selective group III mGlu receptor agonist is associated with reversal of subthalamonigral overactivity. *Neurobiol Dis* 2012;**46**:69–77.
189. Palucha A, Tatarczynska E, Branski P, Szewczyk B, Wieronska JM, Klak K, et al. Group III mGlu receptor agonists produce anxiolytic- and antidepressant-like effects after central administration in rats. *Neuropharmacology* 2004;**46**:151–9.
190. Goudet C, Chapuy E, Alloui A, Acher F, Pin JP, Eschalier A. Group III metabotropic glutamate receptors inhibit hyperalgesia in animal models of inflammation and neuropathic pain. *Pain* 2008;**137**:112–24.
191. Marino MJ, Williams Jr DL, O'Brien JA, Valenti O, McDonald TP, Clements MK, et al. Allosteric modulation of group III metabotropic glutamate receptor 4: a potential approach to Parkinson's disease treatment. *Proc Natl Acad Sci USA* 2003;**100**:13668–73.
192. Flor PJ, Battaglia G, Nicoletti F, Gasparini F, Bruno V. Neuroprotective activity of metabotropic glutamate receptor ligands. *Adv Exp Med Biol* 2002;**513**:197–223.
193. Battaglia G, Busceti CL, Molinaro G, Biagioni F, Traficante A, Nicoletti F, et al. Pharmacological activation of mGlu4 metabotropic glutamate receptors reduces nigrostriatal degeneration in mice treated with 1-methyl-4-phenyl-1,2,3,6-tetrahydropyridine. *J Neurosci* 2006;**26**:7222–9.
194. Betts MJ, O'Neill MJ, Duty S. Allosteric modulation of the group III mGlu receptor 4 provides functional neuroprotection in the 6-OHDA rat model of Parkinson's disease. *Br J Pharmacol* 2012;**166**:2317–30.
195. Williams R, Johnson KA, Gentry PR, Niswender CM, Weaver CD, Conn PJ, et al. Synthesis and SAR of a novel positive allosteric modulator (PAM) of the metabotropic glutamate receptor 4 (mGluR4). *Bioorg Med Chem Lett* 2009;**19**:4967–70.
196. Jones CK, Engers DW, Thompson AD, Field JR, Blobaum AL, Lindsley SR, et al. Discovery, synthesis, and structure-activity relationship development of a series of N-4-(2,5-dioxopyrrolidin-1-yl)phenylpicolinamides (VU0400195, ML182): characterization of a novel positive allosteric modulator of the metabotropic glutamate receptor 4 (mGlu(4)) with oral efficacy in an antiparkinsonian animal model. *J Med Chem* 2011;**54**:7639–47.
197. Jones CK, Bubser M, Thompson AD, Dickerson JW, Turle-Lorenzo N, Amalric M, et al. The metabotropic glutamate receptor 4-positive allosteric modulator VU0364770 produces efficacy alone and in combination with L-DOPA or an adenosine 2A

antagonist in preclinical rodent models of Parkinson's disease. *J Pharmacol Exp Ther* 2012;**340**:404–21.
198. Jimenez HN, Liu KG, Hong SP, Reitman MS, Uberti MA, Bacolod MD, et al. 4-(1-Phenyl-1H-pyrazol-4-yl)quinolines as novel, selective and brain penetrant metabotropic glutamate receptor 4 positive allosteric modulators. *Bioorg Med Chem Lett* 2012;**22**:3235–9.
199. Hong SP, Liu KG, Ma G, Sabio M, Uberti MA, Bacolod MD, et al. Tricyclic thiazolopyrazole derivatives as metabotropic glutamate receptor 4 positive allosteric modulators. *J Med Chem* 2011;**54**:5070–81.
200. Engers DW, Niswender CM, Weaver CD, Jadhav S, Menon UN, Zamorano R, et al. Synthesis and evaluation of a series of heterobiarylamides that are centrally penetrant metabotropic glutamate receptor 4 (mGluR4) positive allosteric modulators (PAMs). *J Med Chem* 2009;**52**:4115–8.
201. East SP, Gerlach K. mGluR4 positive allosteric modulators with potential for the treatment of Parkinson's disease: WO09010455. *Expert Opin Ther Pat* 2010;**20**:441–5.
202. Bennouar KE, Uberti MA, Melon C, Bacolod MD, Jimenez HN, Cajina M, et al. Synergy between l-DOPA and a novel positive allosteric modulator of metabotropic glutamate receptor 4: implications for Parkinson's disease treatment and dyskinesia. *Neuropharmacology* 2013;**66**:158–69.
203. Slawinska A, Wieronska JM, Stachowicz K, Palucha-Poniewiera A, Uberti MA, Bacolod MA, et al. Anxiolytic- but not antidepressant-like activity ofLu AF21934, a novel, selective positive allosteric modulator of the mGlu(4) receptor. *Neuropharmacology* 2012;.
204. Celanire S, Bolea C, Bruckner S, Liverton N, Charvin D, Hess F, et al. Discovery and characterization of novel metabotropic glutamate receptor 4 (mGluR4) positive allosteric modulators, In: *7th International meeting on metabotropic glutamate receptors*, Taormina, Italy; 2011.
205. Neugebauer V. Metabotropic glutamate receptors—important modulators of nociception and pain behavior. *Pain* 2002;**98**:1–8.
206. Azkue JJ, Murga M, Fernandez-Capetillo O, Mateos JM, Elezgarai I, Benitez R, et al. Immunoreactivity for the group III metabotropic glutamate receptor subtype mGluR4a in the superficial laminae of the rat spinal dorsal horn. *J Comp Neurol* 2001;**430**:448–57.
207. Neugebauer V, Chen PS, Willis WD. Groups II and III metabotropic glutamate receptors differentially modulate brief and prolonged nociception in primate STT cells. *J Neurophysiol* 2000;**84**:2998–3009.
208. Han JS, Bird GC, Neugebauer V. Enhanced group III mGluR-mediated inhibition of pain-related synaptic plasticity in the amygdala. *Neuropharmacology* 2004;**46**:918–26.
209. Fallarino F, Volpi C, Fazio F, Notartomaso S, Vacca C, Busceti C, et al. Metabotropic glutamate receptor-4 modulates adaptive immunity and restrains neuroinflammation. *Nat Med* 2010;**16**:897–902.
210. Romano C, Sesma MA, McDonald CT, O'Malley K, Van den Pol AN, Olney JW. Distribution of metabotropic glutamate receptor mGluR5 immunoreactivity in rat brain. *J Comp Neurol* 1995;**355**:455–69.
211. Marino MJ, Conn PJ. Glutamate-based therapeutic approaches: allosteric modulators of metabotropic glutamate receptors. *Curr Opin Pharmacol* 2006;**6**:98–102.
212. Chojnacka-Wojcik E, Tatarczynska E, Pilc A. The anxiolytic-like effect of metabotropic glutamate receptor antagonists after intrahippocampal injection in rats. *Eur J Pharmacol* 1997;**319**:153–6.
213. Hayashi Y, Sekiyama N, Nakanishi S, Jane DE, Sunter DC, Birse EF, et al. Analysis of agonist and antagonist activities of phenylglycine derivatives for different cloned metabotropic glutamate receptor subtypes. *J Neurosci* 1994;**14**:3370–7.

214. Varney MA, Cosford ND, Jachec C, Rao SP, Sacaan A, Lin FF, et al. SIB-1757 and SIB-1893: selective, noncompetitive antagonists of metabotropic glutamate receptor type5. *J Pharmacol Exp Ther* 1999;**290**:170–81.
215. Gasparini F, Lingenhohl K, Stoehr N, Flor PJ, Heinrich M, Vranesic I, et al. 2-Methyl-6-(phenylethynyl)-pyridine (MPEP), a potent, selective and systemically active mGlu5 receptor antagonist. *Neuropharmacology* 1999;**38**:1493–503.
216. Cosford ND, Tehrani L, Roppe J, Schweiger E, Smith ND, Anderson J, et al. 3-[(2-Methyl-1,3-thiazol-4-yl)ethynyl]-pyridine: a potent and highly selective metabotropic glutamate subtype 5 receptor antagonist with anxiolytic activity. *J Med Chem* 2003;**46**:204–6.
217. Tatarczynska E, Klodzinska A, Kroczka B, Chojnacka-Wojcik E, Pilc A. The antianxiety-like effects of antagonists of group I and agonists of group II and III metabotropic glutamate receptors after intrahippocampal administration. *Psychopharmacology (Berl)* 2001;**158**:94–9.
218. Spooren WP, Vassout A, Neijt HC, Kuhn R, Gasparini F, Roux S, et al. Anxiolytic-like effects of the prototypical metabotropic glutamate receptor 5 antagonist 2-methyl-6-(phenylethynyl)pyridine in rodents. *J Pharmacol Exp Ther* 2000;**295**:1267–75.
219. Spooren WP, Schoeffter P, Gasparini F, Kuhn R, Gentsch C. Pharmacological and endocrinological characterisation of stress-induced hyperthermia in singly housed mice using classical and candidate anxiolytics (LY314582, MPEP and NKP608). *Eur J Pharmacol* 2002;**435**:161–70.
220. Schulz B, Fendt M, Gasparini F, Lingenhohl K, Kuhn R, Koch M. The metabotropic glutamate receptor antagonist 2-methyl-6-(phenylethynyl)-pyridine (MPEP) blocks fear conditioning in rats. *Neuropharmacology* 2001;**41**:1–7.
221. Rodrigues SM, Bauer EP, Farb CR, Schafe GE, LeDoux JE. The group I metabotropic glutamate receptor mGluR5 is required for fear memory formation and long-term potentiation in the lateral amygdala. *J Neurosci* 2002;**22**:5219–29.
222. Porter RH, Jaeschke G, Spooren W, Ballard TM, Buttelmann B, Kolczewski S, et al. Fenobam: a clinically validated nonbenzodiazepine anxiolytic is a potent, selective, and noncompetitive mGlu5 receptor antagonist with inverse agonist activity. *J Pharmacol Exp Ther* 2005;**315**:711–21.
223. Patel JB, Martin C, Malick JB. Differential antagonism of the anticonflict effects of typical and atypical anxiolytics. *Eur J Pharmacol* 1982;**86**:295–8.
224. Goldberg ME, Salama AI, Patel JB, Malick JB. Novel non-benzodiazepine anxiolytics. *Neuropharmacology* 1983;**22**:1499–504.
225. Pecknold JC, McClure DJ, Appeltauer L, Wrzesinski L, Allan T. Treatment of anxiety using fenobam (a nonbenzodiazepine) in a double-blind standard (diazepam) placebo-controlled study. *J Clin Psychopharmacol* 1982;**2**:129–33.
226. Steckler T, Oliveira AF, Van Dyck C, Van Craenendonck H, Mateus AM, Langlois X, et al. Metabotropic glutamate receptor 1 blockade impairs acquisition and retention in a spatial Water maze task. *Behav Brain Res* 2005;**164**:52–60.
227. Steckler T, Lavreysen H, Oliveira AM, Aerts N, Van Craenendonck H, Prickaerts J, et al. Effects of mGlu1 receptor blockade on anxiety-related behaviour in the rat lick suppression test. *Psychopharmacology (Berl)* 2005;**179**:198–206.
228. Pietraszek M, Sukhanov I, Maciejak P, Szyndler J, Gravius A, Wislowska A, et al. Anxiolytic-like effects of mGlu1 and mGlu5 receptor antagonists in rats. *Eur J Pharmacol* 2005;**514**:25–34.
229. Gravius A, Pietraszek M, Schafer D, Schmidt WJ, Danysz W. Effects of mGlu1 and mGlu5 receptor antagonists on negatively reinforced learning. *Behav Pharmacol* 2005;**16**:113–21.
230. Pisani A, Gubellini P, Bonsi P, Conquet F, Picconi B, Centonze D, et al. Metabotropic glutamate receptor 5 mediates the potentiation of N-methyl-D-aspartate responses in medium spiny striatal neurons. *Neuroscience* 2001;**106**:579–87.

231. Doherty AJ, Palmer MJ, Bortolotto ZA, Hargreaves A, Kingston AE, Ornstein PL, et al. A novel, competitive mGlu(5) receptor antagonist (LY344545) blocks DHPG-induced potentiation of NMDA responses but not the induction of LTP in rat hippocampal slices. *Br J Pharmacol* 2000;**131**:239–44.
232. Awad H, Hubert GW, Smith Y, Levey AI, Conn PJ. Activation of metabotropic glutamate receptor 5 has direct excitatory effects and potentiates NMDA receptor currents in neurons of the subthalamic nucleus. *J Neurosci* 2000;**20**:7871–9.
233. Li X, Need AB, Baez M, Witkin JM. Metabotropic glutamate 5 receptor antagonism is associated with antidepressant-like effects in mice. *J Pharmacol Exp Ther* 2006;**319**:254–9.
234. Hughes ZA, Neal SJ, Smith DL, Sukoff Rizzo SJ, Pulicicchio CM, Lotarski S, et al. Negative allosteric modulation of metabolic glutamate receptor 5 results in broad spectrum activity relevant to treatment resistant depression. *Neuropharmacology* 2013;**66**:202–14.
235. Belozertseva IV, Kos T, Popik P, Danysz W, Bespalov AY. Antidepressant-like effects of mGluR1 and mGluR5 antagonists in the rat forced swim and the mouse tail suspension tests. *Eur Neuropsychopharmacol* 2007;**17**:172–9.
236. Garber KB, Visootsak J, Warren ST. Fragile X syndrome. *Eur J Hum Genet* 2008;**16**:666–72.
237. Crawford DC, Acuna JM, Sherman SL. FMR1 and the fragile X syndrome: human genome epidemiology review. *Genet Med* 2001;**3**:359–71.
238. Pieretti M, Zhang FP, Fu YH, Warren ST, Oostra BA, Caskey CT, et al. Absence of expression of the FMR-1 gene in fragile X syndrome. *Cell* 1991;**66**:817–22.
239. Bear MF, Dolen G, Osterweil E, Nagarajan N. Fragile X: translation in action. *Neuropsychopharmacology* 2008;**33**:84–7.
240. Lu YM, Jia Z, Janus C, Henderson JT, Gerlai R, Wojtowicz JM, et al. Mice lacking metabotropic glutamate receptor 5 show impaired learning and reduced CA1 long-term potentiation (LTP) but normal CA3 LTP. *J Neurosci* 1997;**17**:5196–205.
241. Huber KM, Roder JC, Bear MF. Chemical induction of mGluR5- and protein synthesis-dependent long-term depression in hippocampal area CA1. *J Neurophysiol* 2001;**86**:321–5.
242. Huang CC, Hsu KS. Sustained activation of metabotropic glutamate receptor 5 and protein tyrosine phosphatases mediate the expression of (S)-3,5-dihydroxyphenylglycine-induced long-term depression in the hippocampal CA1 region. *J Neurochem* 2006;**96**:179–94.
243. Faas GC, Adwanikar H, Gereau 4th RW, Saggau P. Modulation of presynaptic calcium transients by metabotropic glutamate receptor activation: a differential role in acute depression of synaptic transmission and long-term depression. *J Neurosci* 2002;**22**:6885–90.
244. Cohen AS, Raymond CR, Abraham WC. Priming of long-term potentiation induced by activation of metabotropic glutamate receptors coupled to phospholipase C. *Hippocampus* 1998;**8**:160–70.
245. Ayala JE, Chen Y, Banko JL, Sheffler DJ, Williams R, Telk AN, et al. mGluR5 positive allosteric modulators facilitate both hippocampal LTP and LTD and enhance spatial learning. *Neuropsychopharmacology* 2009;**34**:2057–71.
246. Huber KM, Gallagher SM, Warren ST, Bear MF. Altered synaptic plasticity in a mouse model of fragile X mental retardation. *Proc Natl Acad Sci USA* 2002;**99**:7746–50.
247. Lauterborn JC, Rex CS, Kramar E, Chen LY, Pandyarajan V, Lynch G, et al. Brain-derived neurotrophic factor rescues synaptic plasticity in a mouse model of fragile X syndrome. *J Neurosci* 2007;**27**:10685–94.
248. Bear MF, Huber KM, Warren ST. The mGluR theory of fragile X mental retardation. *Trends Neurosci* 2004;**27**:370–7.
249. Yan QJ, Rammal M, Tranfaglia M, Bauchwitz RP. Suppression of two major Fragile X Syndrome mouse model phenotypes by the mGluR5 antagonist MPEP. *Neuropharmacology* 2005;**49**:1053–66.

250. Dolen G, Osterweil E, Rao BS, Smith GB, Auerbach BD, Chattarji S, et al. Correction of fragile X syndrome in mice. *Neuron* 2007;**56**:955–62.
251. Michalon A, Sidorov M, Ballard TM, Ozmen L, Spooren W, Wettstein JG, et al. Chronic pharmacological mGlu5 inhibition corrects fragile X in adult mice. *Neuron* 2012;**74**:49–56.
252. Berry-Kravis E, Hessl D, Coffey S, Hervey C, Schneider A, Yuhas J, et al. A pilot open label, single dose trial of fenobam in adults with fragile X syndrome. *J Med Genet* 2009;**46**:266–71.
253. Silverman JL, Smith DG, Rizzo SJ, Karras MN, Turner SM, Tolu SS, et al. Negative allosteric modulation of the mGluR5 receptor reduces repetitive behaviors and rescues social deficits in mouse models of autism. *Sci Transl Med* 2012;**4**:131ra151.
254. Campbell UC, Lalwani K, Hernandez L, Kinney GG, Conn PJ, Bristow LJ. The mGluR5 antagonist 2-methyl-6-(phenylethynyl)-pyridine (MPEP) potentiates PCP-induced cognitive deficits in rats. *Psychopharmacology (Berl)* 2004;**175**:310–8.
255. Itil TM, Seaman BA, Huque M, Mukhopadhyay S, Blasucci DNKT, Ciccone PE. The clinical and quantitative EEG effects and plasma levels of fenobam (McN-3377) in subjects with anxiety: an open rising dose tolerance and efficacy study. *Curr Ther Res* 1978;**24**:708–24.
256. Friedmann CTH, Davis LJ, Ciccone PE, Rubin RT. Phase II double blind controlled study of a new anxiolytic, fenobam (McN-3377) vs placebo. *Curr Ther Res* 1980;**27**:144–51.
257. Zerbib F, Keywood C, Strabach G. Efficacy, tolerability and pharmacokinetics of a modified release formulation of ADX10059, a negative allosteric modulator of metabotropic glutamate receptor 5: an esophageal pH-impedance study in healthy subjects. *Neurogastroenterol Motil* 2010;**22**:859–65 e231.
258. Keywood C, Wakefield M, Tack J. A proof-of-concept study evaluating the effect of ADX10059, a metabotropic glutamate receptor-5 negative allosteric modulator, on acid exposure and symptoms in gastro-oesophageal reflux disease. *Gut* 2009;**58**:1192–9.
259. Chiamulera C, Epping-Jordan MP, Zocchi A, Marcon C, Cottiny C, Tacconi S, et al. Reinforcing and locomotor stimulant effects of cocaine are absent in mGluR5 null mutant mice. *Nat Neurosci* 2001;**4**:873–4.
260. Popik P, Wrobel M. Morphine conditioned reward is inhibited by MPEP, the mGluR5 antagonist. *Neuropharmacology* 2002;**43**:1210–7.
261. Lominac KD, Kapasova Z, Hannun RA, Patterson C, Middaugh LD, Szumlinski KK. Behavioral and neurochemical interactions between Group 1 mGluR antagonists and ethanol: potential insight into their anti-addictive properties. *Drug Alcohol Depend* 2006;**85**:142–56.
262. Herzig V, Capuani EM, Kovar KA, Schmidt WJ. Effects of MPEP on expression of food-, MDMA- or amphetamine-conditioned place preference in rats. *Addict Biol* 2005;**10**:243–9.
263. Tessari M, Pilla M, Andreoli M, Hutcheson DM, Heidbreder CA. Antagonism at metabotropic glutamate 5 receptors inhibits nicotine- and cocaine-taking behaviours and prevents nicotine-triggered relapse to nicotine-seeking. *Eur J Pharmacol* 2004;**499**:121–33.
264. Martin-Fardon R, Baptista MA, Dayas CV, Weiss F. Dissociation of the effects of MTEP [3-[(2-methyl-1,3-thiazol-4-yl)ethynyl]piperidine] on conditioned reinstatement and reinforcement: comparison between cocaine and a conventional reinforcer. *J Pharmacol Exp Ther* 2009;**329**:1084–90.
265. Gass JT, Osborne MP, Watson NL, Brown JL, Olive MF. mGluR5 antagonism attenuates methamphetamine reinforcement and prevents reinstatement of methamphetamine-seeking behavior in rats. *Neuropsychopharmacology* 2009;**34**:820–33.

266. Backstrom P, Bachteler D, Koch S, Hyytia P, Spanagel R. mGluR5 antagonist MPEP reduces ethanol-seeking and relapse behavior. *Neuropsychopharmacology* 2004;**29**:921–8.
267. Cleva RM, Olive MF. mGlu receptors and drug addiction. *Wiley Interdiscip Rev Membr Transp Signal* 2012;**1**:281–95.
268. Schools GP, Kimelberg HK. mGluR3 and mGluR5 are the predominant metabotropic glutamate receptor mRNAs expressed in hippocampal astrocytes acutely isolated from young rats. *J Neurosci Res* 1999;**58**:533–43.
269. Cai Z, Schools GP, Kimelberg HK. Metabotropic glutamate receptors in acutely isolated hippocampal astrocytes: developmental changes of mGluR5 mRNA and functional expression. *Glia* 2000;**29**:70–80.
270. Oliet SH, Piet R, Poulain DA. Control of glutamate clearance and synaptic efficacy by glial coverage of neurons. *Science* 2001;**292**:923–6.
271. Anderson CM, Swanson RA. Astrocyte glutamate transport: review of properties, regulation, and physiological functions. *Glia* 2000;**32**:1–14.
272. Schousboe A. Role of astrocytes in the maintenance and modulation of glutamatergic and GABAergic neurotransmission. *Neurochem Res* 2003;**28**:347–52.
273. Araque A, Sanzgiri RP, Parpura V, Haydon PG. Astrocyte-induced modulation of synaptic transmission. *Can J Physiol Pharmacol* 1999;**77**:699–706.
274. Bradley SJ, Watson JM, Challiss RA. Effects of positive allosteric modulators on single-cell oscillatory Ca2+ signaling initiated by the type 5 metabotropic glutamate receptor. *Mol Pharmacol* 2009;**76**:1302–13.
275. Fellin T, Gomez-Gonzalo M, Gobbo S, Carmignoto G, Haydon PG. Astrocytic glutamate is not necessary for the generation of epileptiform neuronal activity in hippocampal slices. *J Neurosci* 2006;**26**:9312–22.
276. Nakahara K, Okada M, Nakanishi S. The metabotropic glutamate receptor mGluR5 induces calcium oscillations in cultured astrocytes via protein kinase C phosphorylation. *J Neurochem* 1997;**69**:1467–75.
277. Pasti L, Volterra A, Pozzan T, Carmignoto G. Intracellular calcium oscillations in astrocytes: a highly plastic, bidirectional form of communication between neurons and astrocytes in situ. *J Neurosci* 1997;**17**:7817–30.
278. Seifert G, Schilling K, Steinhauser C. Astrocyte dysfunction in neurological disorders: a molecular perspective. *Nat Rev Neurosci* 2006;**7**:194–206.
279. Ding S, Fellin T, Zhu Y, Lee SY, Auberson YP, Meaney DF, et al. Enhanced astrocytic Ca2+ signals contribute to neuronal excitotoxicity after status epilepticus. *J Neurosci* 2007;**27**:10674–84.
280. Bruno V, Battaglia G, Casabona G, Copani A, Caciagli F, Nicoletti F. Neuroprotection by glial metabotropic glutamate receptors is mediated by transforming growth factor-beta. *J Neurosci* 1998;**18**:9594–600.
281. Bruno V, Battaglia G, Copani A, D'Onofrio M, Di Iorio P, De Blasi A, et al. Metabotropic glutamate receptor subtypes as targets for neuroprotective drugs. *J Cereb Blood Flow Metab* 2001;**21**:1013–33.
282. D'Onofrio M, Cuomo L, Battaglia G, Ngomba RT, Storto M, Kingston AE, et al. Neuroprotection mediated by glial group-II metabotropic glutamate receptors requires the activation of the MAP kinase and the phosphatidylinositol-3-kinase pathways. *J Neurochem* 2001;**78**:435–45.
283. Nedergaard M, Rodriguez JJ, Verkhratsky A. Glial calcium and diseases of the nervous system. *Cell Calcium* 2010;**47**:140–9.
284. Rossi D, Volterra A. Astrocytic dysfunction: insights on the role in neurodegeneration. *Brain Res Bull* 2009;**80**:224–32.
285. Slassi A, Isaac M, Edwards L, Minidis A, Wensbo D, Mattsson J, et al. Recent advances in non-competitive mGlu5 receptor antagonists and their potential therapeutic applications. *Curr Top Med Chem* 2005;**5**:897–911.

286. Breysse N, Amalric M, Salin P. Metabotropic glutamate 5 receptor blockade alleviates akinesia by normalizing activity of selective basal-ganglia structures in parkinsonian rats. *J Neurosci* 2003;**23**:8302–9.
287. Breysse N, Baunez C, Spooren W, Gasparini F, Amalric M. Chronic but not acute treatment with a metabotropic glutamate 5 receptor antagonist reverses the akinetic deficits in a rat model of parkinsonism. *J Neurosci* 2002;**22**:5669–78.
288. Ossowska K, Konieczny J, Wolfarth S, Pilc A. MTEP, a new selective antagonist of the metabotropic glutamate receptor subtype 5 (mGluR5), produces antiparkinsonian-like effects in rats. *Neuropharmacology* 2005;**49**:447–55.
289. Ossowska K, Wardas J, Smialowska M, Kuter K, Lenda T, Wieronska JM, et al. A slowly developing dysfunction of dopaminergic nigrostriatal neurons induced by long-term paraquat administration in rats: an animal model of preclinical stages of Parkinson's disease? *Eur J Neurosci* 2005;**22**:1294–304.
290. Rylander D, Iderberg H, Li Q, Dekundy A, Zhang J, Li H, et al. A mGluR5 antagonist under clinical development improves L-DOPA-induced dyskinesia in parkinsonian rats and monkeys. *Neurobiol Dis* 2010;**39**:352–61.
291. Lisman JE, Coyle JT, Green RW, Javitt DC, Benes FM, Heckers S, et al. Circuit-based framework for understanding neurotransmitter and risk gene interactions in schizophrenia. *Trends Neurosci* 2008;**31**:234–42.
292. Marek GJ, Behl B, Bespalov AY, Gross G, Lee Y, Schoemaker H. Glutamatergic (N-methyl-D-aspartate receptor) hypofrontality in schizophrenia: too little juice or a miswired brain? *Mol Pharmacol* 2010;**77**:317–26.
293. Hirose K, Chan PH. Blockade of glutamate excitotoxicity and its clinical applications. *Neurochem Res* 1993;**18**:479–83.
294. Alagarsamy S, Marino MJ, Rouse ST, Gereau 4th RW, Heinemann SF, Conn PJ. Activation of NMDA receptors reverses desensitization of mGluR5 in native and recombinant systems. *Nat Neurosci* 1999;**2**:234–40.
295. Alagarsamy S, Saugstad J, Warren L, Mansuy IM, Gereau 4th RW, Conn PJ. NMDA-induced potentiation of mGluR5 is mediated by activation of protein phosphatase 2B/calcineurin. *Neuropharmacology* 2005;**49**(Suppl. 1):135–45.
296. Attucci S, Carla V, Mannaioni G, Moroni F. Activation of type 5 metabotropic glutamate receptors enhances NMDA responses in mice cortical wedges. *Br J Pharmacol* 2001;**132**:799–806.
297. Ehlers MD. Synapse structure: glutamate receptors connected by the shanks. *Curr Biol* 1999;**9**:R848–R850.
298. Mannaioni G, Marino MJ, Valenti O, Traynelis SF, Conn PJ. Metabotropic glutamate receptors 1 and 5 differentially regulate CA1 pyramidal cell function. *J Neurosci* 2001;**21**:5925–34.
299. Marino MJ, Conn PJ. Direct and indirect modulation of the N-methyl D-aspartate receptor. *Curr Drug Targets CNS Neurol Disord* 2002;**1**:1–16.
300. Pisani A, Bonsi P, Centonze D, Bernardi G, Calabresi P. Functional coexpression of excitatory mGluR1 and mGluR5 on striatal cholinergic interneurons. *Neuropharmacology* 2001;**40**:460–3.
301. Ugolini A, Corsi M, Bordi F. Potentiation of NMDA and AMPA responses by the specific mGluR5 agonist CHPG in spinal cord motoneurons. *Neuropharmacology* 1999;**38**:1569–76.
302. Balschun D, Wetzel W. Inhibition of mGluR5 blocks hippocampal LTP in vivo and spatial learning in rats. *Pharmacol Biochem Behav* 2002;**73**:375–80.
303. Manahan-Vaughan D, Schuetz K. Differential participation of metabotropic glutamate receptor mGlu1 and mGlu5 in spatial learning and hippocampal long-term potentiation in vivo. *Neuropharmacology* 2002;**43**:297.

304. Lindsley CW, Wisnoski DD, Leister WH, O'Brien JA, Lemaire W, Williams Jr DL, et al. Discovery of positive allosteric modulators for the metabotropic glutamate receptor subtype 5 from a series of N-(1,3-diphenyl-1H- pyrazol-5-yl)benzamides that potentiate receptor function in vivo. *J Med Chem* 2004;**47**:5825–8.
305. Horio M, Fujita Y, Hashimoto K. Therapeutic effects of metabotropic glutamate receptor 5 positive allosteric modulator CDPPB on phencyclidine-induced cognitive deficits in mice. *Fundam Clin Pharmacol* 2012 [Epub ahead of print].
306. Vardigan JD, Huszar SL, McNaughton CH, Hutson PH, Uslaner JM. MK-801 produces a deficit in sucrose preference that is reversed by clozapine, D-serine, and the metabotropic glutamate 5 receptor positive allosteric modulator CDPPB: relevance to negative symptoms associated with schizophrenia? *Pharmacol Biochem Behav* 2010;**95**:223–9.
307. Stauffer SR. Progress toward Positive Allosteric Modulators of the Metabotropic Glutamate Receptor Subtype 5 (mGlu5). *ACS Chem Neurosci* 2011;**2**:450–70.
308. Zhao Z, Wisnoski DD, O'Brien JA, Lemaire W, Williams Jr DL, Jacobson MA, et al. Challenges in the development of mGluR5 positive allosteric modulators: the discovery of CPPHA. *Bioorg Med Chem Lett* 2007;**17**:1386–91.
309. Balschun D, Zuschratter W, Wetzel W. Allosteric enhancement of metabotropic glutamate receptor 5 function promotes spatial memory. *Neuroscience* 2006;**142**:691–702.
310. Gass JT, Olive MF. Positive allosteric modulation of mGluR5 receptors facilitates extinction of a cocaine contextual memory. *Biol Psychiatry* 2009;**65**:717–20.
311. Lecourtier L, Homayoun H, Tamagnan G, Moghaddam B. Positive allosteric modulation of metabotropic glutamate 5 (mGlu5) receptors reverses N-Methyl-D-aspartate antagonist-induced alteration of neuronal firing in prefrontal cortex. *Biol Psychiatry* 2007;**62**:739–46.
312. Stefani MR, Moghaddam B. Activation of type 5 metabotropic glutamate receptors attenuates deficits in cognitive flexibility induced by NMDA receptor blockade. *Eur J Pharmacol* 2010;**639**:26–32.
313. Uslaner JM, Parmentier-Batteur S, Flick RB, Surles NO, Lam JS, McNaughton CH, et al. Dose-dependent effect of CDPPB, the mGluR5 positive allosteric modulator, on recognition memory is associated with GluR1 and CREB phosphorylation in the prefrontal cortex and hippocampus. *Neuropharmacology* 2009;**57**:531–8.
314. Spear N, Gadient RA, Wilkins DE, Do M, Smith JS, Zeller KL, et al. Preclinical profile of a novel metabotropic glutamate receptor 5 positive allosteric modulator. *Eur J Pharmacol* 2011;**659**:146–54.
315. Schlumberger C, Pietraszek M, Gravius A, Danysz W. Effects of a positive allosteric modulator of mGluR5 ADX47273 on conditioned avoidance response and PCP-induced hyperlocomotion in the rat as models for schizophrenia. *Pharmacol Biochem Behav* 2010;**95**:23–30.
316. Schlumberger C, Pietraszek M, Gravius A, Klein KU, Greco S, More L, et al. Comparison of the mGlu(5) receptor positive allosteric modulator ADX47273 and the mGlu (2/3) receptor agonist LY354740 in tests for antipsychotic-like activity. *Eur J Pharmacol* 2009;**623**:73–83.
317. Tsai P, Sahin M. Mechanisms of neurocognitive dysfunction and therapeutic considerations in tuberous sclerosis complex. *Curr Opin Neurol* 2011;**24**:106–13.
318. Auerbach BD, Osterweil EK, Bear MF. Mutations causing syndromic autism define an axis of synaptic pathophysiology. *Nature* 2011;**480**:63–8.
319. Chevere-Torres I, Kaphzan H, Bhattacharya A, Kang A, Maki JM, Gambello MJ, et al. Metabotropic glutamate receptor-dependent long-term depression is impaired due to elevated ERK signaling in the DeltaRG mouse model of tuberous sclerosis complex. *Neurobiol Dis* 2012;**45**:1101–10.

320. Cryan JF, Kelly PH, Neijt HC, Sansig G, Flor PJ, van Der Putten H. Antidepressant and anxiolytic-like effects in mice lacking the group III metabotropic glutamate receptor mGluR7. *Eur J Neurosci* 2003;**17**:2409–17.
321. Stachowicz K, Branski P, Klak K, van der Putten H, Cryan JF, Flor PJ, et al. Selective activation of metabotropic G-protein-coupled glutamate 7 receptor elicits anxiolytic-like effects in mice by modulating GABAergic neurotransmission. *Behav Pharmacol* 2008;**19**:597–603.
322. McCallum J, Kim JH, Richardson R. Impaired extinction retention in adolescent rats: effects of D-cycloserine. *Neuropsychopharmacology* 2010;**35**:2134–42.
323. Myers KM, Carlezon Jr WA. Extinction of drug- and withdrawal-paired cues in animal models: relevance to the treatment of addiction. *Neurosci Biobehav Rev* 2010;**35**: 285–302.
324. Myers KM, Davis M. Behavioral and neural analysis of extinction. *Neuron* 2002;**36**:567–84.
325. Toth I, Dietz M, Peterlik D, Huber SE, Fendt M, Neumann ID, et al. Pharmacological interference with metabotropic glutamate receptor subtype 7 but not subtype 5 differentially affects within- and between-session extinction of Pavlovian conditioned fear. *Neuropharmacology* 2012;**62**:1619–26.
326. Broadstock M, Austin PJ, Betts MJ, Duty S. Antiparkinsonian potential of targeting group III metabotropic glutamate receptor subtypes in the rodent substantia nigra pars reticulata. *Br J Pharmacol* 2012;**165**:1034–45.
327. Dolan S, Gunn MD, Biddlestone L, Nolan AM. The selective metabotropic glutamate receptor 7 allosteric agonist AMN082 inhibits inflammatory pain-induced and incision-induced hypersensitivity in rat. *Behav Pharmacol* 2009;**20**:596–604.
328. Dolan S, Gunn MD, Crossan C, Nolan AM. Activation of metabotropic glutamate receptor 7 in spinal cord inhibits pain and hyperalgesia in a novel formalin model in sheep. *Behav Pharmacol* 2011;**22**:582–8.
329. Palazzo E, Fu Y, Ji G, Maione S, Neugebauer V. Group III mGluR7 and mGluR8 in the amygdala differentially modulate nocifensive and affective pain behaviors. *Neuropharmacology* 2008;**55**:537–45.
330. Palucha-Poniewiera A, Branski P, Lenda T, Pilc A. The antidepressant-like action of metabotropic glutamate 7 receptor agonist N, N′-bis(diphenylmethyl)-1,2-ethanediamine (AMN082) is serotonin-dependent. *J Pharmacol Exp Ther* 2010;**334**: 1066–74.
331. Hikichi H, Murai T, Okuda S, Maehara S, Satow A, Ise S, et al. Effects of a novel metabotropic glutamate receptor 7 negative allosteric modulator, 6-(4-methoxyphenyl)-5-methyl-3-pyridin-4-ylisoxazonolo[4,5-c]pyridin-4(5H)-one (MMPIP), on the central nervous system in rodents. *Eur J Pharmacol* 2010;**639**:106–14.
332. Bahi A. The selective metabotropic glutamate receptor 7 allosteric agonist AMN082 prevents reinstatement of extinguished ethanol-induced conditioned place preference in mice. *Pharmacol Biochem Behav* 2012;**101**:193–200.
333. Bahi A, Fizia K, Dietz M, Gasparini F, Flor PJ. Pharmacological modulation of mGluR7 with AMN082 and MMPIP exerts specific influences on alcohol consumption and preference in rats. *Addict Biol* 2012;**17**:235–47.
334. Li X, Li J, Gardner EL, Xi ZX. Activation of mGluR7s inhibits cocaine-induced reinstatement of drug-seeking behavior by a nucleus accumbens glutamate-mGluR2/3 mechanism in rats. *J Neurochem* 2010;**114**:1368–80.
335. Sukoff Rizzo SJ, Leonard SK, Gilbert A, Dollings P, Smith DL, Zhang MY, et al. The metabotropic glutamate receptor 7 allosteric modulator AMN082: a monoaminergic agent in disguise? *J Pharmacol Exp Ther* 2011;**338**:345–52.

336. Johnson KA, Jones CK, Tantawy MN, Bubser M, Marvanova M, Ansari MS, et al. The metabotropic glutamate receptor 8 agonist (S)-3,4-DCPG reverses motor deficits in prolonged but not acute models of Parkinson's disease. *Neuropharmacology* 2013;**66**: 187–95.
337. Lopez S, Turle-Lorenzo N, Acher F, De Leonibus E, Mele A, Amalric M. Targeting group III metabotropic glutamate receptors produces complex behavioral effects in rodent models of Parkinson's disease. *J Neurosci* 2007;**27**:6701–11.
338. Ballesteros JA, Weinstein H. Integrated methods for the construction of three dimensional models and computational probing of structure-function relations in G-protein coupled receptors. *Methods Neurosci* 1995;**25**:366–428.

CHAPTER THREE

Mining the Potential of Label-Free Biosensors for Seven-Transmembrane Receptor Drug Discovery

Magalie Rocheville*, Julio Martin†, Jeff Jerman‡, Evi Kostenis§
*GlaxoSmithKline, Medicines Research Centre, Stevenage, United Kingdom
†GlaxoSmithKline, Centro de Investigacion Basica, Parque Tecnologico de Madrid, Tres Cantos, Spain
‡MRC Technology, Centre for Therapeutics Discovery, London, United Kingdom
§Molecular-, Cellular-, and Pharmacobiology Section, Institute of Pharmaceutical Biology, University of Bonn, Bonn, Germany

Contents

Abstract

Label-free is a broad term used to describe a number of cutting-edge biosensor technologies that have attracted considerable attention in the area of drug discovery for seven-transmembrane G protein-coupled receptors (GPCRs). Label-free biosensors resolve receptor-mediated responses noninvasively in real time and living cells and do so with high textural information and broad signaling-pathway coverage. They should facilitate studies of the receptor's integrated signal transduction biology intractable to classical assays with single pathway focus. Label-free occupies a privilege niche with respect to mechanistic studies in human native cells—healthy or disease-relevant—and the probing of context-dependent pharmacology in relation to whole

Progress in Molecular Biology and Translational Science, Volume 115
ISSN 1877-1173
http://dx.doi.org/10.1016/B978-0-12-394587-7.00003-8

biological system efficacy. It is expected that implementation of label-free approaches into the drug discovery process will improve clinical predictability of drug candidates at early stages of discovery research by their exquisite capability to sense whole cellular responses akin to tissue bioassays. Here, we present an overview of promises and challenges this rapidly evolving technology offers to drug screening and we also discuss the prospect of advancing drug discovery.

1. INTRODUCTION

G protein-coupled receptors (GPCRs) are membrane-bound integral proteins linked to a plethora of disease paradigms, both peripheral and centrally mediated, and have a rich history of therapeutic intervention for drug discovery. They have led to numerous breakthrough medicines, making them a highly tractable target class occupying ~50% market share of prescribed medicines.[1,2] Yet, recent business analysis shows a lack of return on investment in overall pharmaceutical research and forces a rethink of drug discovery models for GPCRs and other target classes.[3] GPCRs remain afflicted by a significant incidence of drug attrition at late stages of development and the issue is costly.[4] Accordingly, there is a strong impetus to improve preclinical predictability and particularly assay relevance at early stages of discovery research by exploiting paradigms amenable to a fuller interrogation of molecule activity. Currently, screening for GPCRs is commonly achieved by running high-capacity and homogenous assays of cellular signaling end points using reconstituted cellular systems overexpressing a receptor of interest.[5–7] In doing so, screening focuses on a receptor species typically isolated from its physiological context but thought to be involved in mechanisms of disease(s). The approach termed target-based discovery has proven valuable and replaced isolated tissues or *in vivo* animal models at the front-end of compound screening. However, it remains a surrogate and simplistic approach to native phenotypes and while powerful, presents some shortcomings by often lacking the desired predictability of intact physio- and/or pathological systems. An improved understanding of biological systems and disease paradigms will undoubtedly help resolving this as the etiology of diseases is complex, likely multifactorial and often patient-specific. A better fit of assay to end-game physiological system relevance as early as possible during screening may also be value-adding to triage molecules.

Understanding, controlling, and improving assay relevance is challenging for GPCRs. One needs to carefully choose assay readouts (that are being measured and how it is detected) and assay systems (cell reagent and species) to employ for screening at defined stages of drug discovery.[5–7] One also needs to consider the greater cellular signaling intricacy uncovered over recent years.[8–13] Signaling is thought to be collateral and intertwined with multiple cascades overlapping in both time and cellular sub-localization. Signaling is elicited via means dependently or independently of G proteins and relies on the characteristics of the system under study. Signaling can also be biased by compounds for a subset of features via receptor binding, a notion termed ligand-bias or functional selectivity. As a result, the observed receptor pharmacology will vary depending on the signaling end point being measured, the cellular context/reagent, and the compound used. Assays commonly employed in screening are focused on single functional end points and so provide a "reductionist" perspective with respect to the diversity of signaling. One cannot rule out the possibility that they may, when considered in isolation, be under-appreciative of compound activity and potentially mislead medicinal chemistry efforts. Multiple vantage points can be gathered by parallel testing of assays but the strategy holds several drawbacks, not least in terms of logistic, resource, and cost. Such effort is likely to be limiting as a simple collation of outputs will not necessarily capture the interplay between readouts (negative, positive, neutral).[5,9] To this end, label-free is of great interest as it captures whole cell responses noninvasively in a manner akin to tissue bioassays.[5,14,15] Its applicability is broad throughout drug discovery but raises several points for consideration that may limit it fulfilling a generic screening solution for GPCRs.

2. PROMISES OF LABEL-FREE SCREENING

Label-free is a broad term used to describe a series of cutting-edge assay technologies rapidly incorporating screening in industrial settings.[14–22] Label-free relies on biosensor-coated microplates to sense-integrated phenotypic changes that originate from application of pharmacological stimuli to living cells. It spans broad applicability across discovery research (Table 3.1). Numerous examples of pharmacological outputs have been reported across types of receptors and cell backgrounds,[23–29] and details of instrumentation have been recently reviewed elsewhere.[16] Here, we focus on aspects that may tackle assay relevance and throughput compatible with large-scale screening.

Table 3.1 The applicability of label-free ranges from target inception to lead optimization and *in vitro* safety profiling

Drug discovery activity	Potential application
Target ID	Receptor panning against immortalized cell lines, primary cultured cells, stem cells
	Orphan screening across array of cell types; catch-all platform to screen for multiple signaling route(s)
Assay development	Direct measure of signaling route(s) traditionally deemed challenging such as Gα12/13 without forcing coupling via promiscuous or chimeric G proteins
	Use of endogenously expressed targets in immortalized cell lines or primary cultures for assay development
	Straight-forward reagent validation without requiring target tagging, including study of preferred coupling route(s) and/or pleiotropic coupling
	Selectivity testing at multiple receptor subtypes, enabling cross-validation of conventional technology outputs versus label-free for further reagent characterization
	Profiling of host cells prior to reagent generation. This aspect may be valuable for ubiquitously expressed receptor families for which identifying null-background cells has proven challenging, for example, lipid receptors
Primary screening	Robust single shot assay platform for hit identification, which may offer the potential of additional chemotypes versus the use of a single linear output
	Potential reduction in technology-based false positives typically associated with existing screening strategies, consequently lower hit rate and follow-up work
	New (Gα12/13) or complementary approach (i.e., non-radioactive Gαi/o) to commonly used HTS assay formats
	Enabler of HTS based on native cells, primary cells, stem cells, and cells isolated from disease patients
Secondary screening & MoA	Support of SAR and MoA using recombinant cell lines or native cell types aligned to disease interests (context-relevant)
	Optimization of lead molecules for subset of signaling pathways deemed physiologically relevant (functional selectivity)

Table 3.1 The applicability of label-free ranges from target inception to lead optimization and *in vitro* safety profiling—cont'd

Drug discovery activity	**Potential application**
	Additional texture on existing chemotypes, providing fingerprint to help define "desired" drug profiles in relation with *in vivo* testing in behavioral animal models
	Downstream native assay; confirmation of recombinant HTS/SAR data in native systems
	Integrated view into cell physiology, for example, cell–cell interactions, chemotaxis, adherence or polarization, membrane permeability, cell differentiation, proliferation, etc.
Safety profiling	Exploitation of native cells such as human stem cells to improve data predictability and physiological relevance

2.1. Sensitivity of detection and compatibility with native cells

Label-free is characterized by an exquisite sensitivity of detection to whole cellular responses achieved independently of target overexpression or tagging. Its ability to do so is impressive in terms of overall power of resolution and extends beyond the realm of light microscopy. Both optical and impedance-based technologies perform comparatively and quantify receptor activation robustly through signals elicited within an estimated 100–500 nm from the biosensor surface upon cycloskeletal changes, dynamic mass redistribution (DMR), or changes in cell adhesion.[16,30–32] Reader capabilities have progressed significantly over the last year, and according to vendors, signals can be resolved in milliseconds to observe rapid cellular events such as cardiomyocyte beating patterns (e.g., Roche xCELLIgence Cardio).

Signals can also be selectively measured for a subset of cells covering the well using a combination of label-free and imaging techniques, therein providing a powerful solution to screen sparsely represented cell types among heterogeneous populations or small cell clusters in nonconfluent wells (Fig. 3.1 and Ref. 33). Label-free embraces a large repertoire of fit-for-purpose cultured cell reagents expressing receptors of interest heterologously or endogenously.[15–22] In particular, the approach has already been applied successfully not only to various primary cell types, as exemplified in Fig. 3.2 for human keratinocytes, but also to muscle, cancer, blood, and neuronal cells.[18,19,23,34] The ability to test native-like cell types coupled with the

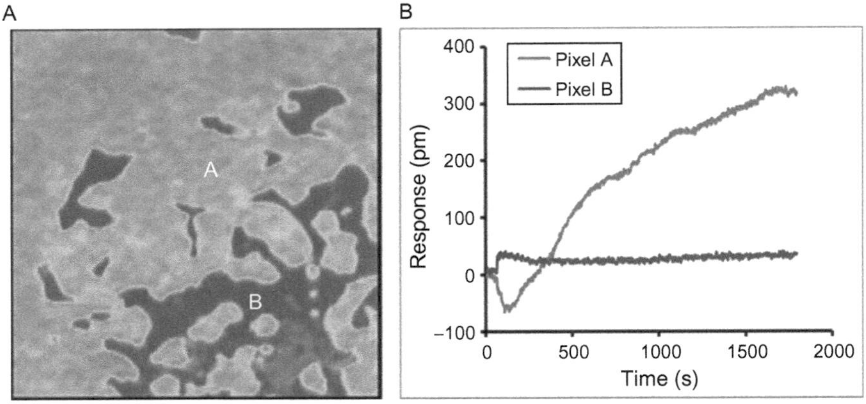

Figure 3.1 Resonant waveguide grating imager for real-time measurement of cell signaling in living cells. (a) False-colored basal resonant wavelength image of nonconfluent A431 cells. Cell adhesion results in higher basal resonant wavelength (λ_{br}) values (green area: adherent cells; blue area: cell-free). (b) Real-time DMR recording in A431 cells treated with 100 nM of the β-adrenergic agonist epinephrine at two distinct pixelated positions shown in (a). The imager used has a spatial resolution of 12 μm. (The Figure is a courtesy of Ye Fang, Corning Inc.; for details on the DMR imager, see Ref. 33) (See Color Insert.)

sensitivity of signals from 384- or 1536-well microplates to minimize cell requirement is truly enabling and can extend beyond primary targets to the profiling of selectivity and liabilities.[35] Preliminary reports also suggest the ability to assay differentiated embryonic stem cells, an emerging area that may provide human-derived cell models and improve relevance during early *in vitro* toxicity screening.[36–38]

2.2. Derivation of signature response profiles

Label-free can qualitatively and temporally distinguish signal transduction pathway(s) involved in cellular responses through an examination of response profiles as recently reviewed by Schröder *et al.*[23,24] Response profiles are informative "signatures" as they hold shape characteristics pertaining to distinct pathways, allowing compounds to be classified and compared. Signatures have been reported for Gαs, Gαi/o, Gαq, and Gα12/13, but it is important to note that such observed profiles are assay system- and technology-dependent and so call for a rigorous approach of deconvolution if mechanistic information is to be extracted.[16,23,24] Figure 3.3 exemplifies that a mere inspection of response profiles is not sufficient to deduct the underlying signaling mechanisms even in a defined cellular background. Nevertheless,

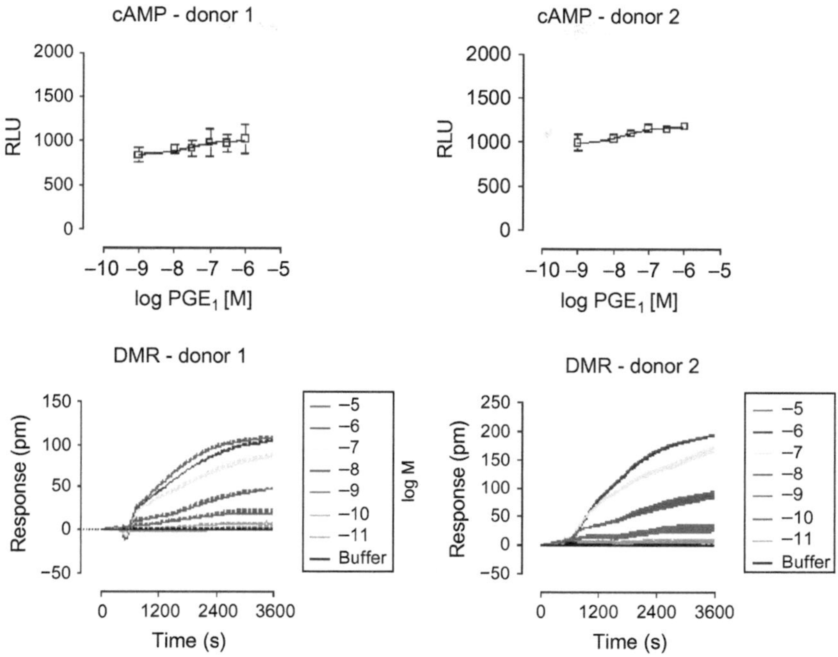

Figure 3.2 DMR enables analysis of GPCR function in primary human keratinocytes. Primary human keratinocytes were obtained from patients who underwent skin surgery (for details, see Ref. 23), and DMR was monitored upon application of prostaglandin E_1 (PGE_1), a lipid mediator known to affect growth and cytokine production of this cell type. Robust and concentration-dependent DMR recordings were observed over time in cells from individual donors. Signal-to-noise ratio and scatter in DMR recordings are clearly superior to classical end point assays, quantifying production of the second messenger cAMP with a competitive immunoassay. RLU, relative light units. (See Color Insert.)

when response profiles are collected for a given receptor in a defined cellular context, signature profiles can be established with endogenous activator molecules or reference ligands and deviations from signature profiles of the reference ligand are then most likely indicative of altered cellular signaling.

The ability to directly assay Gα12/13 or Gαi/o without either the presence of forskolin and phosphodiesterase inhibitors or the need of adding chimeric G proteins for forced G protein coupling is particularly enabling, as shown for the Gαi-linked prostaglandin D2 receptor CRTH2, the Gαi-sensitive dopamine receptor DRD2, the Gα12/13-sensitive GPR55, and other receptors.[16,23,24,39–41] Signal specificity can be tested using receptor antagonists, signaling-pathway blockers, or silent RNA delivery. However, careful validation of such tools is imperative as specificity may not translate

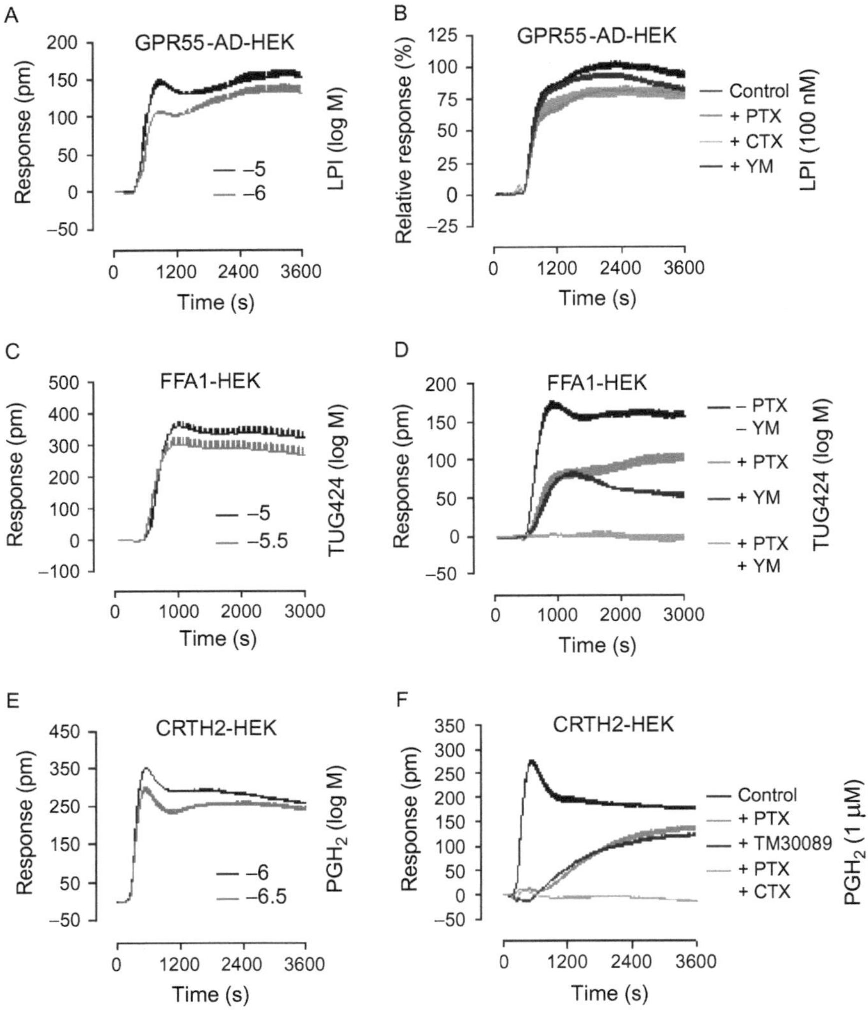

Figure 3.3 Cellular response profiles cannot be linked automatically to defined signaling routes. (A) GPR55-HEK cells were stimulated with the indicated concentrations of lyso-phosphatidylinositol (LPI) and DMR is recorded over time. (B) LPI-mediated GPR55 activation is insensitive to pretreatment of cells with choleratoxin (CTX) to mask Gαs signaling, pertussis toxin (PTX) to abolish Gαi signaling, and YM to inhibit Gαq signaling. (C) FFA1-HEK cells were stimulated with the small-molecule FFA1 agonist TUG424 and DMR was monitored over time. (D) TUG424-mediated FFA1 activation is sensitive to pretreatment of cells with PTX and YM. (E) CRTH2-HEK cells were stimulated with the lipid mediator prostaglandin H_2 (PGH_2) and cell response was monitored over time. (F) PGH_2-mediated signals are sensitive to pretreatment of cells with PTX and CTX. Note that PGH_2-induced DMR in a HEK cell background is triggered by the Gαi-linked CRTH2 receptor and endogenous Gαs-coupled EP2/4 receptors. (See Color Insert.)

from other test systems,[16,23,24] and pathway-perturbing agents frequently used in other assays often display prominent effects on cellular function on their own making it hard to interpret sensitivity to the GPCR agonist response.

2.3. Integration of signaling responses

Label-free resolves signaling in a cumulative manner into a single measurable function and does so without any bias or exclusion. Consequently, it holds a privilege and arguably unique position from where to observe pleiotropic signaling and functional selectivity in plate-based assays.[5,15,16,23,24,29,30] These areas have benefited from increasing therapeutic interest and are of theoretically high value to drug discovery and the design of safer drugs.[11–13] In practical terms, however, they are very difficult to prosecute at the early end of drug discovery. This may be attributed, at least in part, to the reliance of multiple distinct "reductionist" assays to witness it. Such cross-analysis is sometimes done using assays configured in different cellular backgrounds or receptor expression levels, all of which may lead to confounding pharmacologies. Moreover, once/if observed, understanding the physiological consequence of functional selectivity from a perspective of a theoretical reintegration of reductionist assay data is extremely challenging. Functional selectivity can manifest itself in different ways, as the underlying responses may be temporally distinct.[42,43] Pathways involved may also operate independently or dependently,[42] emphasizing the need to focus on integrated system responses. Studying the same system through the "eyes" of label-free may therefore be an interesting catch-all strategy, whether for high-throughput screening (HTS), structure–activity relationship (SAR), or mode-of-action (MoA) studies. This is illustrated by GPR55, which engages in multiple signaling routes, differentially so in response to distinct ligands as shown by reductionist and integrated readouts.[39]

2.4. Application of label-free to study complex cellular phenomena

Recently, label-free DMR has been applied in an elegant experimental approach to study assembly and disassembly of purinosomes, a multienzyme complex required for biosynthesis of the purine nucleosides adenosine and guanosine and important for cell replication.[28,44] Purinosome formation has so far been studied with methods such as fluorescence live-cell imaging. By linking purinosome assembly and disassembly to discrete alterations of cellular DMR, Verrier *et al.* examined a set of GPCR modulators for their effects on purinosome formation in HeLa cells and identified the adrenergic α2a receptor and its associated Gαi signaling pathway as the specific receptor mediating the cellular response.[28,44] Although the molecular details linking Gαi signaling with purinosome formation still remain to be unraveled, these data offer the intriguing possibility to use label-free DMR technology to search for new therapeutics to regulate mitogenic GPCR signaling.

3. CHALLENGES OF LABEL-FREE SCREENING

The use of label-free in screening is associated with perceived limitations that are briefly summarized here below. But first, it is important to qualify the emergence of label-free in context of the increasing demands of screening throughput for drug discovery over the past decade. While improving assay relevance is undoubtedly an area of intense focus, the cultural aspect of running signaling end point assays at the stages of HTS and primary SAR is still deeply engrained. Label-free is a young field of investigation for GPCRs and yet unproven in terms of exploitability of outputs to discover "better drugs" and so comes at a short-term cost for screening organizations, even if it may ultimately lead to longer-term gain, for example, reduced attrition.

3.1. Data analysis

Label-free responses are qualitatively informative but analysis of outputs can be highly challenging (Fig. 3.4). They entail a significant degree of complexity linked to the dynamics of cell signaling, which can be observationally differentiated by response magnitude and signal orientation over the defined reading time. While some responses are simple and monophasic, others appear multiphasic (negative and/or positive), and this bears significance for screening.[16,23,24,30,33] Depending on the part of the response chosen, a compound may display differential potency and/or efficacy values, making it hard to define what is most meaningful to drive medicinal chemistry. Kinetic profiles may also be unique to distinct compound concentrations, even for closely related ligands acting via the same receptor, which complicates fitting of concentration–effect curves further (Fig. 3.5). When profiling related chemotypes during SAR, it is also possible to witness differential modulation in aspects of response signatures as a consequence of ligand-bias or multiple signaling pathways, each of which may be distinct from responses of native ligand(s) (Fig. 3.6). For example, appreciative yet modest changes to response profiles were picked up across closely related compounds as seen for A431 cells and β2-ligands.[46] Label-free can also detect additional pharmacologies from classical single end point readouts,[16,23,24] whether desirable

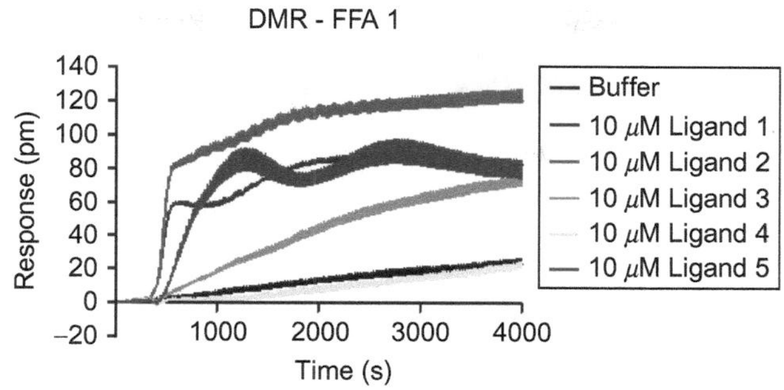

Figure 3.4 Small-molecule ligands targeting the long-chain fatty acid receptor FFA1 do not trigger uniform signature profiles. FFA1-HEK cells were stimulated with the indicated small-molecule agonists, each applied at a maximally effective concentration and DMR was recorded over time as a measure of FFA1 activity. If DMR assays are performed in an end point mode, throughput and screening logistics may benefit; however, mechanistic differences between individual chemotypes may be lost. Ligand 2 is identical to GW9508, Ligand 1 corresponds to TUG20 in Ref. 45, and Ligands 3–5 are unpublished so far. Ligand 4 appears inactive in DMR assays but stimulates inositolphosphate production in IP1 accumulation assays in a FFA1-dependent manner. (See Color Insert.)

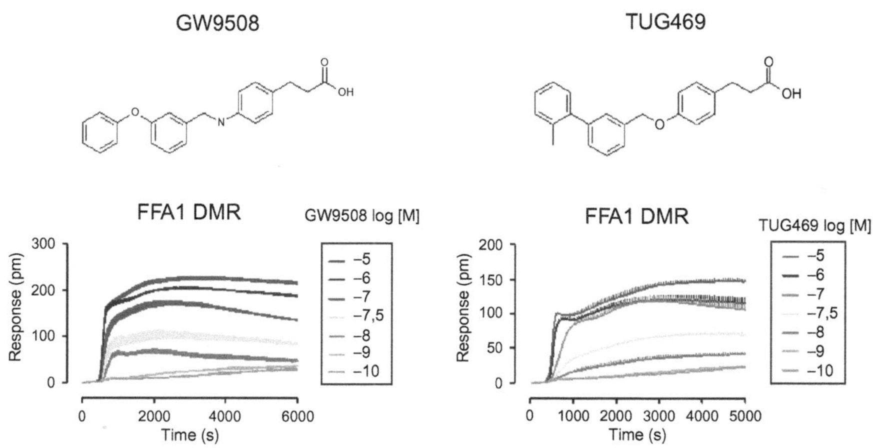

Figure 3.5 Label-free DMR recordings of small-molecule FFA1 agonists are concentration-dependent and chemotype-specific. FFA1-HEK cells were stimulated with the indicated concentrations of the small-molecule agonists GW9508 and TUG469,[45] and DMR was monitored in real time as a measure of receptor activation. Please note that agonist traces are concentration-dependent, chemotype-dependent, and additionally differ kinetically in the time-to-peak profile. By measuring a set time point, "profile-distinct" hits or changes in the subset of the profiles may be missed or overlooked, which is not desirable. (See Color Insert.)

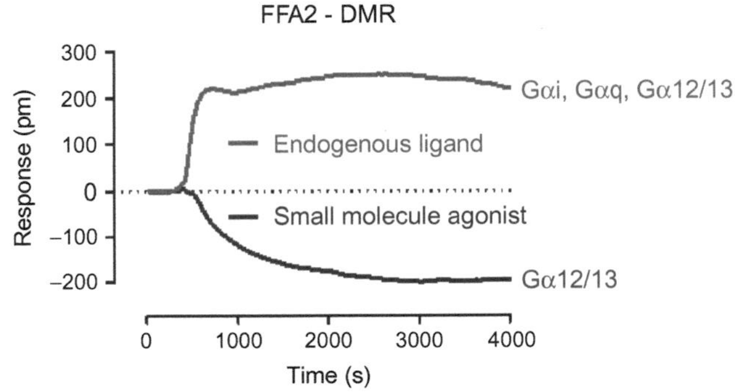

Figure 3.6 Label-free DMR recordings of FFA2 modulators differ depending on signaling pathway usage. FFA2-HEK cells were stimulated with 300 μM of the endogenous agonist propionic acid or 100 μM of a small-molecule activator and DMR was monitored over time. The signature profile of the endogenous agonist propionic acid—applied in a maximally effective concentration—is totally distinct from that observed with a small-molecule ligand biased toward Gα12/13 signaling (J. Schmidt and E. Kostenis, unpublished observations). (See Color Insert.)

or not for therapeutic efficacy, and analysis methods need to evolve to allow on-demand detailed analysis and pattern recognition to be performed when necessary. In fact, elegant examples for pattern recognition using DMR analysis have recently emerged for drugs acting via the adrenergic β2 receptor and the mu opioid receptor.[47,48]

3.2. Data interpretation

An area of challenge pertains to the cumulative overlay of distinct pathways. The integrated trace alone may be insufficient to drive decision making on compounds if the changes observed through SAR are subtle or possibly masked by dominant coupling events. Conventional assays may need to be used in parallel to independently corroborate findings of individual pathways. However, one needs to be aware that assay formats may not be equivalent in their pharmacological outputs as they measure distinct signaling end points. This is exemplified by CRTH2 and a receptor truncate profiled in label-free versus GTPγS, inositol phosphate, and cAMP assays.[40] Studies using pathway blockers can be useful; but in other cases, may lead to the uncovering of novel features of signaling that were otherwise not dominant, making outputs hard to interpret.[16] Assessing compound selectivity can prove difficult as label-free captures cellular signaling beyond the target of interest and more often than not, compound optimization stages involve chemotypes with some degree of panactivity.[23] Screening with compounds that have unappreciated activities at related or unrelated receptors may result in conflicting pharmacologies or inappropriate conclusions.

3.3. Signal specificity

Label-free is a "black-box" assay readout that may not be easily amenable to simple deconvolution. This is particularly true for endogenously expressed targets when selective pharmacological tools are not necessarily available to probe for receptor specificity. Signals have been modeled in terms of output parameters and are likely dependent on actin re-arrangement although the exact details underpinning label-free responses are not yet resolved.[16] The relevance of such changes to cell physiology is not fully understood and may vary upon cell types used. Label-free works on the assumption that it offers a "catch-all" approach to cell signaling, but how universal sensitivity may be across GPCR types, cellular systems, and signaling events is currently unknown. Also unknown is the capability of label-free to detect G protein-independent but arrestin-dependent events, although the relevance of arrestin as signal transducer in its own right has recently been questioned at least for the angiotensin II AT1a receptor.[42] Cellular adherence is a parameter of prime importance believed to dictate on signal magnitude and several surface chemistries and coating agents have been explored (including cell–cell matrix) in which to favor or limit cellular attachment in a cellular background-specific manner. However, this can theoretically introduce some level of bias on cellular signaling, especially considering functional selectivity for distinct route(s) over others (Fig. 3.7).

3.4. Assay throughput and cost

Cellular responses are resolved in real time, and each plate can typically take minutes up to an hour to read depending on the type of experiment. The capture of full temporal profiles therefore comes at a significant cost in throughput and resource, which may be prohibitive to fulfill large screening demands such as HTS. To circumvent this issue, data acquisition can be reduced to one or more discrete time points using a screen configured with known native or standard tool ligand(s) as exemplified for the β2 receptor.[48] However, such an approach does not preclude the possibility that hits may be missed or classified inappropriately, especially so for agonists or positive allosteric modulators. For SAR, the implication around throughput is less critical as it typically involves fewer plates so may withstand more flexibility

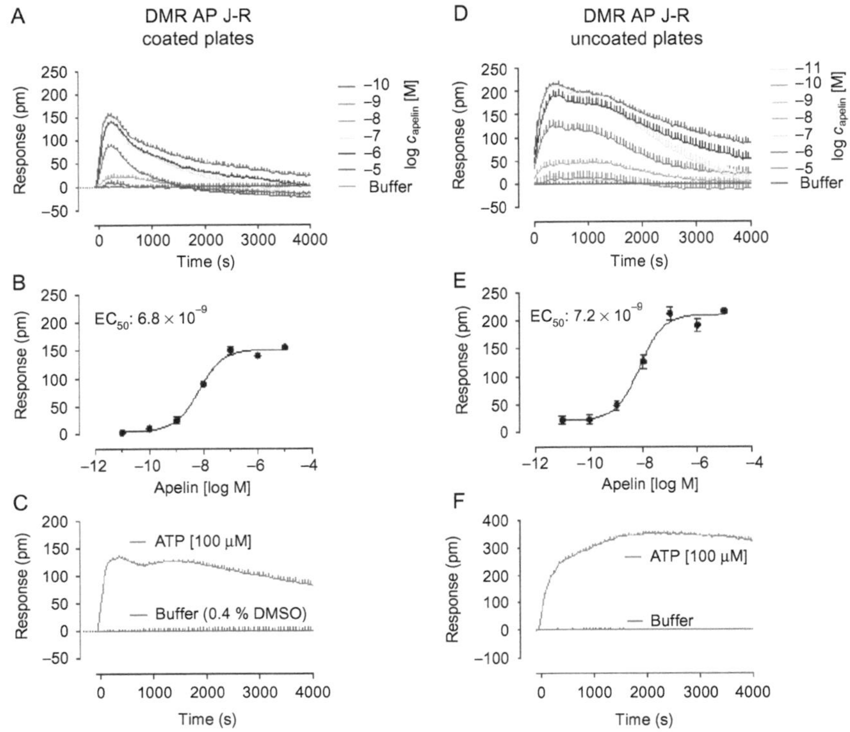

Figure 3.7 Label-free DMR recordings of apelin activity on the APJ receptor. (A, B) CHO-K1 cells stably expressing the apelin receptor APJ were stimulated with the indicated concentrations of apelin, and DMR was monitored over time as a measure of receptor activity on coated DMR plates (left panels) and uncoated DMR plates (right panels). (C, D) concentration–effect curves derived from the data in A and B. (E, F) ATP traces obtained on CHO-APJ cells using coated (E) and uncoated (F) plates, respectively. Surface coating affects the signaling profiles of apelin acting via the APJ receptor and ATP-stimulating endogenous nucleotide receptors (H. Dargatz and E. Kostenis, unpublished data). (See Color Insert.)

in screening run times. In addition to throughput, label-free's operating costs per well are significantly higher than several other commonly employed GPCR assay technologies, prompting users to be more critical as to what level of screening support can be sustained with respect to budget limitations, for example, weekly, monthly, or spot-checks.

4. THE VALUE PROPOSITION OF LABEL-FREE TO DRUG DISCOVERY

Selection of the "right" molecules to progress through the life of a drug discovery program is a major challenge faced by the pharmaceutical industry. Attention is not solely focused on developing "good" assays, so-called fit for purpose, but also "relevant" in terms of biological/disease output. This prompts questions such as which system(s)/reagent(s) should be used and what output should be measured to best capitalize and integrate with screening strategies. In this context, the advent of assay technologies that enable phenotypic pharmacology is very appealing. First, label-free presents a means to *interrogate what molecules do (or do not do)* by exploiting the enhanced biological texture of outputs. It infers complementary information on the dynamic nature of cell signaling beyond the quantification of receptor activation *per se* at a single point in time. The response profile lends itself to classification of compound responses (same or distinct), which may be subsequently followed up for signal deconvolution. It offers "activity fingerprints" as integrated responses are captured irrespectively of pathway or target class (on- and off-target effects), akin to tissue bioassays. Old drugs and failed candidates can be screened in order to build an understanding of assay relevance to suit desired therapeutic profiles and can potentially lead to repositioning within clinical settings. Examples of biological texture include receptor internalization, kinetic of binding (on/off rates), persistent agonism, surmountable versus insurmountable antagonism or else and thought to be key to drug efficacy in preclinical animal models or even clinical settings.[5,6,49–51] Label-free may offer a tangible opportunity to fill gaps in current screening paradigms. Late-stage drug attrition calls for discovery research to evolve toward a more comprehensive assessment of compound action. Conceptually, this endeavor is very challenging as existing screening strategies are mostly implemented on the basis of historical knowledge and are less geared toward exploring the unknown.[52] Also, rarely are conventional screenings resolved kinetically. Hence, screening does not typically lend itself to incorporation of assay systems perceived to be "black boxes" when this is arguably what makes label-free most appealing.

Second, label-free offers *a discovery platform to identify hits independently of any a priori bias of receptor signaling pathways* described by traditional means. Label-free can be described as a "catch-all" approach that may uncover novelties of ligands within the confined chemical space coverage of existing compound libraries.[23,24] A recent publication from Dodgson *et al.* suggests that the hit rate and incidence of technology false-positives in HTS mode are both low for label-free versus conventional detection technologies and proves successful at identifying multiple types of compounds for a given receptor, including some that are putatively novel.[53] Diversity in molecule MoAs can be of high value to drug discovery programs as it enables the testing of biological hypotheses, therein gaining a better understanding of disease paradigms. Moreover, biased signaling can be assayed during lead optimization, thereby offering an all-encompassing screening readout of cellular phenotype.[14–16,23,30] The concept of functional selectivity is a very attractive one for drug discovery, as it indicates that compounds may be designed to retain desirable functionalities at the expense of putative side-effect or undesirable features.[11–13] It is therefore imperative to consider the need for a fuller interrogation of signaling during early phase of drug discovery. It can be argued that if a screen is thought to be predictive of *in vivo* efficacy, then subcellular signaling resolution may be achieved subsequently. While this has merit, it is also true that distinction of profiles at early phases can allow grouping of preferred characteristics and aid elucidation of cause (desired or undesired activities) at later stage.

Third, label-free enables *the integration of native biological systems at a higher level of throughput* than previously possible. It therefore supports translational pharmacology across discovery research and preclinical development. It pushes screening a step closer to assaying intact systems, a strategy that has in the past been fruitful to the discovery of drugs at GPCRs. The idea of probing for cell phenotypes as a screening approach is far from new. Indeed, this has formed the bases of most pregenetically engineered target initiatives of drug discovery. The prospect of profiling compounds using healthy or therapeutically relevant cell phenotypes has always been very attractive. Examples include the release of cytokines or chemotaxis from immune cells, the propagation of action potentials from cardiomyocytes or membrane permeability from endothelial cells. However, this is rarely feasible at scale, especially when considering human-derived tissues (healthy or disease-relevant donors) due to the limited nature of cell supply, assay costs, or lack of suitable detection methodology. Studying cells unaltered by label-free offers clear value in this arena. Questions relating to cell physiology can be probed by a holistic assessment of drug activity, as label-free captures complicated biochemical networks and processes influenced or controlled by receptor modulation. Where this may be important to drug discovery is that such network interactions can be partially or fully redundant within the cells and play a role in the overall efficacy profile of a drug; conversely, the engagement of multiple targets may be required for desired drug activity.[54–56] Consequently, a system-based approach to efficacy and safety may be worthwhile as it embraces the dynamic interplay of macromolecular targets. Such paradigm contrasts significantly from single-analyte screens typically used in the recent past. It can be postulated that systems "as a whole" may help delivering the next generation of therapeutics. In support of this, recent evidence highlights the benefit of agents acting via multiple points of intervention within biological mechanism(s) underlying the onset and development of diseases. This is exemplified by an analysis of clinical antipsychotics for schizophrenia and mood disorders, which links drug effectiveness to receptor signatures or poly-pharmacology.[57]

5. CONCLUDING REMARKS

As any new technology unfolds such as label-free detection, it raises fundamental questions on the potential benefits it will provide. While the applicability of label-free is broad and holds tantalizing prospects to transform drug discovery, its uptake in the "real world" appears slow. This may be a reflection, at least in part, of the current economical situation limiting research budgets globally, but it may also be cultural as label-free is a black-box approach. It is therefore not a replacement technology, a gain in process efficiency, or a reduction in operational costs for GPCR screening. The introduction of label-free gives new opportunities to screen whole cellular systems as it integrates multiple independent activities to give a holistic assessment of drug activity. In many ways, it is reminiscent of tissue/organ-based experimentation. As it is of relatively large throughput, label-free may improve focus of precious resource on molecules ultimately successful toward the clinic. This is, however, aspirational and only time will tell if label-free can deliver. An area of immediate impact for label-free relates to functional selectivity using human native cells. The cost incurred by functional selectivity of chemotypes progressed through the discovery phase is currently unknown, and addressing this issue requires the ability to interrogate it and modify screening cascades accordingly.[8,10] In the years to come, the exploration of the spatiotemporal dynamics of cell signaling with other biophysical techniques may be powerful to complement label-free and kinetically resolve discrete events underlying receptor activation.[58] In addition, advances in label-free will likely include the deciphering of kinetic traces to a subset of distinct signaling event(s) in order to maximize the exploitability of label-free's high mechanistic content. In this regard, a recent study presents a compelling approach to tackle response complexity.[23] Safety profiling and assessment of cardiomyocyte safety/toxicity and function with human cells is yet another emerging area of great promise that may be transformational. As drug discovery takes on label-free methods, it raises the question of whether it is advancing or more simply moving "back to the future" in the days of whole systems research.

Declaration of interests. The authors report no conflict of interest.

REFERENCES

1. Overington JP, et al. How many drug targets are there? *Nat Rev Drug Discov* 2006;**5**:993–6.
2. Rask-Andersen M, et al. Trends in the exploitation of novel drug targets. *Nat Rev Drug Discov* 2011;**10**(8):579–90.
3. Munos B. Lessons from 60 years of pharmaceutical innovation. *Nat Rev Drug Discov* 2009;**8**:959–68.
4. Kola I, Landis J. Can the pharmaceutical industry reduce attrition rates? *Nat Rev Drug Discov* 2004;**3**:711–5.
5. Kenakin T. Cellular assays as portals to seven-transmembrane receptor-based drug discovery. *Nat Rev Drug Discov* 2009;**8**:617–26.
6. Williams C, Hill SJ. GPCR signaling: understanding the pathway to successful drug discovery. *Methods Mol Biol* 2009;**552**:39–50.
7. Heilker R, et al. G-protein-coupled receptor-focused drug discovery using a target class platform approach. *Drug Discov Today* 2009;**14**:231–40.
8. Mailman RB, Murthy V. Ligand functional selectivity advances our understanding of drug mechanisms and drug discovery. *Neuropharmacology* 2010;**35**:345–6.
9. Clarke WP, Berg KA. Use of functional assays to detect and quantify functional selectivity. *Drug Discov Today (Technol)* 2010;**7**:e31–e36.
10. Kenakin T. Interrogating 7TM receptors: does texture in the question yield greater texture in the answer? *J Recept Signal Transduct Res* 2009;**29**:132–9.
11. Kenakin T, Miller LJ. Seven transmembrane receptors as shapeshifting proteins: the impact of allosteric modulation and functional selectivity on new drug discovery. *Pharmacol Rev* 2010;**62**:265–304.
12. Kenakin T. Collateral efficacy in drug discovery: taking advantage of the good (allosteric) nature of 7TM receptors. *Trends Pharmacol Sci* 2007;**28**:390–6.
13. Galadrin S, et al. The evasive nature of drug efficacy: implications for drug discovery. *Trends Pharmacol Sci* 2007;**28**:423–30.
14. Kenakin T. A holistic view of GPCR signaling. *Nat Biotechnol* 2010;**28**:928–9.
15. Rocheville M, Jerman JC. 7TM pharmacology measured by label-free: a holistic approach to cell signaling. *Curr Opin Pharmacol* 2009;**9**:643–9.
16. Scott CW, Peters MF. Label-free whole cell assays: expanding the scope of GPCR screening. *Drug Discov Today* 2010;**15**:704–16.
17. Minor LK. Cell-based assays for GPCR drug discovery: focus on label-free. *Int Drug Discov* 2010;**5**:32–6.
18. Martin J, et al. Addressing attrition in early drug discovery by label-free methodologies: receptor pharmacology in native cells. *Int Drug Discov* 2010;**5**:38–45.
19. McGuiness R. Label-free cell-based assays bring greater biorelevance to early drug discovery. *Assay Drug Dev Technol* 2009;**7**:191–3.
20. Fang Y, et al. Label-free cell-based assays for GPCR screening. *Comb Chem High Throughput Screen* 2008;**11**:357–69.
21. Xi B, et al. The application of cell-based label-free technology in drug discovery. *Biotechnol J* 2008;**3**:484–95.
22. Shiau AK, et al. Back to basics: label-free technologies for small molecule screening. *Comb Chem High Throughput Screen* 2008;**11**:231–7.
23. Schröder R, et al. Deconvolution of complex G protein-coupled receptor signaling in living cells using dynamic mass redistribution measurements. *Nat Biotechnol* 2010;**28**:943–9.
24. Schröder R, et al. Applying label-free dynamic mass redistribution technology to frame signaling of G protein-coupled receptors noninvasively in living cells. *Nat Protoc* 2011;**6**(11):1748–60.

25. Schmidt J, et al. Selective orthosteric free fatty acid receptor 2 (FFA2) agonists: identification of the structural and chemical requirements for selective activation of FFA2 versus FFA3. *J Biol Chem* 2011;**286**(12):10628–40.
26. Lee P, et al. Evaluation of dynamic mass redistribution technology for pharmacological studies of recombinant and endogenously expressed G protein-coupled receptors. *Assay Drug Dev Technol* 2008;**6**:83–94.
27. Peters MF, Scott CW. Evaluating cellular impedance assays for detection of GPCR pleiotropic signaling and functional selectivity. *J Biomol Screen* 2009;**14**:246–55.
28. Verrier F, et al. GPCRs regulate the assembly of a multienzyme complex for purine biosynthesis. *Nat Chem Biol* 2011;**7**(12):909–15.
29. Blättermann S, et al. A biased ligand for OXE-R uncouples Gα and Gβγ signaling within a heterotrimer. *Nat Chem Biol* 2012;**8**:631–8.
30. Fang Y, et al. Resonant waveguide grating biosensor for whole-cell GPCR assays. *Methods Mol Biol* 2009;**552**:239–52.
31. Atienza M, et al. Dynamic and label-free cell-based assays using the real-time cell electronic sensing system. *Assay Drug Dev Technol* 2006;**4**:597–607.
32. Yu N, et al. Real-time monitoring of morphological changes in living cells by electronic cell sensor arrays: an approach to study G protein-coupled receptors. *Anal Chem* 2006;**78**:35–43.
33. Ferrie AM, et al. High resolution resonant waveguide grating imager for cell cluster analysis under physiological condition. *Appl Phys Lett* 2012;**100**:223701-1–223701-4.
34. Ona T, Shibata J. Advanced dynamic monitoring of cellular status using label-free and non-invasive cell-based sensing technology for the prediction of anticancer drug efficacy. *Anal Bioanal Chem* 2010;**398**(6):2505–33.
35. McGuinness R, et al. Enhanced selectivity screening of GPCR ligands using a label-free cell based assay technology. *Comb Chem High Throughput Screen* 2009;**12**:812–23.
36. Pouton CW, Haynes JM. Embryonic stem cells as a source of models for drug discovery. *Nat Rev Drug Discov* 2007;**6**:605–16.
37. Dick E, et al. Evaluating the utility of cardiomyocytes from human pluripotent stem cells for drug screening. *Biochem Soc Trans* 2010;**38**:1037–42.
38. McKim J. Building a tiered approach to in vitro predictive toxicity screening: a focus on assays with in vivo relevance. *Comb Chem High Throughput Screen* 2010;**13**:188–206.
39. Henstridge CM, et al. GPR55 ligands promote receptor coupling to multiple signaling pathways. *Br J Pharmacol* 2010;**160**:604–14.
40. Schröder R, et al. The C-terminal tail of CRTH2 is a key molecular determinant that constrains Galphai and downstream signaling cascade activation. *J Biol Chem* 2009;**284**:1324–36.
41. Peters MF, et al. Evaluation of cellular dielectric spectroscopy, a whole-cell, label-free technology for drug discovery on Gi-coupled receptors. *J Biomol Screen* 2007;**12**:312–9.
42. Sauliere A, et al. Deciphering biased-agonism complexity reveals a new active AT(1) receptor entity. *Nat Chem Biol* 2012;**8**:622–30.
43. Ayoub MA, et al. Differential association modes of the thrombin receptor PAR1 with Galphail, Galpha12, and beta-arrestin 1. *FASEB J* 2010;**24**(9):3522–35.
44. Mohr K, Kostenis E. Signaling pathways: purinosomes under GPCR control. *Nat Chem Biol* 2011;**7**(12):860–1.
45. Christiansen ME, et al. Structure–activity study of dihydrocinnamic acids and discovery of the potent FFA1 (GPR40) agonist TUG-469. *ACS Med Chem Lett* 2010;**1**(7):345–9.
46. Fang Y, Ferrie AM. Label-free optical biosensor for ligand-directed functional selectivity acting on beta(2) adrenoceptor in living cells. *FEBS Lett* 2008;**582**:558–64.
47. Morse M, et al. Ligand-directed functional selectivity at the mu opioid receptor revealed by label-free integrative pharmacology on-target. *PLoS One* 2011;**6**(10):e25643.

48. Ferrie AM, et al. Label-free integrative pharmacology on-target of drugs at the β(2)-adrenergic receptor. *Sci Rep* 2011;**1**:33.
49. Swinney DC. The role of binding kinetics in therapeutically useful drug action. *Curr Opin Drug Discov Dev* 2009;**12**:31–9.
50. Calebiro D, et al. Signaling by internalised G-protein coupled receptors. *Trends Pharmacol Sci* 2010;**31**:221–8.
51. Müllerhausen F, et al. Persistent signaling induced by FTY720-phosphate is mediated by internalised S1P1 receptors. *Nat Chem Biol* 2009;**5**:428–34.
52. Merino A, et al. Drug profiling: knowing where it hits. *Drug Discov Today* 2010;**15**:749–55.
53. Dodgson K, et al. A 100K well screen for a muscarinic receptor using the Epic label-free system—a reflection on the benefits of the label-free approach to screening seven-transmembrane receptors. *J Recept Signal Transduct Res* 2009;**29**:163–72.
54. Pujol A, et al. Unveiling the role of network and systems biology in drug discovery. *Trends Pharmacol Sci* 2010;**31**(3):115–23.
55. Heitzler D, et al. Towards a systems biology approach of G protein-coupled receptor signaling: challenges and expectations. *C R Biol* 2009;**332**:947–57.
56. Butcher EC, et al. Systems biology in drug discovery. *Nat Biotechnol* 2004;**22**:1253–9.
57. Wong EH, et al. The effectiveness of multi-target agents in schizophrenia and mood disorders: relevance of receptor signature to clinical action. *Pharmacol Ther* 2010;**126**:173–85.
58. Vilardaga JP, et al. GPCR and G proteins: drug efficacy and activation in live cells. *Mol Endocrinol* 2009;**23**:590–9.

CHAPTER FOUR

G Protein-Coupled Receptors in Cancer: Biochemical Interactions and Drug Design

Yves Audigier, François-Xavier Picault, Carline Chaves-Almagro, Bernard Masri

Cancer Research Center of Toulouse, INSERM U1037—Université Paul Sabatier Toulouse III, Toulouse, France

Contents

Abstract

G Protein-Coupled Receptors (GPCRs) share the same topology made of seven-transmembrane segments and represent the largest family of membrane receptors. Initially associated with signal transduction in differentiated cells, GPCRs and heterotrimeric G proteins were shown to behave as proto-oncogenes whose overexpression or activating mutations confer transforming properties. The first part of this review focuses on the link between biochemical interactions of a GPCR with other receptors, such as dimerization or multiprotein complexes, and their oncogenic properties. Alteration of these interactions or deregulation of transduction cascades can promote uncontrolled cell proliferation or cell transformation that leads to tumorigenicity and

Progress in Molecular Biology and Translational Science, Volume 115
ISSN 1877-1173
http://dx.doi.org/10.1016/B978-0-12-394587-7.00004-X

malignancy. The second part concerns the design of drugs specifically targeting these complex interactions and their promise in cancer therapy.

1. GENERAL INTRODUCTION

Before 1986, cell proliferation and oncogenic activity were thought to be essentially linked to growth factors and their tyrosine kinase receptors (TKRs). Three pioneer studies have challenged this specificity of TKRs and revealed that G protein-coupled receptors (GPCRs) were also able to behave as proto-oncogenes.[1–3]

1.1. GPCRs as proto-oncogenes

First, the heptahelical protein encoded by *mas* gene was shown to induce the formation of foci in transfected NIH 3T3 fibroblasts and to promote tumorigenesis in nude mice.[1] Second, nonneural expression of serotonin 5HT1c receptors in fibroblasts and their continuous activation result at high frequency in the formation of transformed foci. Moreover, injection of cells derived from these foci induces the formation of tumors in nude mice.[2] Similar results were obtained with muscarinic acetylcholine receptor subtypes, which become oncogenic when stimulated by an agonist.[3] Accordingly, several G_q-coupled receptors can induce a transformed-like phenotype, which requires the continued presence of agonist.[2,3]

These last findings suggested that activation of a GPCR receptor by its endogenous agonist was a crucial step in transforming properties, opening the way to the oncogenic potential of receptor overexpression. Indeed, overexpression of GPCRs caused by genomic amplification or transcriptional upregulation is often observed in diverse types of cancer tissues.[4] In parallel, the study of GPCR functional domains favored the characterization of mutations in the coding sequence that induced a ligand-independent receptor activation (for review, see Ref. 5) and led to the general concept of constitutive activating mutations.[6] Moreover, these mutations also enhance mitogenicity and tumorigenicity.[7] As could be predicted from the transformed phenotype induced by these engineered mutations, similar activating mutations can also occur naturally and their presence is often associated with a number of human diseases,[8–11] including cancer.[12]

The first characterized GPCR proto-oncogene was the receptor of pituitary thyroid-stimulating hormone (TSH). In 30% of hyperfunctioning thyroid adenomas, mutations in the TSH receptor induced constitutive G_s stimulation and increased cAMP synthesis.[12] The somatic mutations were located in the third cytoplasmic loop and restricted to two different residues, aspartic acid at position 619 and alanine at position 623. It must be pointed out that this intracellular loop is often involved in catalyzing GDP dissociation from the α-subunit of G protein. Interestingly, several DNA viruses such as Kaposi's sarcoma-associated herpes virus (KSHV) have hijacked eukaryotic GPCR sequences (for review, see Ref. 13) and selected the emergence of constitutively activating mutations (for review, see Ref. 14). The constitutively active GPCR encoded by this viral genome is related to chemokine receptors and displays both transforming and proangiogenic properties (for review, see Ref. 15).

Overexpression or constitutive activation of GPCRs at the surface of tumor cells can participate in the different phases of tumor development (for review, see Ref. 16). In addition, GPCR expressed by endothelial cells are involved in the formation of tumor blood vessels[17] when activated by angiogenic factors released either by tumor cells or by stromal and immune cells.[18] Finally, chemoattractant GPCRs play a central role in cancer metastasis, as release of chemokines from the secondary organs drives the selective metastasis of tumor cells expressing the corresponding chemokine receptors.[19]

1.2. Heterotrimeric G proteins as proto-oncogenes

As it could be anticipated, the heterotrimeric G proteins may also undergo activating mutations that can lead to cell proliferation and transformation (for review, see Ref. 20). In contrast to GPCR, naturally occurring mutations conferring oncogenicity[21] were characterized concomitantly with engineered mutations.[22] On one hand, reduced GTPase activity of G_s α-subunit was obtained by engineering of point mutations analogous to those described in p21ras.[22] On the other hand, somatic mutations of G_s protein were identified in growth hormone (GH)-secreting human pituitary tumors. The two amino acid substitutions were located in the GTPase domain where they freeze the protein in an activated state, thereby leading to constitutive activation of G_s α-subunit and to overproduction of cAMP second messenger.[21] In addition, a high frequency of the same activating mutations was observed in GH-secreting human pituitary tumors[23] and in human thyroid tumors.[24,25] Finally, in the McCune–Albright syndrome, these mutations can be associated with pituitary adenomas and adrenal hyperplasia.[26] Analogous mutations converting α-subunit of G_i protein into oncogene were also reported in human endocrine tumors.[27]

As far as G_q family is concerned, no mutated subunits have been identified in any tumors. Conflicting results have been obtained with engineered GTPase-deficient G_q α-subunits, leading to either cell death[28] or decreased cell proliferation[29] or cell transformation.[30,31] Nevertheless, as previously mentioned, G_q-coupled receptors are able to stimulate DNA synthesis and transformation of NIH3T3 cells.[2,3]

In the $G_{12/13}$ family, GTPase-deficient α-subunits can transform NIH3T3 cells,[32–34] albeit the transformed phenotype is less severe for G_{13} α-subunit mutants. However, no mutant form of the two α-subunits was found in human tumor cell lines.[35] On the other hand, several receptors coupled to these G proteins, such as lysophosphatidic acid receptor (LPAR)[36,37] or thrombin protease-activated receptor-1 (PAR1)[38] are clearly involved in the migration of cancer cells and tumor angiogenesis. In addition, the ability of $G_{12/13}$ proteins to decrease the adhesive functions of cadherins[39] and to induce vascular endothelial growth factor (VEGF) translation and secretion[40,41] may also participate in the aforementioned properties.

Interestingly, G protein $\beta\gamma$ subunits were also found to have a prominent role in regulating gene expression by shuttling between plasma membrane and nucleus. The $G_{\beta 5}$ subunit colocalizes with regulator of G protein signaling 7 (RGS7) in the nucleus in a palmitoylation-dependent manner.[42] Following GPCR activation, G_β subunit interacts with the glucocorticoid receptor and suppresses its activity in the nucleus.[43] Similarly, the $G_{\gamma 5}$ isoform binds to AEBP transcriptional repressor and inhibits its function.[44] In addition, association of $G_{\beta\gamma}$ subunits and histone deacetylase 5 (HDAC5) participates in the formation of a large protein complex in which inhibition of the HDAC5 catalytic activity occurs in an allosteric manner.[45] Accordingly, this direct link between $G_{\beta\gamma}$ subunits and nuclear transcription factors should also participate in the different cell responses associated with cell proliferation or transformation.

Other GTPase regulators, RGS proteins, increase GTPase activity of heterotrimeric G proteins, thereby accelerating the deactivation of these proteins. Some RGS proteins behave as tumor-suppressor genes[46] and can regulate vessel remodeling during carcinogenesis.[47] Indeed, mutations in RGS genes decrease the risk to develop lung and bladder cancers.

2. BIOCHEMICAL INTERACTIONS

In most cases, GPCR signaling is more than the activation of a single receptor and very frequently involves interaction with other membrane proteins. These interactions lead to the formation of heterodimers or even multiprotein complexes that create a structural and functional microdomain named signalosome.

In Wnt and hedgehog signaling, the main consequence of heterodimerization is the activation of transcription factors whose target genes are the mediators of oncogenicity. For the adhesion family of GPCR, a post-translational cleavage generates a constitutive heterodimer made of an adhesion subunit and a transducing subunit whose interplay may be important in the loss of cell adhesion during invasiveness and metastasis. Finally, a central role is devoted to reciprocal interactions between GPCRs and TKRs, which can lead to opposite effects, activation or inhibition of mitogenic genes and oncogenic proteins. The existence of multiple and complex interactions has favored the emergence of the signalosome concept which, by creating a signaling microdomain, increases the functional efficiency of the signaling machinery.

2.1. Wnt and hedgehog signaling

Deregulation of two signaling pathways, Wnt and hedgehog signaling that are strongly associated with specification of cell fate and cell differentiation, can induce cell dedifferentiation and transformation, which ultimately lead to tumorigenicity and malignancy.

The *Wnt* gene family codes for secreted proteins that bind to a heterodimer composed of frizzled GPCR and a single transmembrane receptor, LRP5 or LRP6.[48] Ligand binding initiates the recruitment of a multicomponent protein complex on the cytoplasmic domain of LRP receptor and induces activation of G proteins downstream of frizzled GPCR. Altogether, it inhibits the β-catenin degradation complex made of adenomatous polyposis coli (APC), Axin, and glycogen synthase kinase 3 (GSK-3β), thereby leading to β-catenin translocation into the nucleus where it interacts with TCF/LEF transcription factors. The transduction downstream of G proteins leads not only to PLC activation and regulation of NFAT transcription factor but also to Rho and Rac activation and increased activity of AP1 transcription factor.

Before characterization of the proteins involved in the transduction of Wnt pathway, *APC* gene was already identified as a tumor-suppressor gene whose deletion or inactivating mutations were associated with colorectal cancers.[49,50] This observation was later extended to various components of the degradation complex and generalized to various cancers, such as lung, breast, gastric, thyroid, and prostate cancers as well as melanomas.[51] Indeed, Wnt ligands are often overexpressed in tumors[52] and their release contribute to tumorigenesis in a paracrine manner[53] and also in an autocrine manner as frizzled GPCRs are overexpressed in various tumors.[54] Very recently, it was shown that the endosomal sequestration of GSK-3 extended the half-life of many other proteins in addition to β-catenin[55] and that Wnt/β-catenin signaling regulates telomerase activity in stem cells and cancer cells.[56]

Hedgehog (Hh) pathway shares many molecular and functional properties with Wnt signaling pathway. Indeed, it uses similar or even identical components whose oligomerization is triggered by activation of smoothened GPCR (SMO). In addition, it is also involved in specification of cell fate and cell differentiation. Moreover, stimulation of Hh pathway can transduce via G_i or G_{12} heterotrimeric proteins and culminate with the activation of Gli transcription factors, which are the mediators of oncogenic effects.[57] As expected, upregulation of Gli expression is often associated with malignancy.[58]

Like Wnt signaling, Hh signaling participates in stem-cell maintenance[59] and aberrant activation of this signaling pathway including SMO mutations contributes to the transformation events leading to basal cell carcinoma and other malignancies.[58,60]

2.2. Adhesion family of GPCRs

Adhesion-GPCRs correspond to a chimeric molecule in which a large extracellular N-terminal cell-adhesion domain is linked to an amino acid sequence similar to that of a seven-transmembrane GPCR.[61,62] In addition, these proteins share a conserved cystein-rich motif that is located in the linker region and contains the site of a proteolytic cleavage occurring in the endoplasmic reticulum. This posttranslational modification generates a heterodimeric receptor composed of an extracellular α-subunit and a seven-transmembrane (TM) β subunit[63,64] and represents an essential maturation step for addressing the receptor to the membrane.[65] Recently, members of this GPCR subfamily have been detected in tumors where they regulate tumor growth and angiogenesis.

CD97 is expressed in many types of tumors[66] and stimulates angiogenesis.[67] Notably, CD97 expression is induced in many epithelial cancers and its expression levels positively correlate with tumor grade and poor prognosis.[68] In prostate cancer cells, activation of this adhesion-GPCR increases Rho-GTP levels through $G\alpha_{12/13}$ proteins and its heterodimerization with LPAR1 enhances LPAR1-induced Rho activation. In addition, coexpression of both receptors is observed in human prostate cancer samples and activation of the heterodimer is strongly correlated with the invasive properties of prostate cancer cells.[69]

BAI1 was the first identified adhesion-GPCR[70] whose expression is downregulated in several cancers and which behaves as an antitumorigenic molecule and an angiogenesis inhibitor.[71] GPR56 has antiangiogenic properties by inhibiting VEGF production[72] but plays a multifaceted role in tumor growth: in melanomas, it acts as a tumor-suppressor gene by impeding tumor growth,[72] whereas in gliomas, it is overexpressed and considered as an oncogenic protein.[73] Another member, GPR110, is overexpressed in lung and prostate cancers.[74] Finally, GPR124 is expressed in endothelial cells of tumor stroma and upregulated during physiological and pathological angiogenesis.[75]

At the present time, little is known about the interplay between the adhesion domain and the signaling domain, as well as about the endogenous ligands. However, it can be assumed that these adhesion-GPCR may represent new pharmacological targets for treating cancer diseases.

2.3. Epidermal growth factor receptor transactivation by GPCRs

Besides the intrinsic oncogenicity of GPCR signaling pathways, cross talk with TKRs such as epidermal growth factor (EGF) and platelet-derived growth factor (PDGF) receptors (for review, see Ref. 76) can transfer the oncogenic activity to the downstream activation of growth factor signaling.

In 1994, a number of G_q- or G_i-coupled receptors were shown to promote mitogen-activated protein kinase (MAPK) activation, which represents a point of convergence for mitogenic signals.[77–79] Although primarily restricted to TKRs, activation of Ras was also shown for chemoattractant GPCRs.[80,81] Indeed, agonist-activated GPCRs promote GTP loading and Ras activation, an event that is required for DNA synthesis and transformation of NIH3T3 cells.[77] Ras activation is mediated mostly by G_i-coupled receptors[82,83] but also by G_i- and G_q-coupled receptors[84,85] and is suppressed by activated G_s-coupled receptors.[86]

One year later, distinct signaling pathways were characterized for G_i- and G_q-mediated MAPK activation. $G_{\beta\gamma}$ subunits are responsible for this effect for G_i-coupled receptors and transduction operates in a Ras-dependent and PKC-independent manner. On the other hand, G_α subunit mediates MAPK activation of G_q-coupled receptors and this activation is Ras-independent and involves PKC.[87] More interestingly, $G_{\beta\gamma}$-mediated MAPK activation by GPCRs displays a common signaling pathway with that of TKRs, that is, tyrosine phosphorylation of Shc.[88] Altogether, these findings establish a link between GPCR stimulation and the previously described transforming properties of this class of membrane receptors. In addition, it suggests a cross communication between the two signaling systems and even physical interactions between the two categories of receptors.

Transduction of the mitogenic signal mediated by some GPCR agonists (endothelin-1, LPA, and thrombin) was later reported to involve tyrosine phosphorylation of EGFR, thereby suggesting an intracellular mechanism for transactivation.[89] These results revealed that TKRs, notably EGF receptors, can behave as downstream mediators in GPCR mitogenic signaling and that their transactivation may proceed in a ligand-independent manner. Indeed, EGFR displays a unique feature in that it does not require ligand binding for dimerization and tyrosine kinase activity. It must be pointed out that EGFRs are frequently overexpressed in various tumors and their overexpression is associated with a poor prognosis.[90]

EGFR transactivation can proceed via three possible mechanisms: matrix metalloproteinase-induced release of EGF,[91] Src-mediated and Ca^{++}-dependent[92] or Src-independent and Ca-independent intracellular transactivation, and even heterodimerization of GPCRs and EGF receptors.[93–96] It is noteworthy that other TKRs, such as PDGFR, can also be transactivated and thus participate in mitogenic signaling.

2.3.1 Ligand-dependent EGFR transactivation

The first described mechanism was a ligand-dependent EGFR transactivation by several GPCR agonists that have a mitogenic activity.[91] Prenzel *et al.* first demonstrated that transactivation depends on the extracellular domain of EGFR and involves the presence of EGF. Then, they observed an increased metalloproteinase activity, which induced release of heparan-bound EGF (HB-EGF) from its membrane-anchored precursor proHB-EGF. Interestingly, batimastat, a metalloproteinase inhibitor, strongly decreased constitutive EGFR phosphorylation and proliferation of PC3 prostate carcinoma cells. In kidney or bladder cell lines, EGFR transactivation can result from the activation of distinct ADAMs (a disintegrin and metalloproteinase) and involves several EGF-like ligands, such as amphiregulin, HB-EGF, and transforming growth factor alpha.[97]

Another similar ligand-dependent EGFR transactivation has been described for ERK activation induced by estradiol and mediated by the receptor GPR30.[98] Indeed, in addition to their genomic effects via the nuclear receptors ERα and β, estrogens also activate a G protein-coupled estrogen receptor (GPR30), which is overexpressed in breast, endometrial and ovarian carcinomas and plays an important role in various cancer cells.[99–102] Accordingly, if they express GPR30 receptors, estrogen receptor

(ER)-negative breast tumors may remain estrogen responsive and a combined therapy of GPR30 and EGFR antagonists should be beneficial for treating this type of tumors.

2.3.2 Ligand-independent EGFR transactivation

The second mechanism is ligand independent and corresponds to an intracellular transactivation that results from a rapid tyrosine phosphorylation of EGFR.[89] Besides transmembrane EGFR, several nonreceptor tyrosine kinases, such as Src[103] and Ca^{++}-regulated focal adhesion kinase Pyk2[104] may also play a role in the link between GPCR stimulation and MAPK activation. Indeed, Pyk2 acts with Src to link G_i- and G_q-coupled receptors with Grb2 and Sos to activate the MAP kinase signaling pathway in PC12 cells.[105] Interestingly, these two types of tyrosine kinase may belong to the same signaling pathway from GPCR to EGFR, as revealed by a role for Pyk2 and Src in tyrosine phosphorylation of EGFR.[106]

However, distinct intracellular effectors may participate in the transduction cascades. For example, m1 muscarinic acetylcholine receptor modulates ion channel activity by a PKC-dependent EGFR transactivation.[107] Alternately, in colonic epithelial cells, carbachol stimulates EGFR transactivation and MAPK in a PKC-independent manner.[108] On the other hand, Ang II-induced transactivation of the EGF receptor is Ca^{++} dependent and serves as a scaffold for preactivated c-Src and for downstream adaptors, leading to MAPK activation.[92] Moreover, EGFR transactivation resulting from activation of angiotensin receptor can also be Ca^{++} independent and not mediated by Pyk2 or Src.[109] In addition to its ability to transactivate EGFR, Ang II can also transactivate PDGFR, essentially PDGFβR,[110,111] which is involved in Ang II-induced ERK activation.[112] The increased PDGFR tyrosine phosphorylation associated with transactivation is directly correlated with intimal hyperplasia.[113] Similarly, the autocrine mitogenic action exerted by endothelins via their ETA receptors in ovarian and prostate carcinoma cells is mediated by a transduction cascade that involves EGFR transactivation.[114]

The signaling mechanisms involved in transactivation are specifically influenced by the activated GPCR and the tumor type in question. The progression of colon, lung, breast, head and neck, and prostate and ovarian cancers have all been reported to be mediated by GPCR–EGFR cross talk.[90]

2.3.3 Oligomerization of GPCR and EGFR

In addition to the mechanisms described before, several activated GPCRs were shown to form protein complexes with transactivated EGFRs.

The first report concerns β2 adrenergic receptors whose stimulation induces EGFR transactivation and formation of a heterodimeric complex containing both activated GPCR and transactivated EGFR.[94] Similarly, β1 adrenergic receptors can form a complex with EGFR in the presence of an adrenergic agonist and the duration of these receptor interactions is dependent on β-arrestin recruitment.[95]

More recently, another β-arrestin-dependent association of a GPCR and transactivated EGFR has been revealed for kisspeptin receptor GPR54. However, this association is already observed in basal conditions and the role of the GPCR agonist is to increase the stability of this receptor complex. This sustained coactivation of both receptors plays a crucial role in the invasion of breast cancer cells.[96]

As mentioned before, estrogen-induced activation of GPR30 promotes EGFR transactivation through an increase of metalloproteinase activity and release of HB-EGF,[98] thereby participating in the proliferation and migration of breast cancer cells.[101] Interestingly, a physical interaction between GPR30 and phosphorylated EGFR has also been observed in breast carcinoma-associated fibroblasts.[115]

An interesting example illustrates the coexistence of the three different mechanisms involved in EGFR transactivation, that is, the upregulated expression of proliferation-related genes following vasopressin V1a receptor activation. Indeed, EGFR transactivation is mediated by HB-EGF release, by ligand-independent activation of Src kinase, and even by vasopressin-induced formation of a complex between V1a receptor and EGFR.[93]

2.4. Inhibition of EGFR transactivation by GPCRs

Although most GPCRs activate EGF receptors, the subfamily of somatostatin (SST) receptors has the opposite effect. Their presence leads to dissociation of homo- and heterodimerization of EGFR subtypes and consequently to inhibition of the mitogenic and oncogenic signaling induced by EGF (for review, see Ref. 116).

A pioneer study first revealed that SSTR subtypes 1 and 5 heterodimerize with EGFR.[117] Such interactions reduce EGFR signaling and counteract EGF-stimulated proliferation, providing an interesting example of cross talk between antiproliferative signaling mediated by these GPCRs and the mitogenic signaling transduced by EGFRs. By itself, expression of the two SSTR subtypes promotes the dissociation of constitutive or EGF-induced heterodimeric complexes of EGFR subtypes, such as EGFR/ErbB2 complexes, thereby controlling EGF-mediated cell survival pathway.[118]

Moreover, the switch to heterodimerization between the SSTR subtype 5 and EGFR is driven in an agonist-dependent manner.[119]

Accordingly, the induced overexpression or activation of SST receptors, which leads to inhibition of EGFR signaling, can be considered as a valuable therapeutic approach in the treatment of ErbB-positive tumors, notably breast cancers. This assumption is strengthened by the observation that a gradual loss of SSTR subtypes correlates with the decrease of pharmacological activity of trastuzumab (a monoclonal antibody antagonizing activation of ErbB2 subtype).[120]

2.5. Transactivation of GPCRs by TKRs

In a reciprocal manner, recent findings have demonstrated that TKR activation can lead to GPCR transactivation, placing GPCR signaling downstream of TKRs (for review, see Ref. 121).

Indeed, the mechanisms of GPCR transactivation somehow reproduce those employed for TKR transactivation. Two modes of transactivation may operate in this GPCR highjack: a ligand-dependent transactivation via transcriptional upregulation of ligand gene expression[122] or enzymatic increase of GPCR ligand synthesis[123] and a ligand-independent transactivation.[124]

The first link between TKRs and GPCR transactivation was suggested by the $G_{\beta\gamma}$ subunit involvement in MAPK activation induced by Type 1 insulin-like growth factor receptor (IGF-1R).[125] Later, the same receptor was found to induce transactivation of the chemokine CCR5 receptor by transcriptional upregulation and secretion of its natural ligand RANTES (regulated and normal T-cell expressed and secreted).[122] As this transactivation is essential for IGF-I-induced cell migration, blockade of CCR5 transduction prevents invasion by tumor cells, whereas RANTES overexpression confers invasive capacity to noninvasive tumor cells. Accordingly, the consequences of this cross talk reveal a connection between tumorigenesis and inflammation.

At the same time was reported an interplay between PDGFR and the EDG-1 subtype of sphingosine-1-phosphate (SP-1-P) receptors, which participates in PDGF-induced chemotaxis.[123] PDGF stimulates the activity of sphingosine kinase leading to an increase in intracellular SP-1-P synthesis. Then SP-1-P secreted in the medium activates its cognate EDG-1 receptor in an autocrine manner. As before, blockade of sphingosine kinase or deletion of EDG-1 receptor suppresses chemotaxis toward PDGF. In addition, both receptors can physically interact and this complex represents a functional platform where both signaling pathways converge and produce stronger mitogenic effects.[126]

Using another mechanism, IGF-I can transactivate the CXCR4 chemokine receptor in a ligand-independent manner and by formation of a multiprotein complex containing IGF-IR, CXCR4, and G_i protein.[124] Again, this TKR-mediated GPCR transactivation plays a crucial role in IGF-I-induced cell migration and metastasis of breast cancer cells.

It must be pointed out that TKR transactivation by GPCR is frequently involved in cell proliferation, whereas GPCR transactivation by TKR usually participates in cell migration.

2.6. GPCRs and signalosome

An important emerging concept is the formation of protein signaling complexes called signalosomes that add an additional layer of complexity to GPCR heterodimerization (for review, see Ref. 127). The biological function of such complexes is to create a signaling microdomain where the close vicinity of the transduction components enables the best efficacy of coupling and signaling. The signalosome can correspond to the formation of a submembrane scaffold of proteins linked to the intracellular domain of GPCRs, such as that of LPA in nonhematopoïetic cells,[128] or to the anchoring of the GPCR C-terminal domain via an adaptor in the cytoskeleton network, such as for the α_{1D}-adrenergic receptor.[129]

It was also reported that the formation of preassembled and constitutively active signalosomes precedes the agonist-induced GPCR stimulation.[130] Indeed, the relaxin family peptide receptor 1 (RXFP1) can form a "membrane" signalosome with A-kinase-anchoring proteins (AKAPs) and adenylylcyclases. Interestingly, increase of cAMP levels induces activation of the phosphodiesterase present in the signaling complex, thereby turning off the signal. The relaxin-like peptide family is expressed in malignancies and their cognate receptors are detected in carcinoma cells and neighboring stromal cells. Activation of their signaling cascades leads to tumor cell proliferation and increased activity of MMPs, both participating in the action of relaxin peptides on tumor cell invasion and metastasis.[131]

3. DRUG DESIGN

GPCRs correspond to the largest family of cell surface receptors with an estimated number of 1000 members and represent the target of approximately 30–40% of marketed drugs. Drug design usually targets extracellular domains of GPCRs, such as ligand-binding sites. However, new strategies are developed that target transmembrane or intracellular domains in order to interfere with GPCR oligomerization or proximal transduction.

The first strategy is to antagonize simultaneously and nonspecifically multiple GPCRs by broad spectrum drugs. Another strategy designed for heterodimeric complexes of GPCRs relies on the development of chimeric molecules made of GPCR agonists or antagonists of each receptor linked together in the same molecule. An additional targeting approach is to simply abrogate GPCR function by preventing their oligomerization. Finally, a new and elegant strategy is related to the use of pepducins that target GPCR intracellular domains and act as agonist, antagonist, or biased agonist in an allosteric manner.

3.1. Broad spectrum drugs: Identification of substance P derivatives

Many tumors including lung tumors not only overproduce peptide hormones but also overexpress their cognate receptors leading to an autocrine stimulation of tumor cell proliferation.[132–134] Inhibition of several of these receptors using drugs selective for a single molecular target has led to cancer regression in preclinical models and in cancer patients.[135,136] Nevertheless, the development of broad spectrum antagonists with pleiotypic actions on several GPCRs becomes interesting to concomitantly antagonize different transduction cascades associated with cell proliferation. Such "selectively nonselective" molecules have been identified for neuropeptide receptors that play an important role in the growth of lung cancers.

Lung cancer is the leading cause of cancer death worldwide and is divided in two main groups: small cell lung cancer (SCLC) and nonsmall cell lung cancer (NSCLC). SCLC represents an aggressive neuroendocrine tumor because tumor cells release a high amount of neuropeptides that induce a stronger activation of their respective GPCRs expressed at the surface of tumor cells.[133]

The first peptide inhibitor reported to be an effective antagonist against several unrelated GPCRs and to induce inhibition of cell proliferation was the substance P antagonist [$DArg^1$, $DPro^2$, $DTrp^{7,9}$, Leu^{11}]substance P. By serendipity, it was found to block the mitogenic effects of both bombesin and vasopressin in Swiss 3T3 cells.[137,138] A modified molecule, [$DArg^1$, $DPhe^5$, $DTrp^{7,9}$, Leu^{11}]substance P derivative (SPD), not only inhibits the action of substance P[139] but also acts at a higher potency and on a larger number of neuropeptide receptors such as bombesin, gastrin-releasing peptide, vasopressin,[140] bradykinin,[141] and endothelin receptors.[142] SPDs are structurally unrelated to bombesin, vasopressin, or bradykinin but may recognize a domain shared by their receptors. Consequently, SPDs have been termed "broad spectrum neuropeptide antagonists." Indeed, SPDs induce apoptosis in SCLC cell lines by blocking activation of $G_{\alpha q}$, protein kinase C activity and calcium mobilization from intracellular stores.[143–147] In addition, they inhibit *in vivo* the tumor growth following injection of SCLC cell lines in nude mice.[145,148]

Nevertheless, SPDs are not merely receptor antagonists since they can induce activation of c-Jun N-terminal kinase (JNK) and the reorganization of the actin cytoskeleton.[143] Altogether, it reveals that SPDs have both agonistic and antagonistic activities.[143] Indeed, SPDs induce stimulation of $G_{\alpha 12,13}$ proteins and downstream events including Jun kinases, caspases, and ultimately trigger apoptosis through a mechanism that depends on the generation of reactive oxygen species.[143,149,150] Accordingly, substance P analogs behave as biased agonists at neuropeptide receptors by inhibiting $G_{\alpha q}$ pathway and calcium mobilization on one hand and by stimulating the $G_{\alpha 12/13}$ pathway on the other hand.

Despite their antimitogenic and proapoptotic activities on tumor cells, substance P analogs inhibit tumor growth via a secondary mechanism related to their antiangiogenic activity. Tumor angiogenesis is a critical step in cancer progression, invasiveness, and metastasis. Growth factors and chemokines released by cancer cells promote tumor angiogenesis by acting on their receptors expressed by endothelial cells. One target of SPDs is the chemokine CXCR2 receptor, which is expressed by endothelial cells. SPD specifically blocked CXCL8-induced angiogenesis in a rat corneal micropocket assay *in vivo* and markedly reduced tumor-associated angiogenesis in a human ductal pancreatic adenocarcinoma cell line (HPAF-II) xenografts.[151]

Thus, a salient feature of SPDs is their activity as biased agonist on different GPCRs such as not only neuropeptide receptors but also chemokine receptors.[143,151] Consequently, their dual action on tumor growth operates through induction of tumor cell apoptosis and blockade of tumor neoangiogenesis.

Nonetheless, high concentrations of SPDs are required to inhibit the growth of SCLC cell lines and their serum half-lives are very short.[143,145,148,152] Thus, search for more potent and stable compounds led to the chemical synthesis and pharmacological characterization of new antagonists targeting bradykinin (BK) receptors. In fact, peptidic antagonists of BK receptors are not efficient at blocking growth of most SCLC cell lines.[135] But surprisingly, when two molecules of these BK antagonists are covalently linked, such as in CU201, they become proliferation inhibitors of both SCLC and NSCLC cell lines *in vitro*. Like SPDs, these "antagonist dimers" trigger apoptosis of lung cancer cells by acting as biased agonists, inhibiting $G_{\alpha q}$ pathway and bradykinin-induced intracellular calcium flux on one hand and stimulating $G_{\alpha 12,13}$ and downstream events like JNK activity on the other hand. Moreover, CU201 was more potent at inducing tumor cell apoptosis compared to SPDs, had a longer serum half-life, and did not alter proliferation of normal lung cells.[152] Such activities of CU201 and related compounds were confirmed not only *in vitro* in breast, prostate, colon, and cervix cancer cell lines but also *in vivo* using tumor cell implantation in nude mice.[153,154] Interestingly, bradykinin antagonist dimers produced synergistic growth inhibition *in vivo* when combined with cytotoxic agents used in therapy such as doxorubicin, etoposide, cisplatin, or paclitaxel.[153] As it could be expected, combination of bradykinin antagonist dimer with SPD conferred additive effects, but intriguingly, a covalently linked heterodimer of SPD and bradykinin antagonist was not more potent at inducing apoptosis of lung cancer cells when compared to SPD monomer alone.[135]

3.2. Chimeric molecules

Despite the unsuccessful use of the antagonist dimer formed by SPD and bradykinin antagonist to block neuroendocrine tumor growth, other dimers composed of GPCR agonists gave remarkable results not only on primary cultures of human pituitary adenoma cells by suppressing GH and prolactin secretion but also in patients with GH/prolactin-secreting adenomas (for review, see Ref. 155).

Like a great number of tumors, pituitary adenomas express both somatostatin receptors (Sst 1-5) and dopamine D2 receptors (D2R) in a stoechiometry that varies from one tumor to the other. These receptors are well expressed and colocalize in several human pituitary adenomas, thereby suggesting possible functional interactions in tumor cell proliferation and control of hormonal hypersecretion. Indeed, in patients with neuroendocrine tumors, monotherapy using somatostatin analogs such as lanreotide or octreotide is not always efficacious and the effect of these compounds decreases with time despite their long half-life.[156] This activity decrease may be associated with receptor desensitization, internalization, and/or degradation. Thus, new somatostatin analogs with a broader spectrum of receptor binding, such as Pasireotide (also called SOM230)[157] or BIM-23244,[158] which bind to multiple somatostatin receptor subtypes, might be effective at blocking cell secretion and proliferation in patients refractory to octreotide or lanreotide.[159]

Dopamine agonists like bromocriptine or cabergoline represent the usual therapeutic indication for the treatment of neuroendocrine disorders. In fact, bromocriptine has been known for many years to effectively block prolactin release in the majority of prolactinomas.[160] On the other hand, cabergoline binds to the D2 receptor with a higher affinity, has a longer plasma half-life, and is effective at decreasing hormonal secretion and inhibiting pituitary tumor growth.[161,162] A combination therapy of somatostatin analogs with dopamine agonists reduces both GH and IGF-1 levels more efficiently than each compound given alone, especially in patients with GH/prolactin-secreting adenomas.[163,164]

As many neuroendocrine cells coexpress both somatostatin and dopamine receptors, it could be hypothesized that these GPCRs may physically interact and work synergistically. Rocheville *et al.* demonstrated a functional heterodimerization of dopamine D2 and Sst 5 receptors that conferred a pharmacology distinct from that of receptor homodimers: a much greater binding affinity for both dopamine and somatostatin analogs and an enhanced coupling to adenylylcyclase.[165] D2R can also heterodimerize with Sst 2 receptor when both receptors are coexpressed in CHO-K1 and HEK-293 cells. This heterodimerization is also associated with an increased affinity of dopamine and somatostatin and a lower EC_{50} in their inhibition of cAMP.[166] Accordingly, agonist-induced hetero-oligomerization of these receptors may induce a novel receptor state that results in enhanced signaling properties.

Based on these observations, new chimeric compounds incorporating both dopaminergic and somatostatinergic agonists into a single molecule, called dopastatin, have been developed. Interestingly, these molecules retain high affinity for both dopamine and somatostatin receptors (for review, see Refs. 159,167,168) and display enhanced potency when compared to activation by dopamine or somatostatin analogs separately.[169] The first chimeric molecule containing structural elements of both somatostatin and dopamine, BIM-23A387, possesses a remarkably enhanced potency in suppressing both GH and prolactin release in primary cultured human pituitary somatotroph adenoma cells when compared to somatostatin and dopamine analogs used either alone or in combination.[170,171] Another chimeric molecule, BIM-23A760, produced a greater decrease of GH and prolactin secretion than octreotide in cultured cells derived from human GH-secreting tumors.[172] Furthermore, as Sst 2 and dopamine D2 receptors are coexpressed at high levels in NSCLC cell line Calu-6, dopastatin activity was evaluated on these cells.[169] Both BIM-23A387 and BIM-23A370 were able to significantly inhibit proliferation of Calu-6 cell line more strongly than the somatostatin and dopamine selective compounds alone, demonstrating additivity of the effects produced by dopastatins.

3.3. GPCRs oligomerization

Although several studies report that some GPCRs can function as monomers, biochemical, biophysical, and functional analyses have demonstrated that other GPCRs such as chemokine receptors are most likely expressed at the cell surface as oligomers (173–175 and for review, see Refs. 176,177). Indeed, receptor oligomerization is an important feature of GPCR structure and function that can be essential in receptor signaling.

The chemokine receptor CXCR4, which plays a crucial role in tumor growth, angiogenesis, and cancer metastasis, forms homodimers[178,179] in the absence of its ligand CXCL12/SDF1α.[174] Indeed, blocking the endogenous ligand or CXCR4 itself by using antibodies raised against these proteins significantly impairs metastasis of breast cancer cells to lymph nodes and lung.[180] Another approach has been developed to block CXCL12/CXCR4 signaling pathway *in vitro* and *in vivo* by promoting dissociation of CXCR4 homodimers. A synthetic peptide containing amino acids of the transmembrane domain 4 (TM4) of CXCR4 reduced the homodimeric interaction and completely abolished CXCL12-induced actin polymerization.[179] Interestingly, TM4 synthetic peptide was able to block chemotaxis of cancer cells but also inhibited migration of normal monocytes toward CXCL12. Although the mechanism of action of TM4 synthetic peptide has yet to be determined, these data suggest that receptor oligomers are appropriate and valuable targets to inhibit tumor cell spread.

3.4. Pepducins

An original and promising strategy targeting GPCRs in cancer is based on the use of synthetic lipopeptide pharmacophores called pepducins. All drugs usually act on GPCRs by interacting with the extracellular part of the receptors and mostly with the orthosteric ligand-binding site. In contrast to this type of targeting, lipidated pepducins interact with intracellular domains and therefore modulate signaling in an allosteric manner.

Pepducin technology was first described in 2002 by Covic *et al.* to target intracellular loops of PAR1, PAR2, and melanocortin-4 receptors.[181] These cell-penetrating lipopeptides can function as allosteric agonists, antagonists, biased ligands or modulators of targeted GPCRs by enhancing either the potency and/or efficacy of endogenous agonists (for review, see Refs. 182,183). The peptide part of pepducins is a 10- to 20-amino acid sequence corresponding to one of the three intracellular cytoplasmic loops or the C-terminal tail of the GPCR, and the lipidic part corresponds to a palmitate or myristate fatty acid chain. This lipid moiety facilitates translocation across cell membrane and tethers the pepducin in the lipid bilayer near the targeted receptor. Although the mechanism of action of pepducins on GPCRs is not completely understood, their effect may rely on the stabilization of the receptors in an active or inactive conformation or even interfere with receptor dimerization.[184]

Such a technology is a highly efficient tool to identify GPCRs' modulators since pepducin library targeting a single receptor can be rapidly designed and easily tested for agonist, antagonist, or biased agonist activity by high-throughput screening.[185,186] Furthermore, pepducins are very effective on cultured cells or animal models where they are well tolerated and display very few side effects (187,188 and for review, see Ref. 189).

At the present time, existing pepducins have been designed to target PAR1/2 and CXCR1/2 chemokine receptors that play a crucial role in cancer progression, invasion, metastasis, and angiogenesis.[190] Antagonistic activity of pepducins targeting intracellular loops of these receptors should constitute a new class of pharmacological agents able to inhibit tumor growth and the spreading of tumor cells. In addition, these lipopeptides could be clinically used as monotherapy or in combination with chemotherapy.

3.4.1 PAR receptors

Like other GPCRs, constitutive overexpression of PAR1 can stimulate focus formation in NIH 3T3 fibroblast cells.[191] Therefore, PAR1 behaves as a proto-oncogene and plays an important role in the invasive and metastatic processes of breast, ovarian, and lung cancers.[192–196]

To examine the *in vivo* efficacy of a pepducin directed against the third intracellular loop (i3) of PAR1 (P1pal-7), different tumor xenograft mouse models have been used to characterize P1pal-7 activity as monotherapy or in combination with chemotherapy. In a mouse model of ovarian cancer, Agarwal *et al.* demonstrated that this pepducin antagonist almost abolished angiogenesis and significantly reduced ascites production in peritoneal ovarian cancer.[192] This antiangiogenic activity of P1pal-7 was confirmed in an athymic nude mouse mammary fat pad injection model using MDA-MB-231 breast tumor cells. In this model, P1pal-7 antagonistic action on PAR1 induced a 60% decrease of the tumor volume when compared to vehicle-treated mice.[194] Combined therapy with P1pal-7 and Taxotere (an antimicrotubule agent used in chemotherapy) reduced by 95% the size of breast tumor xenografts and almost abrogated lung metastasis of breast cancer cells.[196] Similarly, P1pal-7 inhibits migration of cultured primary human lung carcinoma cells and reduces tumor growth by 75% in mouse lung cancer xenograft models.[197] Interestingly, the effect on tumor growth of P1pal-7 was as efficient as Bevacizumab, an inhibitor of the angiogenic factor VEGF used in clinics, suggesting that targeting PAR1 might be as effective as VEGF to treat certain cancers.

When comparing the activity of pepducins targeting either i3 or i1 loops of PAR1 in lung adenocarcinoma cell lines, Cisowski *et al.* identified a pepducin that exhibits biased agonism, delineating divergent signaling pathways mediated by i1 versus i3 loops of PAR1.[197] Indeed, the pepducin-targeting i3 loop of PAR1 completely abolished ERK1/2 activation and cell motility *in vitro*. On the other hand, the pepducin-targeting i1 loop inhibited cell motility but was not able to block ERKs activation. Interestingly, divergent pharmacological effects of these pepducins were also observed in a mouse lung cancer xenograft model: pepducin-targeting i3 loop was efficient at blocking lung tumor growth, whereas pepducin-targeting i1 loop was ineffective. Accordingly, this discrepancy observed between pepducins targeting different intracellular loops of a same receptor illustrates the need to understand the downstream PAR1 signaling pathways and their contribution to lung tumor pathogenesis. Furthermore, such an approach could lead to the appropriate choice of the targeted intracellular loop of a GPCR and a better design of new anticancer drugs.

Antagonistic pepducins were also designed for PAR2 receptor whose stimulation induces proliferation, migration, and invasion of tumor cells in colon, hepatocellular, and pancreatic carcinomas.[198–200] Until recently, no antagonist acting at the ligand-binding site was identified for this receptor whereas a PAR2-selective pepducin antagonist, P2-pal21, was developed and shown to completely block the PAR2-dependent invasion of human hepatocellular carcinoma.[199]

3.4.2 Chemokine receptors

Other GPCRs such as chemokine receptors play important roles in carcinogenesis and modulate cell interactions between the tumor and its stromal microenvironment.[201] CXCR1 and CXCR2 and their respective ligands CXCL8 and CXCL1 are important mediators of immune reaction and play pivotal functions in tumor growth, angiogenesis, and metastasis.[202,203] CXCL8 and CXCL1, whose secretion is positively regulated following

PAR1 activation in ovarian carcinoma cells, promote tumor angiogenesis by stimulating their cognate receptors expressed by endothelial cells. An antagonistic pepducin, which targets the conserved third intracellular loop (i3) of both CXCR1 and CXCR2 receptors, X1/2pal-i3, inhibits tumor angiogenesis and thus blocks ovarian tumor growth.[193]

Altogether, pepducin technology represents a new approach to modulate GPCR activity by acting as allosteric modulators. The lipid moiety and the amino acid side chains can be modified in order to increase the potency and the biological half-life of these compounds. However, a better characterization of the receptor intracellular loops involved in stimulation of the different transduction cascades is crucial to identify the appropriate pepducins that will specifically target one transduction cascade. Interestingly, pharmacokinetic studies of pepducins revealed that these molecules are widely distributed throughout the body and that they are well tolerated in animals. Accordingly, pepducins represent a promising class of therapeutic compounds to treat different diseases such as cancer.

REFERENCES

1. Young D, Waitches G, Birchmeier C, Fasano O, Wigler M. Isolation and characterization of a new cellular oncogene encoding a protein with multiple potential transmembrane domains. *Cell* 1986;**45**:711–9.
2. Julius D, Livelli TJ, Jessell TM, Axel R. Ectopic expression of the serotonin 1c receptor and the triggering of malignant transformation. *Science* 1989;**244**:1057–62.
3. Gutkind JS, Novotny EA, Brann MR, Robbins KC. Muscarinic acetylcholine receptor subtypes as agonist-dependent oncogenes. *Proc Natl Acad Sci USA* 1991;**88**:4703–7.
4. Li S, Huang S, Peng SB. Overexpression of G protein-coupled receptors in cancer cells: involvement in tumor progression. *Int J Oncol* 2005;**27**:1329–39.
5. Marinissen MJ, Gutkind JS. G-protein-coupled receptors and signaling networks: emerging paradigms. *Trends Pharmacol Sci* 2001;**22**:368–76.
6. Cotecchia S, Exum S, Caron MG, Lefkowitz RJ. Regions of the alpha 1-adrenergic receptor involved in coupling to phosphatidylinositol hydrolysis and enhanced sensitivity of biological function. *Proc Natl Acad Sci USA* 1990;**87**:2896–900.
7. Allen LF, Lefkowitz RJ, Caron MG, Cotecchia S. G-protein-coupled receptor genes as protooncogenes: constitutively activating mutation of the alpha 1B-adrenergic receptor enhances mitogenesis and tumorigenicity. *Proc Natl Acad Sci USA* 1991;**88**:11354–8.
8. Gromoll J, Simoni M, Nieschlag E. An activating mutation of the follicle-stimulating hormone receptor autonomously sustains spermatogenesis in a hypophysectomized man. *J Clin Endocrinol Metab* 1996;**81**:1367–70.
9. Rao VR, Cohen GB, Oprian DD. Rhodopsin mutation G90D and a molecular mechanism for congenital night blindness. *Nature* 1994;**367**:639–42.
10. Schipani E, Kruse K, Juppner H. A constitutively active mutant PTH-PTHrP receptor in Jansen-type metaphyseal chondrodysplasia. *Science* 1995;**268**:98–100.
11. Shenker A, Laue L, Kosugi S, Merendino Jr JJ, Minegishi T, Cutler Jr GB. A constitutively activating mutation of the luteinizing hormone receptor in familial male precocious puberty. *Nature* 1993;**365**:652–4.

12. Parma J, Duprez L, Van Sande J, Cochaux P, Gervy C, Mockel J, et al. Somatic mutations in the thyrotropin receptor gene cause hyperfunctioning thyroid adenomas. *Nature* 1993;**365**:649–51.
13. Sodhi A, Montaner S, Gutkind JS. Viral hijacking of G-protein-coupled-receptor signalling networks. *Nat Rev Mol Cell Biol* 2004;**5**:998–1012.
14. Vischer HF, Hulshof JW, de Esch IJ, Smit MJ, Leurs R. Virus-encoded G-protein-coupled receptors: constitutively active (dys)regulators of cell function and their potential as drug target. *Ernst Schering foundation symposium proceedings*; 2006. p. 187–209.
15. Lappano R, Maggiolini M. G protein-coupled receptors: novel targets for drug discovery in cancer. *Nat Rev Drug Discov* 2011;**10**:47–60.
16. Lee HJ, Wall B, Chen S. G-protein-coupled receptors and melanoma. *Pigment Cell Melanoma Res* 2008;**21**:415–28.
17. Folkman J. Tumor angiogenesis: therapeutic implications. *N Engl J Med* 1971;**285**:1182–6.
18. Richard DE, Vouret-Craviari V, Pouyssegur J. Angiogenesis and G-protein-coupled receptors: signals that bridge the gap. *Oncogene* 2001;**20**:1556–62.
19. Ben-Baruch A. Organ selectivity in metastasis: regulation by chemokines and their receptors. *Clin Exp Metastasis* 2008;**25**:345–56.
20. Dhanasekaran N, Heasley LE, Johnson GL. G protein-coupled receptor systems involved in cell growth and oncogenesis. *Endocr Rev* 1995;**16**:259–70.
21. Landis CA, Masters SB, Spada A, Pace AM, Bourne HR, Vallar L. GTPase inhibiting mutations activate the alpha chain of Gs and stimulate adenylyl cyclase in human pituitary tumours. *Nature* 1989;**340**:692–6.
22. Graziano MP, Gilman AG. Synthesis in Escherichia coli of GTPase-deficient mutants of Gs alpha. *J Biol Chem* 1989;**264**:15475–82.
23. Clementi E, Malgaretti N, Meldolesi J, Taramelli R. A new constitutively activating mutation of the Gs protein alpha subunit-gsp oncogene is found in human pituitary tumours. *Oncogene* 1990;**5**:1059–61.
24. O'Sullivan C, Barton CM, Staddon SL, Brown CL, Lemoine NR. Activating point mutations of the gsp oncogene in human thyroid adenomas. *Mol Carcinog* 1991;**4**:345–9.
25. Suarez HG, du Villard JA, Caillou B, Schlumberger M, Parmentier C, Monier R. gsp mutations in human thyroid tumours. *Oncogene* 1991;**6**:677–9.
26. Weinstein LS, Shenker A, Gejman PV, Merino MJ, Friedman E, Spiegel AM. Activating mutations of the stimulatory G protein in the McCune-Albright syndrome. *N Engl J Med* 1991;**325**:1688–95.
27. Lyons J, Landis CA, Harsh G, Vallar L, Grunewald K, Feichtinger H, et al. Two G protein oncogenes in human endocrine tumors. *Science* 1990;**249**:655–9.
28. Wu DQ, Lee CH, Rhee SG, Simon MI. Activation of phospholipase C by the alpha subunits of the Gq and G11 proteins in transfected Cos-7 cells. *J Biol Chem* 1992;**267**:1811–7.
29. Qian NX, Russell M, Buhl AM, Johnson GL. Expression of GTPase-deficient G alpha 16 inhibits Swiss 3T3 cell growth. *J Biol Chem* 1994;**269**:17417–23.
30. De Vivo M, Chen J, Codina J, Iyengar R. Enhanced phospholipase C stimulation and transformation in NIH-3T3 cells expressing Q209LGq-alpha-subunits. *J Biol Chem* 1992;**267**:18263–6.
31. Kalinec G, Nazarali AJ, Hermouet S, Xu N, Gutkind JS. Mutated alpha subunit of the Gq protein induces malignant transformation in NIH 3T3 cells. *Mol Cell Biol* 1992;**12**:4687–93.
32. Jiang H, Wu D, Simon MI. The transforming activity of activated G alpha 12. *FEBS Lett* 1993;**330**:319–22.
33. Vara Prasad MV, Shore SK, Dhanasekaran N. Activated mutant of G alpha 13 induces Egr-1, c-fos, and transformation in NIH 3T3 cells. *Oncogene* 1994;**9**:2425–9.

34. Xu N, Bradley L, Ambdukar I, Gutkind JS. A mutant alpha subunit of G12 potentiates the eicosanoid pathway and is highly oncogenic in NIH 3T3 cells. *Proc Natl Acad Sci USA* 1993;**90**:6741–5.
35. Chan AM, Fleming TP, McGovern ES, Chedid M, Miki T, Aaronson SA. Expression cDNA cloning of a transforming gene encoding the wild-type G alpha 12 gene product. *Mol Cell Biol* 1993;**13**:762–8.
36. Bian D, Mahanivong C, Yu J, Frisch SM, Pan ZK, Ye RD, et al. The G12/13-RhoA signaling pathway contributes to efficient lysophosphatidic acid-stimulated cell migration. *Oncogene* 2006;**25**:2234–44.
37. Xu X, Prestwich GD. Inhibition of tumor growth and angiogenesis by a lysophosphatidic acid antagonist in an engineered three-dimensional lung cancer xenograft model. *Cancer* 2010;**116**:1739–50.
38. Even-Ram S, Uziely B, Cohen P, Grisaru-Granovsky S, Maoz M, Ginzburg Y, et al. Thrombin receptor overexpression in malignant and physiological invasion processes. *Nat Med* 1998;**4**:909–14.
39. Meigs TE, Fedor-Chaiken M, Kaplan DD, Brackenbury R, Casey PJ. Galpha12 and Galpha13 negatively regulate the adhesive functions of cadherin. *J Biol Chem* 2002;**277**:24594–600.
40. Richard DE, Berra E, Pouyssegur J. Nonhypoxic pathway mediates the induction of hypoxia-inducible factor 1alpha in vascular smooth muscle cells. *J Biol Chem* 2000; **275**:26765–71.
41. Williams B, Baker AQ, Gallacher B, Lodwick D. Angiotensin II increases vascular permeability factor gene expression by human vascular smooth muscle cells. *Hypertension* 1995;**25**:913–7.
42. Drenan RM, Doupnik CA, Boyle MP, Muglia LJ, Huettner JE, Linder ME, et al. Palmitoylation regulates plasma membrane-nuclear shuttling of R7BP, a novel membrane anchor for the RGS7 family. *J Cell Biol* 2005;**169**:623–33.
43. Kino T, Tiulpakov A, Ichijo T, Chheng L, Kozasa T, Chrousos GP. G protein beta interacts with the glucocorticoid receptor and suppresses its transcriptional activity in the nucleus. *J Cell Biol* 2005;**169**:885–96.
44. Park JG, Muise A, He GP, Kim SW, Ro HS. Transcriptional regulation by the gamma5 subunit of a heterotrimeric G protein during adipogenesis. *EMBO J* 1999;**18**:4004–12.
45. Spiegelberg BD, Hamm HE. G betagamma binds histone deacetylase 5 (HDAC5) and inhibits its transcriptional co-repression activity. *J Biol Chem* 2005;**280**:41769–76.
46. Hurst JH, Hooks SB. Regulator of G-protein signaling (RGS) proteins in cancer biology. *Biochem Pharmacol* 2009;**78**:1289–97.
47. Manzur M, Hamzah J, Ganss R. Modulation of g protein signaling normalizes tumor vessels. *Cancer Res* 2009;**69**:396–9.
48. Nusse R. Wnt signaling in disease and in development. *Cell Res* 2005;**15**:28–32.
49. Joslyn G, Carlson M, Thliveris A, Albertsen H, Gelbert L, Samowitz W, et al. Identification of deletion mutations and three new genes at the familial polyposis locus. *Cell* 1991;**66**:601–13.
50. Nishisho I, Nakamura Y, Miyoshi Y, Miki Y, Ando H, Horii A, et al. Mutations of chromosome 5q21 genes in FAP and colorectal cancer patients. *Science* 1991;**253**:665–9.
51. Behrens J, Lustig B. The Wnt connection to tumorigenesis. *Int J Dev Biol* 2004;**48**: 477–87.
52. Revet I, Huizenga G, Koster J, Volckmann R, van Sluis P, Versteeg R, et al. MSX1 induces the Wnt pathway antagonist genes DKK1, DKK2, DKK3, and SFRP1 in neuroblastoma cells, but does not block Wnt3 and Wnt5A signalling to DVL3. *Cancer Lett* 2010;**289**:195–207.
53. Cadigan KM, Liu YI. Wnt signaling: complexity at the surface. *J Cell Sci* 2006;**119**: 395–402.

54. Vincan E, Darcy PK, Farrelly CA, Faux MC, Brabletz T, Ramsay RG. Frizzled-7 dictates three-dimensional organization of colorectal cancer cell carcinoids. *Oncogene* 2007;**26**:2340–52.
55. Taelman VF, Dobrowolski R, Plouhinec JL, Fuentealba LC, Vorwald PP, Gumper I, et al. Wnt signaling requires sequestration of glycogen synthase kinase 3 inside multivesicular endosomes. *Cell* 2010;**143**:1136–48.
56. Hoffmeyer K, Raggioli A, Rudloff S, Anton R, Hierholzer A, Del Valle I, et al. Wnt/beta-catenin signaling regulates telomerase in stem cells and cancer cells. *Science* 2012;**336**:1549–54.
57. Kasper M, Regl G, Frischauf AM, Aberger F. GLI transcription factors: mediators of oncogenic Hedgehog signalling. *Eur J Cancer* 2006;**42**:437–45.
58. Katoh Y, Katoh M. Hedgehog target genes: mechanisms of carcinogenesis induced by aberrant hedgehog signaling activation. *Curr Mol Med* 2009;**9**:873–86.
59. Jiang J, Hui CC. Hedgehog signaling in development and cancer. *Dev Cell* 2008;**15**: 801–12.
60. Lum L, Beachy PA. The Hedgehog response network: sensors, switches, and routers. *Science* 2004;**304**:1755–9.
61. Stacey M, Lin HH, Gordon S, McKnight AJ. LNB-TM7, a group of seven-transmembrane proteins related to family-B G-protein-coupled receptors. *Trends Biochem Sci* 2000;**25**:284–9.
62. Yona S, Lin HH, Siu WO, Gordon S, Stacey M. Adhesion-GPCRs: emerging roles for novel receptors. *Trends Biochem Sci* 2008;**33**:491–500.
63. Chang GW, Stacey M, Kwakkenbos MJ, Hamann J, Gordon S, Lin HH. Proteolytic cleavage of the EMR2 receptor requires both the extracellular stalk and the GPS motif. *FEBS Lett* 2003;**547**:145–50.
64. Krasnoperov V, Lu Y, Buryanovsky L, Neubert TA, Ichtchenko K, Petrenko AG. Post-translational proteolytic processing of the calcium-independent receptor of alpha-latrotoxin (CIRL), a natural chimera of the cell adhesion protein and the G protein-coupled receptor. Role of the G protein-coupled receptor proteolysis site (GPS) motif. *J Biol Chem* 2002;**277**:46518–26.
65. Lin HH, Chang GW, Davies JQ, Stacey M, Harris J, Gordon S. Autocatalytic cleavage of the EMR2 receptor occurs at a conserved G protein-coupled receptor proteolytic site motif. *J Biol Chem* 2004;**279**:31823–32.
66. Aust G, Steinert M, Schutz A, Boltze C, Wahlbuhl M, Hamann J, et al. CD97, but not its closely related EGF-TM7 family member EMR2, is expressed on gastric, pancreatic, and esophageal carcinomas. *Am J Clin Pathol* 2002;**118**:699–707.
67. Wang T, Ward Y, Tian L, Lake R, Guedez L, Stetler-Stevenson WG, et al. CD97, an adhesion receptor on inflammatory cells, stimulates angiogenesis through binding integrin counterreceptors on endothelial cells. *Blood* 2005;**105**:2836–44.
68. Mustafa T, Eckert A, Klonisch T, Kehlen A, Maurer P, Klintschar M, et al. Expression of the epidermal growth factor seven-transmembrane member CD97 correlates with grading and staging in human oral squamous cell carcinomas. *Cancer Epidemiol Biomarkers Prev* 2005;**14**:108–19.
69. Ward Y, Lake R, Yin JJ, Heger CD, Raffeld M, Goldsmith PK, et al. LPA receptor heterodimerizes with CD97 to amplify LPA-initiated RHO-dependent signaling and invasion in prostate cancer cells. *Cancer Res* 2011;**71**:7301–11.
70. Nishimori H, Shiratsuchi T, Urano T, Kimura Y, Kiyono K, Tatsumi K, et al. A novel brain-specific p53-target gene, BAI1, containing thrombospondin type 1 repeats inhibits experimental angiogenesis. *Oncogene* 1997;**15**:2145–50.
71. Fukushima Y, Oshika Y, Tsuchida T, Tokunaga T, Hatanaka H, Kijima H, et al. Brain-specific angiogenesis inhibitor 1 expression is inversely correlated with vascularity and distant metastasis of colorectal cancer. *Int J Oncol* 1998;**13**:967–70.

72. Yang L, Chen G, Mohanty S, Scott G, Fazal F, Rahman A, et al. GPR56 Regulates VEGF production and angiogenesis during melanoma progression. *Cancer Res* 2011; **71**:5558–68.
73. Shashidhar S, Lorente G, Nagavarapu U, Nelson A, Kuo J, Cummins J, et al. GPR56 is a GPCR that is overexpressed in gliomas and functions in tumor cell adhesion. *Oncogene* 2005;**24**:1673–82.
74. Lum AM, Wang BB, Beck-Engeser GB, Li L, Channa N, Wabl M. Orphan receptor GPR110, an oncogene overexpressed in lung and prostate cancer. *BMC Cancer* 2010;**10**:40.
75. Carson-Walter EB, Watkins DN, Nanda A, Vogelstein B, Kinzler KW, St Croix B. Cell surface tumor endothelial markers are conserved in mice and humans. *Cancer Res* 2001;**61**:6649–55.
76. Saito Y, Berk BC. Transactivation: a novel signaling pathway from angiotensin II to tyrosine kinase receptors. *J Mol Cell Cardiol* 2001;**33**:3–7.
77. Crespo P, Xu N, Simonds WF, Gutkind JS. Ras-dependent activation of MAP kinase pathway mediated by G-protein beta gamma subunits. *Nature* 1994;**369**:418–20.
78. Faure M, Voyno-Yasenetskaya TA, Bourne HR. cAMP and beta gamma subunits of heterotrimeric G proteins stimulate the mitogen-activated protein kinase pathway in COS-7 cells. *J Biol Chem* 1994;**269**:7851–4.
79. Koch WJ, Hawes BE, Allen LF, Lefkowitz RJ. Direct evidence that Gi-coupled receptor stimulation of mitogen-activated protein kinase is mediated by G beta gamma activation of p21ras. *Proc Natl Acad Sci USA* 1994;**91**:12706–10.
80. Buhl AM, Avdi N, Worthen GS, Johnson GL. Mapping of the C5a receptor signal transduction network in human neutrophils. *Proc Natl Acad Sci USA* 1994;**91**:9190–4.
81. Worthen GS, Avdi N, Buhl AM, Suzuki N, Johnson GL. FMLP activates Ras and Raf in human neutrophils. Potential role in activation of MAP kinase. *J Clin Invest* 1994;**94**:815–23.
82. Alblas J, van Corven EJ, Hordijk PL, Milligan G, Moolenaar WH. Gi-mediated activation of the p21ras-mitogen-activated protein kinase pathway by alpha 2-adrenergic receptors expressed in fibroblasts. *J Biol Chem* 1993;**268**:22235–8.
83. Winitz S, Russell M, Qian NX, Gardner A, Dwyer L, Johnson GL. Involvement of Ras and Raf in the Gi-coupled acetylcholine muscarinic m2 receptor activation of mitogen-activated protein (MAP) kinase kinase and MAP kinase. *J Biol Chem* 1993;**268**: 19196–9.
84. Howe LR, Marshall CJ. Lysophosphatidic acid stimulates mitogen-activated protein kinase activation via a G-protein-coupled pathway requiring p21ras and p74raf-1. *J Biol Chem* 1993;**268**:20717–20.
85. van Corven EJ, Hordijk PL, Medema RH, Bos JL, Moolenaar WH. Pertussis toxin-sensitive activation of p21ras by G protein-coupled receptor agonists in fibroblasts. *Proc Natl Acad Sci USA* 1993;**90**:1257–61.
86. Chen J, Iyengar R. Suppression of Ras-induced transformation of NIH 3T3 cells by activated G alpha s. *Science* 1994;**263**:1278–81.
87. Hawes BE, van Biesen T, Koch WJ, Luttrell LM, Lefkowitz RJ. Distinct pathways of Gi- and Gq-mediated mitogen-activated protein kinase activation. *J Biol Chem* 1995;**270**:17148–53.
88. van Biesen T, Hawes BE, Luttrell DK, Krueger KM, Touhara K, Porfiri E, et al. Receptor-tyrosine-kinase- and G beta gamma-mediated MAP kinase activation by a common signalling pathway. *Nature* 1995;**376**:781–4.
89. Daub H, Weiss FU, Wallasch C, Ullrich A. Role of transactivation of the EGF receptor in signalling by G-protein-coupled receptors. *Nature* 1996;**379**:557–60.
90. Bhola NE, Grandis JR. Crosstalk between G-protein-coupled receptors and epidermal growth factor receptor in cancer. *Front Biosci* 2008;**13**:1857–65.

91. Prenzel N, Zwick E, Daub H, Leserer M, Abraham R, Wallasch C, et al. EGF receptor transactivation by G-protein-coupled receptors requires metalloproteinase cleavage of proHB-EGF. *Nature* 1999;**402**:884–8.
92. Eguchi S, Numaguchi K, Iwasaki H, Matsumoto T, Yamakawa T, Utsunomiya H, et al. Calcium-dependent epidermal growth factor receptor transactivation mediates the angiotensin II-induced mitogen-activated protein kinase activation in vascular smooth muscle cells. *J Biol Chem* 1998;**273**:8890–6.
93. Fuentes LQ, Reyes CE, Sarmiento JM, Villanueva CI, Figueroa CD, Navarro J, et al. Vasopressin up-regulates the expression of growth-related immediate-early genes via two distinct EGF receptor transactivation pathways. *Cell Signal* 2008;**20**:1642–50.
94. Maudsley S, Pierce KL, Zamah AM, Miller WE, Ahn S, Daaka Y, et al. The beta(2)-adrenergic receptor mediates extracellular signal-regulated kinase activation via assembly of a multi-receptor complex with the epidermal growth factor receptor. *J Biol Chem* 2000;**275**:9572–80.
95. Tilley DG, Kim IM, Patel PA, Violin JD, Rockman HA. beta-Arrestin mediates beta1-adrenergic receptor-epidermal growth factor receptor interaction and downstream signaling. *J Biol Chem* 2009;**284**:20375–86.
96. Zajac M, Law J, Cvetkovic DD, Pampillo M, McColl L, Pape C, et al. GPR54 (KISS1R) transactivates EGFR to promote breast cancer cell invasiveness. *PLoS One* 2011;**6**:e21599.
97. Schafer B, Gschwind A, Ullrich A. Multiple G-protein-coupled receptor signals converge on the epidermal growth factor receptor to promote migration and invasion. *Oncogene* 2004;**23**:991–9.
98. Filardo EJ, Quinn JA, Bland KI, Frackelton Jr AR. Estrogen-induced activation of Erk-1 and Erk-2 requires the G protein-coupled receptor homolog, GPR30, and occurs via trans-activation of the epidermal growth factor receptor through release of HB-EGF. *Mol Endocrinol* 2000;**14**:1649–60.
99. Albanito L, Madeo A, Lappano R, Vivacqua A, Rago V, Carpino A, et al. G protein-coupled receptor 30 (GPR30) mediates gene expression changes and growth response to 17beta-estradiol and selective GPR30 ligand G-1 in ovarian cancer cells. *Cancer Res* 2007;**67**:1859–66.
100. Chan QK, Lam HM, Ng CF, Lee AY, Chan ES, Ng HK, et al. Activation of GPR30 inhibits the growth of prostate cancer cells through sustained activation of Erk1/2, c-jun/c-fos-dependent upregulation of p21, and induction of G(2) cell-cycle arrest. *Cell Death Differ* 2010;**17**:1511–23.
101. Pandey DP, Lappano R, Albanito L, Madeo A, Maggiolini M, Picard D. Estrogenic GPR30 signalling induces proliferation and migration of breast cancer cells through CTGF. *EMBO J* 2009;**28**:523–32.
102. Vivacqua A, Bonofiglio D, Recchia AG, Musti AM, Picard D, Ando S, et al. The G protein-coupled receptor GPR30 mediates the proliferative effects induced by 17beta-estradiol and hydroxytamoxifen in endometrial cancer cells. *Mol Endocrinol* 2006;**20**:631–46.
103. Simonson MS, Wang Y, Herman WH. Nuclear signaling by endothelin-1 requires Src protein-tyrosine kinases. *J Biol Chem* 1996;**271**:77–82.
104. Lev S, Moreno H, Martinez R, Canoll P, Peles E, Musacchio JM, et al. Protein tyrosine kinase PYK2 involved in Ca(2+)-induced regulation of ion channel and MAP kinase functions. *Nature* 1995;**376**:737–45.
105. Dikic I, Tokiwa G, Lev S, Courtneidge SA, Schlessinger J. A role for Pyk2 and Src in linking G-protein-coupled receptors with MAP kinase activation. *Nature* 1996;**383**: 547–50.
106. Andreev J, Galisteo ML, Kranenburg O, Logan SK, Chiu ES, Okigaki M, et al. Src and Pyk2 mediate G-protein-coupled receptor activation of epidermal growth factor

receptor (EGFR) but are not required for coupling to the mitogen-activated protein (MAP) kinase signaling cascade. *J Biol Chem* 2001;**276**:20130–5.
107. Tsai W, Morielli AD, Peralta EG. The m1 muscarinic acetylcholine receptor transactivates the EGF receptor to modulate ion channel activity. *EMBO J* 1997;**16**: 4597–605.
108. Keely SJ, Uribe JM, Barrett KE. Carbachol stimulates transactivation of epidermal growth factor receptor and mitogen-activated protein kinase in T84 cells. Implications for carbachol-stimulated chloride secretion. *J Biol Chem* 1998;**273**:27111–7.
109. Wang D, Yu X, Cohen RA, Brecher P. Distinct effects of N-acetylcysteine and nitric oxide on angiotensin II-induced epidermal growth factor receptor phosphorylation and intracellular Ca(2+) levels. *J Biol Chem* 2000;**275**:12223–30.
110. Heeneman S, Haendeler J, Saito Y, Ishida M, Berk BC. Angiotensin II induces transactivation of two different populations of the platelet-derived growth factor beta receptor. Key role for the p66 adaptor protein Shc. *J Biol Chem* 2000;**275**:15926–32.
111. Linseman DA, Benjamin CW, Jones DA. Convergence of angiotensin II and platelet-derived growth factor receptor signaling cascades in vascular smooth muscle cells. *J Biol Chem* 1995;**270**:12563–8.
112. Mondorf UF, Geiger H, Herrero M, Zeuzem S, Piiper A. Involvement of the platelet-derived growth factor receptor in angiotensin II-induced activation of extracellular regulated kinases 1 and 2 in human mesangial cells. *FEBS Lett* 2000;**472**:129–32.
113. Abe J, Deguchi J, Matsumoto T, Takuwa N, Noda M, Ohno M, et al. Stimulated activation of platelet-derived growth factor receptor in vivo in balloon-injured arteries: a link between angiotensin II and intimal thickening. *Circulation* 1997;**96**:1906–13.
114. Vacca F, Bagnato A, Catt KJ, Tecce R. Transactivation of the epidermal growth factor receptor in endothelin-1-induced mitogenic signaling in human ovarian carcinoma cells. *Cancer Res* 2000;**60**:5310–7.
115. Madeo A, Maggiolini M. Nuclear alternate estrogen receptor GPR30 mediates 17beta-estradiol-induced gene expression and migration in breast cancer-associated fibroblasts. *Cancer Res* 2010;**70**:6036–46.
116. Kumar U. Cross-talk and modulation of signaling between somatostatin and growth factor receptors. *Endocrine* 2011;**40**:168–80.
117. Watt HL, Kharmate GD, Kumar U. Somatostatin receptors 1 and 5 heterodimerize with epidermal growth factor receptor: agonist-dependent modulation of the downstream MAPK signalling pathway in breast cancer cells. *Cell Signal* 2009;**21**:428–39.
118. Kharmate G, Rajput PS, Watt HL, Somvanshi RK, Chaudhari N, Qiu X, et al. Role of somatostatin receptor 1 and 5 on epidermal growth factor receptor mediated signaling. *Biochim Biophys Acta* 2011;**1813**:1172–89.
119. Kharmate G, Rajput PS, Watt HL, Somvanshi RK, Chaudhari N, Qiu X, et al. Dissociation of epidermal growth factor receptor and ErbB2 heterodimers in the presence of somatostatin receptor 5 modulate signaling pathways. *Endocrinology* 2011;**152**: 931–45.
120. Berns K, Horlings HM, Hennessy BT, Madiredjo M, Hijmans EM, Beelen K, et al. A functional genetic approach identifies the PI3K pathway as a major determinant of trastuzumab resistance in breast cancer. *Cancer Cell* 2007;**12**:395–402.
121. Delcourt N, Bockaert J, Marin P. GPCR-jacking: from a new route in RTK signalling to a new concept in GPCR activation. *Trends Pharmacol Sci* 2007;**28**:602–7.
122. Mira E, Lacalle RA, Gonzalez MA, Gomez-Mouton C, Abad JL, Bernad A, et al. A role for chemokine receptor transactivation in growth factor signaling. *EMBO Rep* 2001;**2**:151–6.
123. Hobson JP, Rosenfeldt HM, Barak LS, Olivera A, Poulton S, Caron MG, et al. Role of the sphingosine-1-phosphate receptor EDG-1 in PDGF-induced cell motility. *Science* 2001;**291**:1800–3.

124. Akekawatchai C, Holland JD, Kochetkova M, Wallace JC, McColl SR. Transactivation of CXCR4 by the insulin-like growth factor-1 receptor (IGF-1R) in human MDA-MB-231 breast cancer epithelial cells. *J Biol Chem* 2005;**280**:39701–8.
125. Luttrell LM, van Biesen T, Hawes BE, Koch WJ, Touhara K, Lefkowitz RJ. G beta gamma subunits mediate mitogen-activated protein kinase activation by the tyrosine kinase insulin-like growth factor 1 receptor. *J Biol Chem* 1995;**270**:16495–8.
126. Alderton F, Rakhit S, Kong KC, Palmer T, Sambi B, Pyne S, et al. Tethering of the platelet-derived growth factor beta receptor to G-protein-coupled receptors. A novel platform for integrative signaling by these receptor classes in mammalian cells. *J Biol Chem* 2001;**276**:28578–85.
127. Halls ML. Constitutive formation of an RXFP1-signalosome: a novel paradigm in GPCR function and regulation. *Br J Pharmacol* 2012;**165**:1644–58.
128. Sun W, Yang J. Molecular basis of lysophosphatidic acid-induced NF-kappaB activation. *Cell Signal* 2010;**22**:1799–803.
129. Lyssand JS, DeFino MC, Tang XB, Hertz AL, Feller DB, Wacker JL, et al. Blood pressure is regulated by an alpha1D-adrenergic receptor/dystrophin signalosome. *J Biol Chem* 2008;**283**:18792–800.
130. Halls ML, Cooper DM. Sub-picomolar relaxin signalling by a pre-assembled RXFP1, AKAP79, AC2, beta-arrestin 2, PDE4D3 complex. *EMBO J* 2010;**29**:2772–87.
131. Klonisch T, Bialek J, Radestock Y, Hoang-Vu C, Hombach-Klonisch S. Relaxin-like ligand-receptor systems are autocrine/paracrine effectors in tumor cells and modulate cancer progression and tissue invasiveness. *Adv Exp Med Biol* 2007;**612**:104–18.
132. Mendelsohn J. Blockade of receptors for growth factors: an anticancer therapy—the fourth annual Joseph H Burchenal American Association of Cancer Research Clinical Research Award Lecture. *Clin Cancer Res* 2000;**6**:747–53.
133. Moody TW. Peptide hormones and lung cancer. *Panminerva Med* 2006;**48**:19–26.
134. Rozengurt E. Neuropeptides as growth factors for normal and cancerous cells. *Trends Endocrinol Metab* 2002;**13**:128–34.
135. Chan DC, Geraci M, Bunn Jr PA. Anti-growth factor therapy for lung cancer. *Drug Resist Update* 1998;**1**:377–88.
136. Cuttitta F, Carney DN, Mulshine J, Moody TW, Fedorko J, Fischler A, et al. Bombesin-like peptides can function as autocrine growth factors in human small-cell lung cancer. *Nature* 1985;**316**:823–6.
137. Corps AN, Rees LH, Brown KD. A peptide that inhibits the mitogenic stimulation of Swiss 3T3 cells by bombesin or vasopressin. *Biochem J* 1985;**231**:781–4.
138. Jensen RT, Jones SW, Folkers K, Gardner JD. A synthetic peptide that is a bombesin receptor antagonist. *Nature* 1984;**309**:61–3.
139. Tsou K, Wu SX, Lu YA, Way EL. Block of the hyoscine-resistant opiate withdrawal contracture of ileum by a new substance P antagonist [D-Arg1, D-Phe5, D- Trp7,9, Leu 11]substance P. *Eur J Pharmacol* 1985;**110**:155–6.
140. Woll PJ, Rozengurt E. [D-Arg1, D-Phe5, D-Trp7,9, Leu11]substance P, a potent bombesin antagonist in murine Swiss 3T3 cells, inhibits the growth of human small cell lung cancer cells in vitro. *Proc Natl Acad Sci USA* 1988;**85**:1859–63.
141. Woll PJ, Rozengurt E. Two classes of antagonist interact with receptors for the mitogenic neuropeptides bombesin, bradykinin, and vasopressin. *Growth Factors* 1988;**1**: 75–83.
142. Gu XH, Casley DJ, Nayler WG. The inhibitory effect of [D-Arg1, D-Phe, D-Try7,9, Leu11]substance P on endothelin-1 binding sites in rat cardiac membranes. *Biochem Biophys Res Commun* 1991;**179**:130–3.
143. Jarpe MB, Knall C, Mitchell FM, Buhl AM, Duzic E, Johnson GL. [D-Arg1, D-Phe5, D-Trp7,9, Leu11]substance P acts as a biased agonist toward neuropeptide and chemokine receptors. *J Biol Chem* 1998;**273**:3097–104.

144. Reeve JG, Bleehen NM. [D-Arg1, D-Phe5, D-Trp7,9, Leu11]substance P induces apoptosis in lung cancer cell lines in vitro. *Biochem Biophys Res Commun* 1994;**199**:1313–9.
145. Seckl MJ, Higgins T, Widmer F, Rozengurt E. [D-Arg1, D-Trp5,7,9, Leu11] substance P: a novel potent inhibitor of signal transduction and growth in vitro and in vivo in small cell lung cancer cells. *Cancer Res* 1997;**57**:51–4.
146. Sethi T, Langdon S, Smyth J, Rozengurt E. Growth of small cell lung cancer cells: stimulation by multiple neuropeptides and inhibition by broad spectrum antagonists in vitro and in vivo. *Cancer Res* 1992;**52**:2737s–2742s.
147. Woll PJ, Rozengurt E. A neuropeptide antagonist that inhibits the growth of small cell lung cancer in vitro. *Cancer Res* 1990;**50**:3968–73.
148. Langdon S, Sethi T, Ritchie A, Muir M, Smyth J, Rozengurt E. Broad spectrum neuropeptide antagonists inhibit the growth of small cell lung cancer in vivo. *Cancer Res* 1992;**52**:4554–7.
149. MacKinnon AC, Armstrong RA, Waters CM, Cummings J, Smyth JF, Haslett C, et al. [Arg6, D-Trp7,9, NmePhe8]-substance P (6–11) activates JNK and induces apoptosis in small cell lung cancer cells via an oxidant-dependent mechanism. *Br J Cancer* 1999;**80**:1026–34.
150. Stewart JM. Bradykinin antagonists as anti-cancer agents. *Curr Pharm Des* 2003;**9**: 2036–42.
151. Guha S, Eibl G, Kisfalvi K, Fan RS, Burdick M, Reber H, et al. Broad-spectrum G protein-coupled receptor antagonist, [D-Arg1, D-Trp5,7,9, Leu11]SP: a dual inhibitor of growth and angiogenesis in pancreatic cancer. *Cancer Res* 2005;**65**:2738–45.
152. Chan D, Gera L, Stewart J, Helfrich B, Verella-Garcia M, Johnson G, et al. Bradykinin antagonist dimer, CU201, inhibits the growth of human lung cancer cell lines by a "biased agonist" mechanism. *Proc Natl Acad Sci USA* 2002;**99**:4608–13.
153. Chan DC, Gera L, Stewart JM, Helfrich B, Zhao TL, Feng WY, et al. Bradykinin antagonist dimer, CU201, inhibits the growth of human lung cancer cell lines in vitro and in vivo and produces synergistic growth inhibition in combination with other antitumor agents. *Clin Cancer Res* 2002;**8**:1280–7.
154. Stewart JM, Gera L, Chan DC, Bunn Jr PA, York EJ, Simkeviciene V, et al. Bradykinin-related compounds as new drugs for cancer and inflammation. *Can J Physiol Pharmacol* 2002;**80**:275–80.
155. Culler MD. Somatostatin-dopamine chimeras: a novel approach to treatment of neuroendocrine tumors. *Horm Metab Res* 2011;**43**:854–7.
156. Hofland LJ, Lamberts SW. The pathophysiological consequences of somatostatin receptor internalization and resistance. *Endocr Rev* 2003;**24**:28–47.
157. Bruns C, Lewis I, Briner U, Meno-Tetang G, Weckbecker G. SOM230: a novel somatostatin peptidomimetic with broad somatotropin release inhibiting factor (SRIF) receptor binding and a unique antisecretory profile. *Eur J Endocrinol* 2002;**146**:707–16.
158. Saveanu A, Gunz G, Dufour H, Caron P, Fina F, Ouafik L, et al. Bim-23244, a somatostatin receptor subtype 2- and 5-selective analog with enhanced efficacy in suppressing growth hormone (GH) from octreotide-resistant human GH-secreting adenomas. *J Clin Endocrinol Metab* 2001;**86**:140–5.
159. Culler MD, Oberg K, Arnold R, Krenning EP, Sevilla I, Diaz JA. Somatostatin analogs for the treatment of neuroendocrine tumors. *Cancer Metastasis Rev* 2011;**30** (Suppl. 1):9–17.
160. Molitch ME, Elton RL, Blackwell RE, Caldwell B, Chang RJ, Jaffe R, et al. Bromocriptine as primary therapy for prolactin-secreting macroadenomas: results of a prospective multicenter study. *J Clin Endocrinol Metab* 1985;**60**:698–705.
161. Colao A, Di Sarno A, Sarnacchiaro F, Ferone D, Di Renzo G, Merola B, et al. Prolactinomas resistant to standard dopamine agonists respond to chronic cabergoline treatment. *J Clin Endocrinol Metab* 1997;**82**:876–83.

162. Pivonello R, Ferone D, de Herder WW, Kros JM, De Caro ML, Arvigo M, et al. Dopamine receptor expression and function in corticotroph pituitary tumors. *J Clin Endocrinol Metab* 2004;**89**:2452–62.
163. Flogstad AK, Halse J, Grass P, Abisch E, Djoseland O, Kutz K, et al. A comparison of octreotide, bromocriptine, or a combination of both drugs in acromegaly. *J Clin Endocrinol Metab* 1994;**79**:461–5.
164. Minniti G, Jaffrain-Rea ML, Baldelli R, Ferretti E, Caracciolo B, Bultrini A, et al. Acute effects of octreotide, cabergoline and a combination of both drugs on GH secretion in acromegalic patients. *Clin Ter* 1997;**148**:601–7.
165. Rocheville M, Lange DC, Kumar U, Patel SC, Patel RC, Patel YC. Receptors for dopamine and somatostatin: formation of hetero-oligomers with enhanced functional activity. *Science* 2000;**288**:154–7.
166. Baragli A, Alturaihi H, Watt HL, Abdallah A, Kumar U. Heterooligomerization of human dopamine receptor 2 and somatostatin receptor 2 co-immunoprecipitation and fluorescence resonance energy transfer analysis. *Cell Signal* 2007;**19**:2304–16.
167. de Bruin C, Feelders RA, Lamberts SW, Hofland LJ. Somatostatin and dopamine receptors as targets for medical treatment of Cushing's syndrome. *Rev Endocr Metab Disord* 2009;**10**:91–102.
168. Saveanu A, Jaquet P. Somatostatin-dopamine ligands in the treatment of pituitary adenomas. *Rev Endocr Metab Disord* 2009;**10**:83–90.
169. Ferone D, Arvigo M, Semino C, Jaquet P, Saveanu A, Taylor JE, et al. Somatostatin and dopamine receptor expression in lung carcinoma cells and effects of chimeric somatostatin-dopamine molecules on cell proliferation. *Am J Physiol Endocrinol Metab* 2005;**289**:E1044–E1050.
170. Ren SG, Kim S, Taylor J, Dong J, Moreau JP, Culler MD, et al. Suppression of rat and human growth hormone and prolactin secretion by a novel somatostatin/dopaminergic chimeric ligand. *J Clin Endocrinol Metab* 2003;**88**:5414–21.
171. Saveanu A, Lavaque E, Gunz G, Barlier A, Kim S, Taylor JE, et al. Demonstration of enhanced potency of a chimeric somatostatin-dopamine molecule, BIM-23A387, in suppressing growth hormone and prolactin secretion from human pituitary somatotroph adenoma cells. *J Clin Endocrinol Metab* 2002;**87**:5545–52.
172. Jaquet P, Gunz G, Saveanu A, Barlier A, Dufour H, Taylor J, et al. BIM-23A760, a chimeric molecule directed towards somatostatin and dopamine receptors, vs universal somatostatin receptors ligands in GH-secreting pituitary adenomas partial responders to octreotide. *J Endocrinol Invest* 2005;**28**:21–7.
173. Bouvier M, Heveker N, Jockers R, Marullo S, Milligan G. BRET analysis of GPCR oligomerization: newer does not mean better. *Nat Methods* 2007;**4**:3–4 author reply 4.
174. Hamatake M, Aoki T, Futahashi Y, Urano E, Yamamoto N, Komano J. Ligand-independent higher-order multimerization of CXCR4, a G-protein-coupled chemokine receptor involved in targeted metastasis. *Cancer Sci* 2009;**100**:95–102.
175. Salahpour A, Masri B. Experimental challenge to a 'rigorous' BRET analysis of GPCR oligomerization. *Nat Methods* 2007;**4**:599–600 author reply 601.
176. Munoz LM, Lucas P, Holgado BL, Barroso R, Vega B, Rodriguez-Frade JM, et al. Receptor oligomerization: a pivotal mechanism for regulating chemokine function. *Pharmacol Ther* 2011;**131**:351–8.
177. Bennett LD, Fox JM, Signoret N. Mechanisms regulating chemokine receptor activity. *Immunology* 2011;**134**:246–56.
178. Babcock GJ, Farzan M, Sodroski J. Ligand-independent dimerization of CXCR4, a principal HIV-1 coreceptor. *J Biol Chem* 2003;**278**:3378–85.
179. Wang J, He L, Combs CA, Roderiquez G, Norcross MA. Dimerization of CXCR4 in living malignant cells: control of cell migration by a synthetic peptide that reduces homologous CXCR4 interactions. *Mol Cancer Ther* 2006;**5**:2474–83.

180. Muller A, Homey B, Soto H, Ge N, Catron D, Buchanan ME, et al. Involvement of chemokine receptors in breast cancer metastasis. *Nature* 2001;**410**:50–6.
181. Covic L, Gresser AL, Talavera J, Swift S, Kuliopulos A. Activation and inhibition of G protein-coupled receptors by cell-penetrating membrane-tethered peptides. *Proc Natl Acad Sci USA* 2002;**99**:643–8.
182. Dimond P, Carlson K, Bouvier M, Gerard C, Xu L, Covic L, et al. G protein-coupled receptor modulation with pepducins: moving closer to the clinic. *Ann N Y Acad Sci* 2011;**1226**:34–49.
183. O'Callaghan K, Kuliopulos A, Covic L. Turning receptors on and off with intracellular pepducins: new insights into G-protein-coupled receptor drug development. *J Biol Chem* 2012;**287**:12787–96.
184. Sevigny LM, Zhang P, Bohm A, Lazarides K, Perides G, Covic L, et al. Interdicting protease-activated receptor-2-driven inflammation with cell-penetrating pepducins. *Proc Natl Acad Sci USA* 2011;**108**:8491–6.
185. Edwards RJ, Moran N, Devocelle M, Kiernan A, Meade G, Signac W, et al. Bioinformatic discovery of novel bioactive peptides. *Nat Chem Biol* 2007;**3**:108–12.
186. Tchernychev B, Ren Y, Sachdev P, Janz JM, Haggis L, O'Shea A, et al. Discovery of a CXCR4 agonist pepducin that mobilizes bone marrow hematopoietic cells. *Proc Natl Acad Sci USA* 2010;**107**:22255–9.
187. Covic L, Misra M, Badar J, Singh C, Kuliopulos A. Pepducin-based intervention of thrombin-receptor signaling and systemic platelet activation. *Nat Med* 2002;**8**:1161–5.
188. Kaneider NC, Agarwal A, Leger AJ, Kuliopulos A. Reversing systemic inflammatory response syndrome with chemokine receptor pepducins. *Nat Med* 2005;**11**:661–5.
189. Tressel SL, Koukos G, Tchernychev B, Jacques SL, Covic L, Kuliopulos A. Pharmacology, biodistribution, and efficacy of GPCR-based pepducins in disease models. *Methods Mol Biol* 2011;**683**:259–75.
190. Dorsam RT, Gutkind JS. G-protein-coupled receptors and cancer. *Nat Rev Cancer* 2007;**7**:79–94.
191. Martin CB, Mahon GM, Klinger MB, Kay RJ, Symons M, Der CJ, et al. The thrombin receptor, PAR-1, causes transformation by activation of Rho-mediated signaling pathways. *Oncogene* 2001;**20**:1953–63.
192. Agarwal A, Covic L, Sevigny LM, Kaneider NC, Lazarides K, Azabdaftari G, et al. Targeting a metalloprotease-PAR1 signaling system with cell-penetrating pepducins inhibits angiogenesis, ascites, and progression of ovarian cancer. *Mol Cancer Ther* 2008;**7**:2746–57.
193. Agarwal A, Tressel SL, Kaimal R, Balla M, Lam FH, Covic L, et al. Identification of a metalloprotease-chemokine signaling system in the ovarian cancer microenvironment: implications for antiangiogenic therapy. *Cancer Res* 2010;**70**:5880–90.
194. Boire A, Covic L, Agarwal A, Jacques S, Sherifi S, Kuliopulos A. PAR1 is a matrix metalloprotease-1 receptor that promotes invasion and tumorigenesis of breast cancer cells. *Cell* 2005;**120**:303–13.
195. Ghio P, Cappia S, Selvaggi G, Novello S, Lausi P, Zecchina G, et al. Prognostic role of protease-activated receptors 1 and 4 in resected stage IB non-small-cell lung cancer. *Clin Lung Cancer* 2006;**7**:395–400.
196. Yang E, Boire A, Agarwal A, Nguyen N, O'Callaghan K, Tu P, et al. Blockade of PAR1 signaling with cell-penetrating pepducins inhibits Akt survival pathways in breast cancer cells and suppresses tumor survival and metastasis. *Cancer Res* 2009;**69**:6223–31.
197. Cisowski J, O'Callaghan K, Kuliopulos A, Yang J, Nguyen N, Deng Q, et al. Targeting protease-activated receptor-1 with cell-penetrating pepducins in lung cancer. *Am J Pathol* 2011;**179**:513–23.

198. Jikuhara A, Yoshii M, Iwagaki H, Mori S, Nishibori M, Tanaka N. MAP kinase-mediated proliferation of DLD-1 carcinoma by the stimulation of protease-activated receptor 2. *Life Sci* 2003;**73**:2817–29.
199. Kaufmann R, Oettel C, Horn A, Halbhuber KJ, Eitner A, Krieg R, et al. Met receptor tyrosine kinase transactivation is involved in proteinase-activated receptor-2-mediated hepatocellular carcinoma cell invasion. *Carcinogenesis* 2009;**30**:1487–96.
200. Shimamoto R, Sawada T, Uchima Y, Inoue M, Kimura K, Yamashita Y, et al. A role for protease-activated receptor-2 in pancreatic cancer cell proliferation. *Int J Oncol* 2004;**24**:1401–6.
201. Waugh DJ, Wilson C. The interleukin-8 pathway in cancer. *Clin Cancer Res* 2008;**14**: 6735–41.
202. Coussens LM, Werb Z. Inflammation and cancer. *Nature* 2002;**420**:860–7.
203. Matsuo Y, Ochi N, Sawai H, Yasuda A, Takahashi H, Funahashi H, et al. CXCL8/IL-8 and CXCL12/SDF-1alpha co-operatively promote invasiveness and angiogenesis in pancreatic cancer. *Int J Cancer* 2009;**124**:853–61.

CHAPTER FIVE

GPCRs in Stem Cell Function

Van A. Doze*, Dianne M. Perez†
*Department of Pharmacology, Physiology and Therapeutics, School of Medicine and Health Sciences, University of North Dakota, Grand Forks, North Dakota, USA
†Department of Molecular Cardiology, Lerner Research Institute, The Cleveland Clinic Foundation, Cleveland, Ohio, USA

Contents

Abstract

Many tissues of the body cannot only repair themselves, but also self-renew, a property mainly due to stem cells and the various mechanisms that regulate their behavior. Stem cell biology is a relatively new field. While advances are slowly being realized, stem cells possess huge potential to ameliorate disease and counteract the aging process, causing its speculation as the next panacea. Amidst public pressure to advance rapidly to clinical

Progress in Molecular Biology and Translational Science, Volume 115
ISSN 1877-1173
http://dx.doi.org/10.1016/B978-0-12-394587-7.00005-1

trials, there is a need to understand the biology of stem cells and to support basic research programs. Without a proper comprehension of how cells and tissues are maintained during the adult life span, clinical trials are bound to fail. This review will cover the basic biology of stem cells, the various types of stem cells, their potential function, and the advantages and disadvantages to their use in medicine. We will next cover the role of G protein-coupled receptors in the regulation of stem cells and their potential in future clinical applications.

ABBREVIATIONS

2-AG 2-arachidonoylglycerol
AEA *N*-arachidonoylethanolamide
AKT protein kinase B
AR adrenergic receptor
ATP adenosine triphosphate
CB cannabinoid
CD cluster of differentiation
CXCR C-X-C chemokine receptor
cysLT cysteinyl-leukotriene
DAMGO [D-Ala^2,N-Me-Phe^4,Gly^5-ol]-enkephalin
DCX doublecortin
Dlx2 distal-less homeobox gene
DOR delta-opioid receptor
ESC embryonic stem cell
ET endothelin
FZD frizzled
GABA gamma-aminobutyric acid
GFAP glial fibrillary acidic protein
GPCR G protein-coupled receptor
hESC human embryonic stem cell
iPS induced pluripotent stem
KO knockout
KOR kappa-opioid receptor
LPA lysophosphatidic acid
MHC major histocompatibility complex
MOR mu-opioid receptors
MSC mesenchymal stem cell
MT melatonin
NCL neuronal ceroid lipofuscinosis
NPY neuropeptide Y
NSC neural stem cell
OPC oligodendrocyte progenitor cell
PAR proteinase-activated receptor
PI3K phosphoinositide-3-kinase

PKC protein kinase C
RMS rostral migratory stream
S1P sphingosine-1-phosphate
SDF-1 stromal derived factor 1
SFRP secreted frizzled-related protein
SGZ subgranular zone
SULT sulfotransferase
SVZ subventricular zone
TAP transient amplifying progenitor
TM transmembrane
VIP vasoactive intestinal polypeptide

1. INTRODUCTION

There are many organs or tissues in the body that are capable of self-renewal, such as the skin, liver, and epithelium. While the lifespan of fully differentiated cells is relatively short, these organs and tissues must contain some resident cells that can self-replicate and produce daughter cells capable of differentiation into the mature cells of that tissue. These multipotent cells are called adult stem cells that differ from embryonic stem cells (ESCs), which have an even greater potential for differentiation. However, ESCs cannot generate a functional organism.[1,2]

There are also several different types of medical problems that could benefit from the use of stem cells to ameliorate or cure the defect. Type 1 diabetes mellitus in which the beta cells of the pancreas are destroyed or attacked can benefit from stem cells to repopulate the pancreas. There are various types of neurological diseases or injuries such as stroke, spinal cord injuries, Parkinson's disease, or multiple sclerosis where particular types of neurons or glia have the potential to be replaced or regenerated. However, a major problem that has slowed their use in clinical settings is the immunological problems of rejection, mutations due to amplification of stem cells, and aneuploidy.

G protein-coupled receptors (GPCRs) are a large class of transmembrane (TM) receptors that transmit extracellular signals into cells by coupling to guanine nucleotide-binding proteins (G proteins). They play a key role in many complex biological processes and development, and our understanding of their role in stem cell regulation is expanding (Table 5.1). A potential benefit of using our own endogenous receptors to activate the stem cell process or to direct the differentiation of cells would avoid many of the current pitfalls in using stem cells, such as rapid *in vitro* differentiation and expansion or rejection in clinical applications.

GPCRs bind and regulate the effects of 80% of all hormones in the body, comprise 3–5% of the human genome, and account for about 20–50% of pharmaceuticals in the current market.[114] While the receptors are conserved

Table 5.1 GPCRs and roles in stem cell regulation

Type of stem cell	Type of GPCR	Function
Embryonic stem cells	PAR1	Embryonic development[3]
	GPR125	Germ-line progenitor marker[4]
	Gi-coupled	Maintains pluripotency[5]
	S1P, LPA	Directs differentiation;[6,7] ESC maintenance[8,9]
	CB_{1-2}	ESC survival[10]
	WNT	ES self-renewal;[11] hES proliferation;[12] differentiation;[13] maintains pluripotency[14]
	mGlu4,5	ES self-renewal;[15,16] embryoid body diifferentiation;[17] GABAergic differentiation[18]
Induced Pluripotent Stem cells	Gi-coupled	PTX affected stem cell colony retraction in reprogramming[19]
	Gs-coupled	Gs activation with cholera toxin does not affect stem cell morphology[20]
Hematopoietic stem cells	Gs	KO mice do not home or engraft in bone marrow[21]
	SDF-1 (CXCL12)	Regulates homing function[22,23]
	CXCR4	Modulates transendothelial migration;[24] migration, survival, development of multiple cell types[25]
	S1P	With LPA synergistically enhances homing function of SDF-1[26–29]
	WNT	Self-renewal and growth[30,31]

Table 5.1 GPCRs and roles in stem cell regulation—cont'd

Type of stem cell	Type of GPCR	Function
Mesenchymal/ stromal stem cells	S1P, LPA	Inhibit human MSC migration[32] but increasing murine MSC migration;[33,34] differentiates MSC to smooth muscle[35]
	S1P	Antiapoptotic[36]
	LPA	Stimulates osteoblast precursor migration[37]
Cardiac stem cells	WNT	Inhibition required for differentiation[38–41]
	WNT-3A	Migration of progenitor cells via RhoA[42]
	Agtrl1b (Apelin)	Required for heart field formation;[43] required for progenitor cell formation[43]
Neural stem cells	α_1-Adrenergic	Increase proliferation of embryonic neuroepithelial cells;[44] stimulate proliferation, migration, and protection of mouse embryonic neural progenitors;[45–47] increases both mouse neurogenesis and gliogenesis *in vitro* and *in vivo* that is developmentally dependent[48]
	α_2-Adrenergic	Decreased proliferation of hippocampal progenitors;[49] enhanced progenitor survival with antagonist dexefaroxan[50]
	CB	Proliferates embryonic and adult NSCs;[51–53] CB1 involved in migration of neural progenitors;[54] neuronal survival[55,56]
	CXCR4	Proliferation and migration of hippocampal granule progenitors;[57,58] differentiation and migration of cerebellar granule progenitors;[59–63] adult neurogenesis[64–66]
	Dopamine	Adult neurogenesis;[67] SVZ precursor proliferation[68–71]
	Endothelin	Inhibits cortical progenitor cell migration;[72] regulates gliogenesis;[73] OPC differentiation[73]
	LPA	Proliferates and differentiates rat NSCs to cholinergic neurons;[74,75] in hESCs, inhibits neuronal but promotes astrocyte differentiation;[76] both S1P and LPA iInhibit oligodendrocyte maturation[77,78]
	S1P	Inhibits migration of OPCs in rodents[79]

Continued

Table 5.1 GPCRs and roles in stem cell regulation—cont'd

Type of stem cell	Type of GPCR	Function
	Opioids	DOR promotes neuronal differentiation in embryonic NSC;[80] MOR and KOR promote proliferation and differentiation into neurons in ESCs from mouse blastocyst;[81] MOR- and KOR-induced oligodendrocyte progenitor differentiation;[82] decrease adult SGZ proliferation, maturation, and survival[83–87]
Neural stem cells (continued)	VIP	Promotes neuronal differentiation of embryonic hippocampal progenitors in culture;[88] KO mice show reduced progenitor survival and SGZ neurogenesis[89]
	NPY	Increases proliferation in SVZ;[90] NPY1 regulates neuroproliferation and differentiation in SVZ and SGZ through ERK[91–95]
	P2Y	P2Y1 associated with developmental neurogenesis;[96] ATP and P2 proliferate v-myc immortalized neural progenitors;[97] proliferate embryonic neural progenitors and P2Y inhibition-permitted differentiation into neurons and glia[98]
	WNT	Regulates adult hippocampal neurogenesis via proliferation and differentiation[99–102]
Cancer stem cells	Purinergic	Regulates proliferation of glioma cell growth via CSC subpopulation;[103,104] alters size of gliomas in rats and human cultures;[105,106] ATP reduces number of tumor spheres and expression of cancer cell markers[106]
	Second messengers	AKT/PI3K reduces glial tumor formation;[107] glioma stem cells sensitive to AKT signaling than nonstem glioma cells[108]
	Melatonin	MT1 suppresses breast cancer cell proliferation;[109,110] MT1 cancer suppression via G_{i2} signaling;[111] increases expression of SULT1, which protects against excessive estrogen actions[110]
	CXCR4	Suggested to promote metastasis to organs that secrete SDF-1[112,113]

mostly in the TM domains, the ligands span a large range of diverse biology, from peptides to waves of light, and represent various evolutionary classes.[115] They are composed of a single polypeptide containing seven regions of 20–28 hydrophobic amino acids that span the TM domains. The TM segments are α-helices oriented approximately perpendicular to the cell surface as shown in the crystal structure of rhodopsin.[116] The amino terminus is located on the extracellular side of the membrane and contains several glycosylation sites. The carboxy terminus is located on the intracellular side and contains sites for phosphorylation, which are used in the regulation of receptor desensitization and internalization. Three intracellular and three extracellular loops link the TM domains and may contain ligand-binding sites. Many GPCRs also have a highly conserved disulfide bond between the cysteines in the second and third extracellular loops. This bond is needed for proper folding of the protein and the regulation of the high-affinity binding site.[117]

The GPCRs bind a ligand on the extracellular side and trigger conformational changes, causing the intracellular loops to bind and activate the heterotrimeric G protein. Once the G protein is activated, it dissociates from the receptor and its subunits (α and βγ) to amplify a second messenger response mediated by effector molecules such as phospholipases, enzymes, or channels. GPCRs may also signal through non-G protein-mediated events that involve scaffolding proteins, transactivation of tyrosine kinase receptors, and/or G protein receptor kinases that regulate the GPCR signal (reviewed in Ref. [118]).

2. TYPES OF STEM CELLS

2.1. Embryonic stem cells

ESCs are pluripotent cells derived from the blastocyst of an early-stage embryo, generally about 4–5 days after fertilization. Their isolation in 1981 by Sir Martin J. Evans and Matthew Kaufman and independently by Gail R. Martin has contributed extensively to our knowledge of pluripotency and differentiation of stem cells.[119,120] In 1998, James Thomson reported the first successful cultivation of human embryonic stem cell (hESC) lines.[121] Isolation of these cells results in the destruction of the embryo, which is the subject of intense ethical debate. The term pluripotent indicates that the stem cell is capable of differentiation into cells from the ectoderm, endoderm, and mesoderm, that is, the three germinal layers in the body. Pluripotency results in the potential to culture approximately 220 different cell types present in the body. Feeder cells provide factors that are necessary to prevent the ESC from differentiating in culture, such as leukemic inhibitory factor and bone morphogenetic proteins.

While the potential to generate unlimited cell types is ground breaking, disadvantages of ESC therapy include the potential to form tumors upon transplantation (called teratomas). There is also the problem of immune-compatibility when the ESCs are derived from a different genetic background than the patient who receives them. While immunosuppressive drugs can be used to minimize these problems, they are not perfect. Thus, the field of induced pluripotent stem (iPS) cells has developed to solve immune-compatibility, where adult cells from the same person can be forced to switch their lineage or become pluripotent through nuclear reprogramming.

2.2. Induced pluripotent stem cells

For human treatment, there is a need for patient-specific or patient-matched ESCs. These cells can then be grafted into the genetically identical host. Many groups use iPS cells in their research in addition to hESCs, as this research faces fewer ethical issues. iPS cells are generated from adult (somatic) tissue; furthermore, they are easier to use in research.[122,123]

Somatic cells can be reprogrammed to a pluripotent state through a variety of methods. Originally, pluripotent cells were created by injecting the nucleus of an adult cell into an enucleated oocyte.[124,125] This leads to reprogramming of the somatic cell nucleus by the host cell. After several cell divisions, the reprogrammed cell forms a blastocyst, which is now genetically matched with the nuclear donor and can be used to obtain stem cells. The cell is not genetically identical to the donor because of the mitochondrial DNA, which is genetically identical to the recipient and present in the oocyte cytoplasm. This process of forming iPS cells, also known as somatic cell nuclear transfer, can achieve the greatest level of induced pluripotency, but is an inefficient process and potentially damaging to the cell. Instead of using oocytes, ESCs can be used to reprogram somatic cells, but this method is even more inefficient and leads to higher rates of tetraploidy.[126]

The iPS cell field advanced in 2006, when Takahashi and Yamanaka found that retroviral transduction of only four genes was sufficient to induce pluripotency in somatic cells.[122] A combination of four regulatory proteins (Oct3/4, Sox2, Myc, and Klf4) can convert fibroblasts into ES-like cells with the ability to differentiate into diverse cell types, although the rate of conversion is still very low. The goal of patient-specific iPS cell therapy is to isolate somatic cells from diseased patients, reprogram and differentiate them to replace diseased cells, and successfully transplant them back into the patient without immune problems.[127–129] Since retroviral delivery integrates the vector into the cell's genome, there is still a potential to induce tumorigenesis with this method.

Some of the more successful studies demonstrating the potential of iPS cells to treat disorders include rescue of sickle cell anemia in mice with a genetically corrected β-globin locus[130] and in a patient whose skin fibroblasts were used to generate patient-specific iPS cells.[131] Parkinson's disease was improved in a rat model using iPS cells[132] and injection of undifferentiated iPS cells into ischemic cardiac muscle-induced repair.[133] In humans, patient-specific iPS cells from Fanconi anemia patients corrected the genetic defect.[134] Of course, the major failures of iPS cell therapy in clinical applications or animal models are not outlined here but serve as a cautionary note that more studies need to be performed before the use of iPS cells becomes accepted as a major treatment option.

2.3. Adult stem cells

In addition to iPS cells, adult stem cells can be used without ethical controversy. Furthermore, adult stem cells are readily available from several organs and tissues. These cells can be isolated from the patient and differentiated *in vitro* or differentiated *in vivo* through pharmacological agents, thus avoiding immune system incompatibility problems. However, the differentiation potential of adult stem cells is limited to cell types within the boundaries of its endogenous tissue. For example, neural stem cells (NSCs) cannot be induced to form cardiomyocytes or vice versa. The following adult stem cell categories are discussed separately.

2.3.1 Hematopoietic stem cells

The most extensively studied adult stem cell system is that of hematopoietic stem cells in the bone marrow and in the umbilical cord blood.[135] Hematopoietic stem cells differentiate into only myeloid and lymphoid lineages, but there are controversial studies suggesting that they can differentiate into nonhematopoietic lineages.[136,137] Transplantation studies to reverse disease were first performed using hematopoietic stem cells and are the most widely available stem cell therapy,[138,139] with allogenic hematopoietic stem cell transplantation a common treatment for bone marrow failure. Hematopoietic stem cell transplantation has been successful because they do not need to be expanded prior to transplantation and no complex structures or organs need to be regenerated. Patients with leukemia, for example, can be irradiated or treated with chemotherapy to destroy the cancerous cells, as well as the rest of their hematopoietic cells. The patient is then transfused with healthy, noncancerous hematopoietic stem cells that repopulate the bone marrow.[140] Of course, this type of treatment can produce immune rejection, but with careful tissue matching and the use of immunosuppressive drugs, the difficulties can be reduced to acceptable levels.

When the cancer is due to a mutation in only a particular type of blood cell, it is possible to treat patients with their own hematopoietic stem cells. In this case, a sample of the bone marrow is isolated before irradiation and sorted to obtain a pure population of cancer-free hematopoietic stem cells that are then transfused back into the patient. However, this autologous technique is not without problems, as patients may develop engraftment syndrome, typically characterized by fever, rash, and pulmonary edema.[141]

Umbilical cord blood was first used as a source of hematopoietic stem cell transplantation in 1989.[142] Since then, umbilical cord blood has increasingly become a popular alternative to bone marrow, especially in children with hematological malignancies[143] as they require less blood. An alternative source of hematopoietic stem cells is the placenta, which has up to 10 times the number of stem cells as umbilical cord blood.[144]

2.3.2 Mesenchymal or stromal stem cells

Mesenchymal stem cells (MSCs) are capable of differentiating into various mesodermal lineages such as adipocytes, fibroblasts, osteoblasts, neuronal cells, and myocytes.[145] Technically, they are stromal-like cells characterized by adherence in culture and a specific set of cluster of differentiation (CD) cell-surface antigens such as CD105+/CD73+/CD90+/CD45−/CD34−/CD11b− or CD14−/CD19− or CD79α−/HLA-DR1−.[146] If the above conditions are not met completely, the term "mesenchymal stromal cells" or "MSC-like" is used.[147] MSCs have the advantage of possessing antimicrobial activity[148] and can produce trophic factors that promote anti-inflammatory effects and healing.[145] Furthermore, MSCs lack major histocompatibility complex (MHC)-II molecules and show only minimal MHC-I expression that renders them allogeneic and immunosuppressive.[149–151] MSCs have the extraordinary ability to migrate toward injuries,[152] and it is postulated that the MSC microenvironment involves similar mechanisms with wound healing.[153] While MSCs reside in many different organs and tissues, they are the most abundant in the bone marrow and umbilical cord blood, but are also found in the amniotic fluid and placenta, contributing to their ease of collection and storage. Since their ability to self-renew has only recently been shown, it is debatable whether MSCs are true stem cells or a pool of partially differentiated progenitors.[154,155]

2.3.3 Cardiac stem cells

There are various stem cell populations that have been proposed to repair the damaged heart. MSCs derived from the bone marrow can home into the heart and establish a temporary niche to aid in the regeneration of myocardium.[156–158] Bone marrow-derived MSCs also have been shown to improve cardiac performance after injury.[159,160]

However, there is considerable evidence for the existence of an actual population of residing cardiac adult stem cells. Substantial heart muscle regeneration can take place after resectioning of the heart,[161,162] which triggers myocyte dedifferentiation and proliferation.[163] Using various surface markers and transcription factors, most cardiac stem cells have been described to be Sca-1 and c-kit positive and to efflux Hoechst 33342 dye, although no marker is specific to any kind of stem cell.[163–165] A population of residing cardiac stem cells which are negative for CD31, an antigen common to bone marrow cells, has been described that can form beating myocytes and acquire an adult phenotype when associated with differentiated myocytes.[166] Cardiac stem cells reside in clusters within niches in the heart that provide a microenvironment to maintain stemness.[167] Niches are normally located in areas with low wall stress (low blood flow) and hemodynamic load.

When cultured, cardiac stem cells can form free-floating aggregates called cardiospheres. Cardiospheres are heterogeneous cells that express c-kit in a fraction of the cells near the center of the sphere, as well as CD105 in the outer layers (bone marrow meschenymal marker).[146] Connexin 43 is expressed within the highly dividing cells and differentiating cells of the sphere.[168] Cells from these cardiospheres can be differentiated into myocytes when cocultured with neonatal myocytes.[169]

2.3.4 Neural stem cells

Along with heart tissue, the brain and spinal cord have little capacity to self-repair. However, in certain parts of the brain, stem cells continually produce new neurons and glia, although this occurs at naturally low rates in humans. A classic example of adult neurogenesis is the rapid turnover that occurs in songbirds as they learn a new song each breeding season.[170] Evidence for

NSCs was first shown by the development of floating balls in culture called neurospheres from dissociated adult mammalian brain cells. These cells can differentiate into neurons and glia when plated onto adherent surfaces with the appropriate serum and/or growth factors.[171] Depending upon the culture conditions and factors added to the media, NSCs can be maintained in the stem cell state or induced into neurons, glia, or a mixture of the two.

NSCs are defined by their ability to self-renew and are multipotent and able to differentiate into three major cell types in the CNS: neurons, astrocytes, and oligodendrocytes (reviewed in Ref. [172]). The particular cell type of the resident adult NSC that can generate new neurons and glia is still debated and includes multiciliated ependymal cells,[173] subependymal cells,[174] or a special type of astrocytic cell termed radial glia.[175] The astrocyte radial glia theory is the most popular, in which NSCs in the subventricular zone (SVZ) represent a subset of slowly dividing glial fibrillary acidic protein (GFAP)+ astrocytic cells with radial projections. These radial glia can proliferate to generate rapidly dividing cells called transient amplifying progenitor (TAP) cells. TAP cells are identified by being $GFAP^-$ but $Dlx2^+$. The distal-less homeobox gene (Dlx2) is a transcription factor that regulates the fate determination of interneurons and oligodendrocytes in embryogenesis but is proneural in adult neurogenesis.[176] TAP cells divide to generate migrating neuroblasts that are distinguished by the neuronal marker polysialylated neural cell adhesion molecule (PSA-NCAM).[177–179]

The SVZ and the subgranular zone (SGZ) of the hippocampus are the two major regions in the adult CNS capable of active adult neurogenesis. The neurons in the SVZ are generated continuously and migrate to the olfactory bulb where they mature into at least five different types of interneurons as well as a minor population of oligodendrocytes.[180] In a similar but distinct manner, residing and quiescent NSCs in the SGZ are a population of radial astrocytic-like cells containing GFAP, Sox2, and Nestin markers.[181] These stem cell bodies are located at the SGZ with their radial processes projecting through the granular cell layer and shorter tangential processes extending along the border of the hilus and granule cell layer.[182] Similar to TAP cells, NSCs in the SGZ generate actively dividing nonradial transient amplifying progenitors. These cells generate neuroblasts that are doublecortin $(DCX)^+$ and migrate into the dentate granule cell layer where they mature,[183] a process that takes about 28 days.[184] New granule neurons in the dentate gyrus are continuously generated locally[182] and may survive for at least 2 years.[185] Adult neurogenesis has also been reported in the neocortex and striatum, but at much lower rates.[186–188]

NSCs, derived from fetal tissue, are currently the only practical source for human neural cell transplantation in their ability to be grafted into the adult brain and integrate to match their new location.[189] The initial preclinical studies using transplanted exogenously generated brain cells were performed in a Parkinson's disease animal model in the early 1970s[190] and then in very limited human clinical trials in the early 1990s[191,192] with reasonable success. A larger-scale trial of 40 Parkinson's disease patients in 2001 found that transplanted dopamine neurons from embryonic fetal tissue survive in the human brain, but only patients under the age of 60 showed clinical improvement over a three-year time period.[193] While the potential to form tumors is reduced in NSC transplantation compared to ESCs in animal models, the risk may not be so limited in humans and there is a need for caution.[194] Since then, NSCs have been used in successful transplantation studies on animals to repair the spinal cord,[195,196] and to treat multiple sclerosis,[197,198] Huntington's disease,[199] and stroke.[200–202]

The first FDA-approved clinical trial for purified NSCs was performed in 2009 to treat neuronal ceroid lipofuscinosis (NCL, Batten's disease) (http://www.stemcellsinc.com/clinicaltrials/clinicaltrials.html). Human NSCs were transplanted into six patients with either infantile or late infantile NCL. Patients were immunosuppressed for 12 months after transplantation. This phase I trial showed a favorable safety profile with transplanted cells, which exhibited survival long-term; however, patients are still being tracked for outcomes. Other phase I trials using purified NSCs are ongoing for Pelizaeus–Merzbacher disease (a myelination disorder) and for chronic spinal cord injury at the same company (StemCells, Inc.). Future areas of research may involve the utilization of endogenous mechanisms that activate adult neurogenesis after various brain injuries.[203]

2.3.5 Cancer stem cells

Cancer stem cells were first described in 1994 when a cell type of low abundance (0.1% of cells) derived from human acute myeloid leukemia caused cancer to develop in nude mice.[204–206] Today, cancer stem cells are derived from most types of hematopoietic cancers including various types of myeloid cancers[207–209] and acute lymphoblastic leukemia.[210] They are also found in solid tumors of the brain,[211,212] breast,[213,214] lung,[215,216] prostate,[217] and more recently in liver[218,219] and colon.[220,221] The presence of cancer stem cells has important implications for treatment, as current therapies may abolish the bulk of tumor cells but miss the cancer stem cells leading to tumor reformation. Therefore, future treatments specifically targeting cancer stem cells may be more successful.

The origin of cancer stem cells is still being debated, particularly whether they represent a stromal stem cell, which has undergone some sort of malignant change, or whether they are differentiated cells that return to a "stem" state as part of the malignant transformation process.[222–224]

Each cancer stem cell has unique phenotypic features and niche environment that may prove to complicate future targeted therapies.[207,208,225–227] However, they do share some common cell-surface markers such as $CD34^{+}$, $CD38^{-}$, and Lin^{-}.[204,206,208,210] Cancer and neural stem cells share the markers CD133, CD15/Lex/Ssea-1, and CD34.[211,228] Tumor-suppressor genes are commonly lacking in many cancer stem cells that may provide their ability to self-renew.[229–231]

3. GPCRs IN EMBRYONIC STEM CELLS

Since GPCRs play a major role in development,[232] it stands to reason that they may also be a major regulator of stem cells (Table 5.1). Signal transduction pathways that are activated during stem cell renewal and differentiation are shared, cross-activated, or synergistic with GPCR stimulation (reviewed in Ref. [233]). The first study to define a major role for GPCRs in stem cell function performed a real-time PCR-based expression profile of 343 different GPCRs in mouse ESCs.[234] Of these, 161 GPCRs were expressed in undifferentiated ESC at a low level, 30 GPCRs were moderately expressed, and 7 were highly expressed. Many GPCRs display a dramatic difference in expression when the cells differentiate.[234]

3.1. Proteinase-activated receptors

Using a GPCR-specific real-time PCR microarray as described in the above studies, two GPCRs from the proteinase-activated receptor family (PAR), F2r (PAR1) and F2rl1, were highly expressed in ESCs[234] and shown in knockout (KO) mice studies to play an important role in early embryonic development.[3] Another important GPCR differentially expressed in ESCs was GPR125, an orphan adhesion-type GPCR[234] that was identified as a marker of germ-line progenitors.[4]

3.2. G_i-coupled receptors

The role of G_i-coupled signaling in pluripotent stem cells is largely unknown, but it has been implicated in the maintenance of pluripotency[5] and directed differentiation[235] of hESCs. The G_s-coupled pathway and the generation of cAMP have been suggested to regulate ESC self-renewal[236] and maintenance of ESC pluripotency.[234]

The secreted factors produced by ESCs may also provide insight into the production of therapeutics for the regeneration and repair of damaged tissues. Direct activation of endogenous progenitor cells as pluripotent hESCs can secrete paracrine signals associated with embryogenesis. Conditioned media from these stem cells was found to induce karyokinesis, cytokinesis, and proliferation in mono- and binucleated cardiomyocytes through the RhoA/Rho-associated kinases (ROCK) 1 and 2 pathways.[237] While RhoA is a small G protein, it serves as an effector of numerous GPCRs such as thrombin, sphingosine-1-phopshate (S1P), and lysophosphatidic acid (LPA)[238] to regulate proliferation, migration, and survival.

3.3. Sphingosine-1-phosphate and lysophosphatidic acid receptors

Particular lipids acting through the binding and activation of GPCRs have been shown to regulate various aspects of ESC biology. Such lipids include the phospholipids S1P and LPA. Both LPA and S1P are positive regulators of ESC maintenance through either ERK or Ca^{+2}-mediated pathways.[8,9] Mouse ESCs express receptor subtypes $LPAR_{1\text{-}3,5}$ and $S1PR_{1\text{-}5}$[5,9] while hESCs express $LPAR_{1\text{-}5}$ and $S1PR_{1\text{-}3}$.[76,239,240] The LPA signaling pathway that effects mouse ES maintenance is the ERK activation of *c-fos*.[241] S1P activation of mouse ESC also signals through ERK, likely mediated by $S1PR_5$[8] utilizing G_i-, PKC-, and c-Src-dependent mechanisms. While the ERK signaling pathway is implicated in mouse ESC proliferation, there is also evidence for a role in its differentiation by the suppression of ERK signals.[6,7] The signaling pathway in hESCs is similar to mouse ESCs and involves G_i- and ERK-dependent mechanisms[5] resulting in the inhibition of apoptosis and activation of prosurvival signals.

3.4. Cannabinoid receptors

The cannabinoid receptors CB_1 and CB_2 are activated by three major groups of ligands: endocannabinoids, plant cannabinoids, and synthetic cannabinoids. Mammalian endocannabinoids include four known compounds anandamide (AEA), anandamide derivatives (2-arachidonoylglycerol (2-AG) and noladin ether), virodhamine, and *N*-arachidonoyl-dopamine.[242] CB_1 and CB_2 are likely involved in mouse ESC survival since specific receptor antagonists AM251 and AM630 promoted cell death.[10] While mouse

ESC express low levels of CB_1 and CB_2, they express high levels of 2-AG and moderate levels of AEA agonists.

3.5. Wnt/frizzled

Wnt signaling has been implicated in ESC self-renewal[11]; however, its exact role in ESC biology remains to be clarified. Wnt proteins bind to receptors of the atypical GPCR frizzled and low-density-lysoprotein-related-protein5/6 (LRP) families on the cell surface. Wnt binding to its receptors induces the nuclear translocation of β-catenin that acts as a transcriptional cofactor of lymphoid enhancer binding factor1 to modify gene transcription (referred to as the "canonical" Wnt pathway). Wnt can also signal in non-canonical pathways independent of β-catenin, leading to either small G protein activation and cytoskeletal changes or through Ca^{2+} signaling with activation of heterotrimeric G proteins, causing various cell responses.[243] hESCs have been shown to express members of the frizzled (FZD) receptor family (FZD1, FZD3, FZD4, FZD5, FZD6) and the secreted frizzled-related protein (SFRP) family (SFRP1, SFRP2, FRZB, SFRP4), encoding soluble Wnt antagonists.[12] Adding recombinant Wnt3a stimulated hESC proliferation, but after 4–5 days of treatment, hESCs that survived were undifferentiated and few could self-renew. β-catenin-mediated transcriptional activation in the canonical Wnt pathway was minimal in undifferentiated hESCs but was greatly upregulated during differentiation induced by Wnt.[13] Thus, canonical Wnt/β-catenin activation does not suffice to maintain the undifferentiated and pluripotent state of hES cells. In contrast, some studies suggest that Wnt3a or the GSK-3 inhibitor, 6-bromoindirubin-3′-oxime, maintains hESCs in the undifferentiated state and maintains pluripotency.[14] The Wnt receptor FZD7 has also been identified as important for hESC maintenance and self-renewal. FZD7 mRNA levels have been found to be 200-fold higher in hESCs compared to differentiated cell types and knockdown of FZD7 induces dramatic changes in the morphology of ESC colonies with a loss of OCT-4 expression, an ESC-specific transcription factor.[244]

3.6. mGlu metabotrophic

The mGlu5 metabotrophic GPCR supports self-renewal in mouse ES cells.[15,16] Different mGlu receptor subtypes, mGlu3 and mGlu5 receptors in particular, are found in neural ESCs.[17] Differentiation of ESCs into embryoid bodies is associated with the induction of mGlu4 receptors, the activation of which drives cell differentiation toward the mesoderm and endoderm lineages.[17] Mouse ESCs expressed mGlu5 receptors throughout the entire differentiation period into neurons. Antagonists of mGlu5 receptors accelerated the appearance of the neuronal marker, β-tubulin, and increased the number of cells expressing glutamate decarboxylase-65/67, a marker of GABAergic neurons.[18]

Other GPCRs are also found in ESCs but there are few studies to indicate their functional role. CXCR4 is expressed in mouse ES cells[245] and endothelin receptors in hESCs.[246,247]

4. GPCRs IN INDUCED PLURIPOTENT STEM CELLS

4.1. G_i-coupled receptors

Reprogramming adult somatic cells to create induced pluripotent stem cells requires dramatic morphological and organizational cell changes. Interestingly, human pluripotent colonies form a flat uniform monolayer, while mouse pluripotent colonies form thicker, multilayered colonies.[19] Since pluripotent colony morphology correlates closely with the maintenance of pluripotency, the mechanisms by which these colonies form and organize may be important for controlling somatic cell reprogramming. It was found that inhibition of G_i-coupled receptor signaling via pertussis toxin retracted stem cell colonies inward, adopting a dense multilayered conformation without affecting proliferation, survival or its pluripotency. Activation of G_s receptor-coupled signaling with cholera toxin did not affect the colony morphology.[20] These results suggest that GPCRs play a role in some aspects of somatic cell reprogramming.

5. GPCRs IN ADULT STEM CELLS

5.1. GPCRs in hematopoietic stem cells

5.1.1 G_s-coupled receptors

Hematopoietic stem and progenitor cells change location during development and circulate in mammals throughout life, moving into and out of the bloodstream to engage bone marrow niches in sequential steps of homing, engraftment, and retention. Hematopoietic stem cell engraftment of bone marrow in fetal development is dependent upon the activity of Gsα.[21] Hematopoietic stem cells from adult Gsα KO mice do not home or engraft in the bone marrow despite their ability to differentiate and undergo chemotaxis, suggesting that pharmacological targeting of this pathway may lead to improved transplantation efficiency.[21]

5.1.2 Chemokine receptors

Chemokines are small proinflammatory chemoattractant cytokines that bind to specific GPCRs present on plasma membranes of target cells and are major regulators of cell trafficking. Stromal-derived factor-1 (SDF-1 or CXCL12), an alpha-chemokine that binds to CXCR4, plays an important and unique role in the mobilization and homing of hematopoietic stem cells.[22,23] CXCR4 modulates transendothelial migration of hematopoietic stem cells and progenitors into the blood stream.[24] Drugs that block CXCR4 cause hematopoietic stem cells to move into the circulation. SDF-1 and its receptor CXCR4 are involved in regulation of migration, survival, and development of multiple cell types, including human hematopoietic CD34+/CD38−/low and stromal STRO-1+ stem cells.[25] The homing function of SDF-1 that primes or sensitizes cell chemotaxis to a concentration gradient is dependent on the cholesterol content in the membrane and the incorporation of both CXCR4 and Rac-1 in lipid rafts facilitated by GTP binding.[248]

5.1.3 Sphingosine-1-phosphate and lysophosphatidic acid receptors

In addition, other cytokines such as granulocyte colony-stimulating factor act partially through the modulation of SDF-1/CXCR4 signaling to induce hematopoietic stem cell mobilization in the bone.[23] SDF-1 can also cross talk with other GPCR signaling pathways, including S1P and LPA. For instance, S1P-activation of $S1PR_3$ leads to the transactivation of CXCR4 in human endothelial progenitor cells to stimulate their functional capacity.[26] Both S1P and LPA also synergistically enhance the chemotactic migratory response of the hematopoietic stem cells to SDF-1.[27–29]

5.1.4 Wnt/frizzled

The Wnt pathway has also been implicated as a self-renewal and growth signal in hematopoietic stem cells.[30,31] Purified Wnt3a protein induced self-renewal of hematopoietic stem cells.[31] Overexpression of activated β-catenin expands the pool of hematopoietic stem cells in long-term cultures by both phenotype and function. Ectopic expression of axin or a frizzled ligand-binding domain, which are inhibitors of the Wnt signaling, leads to inhibition of hematopoietic stem cell growth *in vitro* and *in vivo*. In addition, activation of Wnt signaling induces increased expression of HoxB4 and Notch1, genes previously implicated in self-renewal of hematopoietic stem cells.[30]

5.1.5 *Cysteinyl-leukotriene receptor*

Hematopoietic stem cells express other GPCRs such as the leukotriene D4 receptor, cysteinyl-leukotriene (cysLT) receptor cysLT1,[249] and lysophospholipid receptors such as $S1PR_1$ that differentially regulate chemotaxis, adhesion, and proliferation.[250] The coactivation of both G_q and G_i by cysLT1 results in stronger proliferation of hematopoietic stem cells than stimulation of G_i by SDF-1 or S1P alone.[250]

5.2. GPCRs in mesenchymal stem cells

5.2.1 *Sphingosine-1-phosphate and lysophosphotidic acid receptors*

Both LPA and S1P play a role in mesenchymal/stromal stem cell recruitment and differentiation. Both GPCR systems inhibit human MSC migration; an effect linked to Rho activation.[32] However, they stimulate murine MSC migration through Rho/ERK signals.[33,34] LPA exerts potent antiapoptotic effects on MSCs through $LPAR_1$ and the activation of G_i/ERK/PI3K/Akt signaling pathway.[36] S1P1 induces differentiation of MSC to smooth muscle cells[35] with S1P1 receptor KO mice lacking vascular maturation caused by a deficiency of MSC-derived-vascular smooth muscle cells and pericytes,[251] causing them to be embryonically lethal. LPA also stimulates migration of osteoblast precursors through ERK phosphorylation[37] and either promotes or inhibits osteogenic differentiation of human MSCs by $LPAR_1$ and $LPAR_4$, respectively.[252]

5.2.2 *Prostaglandin receptor*

Prostaglandins are lipids belonging to the eicosanoid family and act through receptor isoforms EP_{1-4}. Studies have demonstrated the importance of prostaglandins on hematopoietic stem/progenitor cells and on MSCs.[253,254] For example, PGE_2 acts as an MSC survival factor through the binding of its receptor EP_4 and has antiapoptotic effect on MSCs that correlates with intracellular S1P synthesis.[255]

5.3. GPCRs in cardiac stem cells

Circulating mobilized bone marrow stem cells can be recruited to damaged myocardial tissue, and this homing behavior is suggested to play a key role for myocardial repair and regeneration.[256] However, there is increasing evidence that an endogenous, self-maintained pool of resident progenitor cells in the myocardium exists.[161,162,257,258]

5.3.1 *Wnt/frizzled*

Multiple signaling pathways, acting in both stimulatory and inhibitory fashions, can act to regulate cardiomyogenesis. These signals include Wnts and Wnt inhibitors.[38] During development, there is evidence that β-catenin-dependent Wnt signaling negatively regulates cardiac differentiation, and that the inhibition of the Wnt pathway is required for cardiac specification.[38–41] On the other hand, β-catenin-independent Wnt signaling is thought to enhance myocardial differentiation.[259] Cardiac progenitor cell migration is mediated via Wnt3A–RhoA activity.[42]

5.3.2 *Apelin receptor*

Grn encodes the GPCR Agtrl1b (APJ receptor). Agtrl1b with its ligand, apelin, is a known regulator of heart field formation.[43] Reduced or excess Agtrl1b or apelin function caused deficiency of cardiac precursors and, subsequently, the heart tissue.[43] In Apelin-deficiency, the cardiac precursors converged inefficiently to the heart fields and showed ectopic distribution, whereas cardiac precursors overexpressing Apelin exhibited abnormal morphology and rostral migration. Ectopic expression of apelin results in the complete absence of cardiomyocytes.[260] Zebrafish *grinch* (*grn*) mutants form a reduced number of myocardial progenitor cells, which results in a profound deficit in cardiomyocyte numbers in the most severe cases.[260]

5.4. GPCRs in neural stem cells

The role of GPCRs in regulating various aspects of adult neurogenesis and gliogenesis is extensive (reviewed in Ref. [261]). The types of GPCRs that regulate adult neurogenesis and/or gliogenesis include adrenergics, cannabinoids, chemokines (SD1/CXCR4), dopamine, LPA and sphingosine, opioids, PACAP, various peptide hormones, neuropeptide Y (NPY), purinergic, and various GPCR-coupled effectors (phospholipase C-β1, phospholipase A2, phosphodiesterase-4D). The major GPCR systems will be briefly summarized here.

5.4.1 *α_1-Adrenergic receptors*

Previous studies indicate that α_1-adrenergic receptors (ARs) increase proliferation of embryonic neuroepithelial cells, suggesting that this subtype might influence adult progenitor proliferation.[44] α_1-ARs stimulate proliferation,[45] migration,[46] and protect against stress-induced death of mouse embryonic brain-derived neural progenitor cells through a caspase 3/7-independent mechanism *in vitro*.[47] We showed that α_{1A}-AR subtype stimulation increased neurogenesis and gliogenesis in adult mice *in vivo* and may be developmentally dependent as embryonic and neonatal neuropheres differentiated into neurons while adult neurospheres differentiated into glia.[48]

5.4.2 α_2-Adrenergic receptors

In the adult, the α_2-AR agonists clonidine and guanabenz decreased proliferation of hippocampal progenitors with no effect on survival or differentiation.[49] Adult hippocampal progenitors *in vitro* expressed all the α_2-AR subtypes that decrease neurosphere frequency and BrdU incorporation, effects that are blocked with the α_2-AR antagonist yohimbine. In contrast, another study reported enhanced progenitor survival following long-duration treatment with the α_2-AR antagonist dexefaroxan.[50]

5.4.3 Cannabinoid receptors

The proliferation of embryonic and adult rodent NSCs is significantly increased by administration of endogenous anandamide (e.g., AEA) and synthetic agonists (e.g., HU210) for the CB_1 and/or CB_2 receptors, respectively.[51–53] Consistent with these results, CB_1 KO mice show impaired proliferation via BrdU incorporation studies of embryonic and adult neural stem/progenitors. A reverse effect is observed in fatty acid amide hydrolase (*Faah*) KO mice, which lack the ability to degrade endogenous AEA.[51–53,262,263] CB_1 KO mice also demonstrated cortical lamination that was significantly perturbed, suggesting that CB1 is involved in migration of neural progenitors.[54] Concomitantly, cortical neuronal migration is accelerated with HU 210, an agonist of CB_1 and CB_2 or URB 597, a fatty acid amide hydrolase inhibitor in a cortical slice culture.[54] PI3K/Akt phosphorlyation is also thought to be involved with neuronal survival functions of endocannabinoids[55,56] through inhibition of glutamate neurotoxicity.[56] Cannabinoid-induced adult neurogenesis mechanisms include opposing the antineurogenic effect of neuronal nitric oxide.[264]

5.4.4 Chemokine receptors

Developing granule cells of the cerebellum and hippocampus express SDF-1 and CXCR4,[265–268] and their CXCR4 signaling is suggested to be involved in the maintenance of the proliferative state and for their migration properties.[57,58] Consistent with this hypothesis, SDF-1 and CXCR4 KO mice show severe malformation in the cerebellum and hippocampus, owing to impaired proliferation and migration of granule progenitor cells.[59,60,265,266] SDF-1/CXCR4 regulates the cell cycle of early granule progenitor cells to help them survive in a quiescent, but proliferative-ready state.[61] Chemokines mediate progenitor proliferation through activation of Akt and concomitant inactivation of FOXO3, a fork head family of transcription factor regulating the cell cycle.[62] There is also evidence for involvement of Shp2 tyrosine phosphatase, a downstream signal of CXCR4, in migration and differentiation of cerebellar granule progenitors, as conditional KO of Shp2 displays cerebellar malformation and explants from these mice fail to migrate in response to SDF-1.[63]

CXCR4 is also suggested to be involved in adult neurogenesis.[64–66] There is a loss of mitotic, nestin+, and $(DCX)^+$ cells in the dentate gyrus when a CXCR4 antagonist, AMD3100, is infused.[66] SDF-1 can also modulate gamma-aminobutyric acid (GABA)-induced maturation of granule progenitor cells and is coreleased with GABA from interneurons onto immature neural progenitor cells.[65]

5.4.5 Dopamine receptors

Ciliary neurotrophic factor (CNTF) regulation of adult neurogenesis in mice is mediated by the D2 receptor.[67] Systemic treatment of either normal or dopamine-depleted rats with the D2-like agonists ropinirole or 7-OH-DPAT significantly increased precursor cell proliferation in the SVZ,[68–70] resulting in increased differentiation of neurons. Destruction of dopaminergic neurons in the substantia nigra and ventral tegmental areas reduced the number of proliferating neural precursors in the SVZ by 40%.[71]

5.4.6 Endothelin receptors

Migration of cortical progenitor cells is inhibited by ET-1 through ET_B signaling coupled to G_q and the JNK pathway.[72] Evidence also suggests a role of ET-1 signaling in gliogenesis,[73] as oligodendrocyte progenitors express ET receptors.[73] Since ET is released by astrocytes, it is possible that oligodendrocyte progenitor cell (OPC) differentiation is regulated through paracrine effects.

5.4.7 Sphingosine-1-phosphate and lysophosphotidic acid receptors

LPA stimulates the differentiation of mouse cortical neuroblasts[74] and the proliferation and differentiation of rat NSCs to cholinergic neurons.[75] In contrast, in hESC-derived neurospheres that expressed all five of the LPA receptors, LPA specifically inhibits the differentiation of NSCs toward neurons, without affecting proliferation, whereas it maintains the differentiation of NSCs toward astrocytes.[76] In humans, hESC-derived NSCs express

$S1P_{1,3}$ where they stimulate proliferation and growth.[76,269] LPA_1 and $S1P_{5>1-3}$ are expressed in embryonic and adult rodent OPCs,[77–79,270] while fetal human OPCs express higher levels of S1P1.[271] Both LPA and S1P act on OPCs and inhibit oligodendrocyte maturation.[77,78] $S1P_5$ is preferentially expressed in OPCs and is responsible for inhibition of their migration in rodents.[79] In human OPCs, $S1P_{5,3}$ agonism (FTY720) first inhibits differentiation via process retraction, but then stimulates this function and increases cell survival in a $S1P_1$-ERK-dependent manner.[271,272]

5.4.8 Opioid receptors

Endogenous opioids and synthetic agonists acting on the mu-opioid receptors (MORs), delta-opioid receptors (DORs), and kappa-opioid receptors (KORs) have been shown to affect proliferation and differentiation in various embryonic neural cell types. In embryonic NSCs, stimulation of DORs using SNC80 promoted neuronal differentiation through PI3K/PKC/CAMKII/MEK, but MOR agonism (DAMGO) or KOR stimulation (U50,488H) had no effect.[80] In ESCs derived from a mouse blastocyst, both MOR and KOR stimulation promoted proliferation and differentiation into neural progenitors through ERK.[81] MOR and KOR functionality was found in neural progenitor-derived oligodendrocytes and induced their differentiation via ERK and p38.[82] Agonism of these opioid receptors inhibited the differentiation of both astrocytes and neurons in retinoic acid-induced neural progenitors.[82] Opiates acting on the adult SVZ and SGZ inhibit progenitor proliferation, maturation, and survival,[83–87] and alter the progenitor cell cycle.[84,273,274]

5.4.9 Vasoactive intestinal polypeptide receptors

Vasoactive intestinal polypeptide (VIP) is a peptide neurotransmitter released by GABAergic interneurons in the dentate gyrus. VIP and its receptors ($VPAC_1$ and $VPAC_2$) are expressed in developing adult dentate gyrus. VIP shortens the cell cycle of embryonic neuroepithelial cells[275] and promotes neuronal differentiation of embryonic hippocampal neurons in culture.[88] Adult *Vipr2*$^{-/-}$ mice showed reduced progenitor survival and SGZ neurogenesis.[89]

5.4.10 Neuropeptide Y receptors

NPY is a 36-amino acid peptide that belongs to a family of peptides that include pancreatic polypeptide and peptide YY. NPY is secreted by the hypothalamus and is widely distributed in the central and peripheral nervous systems. In an autocrine or paracrine fashion, NPY increases proliferation in the SVZ through an ERK pathway but does not affect the self-renewal of

NSCs.[90] Using KO mice *in vivo* or cell culture with specific NPY receptor agonists and antagonists, the neuroproliferative and neuronal differentiating effects of NPY in the SVZ and SGZ were found to be mediated by the Y1 receptor subtype through ERK-mediated signaling.[91–95]

5.4.11 Purinergic receptors

P2Y receptors are a family of purinergic GPCRs stimulated by nucleotides such as ATP, ADP, UTP, UDP, and UDP-glucose. ATP-mediated purinergic signaling through the P2Y1 receptor has been associated with developmental neurogenesis.[96] In human NSCs, ATP and P2 receptors have been identified as a mitogen for v-myc immortalized neural progenitor cells through calcium release and PI3K.[97] Embryonic neural progenitors expressed P2Y purinergic receptors and release ATP themselves in burst events to mobilize intracellular calcium and increase progenitor proliferation. Receptor antagonists suppressed proliferation and permitted differentiation into neurons and glia *in vitro*, while subsequent removal of purinergic inhibition restored progenitor cell expansion.[98] Neurospheres isolated from P2Y1-deficient mice exhibited reduced proliferation.[276]

5.4.12 Wnt/frizzled receptors

Wnt receptors are also implicated in adult hippocampal neurogenesis by regulating proliferation and differentiation.[99–102] The Wnt3 receptor is expressed and β-catenin pathway is active in the hippocampal SGZ. Overexpression of Wnt3 is sufficient to increase neurogenesis in the SGZ *in vitro* and *in vivo*. By contrast, blockade of Wnt signaling abolishes neurogenesis almost completely *in vivo*.[99] LRP6 mutant mice had reduced production and proliferation of dentate granule neurons and abnormalities in the radial glial scaffolding in the dentate gyrus.[100] In a series of *in vitro* studies utilizing adult mouse neurospheres, both Wnt3 and Wnt5a promoted neurogenesis, but only the Wnt5a neurogenesis was blocked by PKC inhibition.[101]

6. GPCRs IN CANCER STEM CELLS

6.1. GPCRs in glia tumors

6.1.1 Purinergic receptors

Glioblastoma is the most common and aggressive tumor in the brain and possibly characterized by having a cancer stem cell subpopulation essential for tumor survival.[277] The purinergic system plays an important role in glioma cell growth, since adenosine triphosphate (ATP) can regulate proliferation[103,104] in these cell types. Nucleotide receptor-mediated cell communication is controlled by ecto-nucleotidases, which hydrolyze ATP into adenosine in the extracellular space. Changes in extracellular ATP degradation by the use of these exogenous nucleotidases dramatically alters the size of gliomas in rats and human glioma cultures.[105,106] Several purinergic receptor mRNAs were differently expressed in tumor spheres containing markers for cancer stem cells when compared to a cell monolayer not containing spheres or the cancer markers. Treatment of human gliomas U87 or U343 as well as rat C6 gliomas with 100 μM of ATP reduced the number of tumor spheres as well as the expression of cancer stem cell markers.[106] The differential expression of purinergic receptors in tumor spheres and the effect of ATP in reducing the number of tumor spheres suggest that the purinergic system affects cancer stem cell biology and that ATP may be a potential agonist for differentiation therapy.

6.1.2 Akt/PI3K

Downstream signaling pathways of GPCRs such as Ras/Raf/MAPK and Akt/PI3K promote cell survival, cell proliferation, cell migration, and angiogenesis in glial tumor cells. AKT/PI3K activation through loss of phosphatase and tensin homolog (PTEN) in combination with constitutively active epidermal growth factor receptor (EGFR) signaling has been shown to induce glial tumor formation.[107] In addition, glioma stem cells appear to be more dependent on AKT signaling than nonstem glioma cells,[108] suggesting that AKT inhibition may used to target the stem cell population in brain tumors. Several inhibitors of mammalian target of rapamycin (mTOR), which constitutes a downstream target of PI3K, are currently being evaluated in phase II clinical trials in patients with malignant glioma.[278–280]

6.2. GPCRs in breast cancer stem cells

6.2.1 Melatonin receptors

Breast carcinoma is the most common cancer form among women worldwide, registering about one-third of all new cancer cases each year.[281] Melatonin suppresses breast cancer cell proliferation by inhibiting the upregulation of estrogen-induced cyclin D1 via its GPCR MT1.[109] Melatonin can downregulate estrogen receptor alpha (ER-α) and blocks the binding of estrogen receptor complexes that induce transcriptional activity on genes that regulate cell growth, proliferation, and survival.[110] The cancer suppression effects of melatonin are mediated through its G_{i2} G protein signaling pathway and the MT1 receptor.[111] The oxidated form of melatonin, 6-hydroxymelatonin, has also been shown to bind selectively to MT1 and have antioxidant properties.[282,283] Some isoforms of sulfotransferase (SULT), such as SULT1, show a high affinity for 6-hydroxymelatonin in breast cancer cells,[284] and SULT1 expression is elevated in malignant breast tissue.[285,286] The enzyme SULT1E1 catalyzes the conjugation of estrogens to sulfate residues, protecting breast cancer cells from excessive estrogenic actions.[110] Since melatonin can increase the expression of SULT1 isoforms,[110] the antiproliferative effects of MT1 receptor stimulation may also act through this mechanism.

6.2.2 Chemokine receptors

While the effects of SDF-1/CXCR4 are well known in hematopoietic stem cells, CXCR4 is expressed on several tumor cells that may metastasize to the organs that secrete/express SDF-1 (e.g., bones, lymph nodes, lung, and liver).[112,113] SDF-1 exerts pleiotropic effects regulating processes essential to tumor metastasis, such as locomotion of malignant cells, their chemoattraction, and adhesion, as well as playing an important role in tumor vascularization.

7. SUMMARY

Stem cells have the potential to be artificially manipulated into various cell types and transplanted into tissues for the treatment of multiple diseases. Hematopoietic stem cells can be transfused into leukemia patients or NSCs can be grafted into the adult brain to produce partial repair. However, ethical issues still need to be resolved in their use and their cell biology still needs to be explored to determine the best course of treatment.

Each type of stem cell has a restricted range of potential differentiated cell types. ESCs can differentiate into any cell type in the body, but they are still restricted because they cannot form trophoblasts that are crucial in formation of many structures needed to complete an organ or organism. The use of patient-specific and inducible stem cells have great promise for tissue repair and avoids the problems of immune rejection since the cells contain the patients' genome.

Stem cells, besides their direct therapeutic potential, also hold great promise as a tool in drug discovery. Large homogenous populations of differentiated cells can be cultured and used to test the effects of chemical libraries in search of new drugs. Improving upon nature's mechanism of tissue repair is a discovery with great promise, providing an exciting time in science and medicine. The role of GPCRs in this process and our ever expanding knowledge of their roles in stem cell functions will make this promise a reality.

REFERENCES

1. Russ AP, Wattler S, Colledge WH, Aparicio SA, Carlton MB, Pearce JJ, et al. Eomesodermin is required for mouse trophoblast development and mesoderm formation. *Nature* 2000;**404**:95–9.
2. Hadjantonakis A, Papaioannou V. The stem cells of early embryos. *Differentiation* 2001;**68**:159–66.
3. Griffin CT, Srinivasan Y, Zheng YW, Huang W, Coughlin SR. A role for thrombin receptor signaling in endothelial cells during embryonic development. *Science* 2001;**293**:1666–70.
4. Seandel M, James D, Shmelkov SV, Falciatori I, Kim J, Chavala S, et al. Generation of functional multipotent adult stem cells from GPR125+ germline progenitors. *Nature* 2007;**449**:346–50.
5. Pebay A, Wong RC, Pitson SM, Wolvetang EJ, Peh GS, Filipczyk A, et al. Essential roles of sphingosine-1-phosphate and platelet-derived growth factor in the maintenance of human embryonic stem cells. *Stem Cells* 2005;**23**:1541–8.
6. Burdon T, Stracey C, Chambers I, Nichols J, Smith A. Suppression of SHP-2 and ERK signalling promotes self-renewal of mouse embryonic stem cells. *Dev Biol* 1999;**210**:30–43.
7. Burdon T, Smith A, Savatier P. Signalling, cell cycle and pluripotency in embryonic stem cells. *Trends Cell Biol* 2002;**12**:432–8.
8. Rodgers A, Mormeneo D, Long JS, Delgado A, Pyne NJ, Pyne S. Sphingosine 1-phosphate regulation of extracellular signal-regulated kinase-1/2 in embryonic stem cells. *Stem Cells Dev* 2009;**18**:1319–30.
9. Todorova MG, Fuentes E, Soria B, Nadal A, Quesada I. Lysophosphatidic acid induces Ca2+ mobilization and c-Myc expression in mouse embryonic stem cells via the phospholipase C pathway. *Cell Signal* 2009;**21**:523–8.
10. Jiang S, Fu Y, Williams J, Wood J, Pandarinathan L, Avraham S, et al. Expression and function of cannabinoid receptors CB1 and CB2 and their cognate cannabinoid ligands in murine embryonic stem cells. *PLoS One* 2007;**2**:e641.
11. Nusse R. Wnt signaling and stem cell control. *Cell Res* 2008;**18**:523–7.
12. Walsh J, Andrews PW. Expression of Wnt and Notch pathway genes in a pluripotent human embryonal carcinoma cell line and embryonic stem cell. *Apmis* 2003;**111**:197–210.
13. Dravid G, Ye Z, Hammond H, Chen G, Pyle A, Donovan P, et al. Defining the role of Wnt/βcatenin signaling in the survival, proliferation, and self-renewal of human embryonic stem cells. *Stem Cells* 2005;**23**:1489–501.
14. Sato N, Meijer L, Skaltsounis L, Greengard P, Brivanlou AH. Maintenance of pluripotency in human and mouse embryonic stem cells through activation of Wnt signaling by a pharmacological GSK-3-specific inhibitor. *Nat Med* 2004;**10**:55–63.
15. Cappuccio I, Spinsanti P, Porcellini A, Desiderati F, De Vita T, Storto M, et al. Endogenous activation of mGlu5 metabotropic glutamate receptors supports self-renewal of cultured mouse embryonic stem cells. *Neuropharmacology* 2005;**49**(Suppl. 1):196–205.
16. Spinsanti P, De Vita T, Di Castro S, Storto M, Formisano P, Nicoletti F, et al. Endogenously activated mGlu5 metabotropic glutamate receptors sustain the increase in c-Myc expression induced by leukaemia inhibitory factor in cultured mouse embryonic stem cells. *J Neurochem* 2006;**99**:299–307.

17. Melchiorri D, Cappuccio I, Ciceroni C, Spinsanti P, Mosillo P, Sarichelou I, et al. Metabotropic glutamate receptors in stem/progenitor cells. *Neuropharmacology* 2007; **53**:473–80.
18. Sarichelou I, Cappuccio I, Ferranti F, Mosillo P, Ciceroni C, Sale P, et al. Metabotropic glutamate receptors regulate differentiation of embryonic stem cells into GABAergic neurons. *Cell Death Differ* 2008;**15**:700–7.
19. Maherali N, Hochedlinger K. Guidelines and techniques for the generation of induced pluripotent stem cells. *Cell Stem Cell* 2008;**3**:595–605.
20. Nakamura K, Salomonis N, Tomoda K, Yamanaka S, Conklin BR. G(i)-coupled GPCR signaling controls the formation and organization of human pluripotent colonies. *PLoS One* 2009;**4**:e7780.
21. Adams GB, Alley IR, Chung UI, Chabner KT, Jeanson NT, Lo Celso C, et al. Haematopoietic stem cells depend on Gαs-mediated signalling to engraft bone marrow. *Nature* 2009;**459**:103–7.
22. Ma Q, Jones D, Springer TA. The chemokine receptor CXCR4 is required for the retention of B lineage and granulocytic precursors within the bone marrow microenvironment. *Immunity* 1999;**10**:463–71.
23. Petit I, Szyper-Kravitz M, Nagler A, Lahav M, Peled A, Habler L, et al. G-CSF induces stem cell mobilization by decreasing bone marrow SDF-1 and up-regulating CXCR4. *Nat Immunol* 2002;**3**:687–94.
24. Mohle R, Bautz F, Rafii S, Moore MA, Brugger W, Kanz L. The chemokine receptor CXCR-4 is expressed on $CD34^+$ hematopoietic progenitors and leukemic cells and mediates transendothelial migration induced by stromal cell-derived factor-1. *Blood* 1998;**91**:4523–30.
25. Dar A, Kollet O, Lapidot T. Mutual, reciprocal SDF-1/CXCR4 interactions between hematopoietic and bone marrow stromal cells regulate human stem cell migration and development in NOD/SCID chimeric mice. *Exp Hematol* 2006;**34**:967–75.
26. Walter DH, Rochwalsky U, Reinhol J, Seeger F, Aicher A, Urbich C, et al. Sphingosine-1-phosphate stimulates the functional capacity of progenitor cells by activation of the CXCR4-dependent signaling pathway via the S1P3 receptor. *Arterioscler Thromb Vasc Biol* 2007;**27**:275–82.
27. Whetton AD, Lu Y, Pierce A, Carney L, Spooncer E. Lysophospholipids synergistically promote primitive hematopoietic cell chemotaxis via a mechanism involving Vav 1. *Blood* 2003;**102**:2798–802.
28. Kimura T, Boehmler AM, Seitz G, Kuçi S, Wiesner T, Brinkmann V, et al. The sphingosine 1-phosphate receptor agonist FTY720 supports CXCR4-dependent migration and bone marrow homing of human $CD34^+$ progenitor cells. *Blood* 2004;**103**:4478–86.
29. Seitz G, Boehmler AM, Kanz L, Mohle R. The role of sphingosine 1-phosphate receptors in the trafficking of hematopoietic progenitor cells. *Ann N Y Acad Sci* 2005;**1044**: 84–9.
30. Reya T, Duncan AW, Ailles L, Domen J, Scherer DC, Willert K, et al. A role for Wnt signalling in self-renewal of haematopoietic stem cells. *Nature* 2003;**423**:409–14.
31. Willert K, Brown JD, Danenberg E, Duncan AW, Weissman IL, Reya T, et al. Wnt proteins are lipid-modified and can act as stem cell growth factors. *Nature* 2003;**423**: 448–52.
32. Jaganathan BG, Ruester B, Dressel L, Stein S, Grez M, Seifried E, et al. Rho inhibition induces migration of mesenchymal stromal cells (MSCs). *Stem Cells* 2007;**25**:1966–74.
33. Annabi B, Thibeault S, Lee YT, Bousquet-Gagnon N, Eliopoulos N, Barrette S, et al. Matrix metalloproteinase regulation of sphingosine-1-phosphate-induced angiogenic properties of bone marrow stromal cells. *Exp Hematol* 2003;**31**:640–9.
34. Meriane M, Duhamel S, Lejeune L, Galipeau J, Annabi B. Cooperation of matrix metalloproteinases with the RhoA/Rho kinase and mitogen-activated protein kinase

kinase-1/extracellular signal-regulated kinase signaling pathways is required for the sphingosine-1-phosphate-induced mobilization of marrow-derived stromal cells. *Stem Cells* 2006;**24**:2557–65.
35. Nincheri N, Luciani P, Squecco R, Donati C, Bernacchioni C, Borgognoni L, et al. Sphingosine 1-phosphate induces differentiation of adipose tissue-derived mesenchymal stem cells towards smooth muscle cells. *Cell Mol Life Sci* 2009;**66**:1741–54.
36. Chen J, Baydoun AR, Xu R, Deng L, Liu X, Zhu W, et al. Lysophosphatidic acid protects mesenchymal stem cells against hypoxia and serum deprivation-induced apoptosis. *Stem Cells* 2008;**26**:135–45.
37. Karagiosis SA, Chrisler WB, Bollinger N, Karin NJ. Lysophosphatidic acid-induced ERK activation and chemotaxis in MC3T3-E1 preosteoblasts are independent of EGF receptor transactivation. *J Cell Physiol* 2009;**219**:716–23.
38. Schneider VA, Mercola M. Wnt antagonism initiates cardiogenesis in Xenopus laevis. *Genes Dev* 2001;**15**:304–15.
39. Marvin MJ, Di Rocco G, Gardiner A, Bush SM, Lassar AB. Inhibition of Wnt activity induces heart formation from posterior mesoderm. *Genes Dev* 2001;**15**:316–27.
40. Tzahor E, Lassar AB. Wnt signals from the neural tube block ectopic cardiogenesis. *Genes Dev* 2001;**15**:255–60.
41. Eisenberg LM, Eisenberg CA. Evaluating the role of Wnt signal transduction in promoting the development of the heart. *Scientific World J* 2007;**2**:161–76.
42. Yue Q, Wagstaff L, Yang Y, Weijer C, Münsterberg A. Wnt3a-mediated chemorepulstion controls movement patterns of cardiac progenitors and requires RhoA function. *Development* 2008;**135**:1029–37.
43. Zeng XX, Wilm TP, Sepich DS, Solnica-Krezel L. Apelin and its receptor control heart field formation during zebrafish gastrulation. *Dev Cell* 2007;**12**:391–402.
44. Kulkarni VA, Jha S, Vaidya VA. Depletion of norepinephrine decreases the proliferation, but does not influence the survival and differentiation, of granule cell progenitors in the adult rat hippocampus. *Eur J Neurosci* 2002;**16**:2008–12.
45. Hiramoto T, Ihara Y, Watanabe Y. α_1-Adrenergic receptors stimulation induces the proliferation of neural progenitor cells in vitro. *Neurosci Lett* 2006;**408**:25–8.
46. Hiramoto T, Satoh Y, Takishima K, Watanabe Y. Induction of cell migration of neural progenitor cells in vitro by α_1-adrenergic receptor and dopamine D1 receptor stimulation. *Neuroreport* 2008;**19**:793–7.
47. Ohashi H, Nishikawa K, Ayukawa K, Hara Y, Nishimoto M, Kudo Y, et al. α_1-Adrenoceptor agonists protect against stress-induced death of neural progenitor cells. *Eur J Pharmacol* 2007;**573**:20–8.
48. Gupta MK, Papay RS, Jurgens CW, Gaivin RJ, Shi T, Doze VA, et al. α_1-Adrenergic receptors regulate neurogenesis and gliogenesis. *Mol Pharmacol* 2009;**76**:314–26.
49. Yanpallewar SU, Fernandes K, Marathe SV, Vadodaria KC, Jhaveri D, Rommelfanger K, et al. α_2-Adrenoceptor blockade accelerates the neurogenic, neurotrophic, and behavioral effects of chronic antidepressant treatment. *J Neurosci* 2010;**30**:1096–109.
50. Rizk P, Salazar J, Raisman-Vozari R, Marien M, Ruberg M, Colpaert F, et al. The α_2-adrenoceptor antagonist dexefaroxan enhances hippocampal neurogenesis by increasing the survival and differentiation of new granule cells. *Neuropsychopharmacology* 2006;**31**:1146–57.
51. Aguado T, Monory K, Palazuelos J, Stell N, Cravatt B, Lutz B, et al. The endocannabinoid system drives neural progenitor proliferation. *FASEB J* 2005;**19**:1704–6.
52. Jiang W, Zhang Y, Xiao L, Van Cleemput J, Ji SP, Bai G, et al. Cannabinoids promote embryonic and adult hippocampus neurogenesis and produce anxiolytic- and antidepressant-like effects. *J Clin Invest* 2005;**115**:3104–16.

53. Molina-Holgado F, Rubio-Araiz A, Garcia-Ovejero D, Williams RJ, Moore JD, Arévalo-Martín A, et al. CB2 cannabinoid receptors promote mouse neural stem cell proliferation. *Eur J Neurosci* 2007;**25**:629–34.
54. Mulder J, Aguado T, Keimpema E, Barabás K, Ballester Rosado CJ, Nguyen L, et al. Endocannabinoid signaling controls pyramidal cell specification and long-range axon patterning. *Proc Natl Acad Sci USA* 2008;**105**:8760–5.
55. Galve-Roperh I, Aguado T, Palazuelos J, Guzmán M. Mechanisms of control of neuron survival by the endocannabinoid system. *Curr Pharm Des* 2008;**14**:2279–88.
56. Maison P, Walker DJ, Walsh FS, Williams G, Doherty P. BDNF regulates neuronal sensitivity to endocannabinoids. *Neurosci Lett* 2009;**467**:90–4.
57. Borrell V, Kaspar BK, Gage FH, Callaway EM. In vivo evidence for radial migration of neurons by long-distance somal translocation in the developing ferret visual cortex. *Cereb Cortex* 2006;**16**:1571–83.
58. Li G, Kataoka H, Coughlin SR, Pleasure SJ. Identification of a transient subpial neurogenic zone in the developing dentate gyrus and its regulation by Cxcl12 and reelin signaling. *Development* 2009;**136**:327–35.
59. Lu M, Grove EA, Miller RJ. Abnormal development of the hippocampal dentate gyrus in mice lacking the CXCR4 chemokine receptor. *Proc Natl Acad Sci USA* 2002;**99**: 7090–5.
60. Bagri A, Gurney T, He X, Zou YR, Littman DR, Tessier-Lavigne M, et al. The chemokine SDF1 regulates migration of dentate granule cells. *Development* 2002;**129**: 4249–60.
61. Krathwohl MD, Kaiser JL. Chemokines promote quiescence and survival of human neural progenitor cells. *Stem Cells* 2004;**22**:109–18.
62. Wu Y, Peng H, Cui M, Whitney NP, Huang Y, Zheng JC. CXCL12 increases human neural progenitor cell proliferation through Akt-1/FOXO3a signaling pathway. *J Neurochem* 2009;**109**:1157–67.
63. Hagihara K, Zhang EE, Ke YH, Liu G, Liu JJ, Rao Y, et al. Shp2 acts downstream of SDF-1alpha/CXCR4 in guiding granule cell migration during cerebellar development. *Dev Biol* 2009;**334**:276–84.
64. Tran PB, Banisadr G, Ren D, Chenn A, Miller RJ. Chemokine receptor expression by neural progenitor cells in neurogenic regions of mouse brain. *J Comp Neurol* 2007;**500**: 1007–33.
65. Bhattacharyya BJ, Banisadr G, Jung H, Ren D, Cronshaw DG, Zou Y, et al. The chemokine stromal cell-derived factor-1 regulates GABAergic inputs to neural progenitors in the postnatal dentate gyrus. *J Neurosci* 2008;**28**:6720–30.
66. Kolodziej A, Schulz S, Guyon A, Wu DF, Pfeiffer M, Odemis V, et al. Tonic activation of CXC chemokine receptor 4 in immature granule cells supports neurogenesis in the adult dentate gyrus. *J Neurosci* 2008;**28**:4488–500.
67. Yang P, Arnold SA, Habas A, Hetman M, Hagg T. Ciliary neurotrophic factor mediates dopamine D2 receptor-induced CNS neurogenesis in adult mice. *J Neurosci* 2008;**28**:2231–41.
68. Hoglinger GU, Rizk P, Muriel MP, Duyckaerts C, Oertel WH, Caille I, et al. Dopamine depletion impairs precursor cell proliferation in Parkinson disease. *Nat Neurosci* 2004;**7**:726–35.
69. Van Kampen JM, Hagg T, Robertson HA. Induction of neurogenesis in the adult rat subventricular zone and neostriatum following dopamine D receptor stimulation. *Eur J Neurosci* 2004;**19**:2377–87.
70. Van Kampen RM, Robertson HA. A possible role for dopamine D3 receptor stimulation in the induction of neurogenesis in the adult rat substantia nigra. *Neuroscience* 2005;**136**:381–6.

71. Baker SA, Baker KA, Hagg T. Dopaminergic nigrostriatal projections regulate neural precursor proliferation in the adult mouse subventricular zone. *Eur J Neurosci* 2004;**20**:575–9.
72. Mizuno N, Kokubu H, Sato M, Nishimura A, Yamauchi J, Kurose H, et al. G protein-coupled receptor signaling through Gq and JNK negatively regulates neural progenitor cell migration. *Proc Natl Acad Sci USA* 2005;**102**:12365–70.
73. Gadea A, Aguirre A, Haydar TF, Gallo V. Endothelin-1 regulates oligodendrocyte development. *J Neurosci* 2009;**29** 10047–62.
74. Fukushima N, Shano S, Moriyama R, Chun J. Lysophosphatidic acid stimulates neuronal differentiation of cortical neuroblasts through the LPA1-G(i/o) pathway. *Neurochem Int* 2007;**50**:302–7.
75. Cui HL, Qiao JT. Promotive action of lysophosphatidic acid on proliferation of rat embryonic neural stem cells and their differentiation to cholinergic neurons in vitro. *Sheng Li Xue Bao* 2006;**58**:547–55.
76. Dottori M, Leung J, Turnley AM, Pebay A. Lysophosphatidic acid inhibits neuronal differentiation of neural stem/progenitor cells derived from human embryonic stem cells. *Stem Cells* 2008;**26**:1146–54.
77. Dawson J, Hotchin N, Lax S, Rumsby M. Lysophosphatidic acid induces process retraction in CG-4 line oligodendrocytes and oligodendrocyte precursor cells but not in differentiated oligodendrocytes. *J Neurochem* 2003;**87**:947–57.
78. Jaillard C, Harrison S, Stankoff B, Aigrot MS, Calver AR, Duddy G, et al. Edg8/S1P5: an oligodendroglial receptor with dual function on process retraction and cell survival. *J Neurosci* 2005;**25**:1459–69.
79. Novgorodov AS, El-Alwani M, Bielawski J, Obeid LM, Gudz TI. Activation of sphingosine-1-phosphate receptor S1P5 inhibits oligodendrocyte progenitor migration. *FASEB J* 2007;**21**:1503–14.
80. Narita M, Kuzumaki N, Miyatake M, Sato F, Wachi H, Seyama Y, et al. Role of delta-opioid receptor function in neurogenesis and neuroprotection. *J Neurochem* 2006;**97**:1494–505.
81. Kim E, Clark AL, Kiss A, Hahn JW, Wesselschmidt R, Coscia CJ, et al. Mu- and kappa-opioids induce the differentiation of embryonic stem cells to neural progenitors. *J Biol Chem* 2006;**281**:33749–60.
82. Hahn JW, Jagwani S, Kim E, Rendell VR, He J, Ezerskiy LA, et al. Mu and kappa opioids modulate mouse embryonic stem cell-derived neural progenitor differentiation via MAP kinases. *J Neurochem* 2010;**112**:143114–41.
83. Vértes Z, Melegh G, Vértes M, Kovács S. Effect of naloxone and D-met2-pro5-enkephalinamide treatment on the DNA synthesis in the developing rat brain. *Life Sci* 1982;**31**:119–26.
84. Arguello AA, Harburg GC, Schonborn JR, Mandyam CD, Yamaguchi M, Eisch AJ. Time course of morphine's effects on adult hippocampal subgranular zone reveals preferential inhibition of cells in S phase of the cell cycle and a subpopulation of immature neurons. *Neuroscience* 2008;**157**:70–9.
85. Eisch AJ, Barrot M, Schad CA, Self DW, Nestler EJ. Opiates inhibit neurogenesis in the adult rat hippocampus. *Proc Natl Acad Sci USA* 2000;**97**:7579–84.
86. Kahn L, Alonso G, Normand E, Manzoni OJ. Repeated morphine treatment alters polysialylated neural cell adhesion molecule, glutamate decarboxylase-67 expression and cell proliferation in the adult rat hippocampus. *Eur J Neurosci* 2005;**21**:493–500.
87. Khurshid N, Hameed LS, Mohanasundaram S, Iyengar S. Opioid modulation of cell proliferation in the ventricular zone of adult zebra finches (*Taenopygia guttata*). *FASEB J* 2010;**24**:3681–95.
88. Blondel O, Collin C, McCarran WJ, Zhu S, Zamostiano R, Gozes I, et al. A glia-derived signal regulating neuronal differentiation. *J Neurosci* 2000;**20**:8012–20.

89. Zaben M, Sheward WJ, Shtaya A, Abbosh C, Harmar AJ, Pringle AK, et al. The neurotransmitter VIP expands the pool of symmetrically dividing postnatal dentate gyrus precursors via VPAC2 receptors or directs them toward a neuronal fate via VPAC1 receptors. *Stem Cells* 2009;**27**:2539–51.
90. Thiriet N, Agasse F, Nicoleau C, Guégan C, Vallette F, Cadet JL, et al. NPY promotes chemokinesis and neurogenesis in the rat subventricular zone. *J Neurochem* 2011;**116**:1018–27.
91. Hansel DE, Eipper BA, Ronnett GV. Neuropeptide Y functions as a neuroproliferative factor. *Nature* 2001;**410**:940–4.
92. Howell OW, Scharfman HE, Herzog H, Sundstrom LE, Beck-Sickinger A, Gray WP. Neuropeptide Y is neuroproliferative for post-natal hippocampal precursor cells. *J Neurochem* 2003;**86**:646–59.
93. Howell OW, Doyle K, Goodman JH, Scharfman HE, Herzog H, Pringle A, et al. Neuropeptide Y stimulates neuronal precursor proliferation in the post-natal and adult dentate gyrus. *J Neurochem* 2005;**93**:560–70.
94. Agasse F, Bernardino L, Kristiansen H, Christiansen SH, Ferreira R, Silva B, et al. Neuropeptide Y promotes neurogenesis in murine subventricular zone. *Stem Cells* 2008;**26**:1636–45.
95. Decressac M, Prestoz L, Veran J, Cantereau A, Jaber M, Gaillard A. Neuropeptide Y stimulates proliferation, migration and differentiation of neural precursors from the subventricular zone in adult mice. *Neurobiol Dis* 2009;**34**:441–9.
96. Weissman T, Riquelme P, Ivic L, Flint A, Kriegstein A. Calcium waves propagate through radial glial cells and modulate proliferaytion in the developing neocortex. *Neuron* 2004;**43**:647–61.
97. Ryu JK, Choi HB, Hatori K, Heisel RL, Pelech SL, McLarnon JG, et al. Adenosine triphosphate induces proliferation of human neural stem cells: role of calcium and p70 ribosomal protein S6 kinase. *J Neurosci Res* 2003;**72**:352–62.
98. Lin JH, Takano T, Arcuino G, Wang X, Hu F, Darzynkiewicz Z, et al. Purinergic signaling regulates neural progenitor cell expansion and neurogenesis. *Dev Biol* 2007;**302**:356–66.
99. Lie DC, Colamarino SA, Song HJ, Désiré L, Mira H, Consiglio A, et al. Wnt signalling regulates adult hippocampal neurogenesis. *Nature* 2005;**437**:1370–5.
100. Zhou CJ, Zhao C, Pleasure SJ. Wnt signaling mutants have decreased dentate granule cell production and radial glial scaffolding abnormalities. *J Neurosci* 2004;**24**:121–6.
101. Yu JM, Kim JH, Song GS, Jung JS. Increase in proliferation and differentiation. *Mol Cell Biochem* 2006;**288**:17–28.
102. Wexler EM, Paucer A, Kornblum HI, Palmer TD, Geschwind DH. Endogenous Wnt signaling maintains neural progenitor cell potency. *Stem Cells* 2009;**27**:1130–41.
103. Ralevic V, Burnstock G. Receptors for purines and pyrimidines. *Pharmacol Rev* 1998;**50**:413–92.
104. Rathbone MP, Middlemiss PJ, Gysbers JW, Andrew A, Herman M, Reed JK, et al. Trophic effects of purines in neurons and glial cells. *Prog Neurobiol* 1999;**59**:663–90.
105. Braganhol E, Huppes D, Bernardi A, Wink MR, Lenz G, Battastini AM. A comparative study of ectonucleotidase and P2 receptor mRNA profiles in C6 cell line cultures and C6 ex vivo glioma model. *Cell Tissue Res* 2009;**335**:331–40.
106. Ledur PF, Villodre ES, Paulus R, Cruz LA, Flores DG, Lenz G. Extracellular ATP reduces tumor sphere growth and cancer stem cell population in glioblastoma cells. *Purinergic Signal* 2012;**8**:39–48.
107. Li L, Dutra A, Pak E, Labrie 3rd JE, Gerstein RM, Pandolfi PP, et al. EGFR*vIII* expression and PTEN loss synergistically induce chromosomal instability and glial tumors. *Neuro Oncol* 2009;**11**:9–21.

108. Eyler CE, Foo WC, LaFiura KM, McLendon RE, Hjelmeland AB, Rich JN. Brain cancer stem cells display preferential sensitivity to Akt inhibition. *Stem Cells* 2008;**26**:3027–36.
109. Rögelsperger O, Ekmekcioglu C, Jäger W, Klimpfinger M, Königsberg R, Krenbek D, et al. Coexpression of the melatonin receptor 1 and nestin in human breast cancer specimens. *J Pineal Res* 2009;**46**:422–32.
110. Gonzalez A, Cos S, Martinez-Campa C, Alonso-Gonzalez C, Sanchez-Mateos S, Mediavilla MD, Sanchez-Barcelo EJ. Selective estrogen enzyme modulator actions of melatonin in human breast cancer cells. *J Pineal Res* 2008;**45**:86–92.
111. Lai L, Yuan L, Chen Q, Dong C, Mao L, Rowan B, et al. The Galphai and Galphaq proteins mediate the effects of melatonin on steroid/thyroid hormone receptor transcriptional activity and breast cancer cell proliferation. *J Pineal Res* 2008;**45**:476–88.
112. Ratajczak MZ, Kucia M, Reca R. Stem cell plasticity revisited: CXCR4-positive cells expressing mRNA for early muscle, liver and neural cells "hide out" in the bone marrow. *Leukemia* 2004;**18**:29–40.
113. Kucia M, Ratajczak J, Ratajczak MZ. Bone marrow as a source of circulating CXCR4^{+} tissue committed stem cells (TCSC). *Biol Cell* 2005;**97**:133–46.
114. Overington JP, Al-Lazikani B, Hopkins AL. How many drug targets are there? *Nat Rev Drug Discov* 2006;**5**:993–6.
115. Fredriksson R, Lagerstrom MC, Lundin LG, Schioth HB. The G-protein-coupled receptors in the human genome form five main families. Phylogenetic analysis, paralogon groups, and fingerprints. *Mol Pharmacol* 2003;**63**:1256–72.
116. Palczewski K, Kumasaka T, Hori T, Behnke CA, Motoshima H, Fox BA, et al. Crystal structure of rhodopsin: a G protein-coupled receptor. *Science* 2000;**289**:739–45.
117. Karnik SS, Khorana HG. Assembly of functional rhodopsin requires a disulfide bond between cysteine residues 110 and 117. *J Biol Chem* 1990;**265**:17520–4.
118. Bockaert J, Marin P, Dumuis A, Fagni L. The 'magic tail' of G protein-coupled receptors: an anchorage for functional protein networks. *FEBS Lett* 2003;**546**:65–72.
119. Evans MJ, Kaufman MH. Establishment in culture of pluripotential cells from mouse embryos. *Nature* 1981;**292**:154–6.
120. Martin GR. Isolation of a pluripotent cell line from early mouse embryos cultured in medium conditioned by teratocarcinoma stem cells. *Proc Natl Acad Sci USA* 1981;**78**:7634–7.
121. Thomson JA, Itskovitz-Eldor J, Shapiro SS, Waknitz MA, Swiergiel JJ, Marshall VS, et al. Embryonic stem cell lines derived from human blastocysts. *Science* 1998;**282**:1145–7.
122. Takahashi K, Yamanaka S. Induction of pluripotent stem cells from mouse embryonic and adult fibroblast cultures by defined factors. *Cell* 2006;**126**:663–76.
123. Takahashi K, Tanabe K, Ohnuki M, Narita M, Ichisaka T, Tomoda K, et al. Induction of pluripotent stem cells from adult human fibroblasts by defined factors. *Cell* 2007;**131**:861–72.
124. French AJ, Adams CA, Anderson LS, Kitchen JR, Hughes MR, Wood SH. Development of human cloned blastocysts following somatic cell nuclear transfer with adult fibroblasts. *Stem Cells* 2008;**26**:485–93.
125. Gurdon JB, Melton DA. Nuclear reprogramming in cells. *Science* 2008;**322**:1811–81.
126. Cowan CA, Atienza J, Melton DA, Eggan K. Nuclear reprogramming of somatic cells after fusion with human embryonic stem cells. *Science* 2005;**309**:1369–73.
127. Dimos JT, Rodolfa KT, Niakan KK, Weisenthal LM, Mitsumoto H, Chung W, et al. Induced pluripotent stem cells generated from patients with ALS can be differentiated into motor neurons. *Science* 2008;**321**:1218–21.
128. Park IH, Arora N, Huo H, Maherali N, Ahfeldt T, Shimamura A, et al. Disease-specific induced pluripotent stem cells. *Cell* 2008;**134**:877–86.

129. Soldner F, Hockemeyer D, Beard C, Gao Q, Bell GW, Cook EG, et al. Parkinson's disease patient-derived induced pluripotent stem cells free of viral reprogramming factors. *Cell* 2009;**136**:964–77.
130. Hanna J, Wernig M, Markoulaki S, Sun CW, Meissner A, Cassady JP, et al. Treatment of sickle cell anemia mouse model with iPS cells generated from autologous skin. *Science* 2007;**318**:1920–3.
131. Ye L, Chang JC, Lin C, Sun X, Yu J, Kan YW. Induced pluripotent stem cells offer new approach to therapy in thalassemia and sickle cell anemia and option in prenatal diagnosis in genetic diseases. *Proc Natl Acad Sci USA* 2009;**106**:9826–30.
132. Wernig M, Zhao JP, Pruszak J, Hedlund E, Fu D, Soldner F, et al. Neurons derived from reprogrammed fibroblasts functionally integrate into the fetal brain and improve symptoms of rats with Parkinson's disease. *Proc Natl Acad Sci USA* 2008;**105**:5856–61.
133. Nelson TJ, Martinez-Fernandez A, Yamada S, Perez-Terzic C, Ikeda Y, Terzic A. Repair of acute myocardial infarction by human stemness factors induced pluripotent stem cells. *Circulation* 2009;**120**:408–16.
134. Raya A, Rodríguez-Pizà I, Guenechea G, Vassena R, Navarro S, Barrero MJ, et al. Disease-corrected haematopoietic progenitors from Fanconi anaemia induced pluripotent stem cells. *Nature* 2009;**460**:53–9.
135. McCulloch EA, Till JE. The radiation sensitivity of normal mouse bone marrow cells, determined by quantitative marrow transplantation into irradiated mice. *Radiat Res* 1960;**13**:115–25.
136. Steidl U, Kronenwett R, Martin S, Haas R. Molecular biology of hematopoietic stem cells. *Vitam Horm* 2003;**66**:1–28.
137. Guthrie SM, Caballero S, Mames RN, Grant MB, Scott EW. Analysis of the vascular potential of hematopoietic stem cells. *Methods Mol Med* 2005;**105**:369–80.
138. Speck B, Sartorius J, Cornu P, Nissen C, Weber W, Gratwohl A. Bone marrow transplantation in severe aplastic anemia and acute leukemia. *Onkologie* 1979;**2**:79–82 [Article in german].
139. Perry AR, Linch DC. The history of bone-marrow transplantation. *Blood Rev* 1996;**10**:215–9.
140. Ashfaq K, Yahaya I, Hyde C, Andronis L, Barton P, Bayliss S, et al. Clinical effectiveness and cost-effectiveness of stem cell transplantation in the management of acute leukaemia: a systematic review. *Health Technol Assess* 2010;**14**:1–141, iii-iv, ix-xi.
141. Koreth J, Biernacki M, Aldridge J, Kim HT, Alyea 3rd EP, Armand P, et al. Syngeneic donor hematopoietic stem cell transplantation is associated with high rates of engraftment syndrome. *Biol Blood Marrow Transplant* 2011;**17**:421–8.
142. Gluckman E, Broxmeyer HA, Auerbach AD, Friedman HS, Douglas GW, Devergie H, et al. Hematopoietic reconstitution in a patient with Fanconi's anemia by means of umbilical cord blood from an HLA-identical sibling. *N Engl J Med* 1989;**321**:1174–8.
143. Bertaina A, Bernardo ME, Caniglia M, Vinti L, Giorgiani G, Locatelli F. Cord blood transplantation in children with haematological malignancies. *Best Pract Res Clin Haematol* 2010;**23**:189–96.
144. Cairo MS, Wagner JE. Placental and/or umbilical cord blood: an alternative source of hematopoietic stem cells for transplantation. *Blood* 1997;**90**:4665–78.
145. Schraufstatter IU, Discipio RG, Khaldoyanidi S. Mesenchymal stem cells and their microenvironment. *Front Biosci* 2011;**17**:2271–88.
146. Dominici M, Le Blanc K, Mueller I, Slaper-Cortenbach I, Marini F, Krause D, et al. Minimal criteria for defining multipotent mesenchymal stromal cells. The International Society for Cellular Therapy position statement. *Cytotherapy* 2006;**8**:315–7.
147. Choi YH, Kurtz A, Stamm C. Mesenchymal stem cells for cardiac cell therapy. *Hum Gene Ther* 2011;**22**:3–17.

148. Krasnodembskaya A, Song Y, Fang X, Gupta N, Serikov V, Lee JW, et al. Antibacterial effect of human mesenchymal stem cells is mediated in part from secretion of the antimicrobial peptide LL-37. *Stem Cells* 2010;**28**:2229–38.
149. Le Blanc K. Immunomodulatory effects of fetal and adult mesenchymal stem cells. *Cytotherapy* 2003;**5**:485–9.
150. Koppula PR, Chelluri LK, Polisetti N, Vemuganti GK. Histocompatibility testing of cultivated human bone marrow stromal cells—a promising step towards pre-clinical screening for allogeneic stem cell therapy. *Cell Immunol* 2009;**259**:61–5.
151. Griffin MD, Ritter T, Mahon BP. Immunological aspects of allogeneic mesenchymal stem cell therapies. *Hum Gene Ther* 2010;**21**:1641–65.
152. Chapel A, Bertho JM, Bensidhoum M, Fouillard L, Young RG, Frick J, et al. Mesenchymal stem cells home to injured tissues when co-infused with hematopoietic cells to treat a radiation-induced multi-organ failure syndrome. *J Gene Med* 2003;**5**:1028–38.
153. Dvorak HF. Tumors: wound that do not heal. Similarities between tumor stroma generation and wound healing. *N Engl J Med* 1986;**315**:1650–9.
154. Smith JR, Pochampally R, Perry A, Hsu SC, Prockop DJ. Isolation of a highly clonogenic and multipotential subfraction of adult stem cells from bone marrow stroma. *Stem Cells* 2004;**22**:823–31.
155. Sarugaserb R, Hanoun L, Keating A, Stanford WL, Davies JE. Human mesenchymal stem cells self-renew and differentiate according to a deterministic hierarchy. *PLoS One* 2009;**4**:e6498.
156. Orlic D, Kajstura J, Chimenti S, Limana F, Jakoniuk I, Quaini F, et al. Mobilized bone marrow cells repair the infarcted heart, improving function and survival. *Proc Natl Acad Sci USA* 2001;**98**:10344–9.
157. Balsam LB, Wagers AJ, Christensen JL, Kofidis T, Weissman IL, Robbins RC. Haematopoietic stem cells adopt mature haematopoietic fates in ischaemic myocardium. *Nature* 2004;**428**:668–73.
158. Rota M, Kajstura J, Hosoda T, Bearzi C, Vitale S, Esposito G, et al. Bone marrow cells adopt the cardiomyogenic fate in vivo. *Proc Natl Acad Sci USA* 2007;**45**:17783–8.
159. Hare JM, Traverse JH, Henry TD, Dib N, Strumpf RK, Schulman SP, et al. A randomized, double-blind, placebo-controlled, dose-escalation study of intravenous adult human mesenchymal stem cells (prochymal) after acute myocardial infarction. *J Am Coll Cardiol* 2009;**54**:2277–86.
160. Trachtenberg B, Velazquez DL, Williams AR, McNiece I, Fishman J, Nguyen K, et al. Rationale and design of the transendocardial injection of autologous human cells (bone marrow or mesenchymal) in chronic ischemic left ventricular dysfunction and heart failure secondary to myocardial infarction (TAC-HFT) trial: a randomized, double-blind, placebo-controlled study of safety and efficacy. *Am Heart J* 2011;**161**:487–93.
161. Poss KD, Wilson LG, Keating MT. Heart regeneration in zebrafish. *Science* 2002;**298**:2188–90.
162. Porrello ER, Mahmoud AI, Simpson E, Hill JA, Richardson JA, Olson EN, et al. Transient regenerative potential of the neonatal mouse heart. *Science* 2011;**331**:1078–80.
163. Jopling C, Sleep E, Raya M, Martí M, Raya A, Belmonte JC. Zebrafish heart regeneration occurs by cardiomyocyte dedifferentiation and proliferation. *Nature* 2010;**464**:606–9.
164. Hierlihy AM, Seale P, Lobe CG, Rudnicki MA, Megeney LA. The post-natal heart contains a myocardial stem cell population. *FEBS Lett* 2002;**530**:239–43.
165. Oh H, Bradfute SB, Gallardo TD, Nakamura T, Gaussin V, Mishina Y, et al. Cardiac progenitor cells from adult myocardium: homing, differentiation, and fusion after infarction. *Proc Natl Acad Sci USA* 2003;**100**:12313–8.
166. Pfister O, Mouquet F, Jain M, Summer R, Helmes M, Fine A, et al. CD31 − but not CD31 + cardiac side population cells exhibit functional cardiomyogenic differentiation. *Circ Res* 2005;**97**:52–61.

167. Spradling A, Drummond-Barbosa D, Kai T. Stem cells find their niche. *Nature* 2001;**414**:98–104.
168. Davis DR, Zhang Y, Smith RR, Cheng K, Terrovitis J, Malliaras K, et al. Validation of the cardiosphere method to culture cardiac progenitor cells from myocardial tissue. *PLoS One* 2009;**4**:e7195.
169. Smith RR, Barile L, Cho HC, Leppo MK, Hare JM, Messina E, et al. Regenerative potential of cardiosphere-derived cells expanded from percutaneous endomyocardial biopsy specimens. *Circulation* 2007;**115**:896–908.
170. Alvarez-Buylla A, Kirn J. Birth, migration, incorporation and death of vocal control neurons in adult songbirds. *J Neurosci* 1997;**33**:585–601.
171. Reynolds BA, Weiss S. Generation of neurons and astrocytes from isolated cells of the adult mammalian central nervous system. *Science* 1992;**255**:1707–10.
172. Gage FH. Mammalian neural stem cells. *Science* 2000;**287**:1433–8.
173. Johansson CB, Svensson M, Wallstedt L, Janson AM, Frisén J. Neural stem cells in the adult human brain. *Exp Cell Res* 1999;**253**:733–6.
174. Morshead CM, Reynolds BA, Craig CG, McBurney MW, Staines WA, Morassutti D, et al. Neural stem cells in the adult mammalian forebrain: a relatively quiescent subpopulation of subependymal cells. *Neuron* 1994;**13**:1071–82.
175. Doetsch F, Caillé I, Lim DA, García-Verdugo JM, Alvarez-Buylla A. Subventricular zone astrocytes are neural stem cells in the adult mammalian brain. *Cell* 1999;**97**:703–16.
176. Panganiban G, Rubenstein JLR. Developmental functions of the Distal-less/Dlx homeobox genes. *Development* 2002;**129**:4371–86.
177. Bonfanti L, Theodosis DT. Expression of polysialylated neural cell adhesion molecule by proliferating cells in the subependymal layer of the adult rat, in its rostral extension and in the olfactory bulb. *Neuroscience* 1994;**62**:291–305.
178. Tomasiewicz H, Ono K, Yee D, Thompson C, Goridis C, Rutishauser U, et al. Genetic deletion of a neural cell adhesion molecule variant (N-CAM) produces distinct defects in the central nervous system. *Neuron* 1993;**11**:1163–74.
179. Cremer H, Lange R, Christoph A, Plomann M, Vopper G, Roes J, et al. Inactivation of the N-CAM gene in mice results in size reduction of the olfactory bulb and deficits in spatial learning. *Nature* 1994;**367**:455–9.
180. Menn B, Garcia-Verdugo JM, Yaschine C, Gonzalez-Perez O, Rowitch D, Alvarez-Buylla A. Origin of oligodendrocytes in the subventricular zone of the adult brain. *J Neurosci* 2006;**26**:7907–18.
181. Suh H, Consiglio A, Ray J, Sawai T, D'Amour KA, Gage FH. In vivo fate analysis reveals the multipotent and self-renewal capacities of $Sox2^{+}$ neural stem cells in the adult hippocampus. *Cell Stem Cell* 2007;**1**:515–28.
182. Ming G-L, Song H. Adult neurogenesis in the mammalian central nervous system. *Annu Rev Neurosci* 2005;**28**:223–50.
183. Mu Y, Lee SW, Gage FH. Signaling in adult neurogenesis. *Curr Opin Neurobiol* 2010;**20**:416–23.
184. Cameron HA, Woolley CS, McEwen BS, Gould E. Differentiation of newly born neurons and glia in the dentate gyrus of the adult rat. *Neuroscience* 1993;**56**:337–44.
185. Eriksson PS, Perfilieva E, Bjork-Eriksson T, Alborn AM, Nordborg C, Peterson DA, et al. Neurogenesis in the adult human hippocampus. *Nat Med* 1998;**4**:1313–7.
186. Gritti A, Frölichsthal-Schoeller P, Galli R, Parati EA, Cova L, Pagano SF, et al. Neurogenesis in the neocortex of adult primates. *Science* 1999;**286**:548–52.
187. Rietze R, Poulin P, Weiss S. Mitotically active cells that generate neurons and astrocytes are present in multiple regions of the adult mouse hippocampus. *J Comp Neurol* 2000;**424**:397–408.
188. Bedard A, Gravel C, Parent A. Chemical characterization of newly generated neurons in the striatum of adult primates. *Exp Brain Res* 2006;**170**:501–12.

189. Daniela F, Vescovi AL, Bottai D. The stem cells as a potential treatment for neurodegeneration. *Methods Mol Biol* 2007;**399**:199–213.
190. Bjorklund A, Stenevi U. Intracerebral neural implants: neuronal replacement and reconstruction of damaged circuitries. *Annu Rev Neurosci* 1984;**7**:279–308.
191. Sawle GV, Bloomfield PM, Bjorklund A, Brooks DJ, Brundin P, Leenders KL, et al. Transplantation of fetal dopamine neurons in Parkinson's disease: PET (^{18}F)-6-L-fluorodopa studies in two patients with putaminal implants. *Ann Neurol* 1992;**31**:166–73.
192. Lindvall O, Widner H, Rehncrona S, Brundin P, Odin P, Gustavii B, et al. Transplantation of fetal dopamine neurons in Parkinson's disease: one year clinical and neurophysiological observations in two patients with putaminal implants. *Ann Neurol* 1992;**31**:155–65.
193. Freed CR, Greene PE, Breeze BR, Tsai WY, DuMouchel W, Kao R, et al. Transplantation of embryonic dopamine neurons for severe Parkinson's disease. *N Engl J Med* 2001;**344**:710–7.
194. Amariglio N, Hirshberg A, Scheithauer BW, Cohen Y, Loewenthal R, Trakhtenbrot L, et al. Donor-derived brain tumor following neural stem cell transplantation in an ataxia telangiectasia patient. *PLoS Med* 2009;**6**:e1000029.
195. Nakamura M, Toyama Y. Transplantation of neural stem cells into spinal cord after injury. *Nihon Rinsho* 2003;**61**:463–8 [Article in Japanese].
196. Pfeifer K, Vroemen M, Blesch A, Weidner N. Adult neural progenitor cells provide a permissive guiding substrate for corticospinal axon growth following spinal cord injury. *Eur J Neurosci* 2004;**20**:1695–704.
197. Kim SU. Human neural stem cells genetically modified for brain repair in neurological disorders. *Neuropathology* 2004;**24**:159–71.
198. Chandran S, Compston A. Neural stem cells as a potential source of oligodendrocytes for myelin repair. *J Neurol Sci* 2005;**233**:179–81.
199. Ryu JK, Kim J, Cho SJ, Hatori K, Nagai A, Choi HB, et al. Proactive transplantation of human neural stem cells prevents degeneration of striatal neurons in a rat model of Huntington disease. *Neurobiol Dis* 2004;**16**:68–77.
200. Arvidsson A, Collin T, Kirik D, Kokaia Z, Lindvall O. Neuronal replacement from endogenous precursors in the adult brain after stroke. *Nat Med* 2002;**8**:963–70.
201. Chu K, Kim M, Jeong SW, Kim SU, Yoon BW. Human neural stem cells can migrate, differentiate, and integrate after intravenous transplantation in adult rats with transient forebrain ischemia. *Neurosci Lett* 2003;**343**:129–33.
202. Jeong SW, Chu K, Jung KH, Kim SU, Kim M, Roh JK. Human neural stem cell transplantation promotes functional recovery in rats with experimental intracerebral hemorrhage. *Stroke* 2003;**34**:2258–63.
203. Nakaguchi K, Masuda H, Kaneko N, Sawamoto K. Strategies for regenerating striatal neurons in the adult brain by using endogenous neural stem cells. *Neurol Res Int* 2011;**2011**:898012.
204. Lapidot T, Sirard C, Vormoor J, Murdoch B, Hoang T, Caceres-Cortes J, et al. A cell initiating human acute myeloid leukaemia after transplantation into SCID mice. *Nature* 1994;**367**:645–8.
205. Sutherland HJ, Blair A, Zapf RW. Characterization of a hierarchy in human acute myeloid leukemia progenitor cells. *Blood* 1996;**87**:4754–61.
206. Bonnet D, Dick JE. Human acute myeloid leukemia is organized as a hierarchy that originates from a primitive hematopoietic cell. *Nat Med* 1997;**3**:730–7.
207. Blair A, Hogge DE, Ailles LE, Lansdorp PM, Sutherland HJ. Lack of expression of Thy-1 (CD90) on acute myeloid leukemia cells with long-term proliferative ability *in vitro* and *in vivo*. *Blood* 1997;**89**:3104–12.
208. Jordan CT, Upchurch D, Szilvassy SJ, Guzman ML, Howard DS, Pettigrew AL, et al. The interleukin-3 receptor α chain is a unique marker for human acute myelogenous leukemia stem cells. *Leukemia* 2000;**14**:1777–84.

209. Cox CV, Diamanti P, Evely RS, Kearns PR, Blair A. Expression of CD133 on leukemia-initiating cells in childhood ALL. *Blood* 2009;**113**:3287–96.
210. George AA, Franklin J, Kerkof K, Shah AJ, Price M, Tsark E, et al. Detection of leukemic cells in the CD34($^+$)CD38($^-$) bone marrow progenitor population in children with acute lymphoblastic leukemia. *Blood* 2001;**97**:3925–30.
211. Singh SK, Clarke ID, Terasaki M, Bonn VE, Hawkins C, Squire J, et al. Identification of a cancer stem cell in human brain tumors. *Cancer Res* 2003;**63**:5821–8.
212. Singh SK, Hawkins C, Clarke ID, Squire JA, Bayani J, Hide T, et al. Identification of human brain tumour initiating cells. *Nature* 2004;**432**:396–401.
213. Al-Hajj M, Wicha MS, Benito-Hernandez A, Morrison SJ, Clarke MF. Prospective identification of tumorigenic breast cancer cells. *Proc Natl Acad Sci USA* 2003;**100**:3983–8.
214. Stingl J, Eirew P, Ricketson I, Shackleton M, Vaillant F, Choi D, et al. Purification and unique properties of mammary epithelial stem cells. *Nature* 2006;**439**:993–7.
215. Higashi K, Ueda Y, Seki H, Yuasa K, Oguchi M, Noguchi T, et al. Fluorine-18-FDG PET imaging is negative in bronchioloalveolar lung carcinoma. *J Nucl Med* 1998;**39**:1016–20.
216. Eramo A, Lotti F, Sette G, Pilozzi E, Biffoni M, Di Virgilio A, et al. Identification and expansion of the tumorigenic lung cancer stem cell population. *Cell Death Differ* 2008;**15**:504–14.
217. Collins AT, Berry PA, Hyde C, Stower MJ, Maitland NJ. Prospective identification of tumorigenic prostate cancer stem cells. *Cancer Res* 2005;**65**:10946–51.
218. Ma S, Chan KW, Hu L, Lee TK, Wo JY, Ng IO, et al. Identification and characterization of tumorigenic liver cancer stem/progenitor cells. *Gastroenterology* 2007;**132**:2542–56.
219. Yang ZF, Ho DW, Ng MN, Lau CK, Yu WC, Ngai P, et al. Significance of CD90$^+$ cancer stem cells in human liver cancer. *Cancer Cell* 2008;**13**:153–66.
220. Ricci-Vitiani L, Lombardi DG, Pilozzi E, Biffoni M, Todaro M, Peschle C, et al. Identification and expansion of human colon-cancer-initiating cells. *Nature* 2007;**445**:111–5.
221. Dalerba P, Dylla SJ, Park IK, Liu R, Wang X, Cho RW, et al. Phenotypic characterization of human colorectal cancer stem cells. *Proc Natl Acad Sci USA* 2007;**104**:10158–63.
222. Bachoo RM, Maher EA, Ligon KL, Sharpless NE, Chan SS, You MJ, et al. Epidermal growth factor receptor and Ink4a/Arf: convergent mechanisms governing terminal differentiation and transformation along the neural stem cell to astrocyte axis. *Cancer Cell* 2002;**1**:269–77.
223. Cozzio A, Passegue E, Ayton PM, Karsunky H, Cleary ML, Weissman IL. Similar MLL-associated leukemias arising from self-renewing stem cells and short-lived myeloid progenitors. *Genes Dev* 2003;**7**:3029–35.
224. Dean M. Cancer stem cells: redefining the paradigm of cancer treatment strategies. *Mol Interv* 2006;**6**:140–8.
225. Blair A, Sutherland HJ. Primitive acute myeloid leukemia cells with long-term proliferative ability *in vitro* and *in vivo* lack surface expression of c-kit (CD117). *Exp Hematol* 2000;**28**:660–71.
226. Jordan CT, Guzman ML. Mechanisms controlling pathogenesis and survival of leukemic stem cells. *Oncogene* 2004;**23**:7178–87.
227. Visvader JE. Cells of origin in cancer. *Nature* 2011;**469**:314–22.
228. Uchida N, Buck DW, He D, Reitsma MJ, Masek M, Phan TV, et al. Direct isolation of human central nervous system stem cells. *Proc Natl Acad Sci USA* 2000;**97**:14720–5.
229. Di Cristofano A, Pandolfi PP. The multiple roles of PTEN in tumor suppression. *Cell* 2000;**100**:387–90.

230. Reya T, Morrison SJ, Clarke MF, Weissman IL. Stem cells, cancer and cancer stem cells. *Nature* 2001;**404**:105–11.
231. Molofsky AV, He S, Bydon M, Morrison SJ, Pardal R. Bmi-1 promotes neural stem cell self-renewal and neural development but not mouse growth and survival by repressing the p16Ink4a and p19Arf senescence pathways. *Genes Dev* 2005;**19**:1432–7.
232. Latronico AC, Hochberg Z. G protein-coupled receptors in child development, growth, and maturation. *Sci Signal* 2010;**3**:re7.
233. Callihan P, Mumaw J, Machacek DW, Stice SL, Hooks SB. Regulation of stem cell pluripotency and differentiation by G protein coupled receptors. *Pharmacol Ther* 2011;**129**:290–306.
234. Layden BT, Newman M, Chen F, Fisher A, Lowe Jr WL. G protein coupled receptors in embryonic stem cells: a role for Gs-alpha signaling. *PLoS One* 2010;**5**:e9105.
235. D'Aniello C, Lonardo E, Iaconis S, Guardiola O, Liguoro AM, Liguori GL, et al. G protein-coupled receptor APJ and its ligand Apelin act downstream of Cripto to specify embryonic stem cells toward the cardiac lineage through extracellular signal-regulated kinase/p70S6 kinase signaling pathway. *Circ Res* 2009;**105**:231–8.
236. Faherty S, Fitzgerald A, Keohan M, Quinlan LR. Self-renewal and differentiation of mouse embryonic stem cells as measured by Oct4 expression: the role of the cAMP/PKA pathway. *In Vitro Cell Dev Biol Anim* 2007;**43**:37–47.
237. LaFramboise WA, Petrosko P, Krill-Burger JM, Morris DR, McCoy AR, Scalise D, et al. Proteins secreted by embryonic stem cells activate cardiomyocytes through ligand binding pathways. *J Proteomics* 2010;**73**:992–1003.
238. Walsh CT, Stupack D, Brown JH. G protein-coupled receptors go extracellular: RhoA integrates the integrins. *Mol Interv* 2008;**8**:165–73.
239. Kleger A, Busch T, Liebau S, Prelle K, Paschke S, Beil M, et al. The bioactive lipid sphingosylphosphorylcholine induces differentiation of mouse embryonic stem cells and human promyelocytic leukaemia cells. *Cell Signal* 2007;**19**:367–77.
240. Inniss K, Moore H. Mediation of apoptosis and proliferation of human embryonic stem cells by sphingosine-1-phosphate. *Stem Cells Dev* 2006;**15**:789–96.
241. Schuck S, Soloaga A, Schratt G, Arthur JS, Nordheim A. The kinase MSK1 is required for induction of c-fos by lysophosphatidic acid in mouse embryonic stem cells. *BMC Mol Biol* 2003;**4**:6.
242. Bosier B, Muccioli GG, Hermans E, Lambert DM. Functionally selective cannabinoid receptor signalling: therapeutic implications and opportunities. *Biochem Pharmacol* 2010;**80**:1–12.
243. Force T, Woulfe K, Koch WJ, Kerkelä R. Molecular scaffolds regulate bidirectional crosstalk between Wnt and classical seven-transmembrane-domain receptor signaling pathways. *Sci STKE* 2007;**397**:pe41.
244. Melchior K, Weiss J, Zaehres H, Kim YM, Lutzko C, Roosta N, et al. The WNT receptor FZD7 contributes to self-renewal signaling of human embryonic stem cells. *Biol Chem* 2008;**389**:897–903.
245. Southgate TD, McGinn OJ, Castro FV, Rutkowski AJ, Al-Muftah M, Marinov G, et al. CXCR4 mediated chemotaxis is regulated by 5T4 oncofetal glycoprotein in mouse embryonic cells. *PLoS One* 2010;**5**:e9982.
246. Laslett AL, Grimmond S, Gardiner B, Stamp L, Lin A, Hawes SM, et al. Transcriptional analysis of early lineage commitment in human embryonic stem cells. *BMC Dev Biol* 2007;**7**:12.
247. Wong RC, Davidson KC, Leung J, Pera MF, Pebay A. Acute effect of endothelins on intracellular communication of human embryonic stem cells. *J Stem Cells* 2009;**4**:47–55.
248. Wysoczynski M, Reca R, Ratajczak J, Kucia M, Shirvaikar N, Honczarenko M, et al. Incorporation of CXCR4 into membrane lipid rafts primes homing-related responses of hematopoietic stem/progenitor cells to an SDF-1 gradient. *Blood* 2005;**105**:40–8.

249. Bautz F, Denzlinger C, Kanz L, Mohle R. Chemotaxis and transendothelial migration of CD34(+) hematopoietic progenitor cells induced by the inflammatory mediator leukotriene D4 are mediated by the 7-transmembrane receptor CysLT1. *Blood* 2001;**97**:3433–40.
250. Xue X, Cai Z, Seitz G, Kanz L, Moehle R, Weisel K. Differential effects of G protein-coupled receptors on hematopoietic progenitor cell growth depends on their signaling capacities. *Ann N Y Acad Sci* 2007;**1106**:180–9.
251. Liu Y, Wada R, Yamashit T, Mi Y, Deng CX, Hobson JP, et al. Edg-1, the G protein-coupled receptor for sphingosine-1-phosphate, is essential for vascular maturation. *J Clin Invest* 2000;**106**:951–61.
252. Liu YB, Kharode Y, Bodine PV, Yaworsky PJ, Robinson JA, Billiard J. LPA induces osteoblast differentiation through interplay of two receptors: LPA1 and LPA4. *J Cell Biochem* 2010;**109**:794–800.
253. Dupuis F, Desplat V, Praloran V, Denizot Y. Effects of lipidic mediators on the growth of human myeloid and erythroid marrow progenitors. *J Lipid Mediat Cell Signal* 1997;**16**:117–25.
254. Lord AM, North TE, Zon LI. Prostaglandin E2: making more of your marrow. *Cell Cycle* 2007;**6**:3054–7.
255. Weinreb M, Shamir D, Machwate M, Rodan GA, Harada S, Keila S. Prostaglandin E2 (PGE2) increases the number of rat bone marrow osteogenic stromal cells (BMSC) via binding the EP4 receptor, activating sphingosine kinase and inhibiting caspase activity. *Prostaglandins Leukot Essent Fatty Acids* 2006;**75**:81–90.
256. Brunner S, Engelmann MG, Franz WM. Stem cell mobilisation for myocardial repair. *Expert Opin Biol Ther* 2008;**8**:1675–90.
257. Beltrami AP, Barlucchi L, Torella D, Baker M, Limana F, Chimenti S, et al. Adult cardiac stem cells are multipotent and support myocardial regeneration. *Cell* 2003;**114**:763–76.
258. Laugwitz KL, Moretti A, Lam J, Gruber P, Chen Y, Woodard S, et al. Postnatal isl1+ cardioblasts enter fully differentiated cardiomyocyte lineages. *Nature* 2005;**433**:647–53 Erratum in: Nature 2007;446:934.
259. Solloway MJ, Harvey RP. Molecular pathways in myocardial development: a stem cell perspective. *Cardiovasc Res* 2003;**58**:264–77.
260. Scott IC, Masri B, D'Amico LA, Jin SW, Jungblut B, Wehman AM, et al. The g protein-coupled receptor agtrl1b regulates early development of myocardial progenitors. *Dev Cell* 2007;**12**:403–13.
261. Perez DM, Doze VA. GPCRs in adult neurogenesis. *Pharmacol Rev* 2012;**64**:645–75.
262. Cravatt BF, Demarest K, Patricelli MP, Bracey MH, Giang DK, Martin BR, et al. Supersensitivity to anandamide and enhanced endogenous cannabinoid signaling in mice lacking fatty acid amide hydrolase. *Proc Natl Acad Sci USA* 2001;**98**:9371–6.
263. Jin K, Xie L, Kim SH, Parmentier-Batteur S, Sun Y, Mao XO, et al. Defective adult neurogenesis in CB1 cannabinoid receptor knockout mice. *Mol Pharmacol* 2004;**66**:204–8.
264. Kim SH, Won SJ, Mao XO, Ledent C, Jin K, Greenberg DA. Role for neuronal nitric-oxide synthase in cannabinoid-induced neurogenesis. *J Pharmacol Exp Ther* 2006;**319**:150–4.
265. Ma Q, Jones D, Borghesani PR, Segal RA, Nagasawa T, Kishimoto T, et al. Impaired B-lymphopoiesis, myelopoiesis, and derailed cerebellar neuron migration in CXCR4- and SDF-1-deficient mice. *Proc Natl Acad Sci USA* 1998;**95**:9448–53.
266. Zou YR, Kottmann AH, Kuroda M, Taniuchi I, Littman DR. Function of the chemokine receptor CXCR4 in haematopoiesis and in cerebellar development. *Nature* 1998;**393**:595–9.

267. Stumm RK, Zhou C, Ara T, Lazarini F, Dubois-Dalcq M, Nagasawa T, et al. CXCR4 regulates interneuron migration in the developing neocortex. *J Neurosci* 2003;**23**:5123–30.
268. Stumm R, Hollt V. CXC chemokine receptor 4 regulates neuronal migration and axonal pathfinding in the developing nervous system: implications for neuronal regeneration in the adult brain. *J Mol Endocrinol* 2007;**38**:377–82.
269. Harada J, Foley M, Moskowitz MA, Waeber C. Sphingosine-1-phosphate induces proliferation and morphological changes of neural progenitor cells. *J Neurochem* 2004;**88**:1026–39.
270. Terai K, Soga T, Takahashi M, Kamohara M, Ohno K, Yatsugi S, et al. Edg-8 receptors are preferentially expressed in oligodendrocyte lineage cells of the rat CNS. *Neuroscience* 2003;**116**:1053–62.
271. Miron VE, Jung CG, Kim HJ, Kennedy TE, Soliven B, Antel JP. FTY720 modulates human oligodendrocyte progenitor process extension and survival. *Ann Neurol* 2008;**63**:61–71.
272. Coelho RP, Payne SG, Bittman R, Spiegel S, Sato-Bigbee C. The immunomodulator FTY720 has a direct cytoprotective effect in oligodendrocyte progenitors. *J Pharmacol Exp Ther* 2007;**323**:626–35.
273. Mandyam CD, Norris RD, Eisch AJ. Chronic morphine induces premature mitosis of proliferating cells in the adult mouse subgranular zone. *J Neurosci Res* 2004;**76**:783–94.
274. Mandyam CD, Harburg GC, Eisch AJ. Determination of key aspects of precursor cell proliferation, cell cycle length and kinetics in the adult mouse subgranular zone. *Neuroscience* 2007;**146**:108–22.
275. Gressens P, Paindaveine B, Hill JM, Evrard P, Brenneman DE. Vasoactive intestinal peptide shortens both G1 and S phases of neural cell cycle in whole postimplantation cultured mouse embryos. *Eur J Neurosci* 1998;**10**:1734–42.
276. Mishra SK, Braun N, Shukla V, Füllgrabe M, Schomerus C, Korf HW, et al. Extracellular nucleotide signaling in adult neural stem cells: synergism with growth factor-mediated cellular proliferation. *Development* 2006;**133**:675–84.
277. Sanai N, Alvarez-Buylla A, Berger MS. Neural stem cells and the origin of gliomas. *N Engl J Med* 2005;**353**:811–22.
278. Galanis E, Buckner JC, Maurer MJ, Kreisberg JI, Ballman K, Boni J, et al. Phase II trial of temsirolimus (CCI-779) in recurrent glioblastoma multiforme: a North Central Cancer Treatment Group Study. *J Clin Oncol* 2005;**23**:5294–304.
279. Chang SM, Wen P, Cloughesy T, Greenberg H, Schiff D, Conrad C, et al. Phase II study of CCI-779 in patients with recurrent glioblastoma multiforme. *Invest New Drugs* 2005;**23**:357–61.
280. Kuhn JG, Chang SM, Wen PY, Cloughesy TF, Greenberg H, Schiff D, et al. Pharmacokinetic and tumor distribution characteristics of temsirolimus in patients with recurrent malignant glioma. *Clin Cancer Res* 2007;**13**:7401–6.
281. Karim-Kos HE, de Vries E, Soerjomataram I, Lemmens V, Siesling S, Coebergh JW. Recent trends of cancer in Europe: a combined approach of incidence, survival and mortality for 17 cancer sites since the 1990s. *Eur J Cancer* 2008;**44**:1345–89.
282. Dubocovich ML, Rivera-Bermudez MA, Gerdin MJ, Masana MI. Molecular pharmacology, regulation and function of mammalian melatonin receptors. *Front Biosci* 2003;**8**: d1093–d1108.
283. Tan DX, Manchester LC, Terron MP, Flores LJ, Reiter RJ. One molecule, many derivatives: a never-ending interaction of melatonin with reactive oxygen and nitrogen species? *J Pineal Res* 2007;**42**:28–42.
284. Aust S, Jaeger W, Klimpfinger M, Mayer K, Baravalle G, Ekmekcioglu C, et al. Biotransformation of melatonin in human breast cancer cell lines: role of sulfotransferase 1A1. *J Pineal Res* 2005;**39**:276–82.

285. Aust S, Obrist P, Klimpfinger M, Tucek G, Jäger W, Thalhammer T. Altered expression of the hormone- and xenobiotic-metabolizing sulfotransferase enzymes 1A2 and 1C1 in malignant breast tissue. *Int J Oncol* 2005;**26**:1079–85.
286. Rögelsperger O, Wlcek K, Ekmekcioglu C, Humpeler S, Svoboda M, Königsberg R, et al. Melatonin receptors, melatonin metabolizing enzymes and cyclin D1 in human breast cancer. *J Recept Signal Transduct Res* 2011;**31**:180–7.

CHAPTER SIX

Application of Receptor Theory to Allosteric Modulation of Receptors

David A. Hall
Fibrosis DPU, GlaxoSmithKline, Stevenage, Herts, United Kingdom

Contents

Abstract

In this chapter, three topics are considered. The allosteric two-state model (ATSM) is compared with explicit multiconformational models. This demonstrates that the ATSM encapsulates the common behaviors of any model with at least two active and two inactive conformations: the "states" of the model represent ensembles of active and inactive conformations. A matrix representation of multiconformational models is introduced to provide a compact notation for models with arbitrarily large numbers of conformations. Allosteric modulation is further explored in the context of an operational model of receptor activity which includes constitutive receptor activity. Fitting this model allows the apparent affinity, intrinsic efficacy, and cooperativity constants of a

Progress in Molecular Biology and Translational Science, Volume 115
ISSN 1877-1173
http://dx.doi.org/10.1016/B978-0-12-394587-7.00006-3

pair of allosteric ligands to be determined. It is also demonstrated that, within certain limits, it is possible to estimate the parameters of the ATSM. Finally, a novel operational model is developed that may allow the analysis of protean ligands. This model requires a nonlinear stimulus function and two parameters to define the efficacy of a ligand. Expressions describing competitive and allosteric interactions under this model are developed and the results of applying null analyses to the data are determined.

1. INTRODUCTION

When I was considering what to present in this chapter, I decided that adding another simple review of receptor theory as it relates to allosteric modulation of pharmacological receptors would not be particularly helpful. There have been a number of such reviews, some quite recent, for example,[1–5] and adding another is unlikely to be particularly illuminating at this time. I decided, instead, to take the opportunity to further develop a small selection of theoretical strands and will assume that the reader is quite familiar with the basic concepts of allosteric modulation. In the first section, I will consider the allosteric two-state model (ATSM)[6] and its interpretation from the perspective of receptor conformational dynamics. This treatment will of necessity to reproduce some of the results of Onaran and coworkers[7–9] but will take a different and initially somewhat simpler approach that I hope will make the mathematics and conclusions more readily accessible. I will also define some limitations on the behavior of pharmacological systems with small numbers of conformational states. This will provide a framework based on the pharmacological behavior of a system to begin to provide lower limits to such questions as how many active states of a receptor are relevant to a particular pharmacological behavior? I will also demonstrate the use of mathematical matrix notation to represent the generalization of these models to large numbers of states. The second section will have a somewhat more practical feel and will deal with the extension of a model that I have recently developed[2,10] to allow the analysis of the interaction of allosteric ligands. There I will demonstrate that, with an appropriate experimental design, it is possible to derive the macroscopic apparent affinity, absolute intrinsic efficacy, and cooperativity constants for a pair of allosterically interacting ligands in a functional assay. I will also derive relationships between the parameters of this operational model and the more mechanistic models of receptor activation discussed in Section 1. Finally, in Section 3, I will continue the theme from Ref. 2 of investigating the "other extreme" of the ternary complex model of G protein-coupled receptor activation (gTCM).[11] Consideration of the gTCM under conditions where the G protein is limiting led to the initial development of the model of Slack and Hall[10] (in Ref. 2). In the final section, I will investigate the cubic ternary complex model (cTCM)[12–14] with the same assumption and use the resulting model to develop an operational model which may be useful to analyze systems that demonstrate protean agonism (ligands that can be agonists or inverse agonists depending on the conditions[15]) and explore the results of standard pharmacological null analysis on these models. All of the models considered in this chapter relate to the behavior of allosteric ligands at equilibrium and primarily in functional assays. The reader is referred to Ref. 16 for a comprehensive treatment of the analysis of allosteric interactions under pre-steady state conditions. I will also concentrate on allosteric regulation, the models I will present will not explicitly consider receptor oligomerization; however, they are applicable to any system in which ligand binding is well described by a Langmuir-binding isotherm irrespective of the underlying subunit structure of the receptor.

At this point I will make one apology, early letters in the Greek alphabet, particularly α, are used frequently as symbols for a variety of the parameters in pharmacological models. Rather than change these symbols in the models presented below, I have generally used the original notation (or on occasion the one I prefer). This means that α represents a number of different quantities in this chapter. In general, the context should make the meaning clear. Where I felt ambiguity was possible, I have provided subscripts to allow the different alphas to be distinguished.

2. THE ALLOSTERIC TWO-STATE MODEL

The ATSM[6] was really the first pharmacological model to explicitly consider the effects of allosteric modulators on the ability of orthosteric ligands to activate constitutively active receptors. For example, the classic work by Ehlert[17] was derived before the phenomenon of constitutive receptor activity[18] was widely accepted at G protein-coupled receptors and the allosteric modulator was assumed to be a neutral antagonist. The ATSM shows a variety of the behaviors that have been observed for allosteric modulators in functional assays (as described in Ref. 6). It also highlighted that the phenomenon of coagonism[19] can be treated as a form of allosteric modulation. In this case, two ligands that are not agonists via their respective binding sites have positive activation cooperativity resulting in a response only when both ligands are bound to the receptor. However, this model has been criticized[20] for being at "an intermediate level of analysis" since it deals with states of the receptor and hence a microscopic representation of the pharmacological system but includes allosteric constants that are a macroscopic phenomenon. This criticism raises the question as to whether the predictions and behavior of the model could be misleading and whether its structure can indeed be justified. The ATSM cannot be used for the analysis of functional responses (since it does not include signal transduction) so the provision of misleading parameter estimates is not a concern (although see Section 2). As a lead-in to addressing this issue, I will first consider the ternary complex model of allosteric modulation (aTCM)[21] (Scheme 6.1) since this provides a relatively simple system to illustrate the mathematical arguments.

2.1. Explicit conformational selectivity in an allosteric ternary complex model

The simplest possible multiconformational model of allosteric modulation of binding is shown in Scheme 6.2. This model has two conformations and is therefore closely analogous to the ATSM. What it lacks is any cooperativity constants since it explicitly considers the effect of the binding of one ligand on the distribution of the receptor conformations and how this may affect the apparent affinity of the second ligand. The expression for the binding isotherm for ligand A in the aTCM is

$$\mathrm{Bound\,A} = \frac{[R]_T K[A](1+\alpha M[B])}{1+M[B]+K[A](1+\alpha M[B])} \qquad [6.1]$$

where the symbols are defined in the caption to Scheme 6.1. As formulated here, $\alpha > 1$ indicates positive cooperativity. The corresponding expression for Scheme 6.2 is

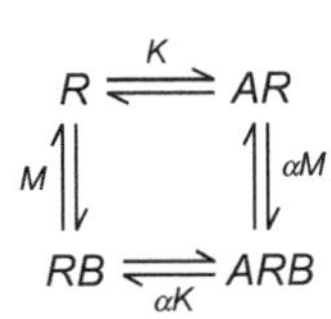

Scheme 6.1 The allosteric ternary complex model. Ligands *A* and *B* bind to distinct sites on the receptor (*R*). *K* and *M* are the association equilibrium-binding constants of *A* and *B*, respectively, and α is the binding cooperativity constant. Unbound ligands have been omitted for clarity. *Reproduced with permission from Ref. 2.*

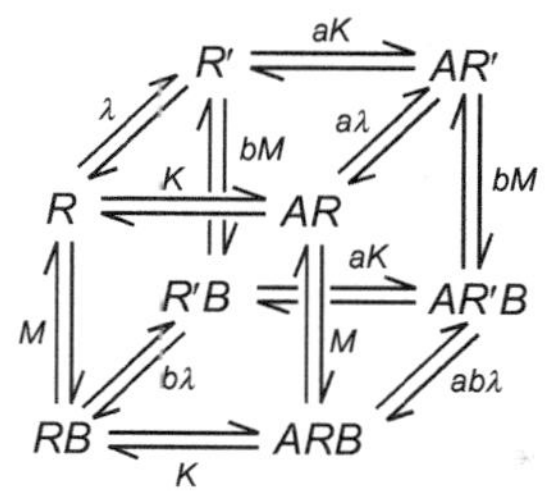

Scheme 6.2 A two-conformation allosteric ternary complex model. The receptor occurs in two conformations in this model, *R* and *R'*. *K* and *M* are the association equilibrium-binding constants of *A* and *B*, respectively, for *R*. The association equilibrium constants for *A* and *B* to *R'* are *aK* and *bM*. The receptor isomerization constants are then λ for the free receptor, $a\lambda$ when *A* is bound, $b\lambda$ when *B* is bound, and $ab\lambda$ when both ligands are bound. Unbound ligands have been omitted for clarity.

$$\text{Bound A} = \frac{[R]_T\kappa[A](1+a\lambda+\mu[B](1+ab\lambda))}{1+\lambda+\mu[B](1+b\lambda)+\kappa[A](1+a\lambda+\mu[B](1+ab\lambda))} \quad [6.2]$$

It is now possible to determine how the parameters of Eqs. (6.1) and (6.2) are related. Initially, it will be assumed that the only constraints on the values of the parameters of Eq. (6.2) are that they be positive and finite. The apparent affinity of *A* from Eq. (6.1) is

$$K_{\text{app1}} = \frac{1+M[B]}{K(1+\alpha M[B])} \quad [6.3]$$

While that from Eq. (6.2) is

$$K_{\text{app2}} = \frac{1+\lambda+\mu[B](1+b\lambda)}{\kappa(1+a\lambda+\mu[B](1+ab\lambda))} \quad [6.4]$$

The right-hand sides of Eqs. (6.3) and (6.4) must be equal. When $[B] = 0$ this gives

$$\frac{1}{K} = \frac{1+\lambda}{\kappa(1+a\lambda)} \Rightarrow K = \frac{\kappa(1+a\lambda)}{1+\lambda} \quad [6.5]$$

and hence, by analogy,

$$M = \frac{\mu(1+b\lambda)}{1+\lambda} \quad [6.6]$$

When $[B] \rightarrow \infty$,

$$\frac{1}{\alpha K} = \frac{1+b\lambda}{\kappa(1+ab\lambda)} \Rightarrow \alpha K = \frac{\kappa(1+ab\lambda)}{1+b\lambda}$$

Again, by analogy,

$$\alpha M = \frac{\mu(1+ab\lambda)}{1+a\lambda}$$

Therefore,

$$\alpha = \frac{\alpha K}{K} = \frac{\kappa(1+ab\lambda)}{1+b\lambda} \times \frac{1+\lambda}{\kappa(1+a\lambda)} = \frac{(1+\lambda)(1+ab\lambda)}{(1+a\lambda)(1+b\lambda)} \quad [6.7]$$

Thus, it is possible to derive the parameters of Scheme 6.1 from those of Scheme 6.2. A further question is, are there any unanticipated constraints applied to the parameters of the aTCM that become apparent from the conformationally explicit treatment? This is equivalent to asking whether it is possible to identify a set of values for the parameters of Scheme 6.2 that correspond to any arbitrary set of values for the parameters of Scheme 6.1 (this need not and generally will not be unique). It is clear that there are no constraints imposed on the values of K and M since for any given values of a, b, and λ we are free to choose the value of κ and μ to achieve any given value of K or M. This argument also implies that a required value of K or M can be achieved from some acceptable combination of values of a, b, κ, λ, and μ.

This is also true of α. First note that, when both a and b are large, Eq. (6.7) simplifies to

$$\alpha \approx \frac{ab\lambda(1+\lambda)}{ab\lambda^2} = \frac{1+\lambda}{\lambda}$$

which immediately shows that all values of α greater than unity are allowed since the limiting values of this expression are λ^{-1} when $\lambda \ll 1$ and 1 when $\lambda \gg 1$ (of course, for any given value of λ, the former defines the maximum possible binding cooperativity). Also, since a, b, and λ are positive, the denominator of Eq. (6.7) cannot be equal to zero so α is defined for all acceptable values of these parameters. Similarly, when both a and b are small, Eq. (6.7) simplifies to $\alpha = 1 + \lambda$ showing again that all values of α greater than unity are allowed. Now let terms in a dominate in Eq. (6.7), this gives

$$\alpha \approx \frac{ab\lambda(1+\lambda)}{a\lambda(1+b\lambda)} = \frac{b(1+\lambda)}{(1+b\lambda)}$$

The limit of this expression when $b \ll 1$ is $b(1+\lambda)$ (the case with $b \gg 1$ has already been considered). Thus, for $a \gg 1$ and $b < 1$, α may take any value in the range $0 < \alpha \leq 1$. A similar argument applies to large b and small a. Hence, when a and b are both large or both small α can be large, while arbitrarily small values of α occur when a and b are of very different magnitudes. This provides an explicit mathematical demonstration that cooperativity is a measure of the concordance (or discordance) of the conformational selectivity of the ligands. Further, it has now been demonstrated that any value of α can be achieved for some combination of a, b, and λ and that each combination of a, b, and λ corresponds to some acceptable value of α. Also, as noted above, since we are free to choose κ and μ, the constraining of a, b, and λ to give a particular value of α does not restrict the values that K and M can take. The two models are, therefore, equivalent since it is possible to map the parameters of either one onto the other. While it may be obvious, it is still worth stating that this shows that the aTCM (Scheme 6.1) must be interpreted as a multiconformational model and that allosteric modulation requires multiple receptor conformations. If only one receptor conformation existed (which is equivalent to letting $\lambda = 0$ in Scheme 6.2), Eq. (6.4) would become

$$K_{\mathrm{app2}} = \frac{1}{\kappa}$$

and there could be no influence of the presence of B on the affinity of A.

If the reference conformation of the receptor, R in this case, is assumed to be the thermodynamically most stable, and hence lowest energy, conformation of the receptor, then it must also be the most abundant conformation at equilibrium in the absence of ligands. This then requires that $\lambda < 1$ (since $[R] = [R]_T/(1+\lambda)$ and $[R'] = \lambda[R]_T/(1+\lambda)$). This will also be true of all of the other conformationally specified models in this chapter, if the reference state is assumed to be of lowest energy. Given the arguments above, this additional constraint does not affect the possible values of α when mapping Scheme 6.2 onto Scheme 6.1.

2.2. Mapping multiconformational models onto the allosteric two-state model

Having defined a conformationally explicit model for the binding interaction of allosteric ligands, it is now possible to extend the model to include active states and to compare the resulting models with the ATSM. One model will be explicitly exemplified along with two of its special cases before a more general result is derived. The first case (Scheme 6.3A) retains the two inactive conformations of Scheme 6.2 and introduces two active states. All these conformations are assumed to be mutually accessible (the justification of the generality of this assumption is given below). A closely analogous

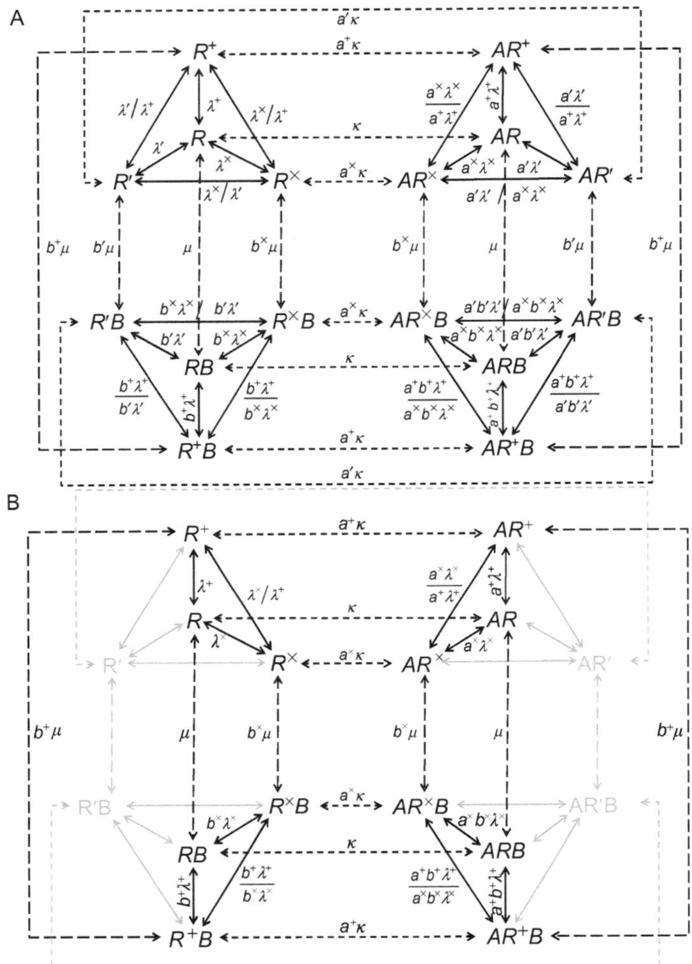

Scheme 6.3 (A) A four-conformation model of allosteric modulation with two active and two inactive conformations. κ and μ are the association equilibrium binding constants of *A* and *B*, respectively, for *R*. The association equilibrium constants of *A* and *B* for R' are $a'\kappa$ and $b'\mu$, for R^+ are $a^+\kappa$ and $b^+\mu$, and for $R^\times$ are $a^\times\kappa$ and $b^\times\mu$. The affinity of a ligand for a given receptor conformation is the same whether or not other ligands are bound. The isomerization equilibrium constants for the *formation* of R', R^+, or $R^\times$ from *R* are λ', λ^+, and $\lambda^\times$, respectively. The isomerization constants for interconversion of the indexed receptor species are ratios of these constants (product as numerator, reactant as denominator so, for example, formation of R^+ from R' has isomerization constant λ^+/λ'). The sense of the equilibrium constants in the state interconversion cycles are conversion from center to periphery, then top to bottom, and finally left to right. The reverse reaction is governed by the reciprocal of the stated isomerization constant. The isomerization constants for ligand receptor complexes are formed by multiplying by the relevant binding selectivity constants (a', a^+, $a^\times$, b', b^+, and $b^\times$) for the bound ligand or ligands. (B) Panel (A) in which $\lambda'=0$. Reactions that can no longer occur have been "grayed out" and their equilibrium constants deleted.

model was considered by Ehlert and Griffin[20]. Indeed, to use the terminology presented by Kenakin and coworkers,[14] these models are part of the same "EQ-equivalence class." However, the assumption that all receptor states are mutually accessible results in mathematical expressions that are of simpler form and highlights the parallels with the work of Onaran and coworkers[7–9] more clearly. The binding isotherm of A from this system is

$$\text{Bound A} = \frac{[R]_T\kappa[A]\begin{pmatrix}1+a'\lambda'+a^{+}\lambda^{+}+a^{\times}\lambda^{\times}\\ +\mu[B]\left(1+a'b'\lambda'+a^{+}b^{+}\lambda^{+}+a^{\times}b^{\times}\lambda^{\times}\right)\end{pmatrix}}{\begin{array}{c}1+\lambda'+\lambda^{+}+\lambda^{\times}+\mu[B]\left(1+b'\lambda'+b^{+}\lambda^{+}+b^{\times}\lambda^{\times}\right)+\\ \kappa[A](1+a'\lambda'+a^{+}\lambda^{+}+a^{\times}\lambda^{\times}+\mu[B](1+a'b'\lambda'+a^{+}b^{+}\lambda^{+}+a^{\times}b^{\times}\lambda^{\times}))\end{array}} \quad [6.8]$$

The analogous expression from the ATSM (from Ref. 6) is

$$\text{Bound A} = \frac{[R]_T K[A](1+\alpha L+\gamma M[B](1+\alpha\beta\delta L))}{1+L+M[B](1+\beta L)+K[A](1+\alpha L+\gamma M[B](1+\alpha\beta\delta L))} \quad [6.9]$$

The apparent affinities (and also EC_{50}s) derived from these expressions are (see Ref. 2 for a discussion of the derivation of the upper and lower asymptotes and midpoint of a linear rational function by inspection)

$$K_{\text{app8}} = \frac{1+\lambda'+\lambda^{+}+\lambda^{\times}+\mu[B]\left(1+b'\lambda'+b^{+}\lambda^{+}+b^{\times}\lambda^{\times}\right)}{\kappa\left(1+a'\lambda'+a^{+}\lambda^{+}+a^{\times}\lambda^{\times}+\mu[B]\left(1+a'b'\lambda'+a^{+}b^{+}\lambda^{+}+a^{\times}b^{\times}\lambda^{\times}\right)\right)} \quad [6.10]$$

$$K_{\text{app9}} = \frac{1+L+M[B](1+\beta L)}{K(1+\alpha L+\gamma M[B](1+\alpha\beta\delta L))} \quad [6.11]$$

The expressions for functional responses from these two models are

$$\frac{E}{[R]_T} = \frac{\begin{array}{c}\lambda^{+}+\lambda^{\times}+\mu[B]\left(b^{+}\lambda^{+}+b^{\times}\lambda^{\times}\right)\\ +\kappa[A]\left(a^{+}\lambda^{+}+a^{\times}\lambda^{\times}+\mu[B]\left(a^{+}b^{+}\lambda^{+}+a^{\times}b^{\times}\lambda^{\times}\right)\right)\end{array}}{\begin{array}{c}1+\lambda'+\lambda^{+}+\lambda^{\times}+\mu[B]\left(1+b'\lambda'+b^{+}\lambda^{+}+b^{\times}\lambda^{\times}\right)+\\ \kappa[A]\left(1+a'\lambda'+a^{+}\lambda^{+}+a^{\times}\lambda^{\times}+\mu[B]\left(1+a'b'\lambda'+a^{+}b^{+}\lambda^{+}+a^{\times}b^{\times}\lambda^{\times}\right)\right)\end{array}} \quad [6.12]$$

and

$$\frac{E}{[R]_T} = \frac{L(1+\beta M[B]+\alpha K[A](1+\beta\gamma\delta M[B]))}{1+L+M[B](1+\beta L)+K[A](1+\alpha L+\gamma M[B](1+\alpha\beta\delta L))} \quad [6.13]$$

Prior to deriving the correspondences between the parameters of the ATSM and Scheme 6.3, the behavior of the system in the presence of a single ligand will be discussed. When $[B]=0$, Eq. (6.8) becomes

$$\text{Bound A} = \frac{[R]_T\kappa[A]\left(1+d'\lambda'+a^+\lambda^+ +a^\times\lambda^\times\right)}{1+\lambda'+\lambda^+ +\lambda^\times +\kappa[A]\left(1+d'\lambda'+a^+\lambda^+ +a^\times\lambda^\times\right)} \qquad [6.14]$$

and Eq. (6.12) becomes

$$\frac{E}{[R]_T} = \frac{\lambda^+ +\lambda^\times +\kappa[A]\left(a^+\lambda^+ +a^\times\lambda^\times\right)}{1+\lambda'+\lambda^+ +\lambda^\times +\kappa[A]\left(1+d'\lambda'+a^+\lambda^+ +a^\times\lambda^\times\right)} \qquad [6.15]$$

The midpoint of these two expressions is

$$K_{\text{app8}} = \text{EC}_{50} = \frac{1+\lambda'+\lambda^+ +\lambda^\times}{\kappa\left(1+d'\lambda'+a^+\lambda^+ +a^\times\lambda^\times\right)}$$

For the present discussion, it will be assumed that the different active conformations of the receptor generate an equivalent response. The implications of differing interactions of the active conformations with transducer proteins will be considered later. The basal activity from Eq. (6.15) is

$$Basal = \frac{[R]_T\left(\lambda^+ +\lambda^\times\right)}{1+\lambda'+\lambda^+ +\lambda^\times}$$

and the maximal response to A (Max_A) is

$$Max_A = \frac{[R]_T\left(a^+\lambda^+ +a^\times\lambda^\times\right)}{1+d'\lambda'+a^+\lambda^+ +a^\times\lambda^\times}$$

Hence, a ligand is an agonist if

$$\frac{[R]_T\left(a^+\lambda^+ +a^\times\lambda^\times\right)}{1+d'\lambda'+a^+\lambda^+ +a^\times\lambda^\times} > \frac{[R]_T\left(\lambda^+ +\lambda^\times\right)}{1+\lambda'+\lambda^+ +\lambda^\times}$$

$$\left(a^+\lambda^+ +a^\times\lambda^\times\right)\left(1+\lambda'+\lambda^+ +\lambda^\times\right) > \left(\lambda^+ +\lambda^\times\right)\left(1+d'\lambda'+a^+\lambda^+ +a^\times\lambda^\times\right)$$

$$\left(a^+\lambda^+ +a^\times\lambda^\times\right)\left(1+\lambda'\right) > \left(\lambda^+ +\lambda^\times\right)\left(1+d'\lambda'\right)$$

$$\frac{\left(a^+\lambda^+ +a^\times\lambda^\times\right)\left(1+\lambda'\right)}{\left(\lambda^+ +\lambda^\times\right)\left(1+d'\lambda'\right)} > 1 \qquad [6.16]$$

In the spirit of mathematical rigor, this statement should be justified. It is true because Eq. (6.12) is a linear rational function, that is, it is of the form

$$y = \frac{a+bx}{c+dx} \qquad [6.17]$$

where c and d cannot both equal zero and a, b, c, and d may be functions of variables other than x. Linear rational functions have no local minima or maxima for finite real x (see below). Hence, the graph of the function is either strictly increasing, strictly decreasing, or it is a horizontal line. In other words, apart from the horizontal case, the graph rises or falls smoothly from *basal* to Max_A. To show this, it is sufficient to show that the first derivative of Eq. (6.17) with respect to x cannot be zero for any x unless it is zero for all x. Using the quotient rule for differentiation,

$$\begin{aligned}\frac{\partial y}{\partial x} &= \frac{b(c+dx)-d(a+bx)}{(c+dx)^2}\\ &= \frac{bc+bdx-ad-bdx}{(c+dx)^2}\\ &= \frac{bc-ad}{(c+dx)^2}\end{aligned}$$

Thus, the gradient of Eq. (6.17) can only equal zero if $bc = ad$ but in this case it is zero for all finite real x and the graph of y against x is a horizontal line (which has no local extrema). Otherwise, the gradient is zero for no finite value of x and there are no local extrema. Hence, a ligand is an agonist if $Max_A > basal$.

The obvious ways in which inequality (6.16) can be satisfied are if $a^+ > 1$, $a^\times > 1$ or $a' < 1$, that is, if the ligand stabilizes an active conformation or destabilizes an inactive conformation. More generally, the overall effect of an agonist ligand must be to favor formation of the active conformations (which is simply the physical interpretation of inequality 6.16). Conversely, therefore, a ligand is an inverse agonist if

$$\frac{(a^+\lambda^+ + a^\times\lambda^\times)(1+\lambda')}{(\lambda^+ + \lambda^\times)(1+a'\lambda')} < 1$$

Neutral antagonism then requires that

$$\frac{(a^+\lambda^+ + a^\times\lambda^\times)(1+\lambda')}{(\lambda^+ + \lambda^\times)(1+a'\lambda')} = 1$$

$$\begin{aligned}(1+a'\lambda') &= \frac{(a^+\lambda^+ + a^\times\lambda^\times)(1+\lambda')}{(\lambda^+ + \lambda^\times)}\\ a'\lambda' &= \frac{(a^+\lambda^+ + a^\times\lambda^\times)(1+\lambda') - (\lambda^+ + \lambda^\times)}{(\lambda^+ + \lambda^\times)}\\ a' &= \frac{\lambda^+(a^+(1+\lambda')-1) + \lambda^\times(a^\times(1+\lambda')-1)}{\lambda'(\lambda^+ + \lambda^\times)}\end{aligned}$$

The numerator of the last expression must not be negative. Both terms in the numerator are guaranteed to be positive if both $a^{+} > (1+\lambda')^{-1}$ and $a^{\times} > (1+\lambda')^{-1}$. An obvious solution to this equation is $a^{+} = a^{\times} = d' = 1$ but this is not the only possibility. Indeed, a solution exists for every combination of $a^{+} \geq 1$ and $a^{\times} \geq 1$ providing an infinite number of combinations of a^{+}, $a^{\times}$, and d' that result in neutral antagonism (even if we apply finite limits to their values). This statement is a little misleading, however, since, if a^{+} and $a^{\times}$ are fixed, only one value of d' can result in neutral antagonism for any fixed λ^{+}, $\lambda^{\times}$, and λ'. Thus, there are many more ways of not being a neutral antagonist than of being one. However, it remains the case that, unlike a simple two-state model (TSM),[22–24] a model with multiple active and inactive conformations allows a ligand that perturbs the equilibrium between the receptor conformations to be a neutral antagonist. Such a ligand could redistribute the inactive states without affecting the active states, enrich an active and an inactive conformation, or enrich one active conformation at the expense of another. The latter possibility provides a mechanism for coagonism among other behaviors (as exemplified below). Indeed, a ligand that does not perturb the conformational equilibrium of the receptor (i.e., for which $a^{+} = a^{\times} = d' = 1$), which will be referred to here as an absolute neutral antagonist, can have no effect on the equilibrium binding of or functional response to any allosteric ligand and cannot itself be allosterically regulated by any other ligand (it is neutrally cooperative with all allosteric ligands). This can be demonstrated by making the relevant substitutions into Eqs. (6.8) and (6.12) (and indeed Eqs. 6.25 and 6.26).

$$\text{Bound A} = \frac{[R]_T\kappa[A]\begin{pmatrix}1+d'\lambda'+a^{+}\lambda^{+}+a^{\times}\lambda^{\times}\\ +\mu[B]\left(1+d'b'\lambda'+a^{+}b^{+}\lambda^{+}+a^{\times}b^{\times}\lambda^{\times}\right)\end{pmatrix}}{\begin{array}{c}1+\lambda'+\lambda^{+}+\lambda^{\times}+\mu[B]\left(1+b'\lambda'+b^{+}\lambda^{+}+b^{\times}\lambda^{\times}\right)+\kappa[A]\\ \left(1+d'\lambda'+a^{+}\lambda^{+}+a^{\times}\lambda^{\times}+\mu[B]\left(1+d'b'\lambda'+a^{+}b^{+}\lambda^{+}+a^{\times}b^{\times}\lambda^{\times}\right)\right)\end{array}}$$

$$= \frac{[R]_T\kappa[A]\left(1+\lambda'+\lambda^{+}+\lambda^{\times}+\mu[B]\left(1+b'\lambda'+b^{+}\lambda^{+}+b^{\times}\lambda^{\times}\right)\right)}{\left(1+\lambda'+\lambda^{+}+\lambda^{\times}+\mu[B]\left(1+b'\lambda'+b^{+}\lambda^{+}+b^{\times}\lambda^{\times}\right)\right)(1+\kappa[A])}$$

$$= \frac{[R]_T\kappa[A]}{1+\kappa[A]}$$

This is simply the binding isotherm of A in the absence of other ligands. So there is no effect of an allosteric ligand on the binding of the absolute neutral antagonist and, by the requirement that the effects of allosteric ligands must be reciprocal, this also demonstrates that the absolute neutral antagonist has no effect on the binding of the allosteric ligand. Such a ligand does bind to the receptor, however, so would be a perfectly effective competitive antagonist. Also,

$$\frac{E}{[R]_T}=\frac{\lambda^{+}+\lambda^{\times}+\mu[B]\left(b^{+}\lambda^{+}+b^{\times}\lambda^{\times}\right)+\kappa[A]\left(a^{+}\lambda^{+}+a^{\times}\lambda^{\times}+\mu[B]\left(a^{+}b^{+}\lambda^{+}+a^{\times}b^{\times}\lambda^{\times}\right)\right)}{1+\lambda'+\lambda^{+}+\lambda^{\times}+\mu[B]\left(1+b'\lambda'+b^{+}\lambda^{+}+b^{\times}\lambda^{\times}\right)+\kappa[A]\left(1+a'\lambda'+a^{+}\lambda^{+}+a^{\times}\lambda^{\times}+\mu[B]\left(1+a'b'\lambda'+a^{+}b^{+}\lambda^{+}+a^{\times}b^{\times}\lambda^{\times}\right)\right)}$$

$$=\frac{\left(\lambda^{+}+\lambda^{\times}+\mu[B]\left(b^{+}\lambda^{+}+b^{\times}\lambda^{\times}\right)\right)(1+\kappa[A])}{\left(1+\lambda'+\lambda^{+}+\lambda^{\times}+\mu[B]\left(1+b'\lambda'+b^{+}\lambda^{+}+b^{\times}\lambda^{\times}\right)\right)(1+\kappa[A])}$$

$$=\frac{\lambda^{+}+\lambda^{\times}+\mu[B]\left(b^{+}\lambda^{+}+b^{\times}\lambda^{\times}\right)}{1+\lambda'+\lambda^{+}+\lambda^{\times}+\mu[B]\left(1+b'\lambda'+b^{+}\lambda^{+}+b^{\times}\lambda^{\times}\right)}$$

This is Eq. (6.15) for ligand B, thus, the presence of the absolute neutral antagonist does not influence responses to an allosteric ligand. Of course, since such a ligand can only arise from a unique set of selectivity constants, the probability that a real ligand will be an absolute neutral antagonist is essentially zero.

Comparing Eqs. (6.12) and (6.13), the response in the absence of A and B is

$$\frac{\left(\lambda^{+}+\lambda^{\times}\right)}{1+\lambda'+\lambda^{+}+\lambda^{\times}}=\frac{L}{1+L}$$

$$\lambda^{+}+\lambda^{\times}+L\left(\lambda^{+}+\lambda^{\times}\right)=L\left(1+\lambda'+\lambda^{+}+\lambda^{\times}\right)$$

$$L(1+\lambda')=\lambda^{+}+\lambda^{\times}$$

$$L=\frac{\lambda^{+}+\lambda^{\times}}{1+\lambda'} \qquad [6.18]$$

The maximal response to A in the absence of B is

$$\frac{a^{+}\lambda^{+}+a^{\times}\lambda^{\times}}{1+a'\lambda'+a^{+}\lambda^{+}+a^{\times}\lambda^{\times}}=\frac{\alpha L}{1+\alpha L}$$

$$a^{+}\lambda^{+}+a^{\times}\lambda^{\times}+\alpha L\left(a^{+}\lambda^{+}+a^{\times}\lambda^{\times}\right)=\alpha L\left(1+a'\lambda'+a^{+}\lambda^{+}+a^{\times}\lambda^{\times}\right)$$

$$a^{+}\lambda^{+}+a^{\times}\lambda^{\times}=\alpha L(1+a'\lambda')$$

$$\alpha L=\frac{a^{+}\lambda^{+}+a^{\times}\lambda^{\times}}{1+a'\lambda'}$$

$$\alpha=\frac{\alpha L}{L}$$

$$\alpha=\frac{\left(a^{+}\lambda^{+}+a^{\times}\lambda^{\times}\right)(1+\lambda')}{\left(\lambda^{+}+\lambda^{\times}\right)(1+a'\lambda')} \qquad [6.19]$$

Note then that inequality (6.16) is equivalent to the TSM criterion for agonism that $\alpha > 1$. It is also consistent with the expression derived by Onaran and Costa.[9] By analogy,

$$\beta = \frac{(b^{+}\lambda^{+} + b^{\times}\lambda^{\times})(1+\lambda')}{(\lambda^{+}+\lambda^{\times})(1+b'\lambda')} \qquad [6.20]$$

The response in the presence of saturating concentrations of both A and B is

$$\frac{a^{+}b^{+}\lambda^{+} + a^{\times}b^{\times}\lambda^{\times}}{1 + a'b'\lambda' + a^{+}b^{+}\lambda^{+} + a^{\times}b^{\times}\lambda^{\times}} = \frac{\alpha\beta\delta L}{1+\alpha\beta\delta L}$$

$$\alpha\beta\delta L = \frac{a^{+}b^{+}\lambda^{+} + a^{\times}b^{\times}\lambda^{\times}}{1+a'b'\lambda'}$$

$$\delta = \frac{\alpha\beta\delta L}{\alpha\beta L}$$

$$\delta = \frac{(\lambda^{+}+\lambda^{\times})(a^{+}b^{+}\lambda^{+} + a^{\times}b^{\times}\lambda^{\times})(1+a'\lambda')(1+b'\lambda')}{(a^{+}\lambda^{+} + a^{\times}\lambda^{\times})(b^{+}\lambda^{+} + b^{\times}\lambda^{\times})(1+\lambda')(1+a'b'\lambda')} \qquad [6.21]$$

In the absence of B the apparent affinity of A is

$$\frac{1+\lambda'+\lambda^{+}+\lambda^{\times}}{\kappa(1+a'\lambda'+a^{+}\lambda^{+}+a^{\times}\lambda^{\times})} = \frac{1+L}{K(1+\alpha L)}$$

$$= \frac{1+\dfrac{\lambda^{+}+\lambda^{\times}}{1+\lambda'}}{K\left(1+\dfrac{a^{+}\lambda^{+}+a^{\times}\lambda^{\times}}{1+a'\lambda'}\right)}$$

$$= \frac{(1+a'\lambda')(1+\lambda'+\lambda^{+}+\lambda^{\times})}{K(1+\lambda')(1+a'\lambda'+a^{+}\lambda^{+}+a^{\times}\lambda^{\times})}$$

$$K = \frac{\kappa(1+a'\lambda')}{(1+\lambda')} \qquad [6.22]$$

Again, by analogy,

$$M = \frac{\mu(1+b'\lambda')}{(1+\lambda')} \qquad [6.23]$$

Thus, with these assumptions the ground state affinity in the TSM actually corresponds to the weighted average affinity of the ligands for the inactive conformations of the receptor. Finally, the apparent affinity of A in the presence of a saturating concentration of B is

$$\frac{1+b'\lambda'+b^{+}\lambda^{+}+b^{\times}\lambda^{\times}}{\kappa\left(1+a'b'\lambda'+a^{+}b^{+}\lambda^{+}+a^{\times}b^{\times}\lambda^{\times}\right)}=\frac{1+\beta L}{\gamma K(1+\alpha\beta\delta L)}$$

$$=\frac{1+\dfrac{b^{+}\lambda^{+}+b^{\times}\lambda^{\times}}{1+b'\lambda'}}{\gamma\dfrac{\kappa(1+a'\lambda')}{(1+\lambda')}\left(1+\dfrac{a^{+}b^{+}\lambda^{+}+a^{\times}b^{\times}\lambda^{\times}}{1+a'b'\lambda'}\right)} \quad [6.24]$$

$$\gamma=\frac{(1+\lambda')(1+a'b'\lambda')}{(1+a'\lambda')(1+b'\lambda')}$$

Thus, we may write Eq. (6.24) as

$$\delta=\frac{\left(\lambda^{+}+\lambda^{\times}\right)\left(a^{+}b^{+}\lambda^{+}+a^{\times}b^{\times}\lambda^{\times}\right)}{\left(a^{+}\lambda^{+}+a^{\times}\lambda^{\times}\right)\left(b^{+}\lambda^{+}+b^{\times}\lambda^{\times}\right)}\gamma^{-1} \quad [6.25]$$

The correspondences listed above are intuitively reasonable if we consider the definitions of the parameters of the ATSM in terms of the concentrations of receptor species (see Table 1 of Ref. 6). For example, L is defined as the ratio of the concentrations of free receptors in the active and inactive states and Eq. (6.18) is compatible with this definition since it is the ratio of the sums of the concentrations of free receptors in the active and inactive conformations.

Now, from the previous discussion of the aTCM and Scheme 6.2, it is clear that the four-conformation model does not apply any constraints on the values of K, L, M, α, β, or γ. It is less clear that the value of δ (the activation cooperativity) is unrestricted, particularly as its definition includes the reciprocal of γ as a factor (as indeed it does in the ATSM). One argument that it is unconstrained is simply that there are more parameters in Scheme 6.3 (11) than in the ATSM (7) and hence there must always be some parameter that can be varied when constraints are placed on the others by the required values of the other ATSM parameters. However, there is still some value in exploring the structure of Eq. (6.25) further since this provides some insight into the mechanisms by which activation cooperativity can be achieved. First consider the behavior of Eq. (6.25) as the terms a^{+} and b^{+} become dominant

$$\delta \to \frac{a^+ b^+ \lambda^+ (\lambda^+ + \lambda^\times)}{a^+ b^+ (\lambda^+)^2} \gamma^{-1} = \frac{\lambda^+ + \lambda^\times}{\lambda^+} \gamma^{-1}$$

This expression can be made arbitrarily large by making λ^+ arbitrarily small and has lower limit γ^{-1} (when λ^+ is large). Thus, strong positive activation cooperativity occurs when the ligands strongly stabilize a relatively rare active conformation of the receptor (this will also be true in models with larger numbers of conformations). Of course, being rare this conformation will have little influence on the basal activity of the system, which must be largely defined by the most abundant active conformation. This provides a method for achieving a required value of δ for given values of L, α, and β since each can be largely controlled by varying a different Scheme 6.3 parameter. Now consider, Eq. (6.25) when a^+ and $b^\times$ become dominant

$$\delta \to \frac{(\lambda^+ + \lambda^\times)(a^+ b^+ \lambda^+ + a^\times b^\times \lambda^\times)}{a^+ b^\times \lambda^+ \lambda^\times} \gamma^{-1} = (\lambda^+ + \lambda^\times)\gamma^{-1} \left(\frac{b^+}{b^\times \lambda^\times} + \frac{a^\times}{a^+ \lambda^+} \right)$$

This allows arbitrarily small values of δ to be achieved. In this case, since it relies on the relative magnitudes of the selectivity constants, negative activation cooperativity is not dependent on a relatively rare state. Again, it is possible to control the efficacies, basal activity, and activation cooperativity primarily with different parameters giving a method to achieve any desired level of negative activation cooperativity with a given level of basal activity and given intrinsic efficacies. Also, it is again apparent that the cooperativity constants measure the concordance of the state selectivity of the ligands. However, strong positive activation cooperativity may only be manifest when the agonists induce the functional response by strongly stabilizing the same rare active conformation of the receptor. Thus Scheme 6.3A has sufficient degrees of freedom not to constrain the derived ATSM parameters. Hence, although the ATSM was derived at an "intermediate level of analysis," its behavior and predictions are sound and it is simply a convenient simplified form of a (family of) more complicated model(s) whose common behavior it encapsulates. Thus, as Onaran and coworkers previously showed,[9] a valid interpretation of the TSM is that the receptor states (inactive and active) each correspond to an ensemble of receptor conformations with the relevant property (an interpretation that has always seemed more natural to the author). It should also be noted that the above arguments show that it is always possible to map the parameters of a multistate model onto those of a model with a simpler structure. It also implies that the inactive state affinity of a ligand derived on the basis of a TSM may in fact be composite and not represent a fundamental thermodynamic constant but a weighted average affinity for multiple inactive conformations. Indeed, if allosteric modulation has been demonstrated for a receptor the latter interpretation must be true unless there is no evidence of binding cooperativity.

As noted above, it is not only possible for a ligand to achieve neutral antagonism by leaving the equilibrium distribution of the receptor conformations unchanged but also by compensating its effects on the levels of one active conformation through its effects on the other and that this is a mechanism for achieving coagonism. As a concrete example, let $\lambda^{+}=0.0001$, $\lambda^{\times}=0.01$, $\lambda'=0.001$, $a^{+}=30$, $a^{\times}=0.71$, $a'=1$, $b^{+}=100$, $b^{\times}=0.01$, and $b'=1$, then the level of activity in the absence of ligands (*basal*) is,

$$basal=\frac{[R]_T(0.0001+0.01)}{1+0.001+0.0001+0.01}=\frac{0.0101}{1.0111}[R]_T$$

The maximal response to A is

$$Max_A=\frac{[R]_T(30\times 0.0001+0.71\times 0.01)}{1+1\times 0.001+30\times 0.0001+0.71\times 0.01}=\frac{0.0101}{1.0111}[R]_T=basal$$

The maximal response to B is

$$Max_B=\frac{[R]_T(100\times 0.0001+0.01\times 0.01)}{1+1\times 0.001+100\times 0.0001+0.01\times 0.01}=\frac{0.0101}{1.0111}[R]_T=basal$$

However, the maximal response to the combination of A and B is

$$\begin{aligned}Max_{AB}&=\frac{[R]_T(30\times 100\times 0.0001+0.71\times 0.01\times 0.01)}{1+1\times 1\times 0.001+30\times 100\times 0.0001+0.71\times 0.01\times 0.01}\\&=\frac{0.300071}{1.301071}[R]_T\approx 0.231[R]_T>basal\end{aligned}$$

The above parameters correspond to the following ATSM parameters: $L\approx 0.010$, $\alpha=1$, $\beta=1$, $\gamma=1$, $\delta=29.71$ (positive activation cooperativity between two neutral antagonists). Hence, the combination of the two ligands is able to induce a functional response while the two ligands individually do not. This mechanistic interpretation of coagonism from Scheme 6.3 does seem more satisfactory than that from the ATSM since the ligands do perturb the equilibrium between the receptor conformations rather than suddenly gaining this ability in combination. Indeed, following on from the discussion above, two absolute neutral antagonists (i.e., $a^+ = a^\times = a' = b^+ = b^\times = b' = 1$) acting at different sites on the receptor cannot give rise to coagonism. However, this does not change the fact that the behaviors of the ATSM do include coagonism.

This, of course, does not imply that this is the only way to achieve coagonism, it simply demonstrates that coagonism is possible when the two ligands are neutral antagonists (and, indeed, is the mechanism proposed in Ref. 19). As noted by Ehlert and Griffin,[20] the two ligands could simply be weakly efficacious, neutrally cooperative agonists that alone cannot induce a measurable response. If this were the case it would impose strict limits on the maximal response to their combination. For example, if $L = 0.01$ (so that the basal activity is approximately 1% of $E_{\max}$) and we assume that it is not possible to detect a response less than 5% of $E_{\max}$, then from the expression for the maximal response, $\alpha < 5.3$ and $\beta < 5.3$. This makes the greatest possible maximal response to the combination (with $\gamma = \delta = 1$) approximately 0.22 $E_{\max}$ (if the basal were 0.1% $E_{\max}$ the upper limit would be ~0.74 $E_{\max}$). Anything larger than this would require positive activation cooperativity. If we do allow activation cooperativity, then one or both of the ligands could also be an inverse agonist and the combination would still result in agonism.

It is instructive to consider the special cases of Scheme 6.3A that have fewer inactive or active conformations. To generate a model with a single active conformation from Scheme 6.3A it is sufficient to let $\lambda' = 0$ (Scheme 6.3B), then the correspondences between Scheme 6.3B and the ATSM become: $K = \kappa$, $M = \mu$, $L = \lambda^+ + \lambda^\times$, $\alpha = \frac{a^+\lambda^+ + a^\times\lambda^\times}{\lambda^+ + \lambda^\times}$, $\beta = \frac{b^+\lambda^+ + b^\times\lambda^\times}{\lambda^+ + \lambda^\times}$, $\gamma = 1$, and $\delta = \frac{(\lambda^+ + \lambda^\times)(a^+b^+\lambda^+ + a^\times b^\times \lambda^\times)}{(a^+\lambda^+ + a^\times\lambda^\times)(b^+\lambda^+ + b^\times\lambda^\times)}$. Thus, there can be no binding cooperativity if there is only one inactive conformation of the receptor, although activation cooperativity is still manifest. The ATSM affinity of the ligands in this case corresponds to the affinity for the single inactive conformation. A model with a single active conformation can be generated from Scheme 6.3 by letting $\lambda^+ = 0$. The correspondences then become: $K = \frac{\kappa(1 + a'\lambda')}{(1 + \lambda')}$, $M = \frac{\mu(1 + b'\lambda')}{(1 + \lambda')}$, $L = \frac{\lambda^\times}{1 + \lambda'}$, $\alpha = \frac{a^\times(1 + \lambda')}{(1 + a'\lambda')}$, $\beta = \frac{b^\times(1 + \lambda')}{(1 + b'\lambda')}$, $\gamma = \frac{(1 + \lambda')(1 + a'b'\lambda')}{(1 + a'\lambda')(1 + b'\lambda')}$, and $\delta = \frac{(1 + a'\lambda')(1 + b'\lambda')}{(1 + \lambda')(1 + a'b'\lambda')} = \gamma^{-1}$. Thus, with a single active conformation there is both binding and activation cooperativity but they are not independent, they become perfectly inversely correlated. With just a single active and inactive conformation ($\lambda' = \lambda^+ = 0$) (again a system considered in Ref. 20), the cooperativities are $\gamma = 1$ and $\delta = 1$, that is, all interactions are neutrally cooperative. Hence, multiple active and inactive conformations are required for independent binding and activation cooperativity between ligands acting at different sites on a receptor.

This conclusion requires some further discussion since the model with a single active and inactive conformation has the same reaction scheme as Scheme 6.2 and it was demonstrated above that this does provide a mechanism for achieving cooperativity. The key difference is in the definition of one of the conformations of the simplified Scheme 6.3 as active. Letting $\lambda' = \lambda^+ = 0$ in Eqs. (6.8) and (6.12) gives,

$$\text{Bound A} = \frac{[R]_T\kappa[A](1 + a^\times\lambda^\times + \mu[B](1 + a^\times b^\times\lambda^\times))}{1 + \lambda^\times + \mu[B](1 + b^\times\lambda^\times) + \kappa[A](1 + a^\times\lambda^\times + \mu[B](1 + a^\times b^\times\lambda^\times))}$$

$$\frac{E}{[R]_T} = \frac{\lambda^\times + b^\times\lambda^\times\mu[B] + a^\times\lambda^\times\kappa[A](1 + b^\times\mu[B])}{1 + \lambda^\times + \mu[B](1 + b^\times\lambda^\times) + \kappa[A](1 + a^\times\lambda^\times + \mu[B](1 + a^\times b^\times\lambda^\times))}$$
$$= \frac{\lambda^\times(1 + a^\times\kappa[A])(1 + b^\times\mu[B])}{1 + \lambda^\times + \mu[B](1 + b^\times\lambda^\times) + \kappa[A](1 + a^\times\lambda^\times + \mu[B](1 + a^\times b^\times\lambda^\times))}$$

The derived ATSM parameters become: $K = \kappa$, $M = \mu$, $L = \lambda^\times$, $\alpha = a^\times$, $\beta = b^\times$, $\gamma = 1$, and $\delta = 1$. The apparent affinity and potency of A in the absence of B is then

$$K_{\text{app}A} = \text{EC}_{50A} = \frac{1 + \lambda^\times}{\kappa(1 + a^\times\lambda^\times)} = \frac{1 + L}{K(1 + \alpha L)}$$

and in the presence of a saturating concentration of B, it is

$$K_{\text{app}A+B} = \text{EC}_{50A+B} = \frac{1 + b^\times\lambda^\times}{\kappa(1 + a^\times b^\times\lambda^\times)} = \frac{1 + \beta L}{K(1 + \alpha\beta L)}$$

The maximal response to A alone becomes

$$Max_A = [R]_T\frac{a^\times\lambda^\times}{1 + a^\times\lambda^\times} = [R]_T\frac{\alpha L}{1 + \alpha L}$$

and in the presence of a saturating concentration of B, it becomes

$$Max_{AB} = [R]_T\frac{a^\times b^\times\lambda^\times}{1 + a^\times b^\times\lambda^\times} = [R]_T\frac{\alpha\beta L}{1 + \alpha\beta L}$$

Thus, the ligands do interact allosterically in this model, but the effect is entirely predictable on the basis of the intrinsic efficacies of the two ligands and the cooperativity is neutral (the concept of cooperativity is generally used to describe the unpredictable part of an interaction). Of course, if it were possible to identify all of the conformations of a receptor and measure the affinities of the ligands for each of these conformations, then it would be possible to specify the behavior of their combination completely, rendering the need to measure cooperativity constants redundant. In the meantime, it does seem useful to make a distinction between the part of an allosteric interaction which is predictable on the basis of the intrinsic efficacies of the ligands and the cooperativities that govern the deviations from this. The overall effect of the combination is then the product of these two factors.

2.3. Generalization to an arbitrary number of states: Matrix representation

It is straightforward to generalize the model in Scheme 6.3 to an arbitrary (but finite) number of receptor conformations. Equations (6.8) and (6.12) can be rewritten more compactly in Σ notation:

$$\text{Bound A} = \frac{[R]_T\kappa[A]\left(\sum_R a^R\lambda^R + \mu[B]\sum_R a^R b^R\lambda^R\right)}{\sum_R \lambda^R + \mu[B]\sum_R b^R\lambda^R + \kappa[A]\left(\sum_R a^R\lambda^R + \mu[B]\sum_R a^R b^R\lambda^R\right)} \quad [6.26]$$

$$\frac{E}{[R]_T} = \frac{\left(\sum_{R^a}\lambda^{R^a} + \mu[B]\sum_{R^a} b^{R^a}\lambda^{R^a} + \kappa[A]\left(\sum_{R^a} a^{R^a}\lambda^{R^a} + \mu[B]\sum_{R^a} a^{R^a}b^{R^a}\lambda^{R^a}\right)\right)}{\sum_R \lambda^R + \mu[B]\sum_R b^R\lambda^R + \kappa[A]\left(\sum_R a^R\lambda^R + \mu[B]\sum_R a^R b^R\lambda^R\right)} \quad [6.27]$$

Here R indicates the set of all conformations and R^a indicates the set of active conformations of the receptor. Also, one of the terms in the summations over R must be 1 since in the formulation presented one of the receptor states (R) has been defined as a "reference" state. Clearly, this generalizes directly to any number of inactive and active conformations by simply adding further terms to the sums to represent these conformations. The structure of the numerator and denominator of Eqs. (6.26) and (6.27) as sums of products lends itself to an even more compact representation using the mathematical notation of vectors and matrices. Let $\boldsymbol{\lambda}=[1\ \lambda_2 \ldots \lambda_n]^T$, $\mathbf{w_a}=[0\ w_2 \ldots w_n]^T$, $\mathbf{w_i}=[1\ 1-w_2 \ldots 1-w_n]^T$, and $\mathbf{w}=\mathbf{w_a}+\mathbf{w_i}=[1\ 1 \ldots 1]^T$ be (column) vectors (the superscript T indicates the transpose, an operation that converts the columns of a matrix or vector into rows and vice versa. It is used in this context to save space when writing out the column vectors). $\boldsymbol{\lambda}$ is the vector of receptor isomerization constants, $\mathbf{w_a}$ is the active conformation selection vector with entries restricted to 0 and 1 where $w_j=1$ when conformation R_j is active and $w_j=0$ when conformation R_j is inactive, $\mathbf{w_i}$ is the inactive conformation selection vector whose entries are 0 where the corresponding entry in $\mathbf{w_a}$ is 1 and 1 where the corresponding entry in $\mathbf{w_a}$ is 0. $\mathbf{w}$ is a vector of n 1s. Also, let $\mathbf{A}$ and $\mathbf{B}$ be the $n \times n$ diagonal matrices of ligand conformational selectivity constants, then $\mathbf{A}$ and $\mathbf{B}$ are

$$\mathbf{A}=\begin{bmatrix} 1 & 0 & \cdots & 0 \\ 0 & a_2 & \cdots & 0 \\ \vdots & \vdots & \ddots & \vdots \\ 0 & 0 & \cdots & a_n \end{bmatrix} \quad \text{and} \quad \mathbf{B}=\begin{bmatrix} 1 & 0 & \cdots & 0 \\ 0 & b_2 & \cdots & 0 \\ \vdots & \vdots & \ddots & \vdots \\ 0 & 0 & \cdots & b_n \end{bmatrix}$$

Of course, it would also be valid to multiply κ and μ into these matrices, respectively, to yield matrices of the affinity constants of A and B for the different receptor conformations. However, one inactive conformation of the receptor will continue to be treated as a reference conformation and conformational selectivity constants defined relative to this. It is then possible to write Eqs. (6.26) and (6.27) as

$$\text{Bound A}=\frac{[R]_T\kappa[A](\mathbf{w}^T\mathbf{A}\boldsymbol{\lambda}+\mu[B]\mathbf{w}^T\mathbf{AB}\boldsymbol{\lambda})}{\mathbf{w}^T\boldsymbol{\lambda}+\mu[B]\mathbf{w}^T\mathbf{B}\boldsymbol{\lambda}+\kappa[A](\mathbf{w}^T\mathbf{A}\boldsymbol{\lambda}+\mu[B]\mathbf{w}^T\mathbf{AB}\boldsymbol{\lambda})} \quad [6.28]$$

$$\frac{E}{[R]_T}=\frac{\mathbf{w_a}^T\boldsymbol{\lambda}+\mu[B]\mathbf{w_a}^T\mathbf{B}\boldsymbol{\lambda}-\kappa[A]\left(\mathbf{w_a}^T\mathbf{A}\boldsymbol{\lambda}+\mu[B]\mathbf{w_a}^T\mathbf{AB}\boldsymbol{\lambda}\right)}{\mathbf{w}^T\boldsymbol{\lambda}+\mu[B]\mathbf{w}^T\mathbf{B}\boldsymbol{\lambda}+\kappa[A](\mathbf{w}^T\mathbf{A}\boldsymbol{\lambda}+\mu[B]\mathbf{w}^T\mathbf{AB}\boldsymbol{\lambda})} \quad [6.29]$$

Note, it is possible (with care) to factorize the common vectors from these equations ($\mathbf{w}^T$ or $\mathbf{w_a}^T$ to the left, $\boldsymbol{\lambda}$ to the right; the order of the matrix multiplications must not be changed), but these common vectors cannot be "canceled" since division is not defined for vectors. Equations (6.18)–(6.25) then become: $L=\frac{\mathbf{w_a}^T\boldsymbol{\lambda}}{\mathbf{w_i}^T\boldsymbol{\lambda}}$, $K=\kappa\frac{\mathbf{w_i}^T\mathbf{A}\boldsymbol{\lambda}}{\mathbf{w_i}^T\boldsymbol{\lambda}}$, $M=\mu\frac{\mathbf{w_i}^T\mathbf{B}\boldsymbol{\lambda}}{\mathbf{w_i}^T\boldsymbol{\lambda}}$, $\alpha=\frac{(\mathbf{w_a}^T\mathbf{A}\boldsymbol{\lambda})(\mathbf{w_i}^T\boldsymbol{\lambda})}{(\mathbf{w_a}^T\boldsymbol{\lambda})(\mathbf{w_i}^T\mathbf{A}\boldsymbol{\lambda})}$, $\beta=\frac{(\mathbf{w_a}^T\mathbf{B}\boldsymbol{\lambda})(\mathbf{w_i}^T\boldsymbol{\lambda})}{(\mathbf{w_a}^T\boldsymbol{\lambda})(\mathbf{w_i}^T\mathbf{B}\boldsymbol{\lambda})}$, $\gamma=\frac{(\mathbf{w_i}^T\boldsymbol{\lambda})(\mathbf{w_i}^T\mathbf{AB}\boldsymbol{\lambda})}{(\mathbf{w_i}^T\mathbf{A}\boldsymbol{\lambda})(\mathbf{w_i}^T\mathbf{B}\boldsymbol{\lambda})}$, and $\delta=\frac{(\mathbf{w_a}^T\boldsymbol{\lambda})(\mathbf{w_a}^T\mathbf{AB}\boldsymbol{\lambda})}{(\mathbf{w_a}^T\mathbf{A}\boldsymbol{\lambda})(\mathbf{w_a}^T\mathbf{B}\boldsymbol{\lambda})}\gamma^{-1}$.

To represent Scheme 6.3, the following expressions would be substituted into Eqs. (6.28) and (6.29) and the matrix forms of Eqs. (6.18)–(6.25): $\boldsymbol{\lambda}=[1\ \lambda'\ \lambda^{+}\ \lambda^{\times}]^T$, $\mathbf{w_a}=[0\ 0\ 1\ 1]^T$, $\mathbf{w_i}=[1\ 1\ 0\ 0]^T$, $\mathbf{w}=[1\ 1\ 1\ 1]^T$,

$$\mathbf{A}=\begin{bmatrix}1 & 0 & 0 & 0\\ 0 & a' & 0 & 0\\ 0 & 0 & a^{+} & 0\\ 0 & 0 & 0 & a^{\times}\end{bmatrix} \text{ and } \mathbf{B}=\begin{bmatrix}1 & 0 & 0 & 0\\ 0 & b' & 0 & 0\\ 0 & 0 & b^{+} & 0\\ 0 & 0 & 0 & b^{\times}\end{bmatrix}$$

The matrix product **AB** is then

$$\mathbf{AB}=\begin{bmatrix}1 & 0 & 0 & 0\\ 0 & a'b' & 0 & 0\\ 0 & 0 & a^{+}b^{+} & 0\\ 0 & 0 & 0 & a^{\times}b^{\times}\end{bmatrix}=\mathbf{BA}$$

(multiplication of diagonal matrices is commutative; this is not a general property of matrix multiplication). In this case, we then calculate L as

$$L=\frac{\mathbf{w_a}^T\boldsymbol{\lambda}}{\mathbf{w_i}^T\boldsymbol{\lambda}}=[0\ \ 0\ \ 1\ \ 1]\begin{bmatrix}1\\ \lambda'\\ \lambda^{+}\\ \lambda^{\times}\end{bmatrix}\div[1\ \ 1\ \ 0\ \ 0]\begin{bmatrix}1\\ \lambda'\\ \lambda^{+}\\ \lambda^{\times}\end{bmatrix}=\frac{\lambda^{+}+\lambda^{\times}}{1+\lambda'}$$

The matrix product $\mathbf{w_i}^T\mathbf{A}\boldsymbol{\lambda}$ is calculated as follows:

$$\mathbf{w_i}^T\mathbf{A}\boldsymbol{\lambda}=[1\ \ 1\ \ 0\ \ 0]\begin{bmatrix}1 & 0 & 0 & 0\\ 0 & a' & 0 & 0\\ 0 & 0 & a^{+} & 0\\ 0 & 0 & 0 & a^{\times}\end{bmatrix}\begin{bmatrix}1\\ \lambda'\\ \lambda^{+}\\ \lambda^{\times}\end{bmatrix}=[1\ \ 1\ \ 0\ \ 0]\begin{bmatrix}1\\ a'\lambda'\\ a^{+}\lambda^{+}\\ a^{\times}\lambda^{\times}\end{bmatrix}$$
$$=1+a'\lambda'$$

although the associativity of matrix multiplication allows this product to be evaluated in either order.

It is necessary to justify that the model given above really is a generalization to any arbitrary number of conformations and that unrestricted interconvertibility of the conformations also encompasses models with restrictions on which conformations can be accessed from the others. In a model with n receptor conformations that are fully interconvertible there are $n-1$ independent receptor isomerization constants governing the conversion of the reference state, R, to the other $n-1$ conformations. Since all conformations can be converted into every other conformation, there is an isomerization cycle involving every set of three conformations. Now, at equilibrium, the free energy change around every closed cycle in the reaction scheme must be zero. Thus, in a cycle involving three conformations, one of the equilibrium constants must be the ratio of the other two and the equilibrium distribution of the receptor conformations is given by

$$[R] = \frac{[R]_T}{1+\lambda_2+\lambda_3}, \quad \left[R^{(2)}\right] = \frac{\lambda_2[R]_T}{1+\lambda_2+\lambda_3}, \quad \text{and} \quad \left[R^{(3)}\right] = \frac{\lambda_3[R]_T}{1+\lambda_2+\lambda_3}$$

This can be summarized more compactly by noting that $[R]_T = [R](1+\lambda_2+\lambda_3)$ and that the concentration of each receptor species is obtained from that of $[R]$ by multiplying the expression for $[R]$ derived from this equation by the appropriate isomerization constant. Thus, the equation $[R]_T = [R](1+\lambda_2+\lambda_3)$ contains all the information necessary to write down the equilibrium distribution of the receptor states. Since all of the three-conformation cycles are connected, this also implies that, of the $n(n-1)/2$ equilibrium constants in the reaction scheme (each of the n conformations can be converted into $n-1$ others and each reaction links two conformations; this is, in fact, an example of the application of the Handshaking Lemma from Graph Theory (e.g., Ref. 25)), only $n-1$ of them are independent and the equilibrium distribution of the states (in the absence of ligands) can therefore be derived from $[R]_T = [R](1+\lambda_2+\cdots+\lambda_n)$. This expression is also true of a system in which the only isomerizations that can take place are from or to the reference state but not directly between any other conformations. If conversions between the conformations are partially restricted (while keeping the same number of conformations), this will not reduce the number of independent isomerization constants and hence would not change the mathematical form of the equilibrium-binding isotherm describing the system, although it may change the interpretation of some of the isomerization constants. This is illustrated in Scheme 6.4 for a four-conformation model with complete interconvertibility and the one

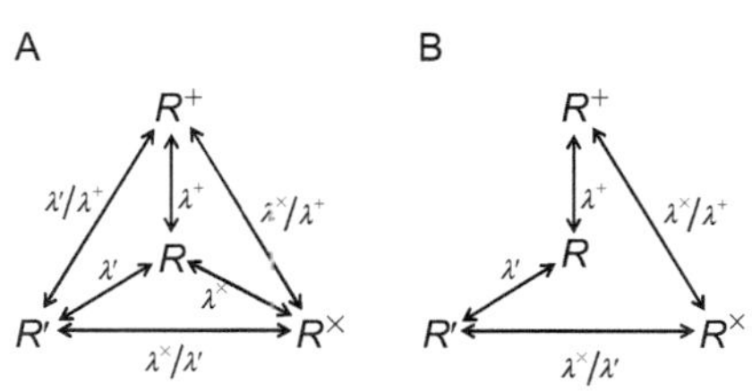

Scheme 6.4 (A) A system with four conformations all of which are mutually interconvertible. (B) A system with four conformations in which interconvertibility is restricted.

in which the $R^{\times}$ is not accessible from R and R^{+} is not accessible from R'. The former can be represented geometrically as a tetrahedron and the latter as a quadrilateral. The equilibrium distribution of receptor conformations for both Schemes can be derived from $[R]_T = [R](1+\lambda'+\lambda^{+}+\lambda^{\times})$. Thus, the equation describing the equilibrium position of both schemes is the same and the interpretation of the isomerization constants is the same. For example, $\lambda^{\times}$ is the equilibrium constant for the formation of $R^{\times}$ from R. In Scheme 6.4A, this is a single step reaction and is defined as $\lambda^{\times} = [R^{\times}]/[R]$. In Scheme 6.4B, the conversion has an intermediate step, however, $[R^{\times}] = \lambda^{\times}[R^{+}]/\lambda^{+} = \lambda^{\times}\lambda^{+}[R]/\lambda^{+} = \lambda^{\times}[R]$ and hence $\lambda^{\times} = [R^{\times}]/[R]$. When ligands are added to the system, it is simply a question of adding a binding reaction with each receptor conformation and an associated equilibrium constant. The binding reactions then define the underlying structure of the linear rational function which describes the behavior of the system (by defining the number terms in the scalar sums in the numerator and denominator) while the isomerization equilibrium defines the number of terms in each of the vector products and hence the structure of the individual terms in those scalar sums.

It is, of course, also possible to render Eqs. (6.2)–(6.7) into this notation and hence generalize Scheme 6.2 to an arbitrary number of receptor conformations although, in this case, there is no requirement for the conformation selection vectors ($\mathbf{w_a}$ and $\mathbf{w_i}$) since this model only considers binding and does not distinguish between active and inactive conformations of the receptor. Equation (6.2) becomes

$$\text{Bound A} = \frac{[R]_T \kappa[A](\mathbf{w}^T\mathbf{A}\boldsymbol{\lambda} + \mu[B]\mathbf{w}^T\mathbf{AB}\boldsymbol{\lambda})}{\mathbf{w}^T\boldsymbol{\lambda} + \mu[B]\mathbf{w}^T\mathbf{B}\boldsymbol{\lambda} + \kappa[A](\mathbf{w}^T\mathbf{A}\boldsymbol{\lambda} + \mu[B]\mathbf{w}^T\mathbf{AB}\boldsymbol{\lambda})}$$

The remaining expressions can then be derived in the same way as above. For example, the Eq. (6.7) is then written

$$\alpha = \frac{(\mathbf{w}^T\boldsymbol{\lambda})(\mathbf{w}^T\mathbf{AB}\boldsymbol{\lambda})}{(\mathbf{w}^T\mathbf{A}\boldsymbol{\lambda})(\mathbf{w}^T\mathbf{B}\boldsymbol{\lambda})}$$

This is again equivalent to the expression derived in Ref. (9). Comparing this with the matrix form of Eq. (6.27) (given below Eq. 6.29) provides a reminder that the binding cooperativity constants in the aTCM and the ATSM represent fundamentally different quantities. In the aTCM, the cooperativity constant is a measure of the effect of the allosteric ligand on the experimentally measured apparent affinity of the orthosteric ligand

(and vice versa). The binding cooperativity constant, γ, in the ATSM measures the effect of the allosteric ligand on the affinity of the orthosteric ligand for the inactive conformations of the receptor. The macroscopic binding cooperativity constant from the ATSM, α_{Mac}, is the ratio of the apparent affinities of A in the absence and presence of a saturating concentration of B, that is,

$$\alpha_{Mac} = \frac{\gamma(1+L)(1+\alpha\beta\delta L)}{(1+\alpha L)(1+\beta L)}$$

If $\gamma = \delta = 1$, consistent with the strict two-conformation model above, this expression becomes

$$\alpha_{Mac} = \frac{(1+L)(1+\alpha\beta L)}{(1+\alpha L)(1+\beta L)}$$

This indicates that the macroscopic binding cooperativity constant would be correlated with the efficacy of the ligands in a strict two-conformation model: positive cooperativity would only occur between two agonists or two inverse agonists; negative cooperativity only between an agonist and an inverse agonist; neutral cooperativity would result if one of the ligands were a neutral antagonist. Indeed, in a system with little constitutive activity ($L \ll 1$), positive cooperativity should only be measurable between two agonists since, when α, β, and $L \ll 1$, all of the terms in the definition of α_{Mac} are approximately equal to unity. Thus, the observation that allosteric modulators of the various muscarinic receptors can be negatively cooperative with both *N*-methylscopolamine (inverse agonist) and acetylcholine (agonist) (e.g., brucine at M_3, M_4, or M_5 receptors[26]) indicates that a strict two-conformation model cannot apply to these receptors. Similarly, eburnamonine has been shown to be positively cooperative with pilopcarpine and negatively cooperative with arecoline at M_2 receptors[27] both of which are agonists. Thus estimates of the affinity of agonists for these receptors for the active and inactive states (based on a TSM) of these receptors (e.g., Ref. 28) must be composite and cannot represent true thermodynamic constants.

The foregoing discussion has focused on heterotypic cooperativity; however, it is equally applicable to systems displaying homotypic cooperativity. All that is required is the replacement of ligand B with a second molecule of ligand A (assuming only two binding sites are involved). If the binding sites are identical, then the reference state affinities and the state selectivity constants should be equal. Since expressions for the binding isotherm and concentration–response curves would then contain terms in $[A]^2$, their behavior can be more varied than that of the heterotypic system (e.g., Refs. 2,29) since such expressions with positive coefficients can have local extrema for positive real values of the variable. Further elaborations (binding sites, ligands, number of sites accessible to each ligand) are clearly possible.

2.4. Interpretation of the active conformations

The interpretation of the multiple active conformations will now be considered. When interest is only in the effects of the ligands on a single signaling pathway, one simple approach is to assume that the basal proportions of the receptor conformations include the influence of the transducer protein. The transducer protein, like any other ligand, will favor those conformations for which it has high affinity. The conformational selectivity constants of the ligands will then implicitly include the influence of the transducer protein. It would then be the case that the activity of various active conformations of the receptor would be equivalent to their concentration and that the total concentration of active conformations could be taken as representative of the functional response (this is justified below). Clearly, this approach is not tenable if the interest is in multiple signaling pathways and modeling-biased agonism (or "agonist trafficking of receptor signals"[30]). In this case, it would be possible to identify subsets of active conformations that interact with different transducer proteins and hence that represent different active states of, for example, a three-state model.[31] This is likely to require rather more than two active conformations of the receptor, particularly if allosteric modulation through the different signaling pathways is to be considered. In terms of Eqs. (6.28) and (6.29), it would be necessary to define multiple active and inactive state selection vectors, a pair for each response (where for each response $\mathbf{w_i} = \mathbf{w} - \mathbf{w_a}$). This is, however, just an extreme case of the alternative approach of defining a set of active conformations and allowing different transducer proteins to interact with these differentially. The latter would require weighting factors in the summations of active conformations to represent the different responses. In either case, given the definitions of γ and δ above, biased agonism leads to the potential for different cooperativity constants to govern the interaction between the same ligands when the effects on different signaling pathways are determined. This is a manifestation of the "permissive pharmacology" of allosteric ligands.[32]

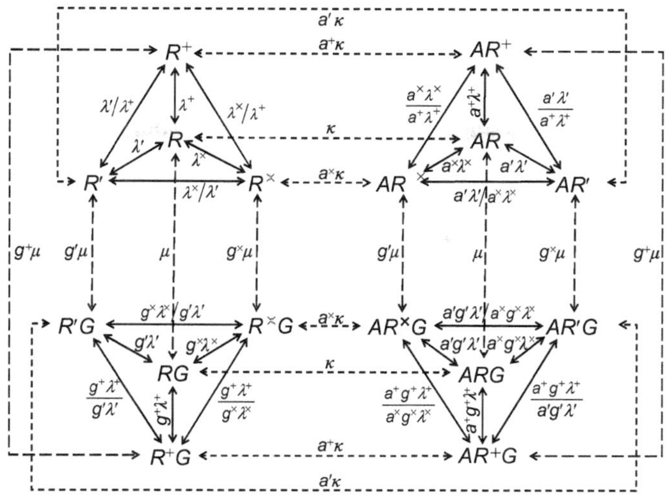

Scheme 6.5 The model of Scheme 6.3 with the allosteric modulator replaced with a transducer protein.

This leads into the consideration of the behavior of Scheme 6.3 when ligand B is taken to be a transducer protein (when it then becomes a conformationally specified cTCM (Scheme 6.5)). The transducer protein will be denoted by G, and the matrix of conformational selectivity constants as

$$\mathbf{G} = \begin{bmatrix} 1 & 0 & \cdots & 0 \\ 0 & g_2 & \cdots & 0 \\ \vdots & \vdots & \ddots & \vdots \\ 0 & 0 & \cdots & g_n \end{bmatrix}$$

The binding isotherm and the concentration–response curve for A are then

$$\text{Bound A} = \frac{[R]_T[A](\mathbf{w}^T\mathbf{A}\boldsymbol{\lambda} + \mu[G]\mathbf{w}^T\mathbf{AG}\boldsymbol{\lambda})}{\mathbf{w}^T\boldsymbol{\lambda} + \mu[G]\mathbf{w}^T\mathbf{G}\boldsymbol{\lambda} + \kappa[A](\mathbf{w}^T\mathbf{A}\boldsymbol{\lambda} + \mu[G]\mathbf{w}^T\mathbf{AG}\boldsymbol{\lambda})} \quad [6.30]$$

$$\frac{E}{[R]_T} = \frac{\mu[G]\left(\mathbf{w}_\mathbf{a}^T\mathbf{G}\boldsymbol{\lambda} + \kappa[A]\mathbf{w}_\mathbf{a}^T\mathbf{AG}\boldsymbol{\lambda}\right)}{\mathbf{w}^T\boldsymbol{\lambda} + \mu[G]\mathbf{w}^T\mathbf{G}\boldsymbol{\lambda} - \kappa[A](\mathbf{w}^T\mathbf{A}\boldsymbol{\lambda} + \mu[G]\mathbf{w}^T\mathbf{AG}\boldsymbol{\lambda})} \quad [6.31]$$

where the response is quantified as the concentration of active receptor conformations bound to the transducer protein and it has been assumed that both A and G are in excess over R. It should be noted then that based on the discussion of the ATSM above, biased agonism will result when a ligand shows concordant selectivity for a rare conformation of the receptor with one but not other G proteins, when there is discordance between the state selectivities of ligands and G proteins for active states of the receptor or when there is either concordance or discordance between inactive state selectivities.

The midpoint of the concentration–response curve to A (Eq. 6.31) and its apparent binding affinity from (Eq. 6.30) is

$$[A]_{50} = \frac{\mathbf{w}^T\boldsymbol{\lambda} + \mu[G]\mathbf{w}^T\mathbf{G}\boldsymbol{\lambda}}{\kappa(\mathbf{w}^T\mathbf{A}\boldsymbol{\lambda} + \mu[G]\mathbf{w}^T\mathbf{A}\mathbf{G}\boldsymbol{\lambda})} \qquad [6.32]$$

The activity in the absence of A is

$$Basal = \frac{[R]_T\mu[G]\mathbf{w}_\mathbf{a}^T\mathbf{G}\boldsymbol{\lambda}}{\mathbf{w}^T\boldsymbol{\lambda} + \mu[G]\mathbf{w}^T\mathbf{G}\boldsymbol{\lambda}} \qquad [6.33]$$

and the maximal response to A is

$$Max_A = \frac{[R]_T\mu[G]\mathbf{w}_\mathbf{a}^T\mathbf{A}\mathbf{G}\boldsymbol{\lambda}}{\mathbf{w}^T\mathbf{A}\boldsymbol{\lambda} + \mu[G]\mathbf{w}^T\mathbf{A}\mathbf{G}\boldsymbol{\lambda}} \qquad [6.34]$$

To give a specific example, it will again be assumed that there are two inactive and two active conformations of the receptor. In that case, the definitions of $\boldsymbol{\lambda}$, $\mathbf{w_a}$, $\mathbf{w_i}$, $\mathbf{w}$, and $\mathbf{A}$ are those given above and

$$\mathbf{G} = \begin{bmatrix} 1 & 0 & 0 & 0 \\ 0 & g' & 0 & 0 \\ 0 & 0 & g^+ & 0 \\ 0 & 0 & 0 & g^\times \end{bmatrix}$$

Equations (6.30) and (6.31) then become

$$\text{Bound A} = \frac{[R]_T\kappa[A]\begin{pmatrix} 1 + a'\lambda' + a^+\lambda^+ + a^\times\lambda^\times \\ +\mu[G](1 + a'g'\lambda' + a^+g^+\lambda^+ + a^\times g^\times\lambda^\times) \end{pmatrix}}{\begin{array}{c} 1 + \lambda' + \lambda^+ + \lambda^\times + \mu[G](1 + g'\lambda' + g^+\lambda^+ + g^\times\lambda^\times) \\ +\kappa[A]\begin{pmatrix} 1 + a'\lambda' + a^+\lambda^+ + a^\times\lambda^\times \\ +\mu[G](1 + a'g'\lambda' + a^+g^+\lambda^+ + a^\times g^\times\lambda^\times) \end{pmatrix} \end{array}} \qquad [6.35]$$

$$\frac{E}{[R]_T} = \frac{\mu[G]\begin{pmatrix} g^+\lambda^+ + g^\times\lambda^\times \\ +\kappa[A](a^+g^+\lambda^+ + a^\times g^\times\lambda^\times) \end{pmatrix}}{\begin{array}{c} 1 + \lambda' + \lambda^+ + \lambda^\times + \mu[G](1 + g'\lambda' + g^+\lambda^+ + g^\times\lambda^\times) \\ +\kappa[A]\begin{pmatrix} 1 + a'\lambda' + a^+\lambda^+ + a^\times\lambda^\times \\ +\mu[G](1 + a'g'\lambda' + a^+g^+\lambda^+ + a^\times g^\times\lambda^\times) \end{pmatrix} \end{array}} \qquad [6.36]$$

The concentration of active species is

$$
\begin{aligned}
[R^a] &= \frac{[R]_T\left(\lambda^+ + \lambda^\times + \mu[G]\left(g^+\lambda^+ + g^\times\lambda^\times\right) + \kappa[A]\left(a^+\lambda^+ + a^\times\lambda^\times + \mu[G]\left(a^+g^+\lambda^+ + a^\times g^\times\lambda^\times\right)\right)\right)}{1+\lambda' + \lambda^+ + \lambda^\times + \mu[G]\left(1+g'\lambda' + g^+\lambda^+ + g^\times\lambda^\times\right) + \kappa[A]\left(1+a'\lambda' + a^+\lambda^+ + a^\times\lambda^\times + \mu[G]\left(1+a'g'\lambda' + a^+g^+\lambda^+ + a^\times g^\times\lambda^\times\right)\right)} \\
&= \frac{[R]_T\left(\lambda^+(1+g^+\mu[G]) + \lambda^\times(1+g^\times\mu[G]) + \kappa[A]\left(a^+\lambda^+(1+g^+\mu[G]) + a^\times\lambda^\times(1+g^\times\mu[G])\right)\right)}{1+\mu[G] + \lambda'(1+g'\mu[G]) + \lambda^+(1+g^+\mu[G]) + \lambda^\times(1+g^\times\mu[G]) + \kappa[A]\begin{pmatrix}1+\mu[G] + a'\lambda'(1+g'\mu[G]) \\ +a^+\lambda^+(1+g^+\mu[G]) + a^\times\lambda^\times(1+g^\times\mu[G])\end{pmatrix}} \\
&= \frac{[R]_T\left(\lambda^+\frac{1+g^+\mu[G]}{1+\mu[G]} + \lambda^\times\frac{1+g^\times\mu[G]}{1+\mu[G]} + \kappa[A]\left(a^+\lambda^+\frac{1+g^+\mu[G]}{1+\mu[G]} + a^\times\lambda^\times\frac{1+g^\times\mu[G]}{1+\mu[G]}\right)\right)}{1+\lambda'\frac{1+g'\mu[G]}{1+\mu[G]} + \lambda^+\frac{1+g^+\mu[G]}{1+\mu[G]} + \lambda^\times\frac{1+g^\times\mu[G]}{1+\mu[G]} + \kappa[A]\left(1+a'\lambda'\frac{1+g'\mu[G]}{1+\mu[G]} + a^+\lambda^+\frac{1+g^+\mu[G]}{1+\mu[G]} + a^\times\lambda^\times\frac{1+g^\times\mu[G]}{1+\mu[G]}\right)} \\
&= \frac{[R]_T\left(\Lambda^+ + \Lambda^\times + \kappa[A]\left(a^+\Lambda^+ + a^\times\Lambda^\times\right)\right)}{1+\Lambda' + \Lambda^+ + \Lambda^\times + \kappa[A]\left(1+a'\Lambda' + a^+\Lambda^+ + a^\times\Lambda^\times\right)}
\end{aligned}
$$

where the various Λ are then functions of $[G]$. This is of the same form as Eqs. (6.18) and (6.26), hence, since the agonist enriches the active conformations by the same factor whether they are bound to G or not, the amount of active state is proportional to the amount of R^aG complex present. This justifies the assumption above that the active conformations in the models with no transducer protein can be summed to quantify the response if only a single pathway is considered.

The equivalences between the terms in Eqs. (6.35) and (6.36) and their cTCM analogues are entirely analogous with those given above for the ATSM and Scheme 6.3, that is, $L=\frac{\lambda^+ + \lambda^\times}{1+\lambda'}$, $K=\kappa\frac{(1+d'\lambda')}{(1+\lambda')}$, $M=\mu\frac{(1+g'\lambda')}{(1+\lambda')}$, $\alpha=\frac{(a^+\lambda^+ + a^\times\lambda^\times)(1+\lambda')}{(\lambda^+ + \lambda^\times)(1+d'\lambda')}$, $\beta=\frac{(g^+\lambda^+ + g^\times\lambda^\times)(1+\lambda')}{(\lambda^+ + \lambda^\times)(1+g'\lambda')}$, $\gamma=\frac{(1+\lambda')(1+d'g'\lambda')}{(1+d'\lambda')(1+g'\lambda')}$, and $\delta=\frac{(\lambda^+ + \lambda^\times)(a^+g^+\lambda^+ + a^\times g^\times\lambda^\times)(1+d'\lambda')(1+g'\lambda')}{(a^+\lambda^+ + a^\times\lambda^\times)(g^+\lambda^+ + g^\times\lambda^\times)(1+\lambda')(1+d'g'\lambda')}$.

The midpoint of Eqs. (6.35) and (6.36) is

$$[A]_{50}=\frac{1+\lambda'+\lambda^+ + \lambda^\times + \mu[G](1+g'\lambda' + g^+\lambda^+ + g^\times\lambda^\times)}{\kappa(1+d'\lambda' + a^+\lambda^+ + a^\times\lambda^\times + \mu[G](1+d'g'\lambda' + a^+g^+\lambda^+ + a^\times g^\times\lambda^\times))} \qquad [6.37]$$

The response in the absence of ligand is

$$Basal=\frac{[R]_T\mu[G](g^+\lambda^+ + g^\times\lambda^\times)}{1+\lambda'+\lambda^+ + \lambda^\times + \mu[G](1+g'\lambda' + g^+\lambda^+ + g^\times\lambda^\times)} \qquad [6.38]$$

and the maximal response to ligand is

$$Max=\frac{[R]_T\mu[G](a^+g^+\lambda^+ + a^\times g^\times\lambda^\times)}{1+d'\lambda' + a^+\lambda^+ + a^\times\lambda^\times + \mu[G](1+d'g'\lambda' + a^+g^+\lambda^+ + a^\times g^\times\lambda^\times)} \qquad [6.39]$$

Thus, in this case, an agonist must satisfy the following criterion

$$\frac{[R]_T\mu[\mathrm{G}]\left(a^+g^+\lambda^+ + a^\times g^\times \lambda^\times\right)}{1 + a'\lambda' + a^+\lambda^+ + a^\times\lambda^\times + \mu[\mathrm{G}]\left(1 + a'g'\lambda' + a^+g^+\lambda^+ + a^\times g^\times \lambda^\times\right)} > \frac{[R]_T\mu[\mathrm{G}]\left(g^+\lambda^+ + g^\times \lambda^\times\right)}{1 + \lambda' + \lambda^+ + \lambda^\times + \mu[\mathrm{G}]\left(1 + g'\lambda' + g^+\lambda^+ + g^\times \lambda^\times\right)}$$

$$\frac{\left(a^+g^+\lambda^+ + a^\times g^\times \lambda^\times\right)}{1 + a'\lambda' + a^+\lambda^+ + a^\times\lambda^\times + \mu[\mathrm{G}]\left(1 + a'g'\lambda' + a^+g^+\lambda^+ + a^\times g^\times \lambda^\times\right)} > \frac{\left(g^+\lambda^+ + g^\times \lambda^\times\right)}{1 + \lambda' + \lambda^+ + \lambda^\times + \mu[\mathrm{G}]\left(1 + g'\lambda' + g^+\lambda^+ + g^\times \lambda^\times\right)}$$

$$\left(a^+g^+\lambda^+ + a^\times g^\times \lambda^\times\right)\left(1 + \lambda' + \lambda^+ + \lambda^\times + \mu[\mathrm{G}](1 + g'\lambda')\right) > \left(g^+\lambda^+ + g^\times \lambda^\times\right)\left(1 + a'\lambda' + a^+\lambda^+ + a^\times\lambda^\times + \mu[\mathrm{G}](1 + a'g'\lambda')\right)$$

$$\frac{\left(a^+g^+\lambda^+ + a^\times g^\times \lambda^\times\right)\left(1 + \lambda' + \lambda^+ + \lambda^\times + \mu[\mathrm{G}](1 + g'\lambda')\right)}{\left(g^+\lambda^+ + g^\times \lambda^\times\right)\left(1 + a'\lambda' + a^+\lambda^+ + a^\times\lambda^\times + \mu[\mathrm{G}](1 + a'g'\lambda')\right)} > 1 \quad [6.40]$$

Unsurprisingly, this is a rearrangement of the criterion for agonism in the cTCM,

$$\delta > \frac{1+\alpha L+\gamma M[G]}{\alpha\gamma(1+L+M[G])} \qquad [6.41]$$

which was derived in Ref. (6), as making the appropriate substitutions into this expression would demonstrate. To continue with the pattern of requiring some quantity to be greater than unity to achieve agonism, Eq. (6.41) can be rearranged to

$$\frac{\alpha\gamma\delta(1+L+M[G])}{1+\alpha L+\gamma M[G]} > 1$$

This is also a rearrangement of the criterion for agonism given in Ref. (14). This criterion also directly demonstrates the possibility of "protean" agonism[15] since inequalities (6.40) and (6.41) change with $[G]$. When the terms in $[G]$ are negligible, the criteria for agonism become

$$\frac{\alpha\gamma\delta(1+L)}{1+\alpha L} > 1 \text{ or equivalently} \frac{\left(a^{+}g^{+}\lambda^{+}+a^{\times}g^{\times}\lambda^{\times}\right)\left(1+\lambda'+\lambda^{+}+\lambda^{\times}\right)}{\left(g^{+}\lambda^{+}+g^{\times}\lambda^{\times}\right)\left(1+a'\lambda'+a^{+}\lambda^{+}+a^{\times}\lambda^{\times}\right)} > 1$$

and when the terms in $[G]$ dominate, they become

$$\alpha\delta > 1 \text{ or} \frac{\left(a^{+}g^{+}\lambda^{+}+a^{\times}g^{\times}\lambda^{\times}\right)(1+g'\lambda')}{\left(g^{+}\lambda^{+}+g^{\times}\lambda^{\times}\right)(1+a'g'\lambda')} > 1$$

So it is possible for inequalities (6.40) and (6.41) to be satisfied for concentrations of transducer protein above but not below some threshold level (or indeed below but not above it since these expressions can be increasing or decreasing functions of $[G]$). A ligand may convert from agonist to inverse agonist or from inverse agonist to agonist as $[G]$ increases. There will be further discussion of protean agonism in Section 3.

3. OPERATIONAL MODELS OF RECEPTOR ACTIVATION WITH CONSTITUTIVE ACTIVITY

3.1. Properties of the model

The discussion of the multiconformational models is highly theoretical and, while it impacts our interpretation of pharmacological data, it does not lend itself to data analysis since the parameters of these models will be extremely difficult to measure experimentally, if this is possible at all. Recently, I derived a novel operational model of agonism that includes constitutive activity.[2,10] In Ref. 10, it was shown that it is possible to measure the intrinsic efficacy of agonists (and inverse agonists) directly if the response to the ligand can be measured in systems with a variety of levels of constitutive activity. It is possible to extend this model to include allosteric modulators and indeed to model the interaction between allosteric modulators (unpublished). One question that must be addressed is whether the extended model can be used to characterize the pharmacological properties of an allosteric modulator given data on its interaction with an orthosteric ligand. This question will be explored in this section using Monte Carlo simulations. However, the properties of the model itself will first be explored in more detail.

The equation for the response to an orthosteric ligand in the presence of an allosteric ligand from the model of Slack and Hall[10] (which generalizes the model derived in Ref. 2 to concentration–response curves with slopes that differ from unity) is

$$E=\frac{E_{\max}\chi^n\left(1+\frac{\varepsilon_B[B]}{K_b}+\frac{\varepsilon_A[A]}{K_a}\left(1+\frac{\varepsilon_{AB}[B]}{\varepsilon_A\alpha K_b}\right)\right)^n}{\left(1+\frac{[B]}{K_b}+\frac{[A]}{K_a}\left(1+\frac{[B]}{\alpha K_b}\right)\right)^n+\chi^n\left(1+\frac{\varepsilon_B[B]}{K_b}+\frac{\varepsilon_A[A]}{K_a}\left(1+\frac{\varepsilon_{AB}[B]}{\varepsilon_A\alpha K_b}\right)\right)^n} \quad [6.42]$$

In this equation, $E_{\max}$ is the maximal response that can be achieved in the system, K_a and K_b are the affinities (dissociation constants) of ligands A and B, respectively, α is the binding cooperativity between A and B (defined in the reciprocal sense to that in Eqs. (6.1–6.7) so $\alpha>1$ indicates negative cooperativity), ε_A and ε_B are the intrinsic efficacies of A and B individually, ε_{AB} is the intrinsic efficacy of the combination of A and B, n is the Hill coefficient of the transducer function (the relationship between receptor density and response), and χ is the coupling efficiency of the signal transduction system. The definition of χ is similar to that of the transducer ratio, τ, of the Operational Model,[33] $\chi=[R]_T/K_e$ (where $[R]_T$ is the receptor density and K_e is the midpoint of the transducer function). However, its role in the model is different since it defines the effects of the free receptor rather than that of ligand:receptor complexes (see below). Note that, in this case, deviations of the Hill coefficient of concentration–response curves from unity are assumed to be due to the transducer function (as assumed in the original Operational Model). When the Hill coefficient of Eq. (6.42) is equal to unity, it simplifies (after rearrangement) to

$$E = \frac{E_{\max}\chi\left(1 + \frac{\varepsilon_B[B]}{K_b} + \frac{\varepsilon_A[A]}{K_a}\left(1 + \frac{\varepsilon_{AB}[B]}{\varepsilon_A \alpha K_b}\right)\right)}{1 + \chi + \frac{[B]}{K_b}(1 + \varepsilon_B \chi) + \frac{[A]}{K_a}\left(1 + \varepsilon_A \chi + \frac{[B]}{\alpha K_b}(1 + \varepsilon_{AB}\chi)\right)} \qquad [6.43]$$

It was also assumed that the ligands are completely specific for the orthosteric (*A*) or allosteric site (*B*). With the above definitions, the activation cooperativity between *A* and *B* is given by $\delta_{AB} = \varepsilon_{AB}/\varepsilon_A\varepsilon_B$. In other words, it is assumed that the intrinsic efficacy of the combination of an orthosteric and an allosteric agonist which exhibits neutral activation cooperativity ($\delta_{AB} = 1$) is the product of their individual intrinsic efficacies. This is consistent with the discussion of multiconformational models in the previous section.

This model is closely related to that presented by Leach *et al.*[3] The concentration–response curve for the Leach *et al.* model is given by

$$E = \frac{E_{\max}\left(\frac{\tau_A[A]}{K_a} + \frac{\alpha\beta\tau_A[A][B]}{K_aK_b} + \frac{\tau_B[B]}{K_b}\right)^n}{\left(1 + \frac{[A]}{K_a} + \frac{[B]}{K_b} + \frac{\alpha[A][B]}{K_aK_b}\right)^n + \left(\frac{\tau_A[A]}{K_a} + \frac{\alpha\beta\tau_A[A][B]}{K_aK_b} + \frac{\tau_B[B]}{K_b}\right)^n}$$

where τ_A and τ_B are the operational model transducer ratios (efficacies) for ligands *A* and *B*, α is the binding cooperativity (defined in the reciprocal sense to that in Eq. 6.42) and $\beta\tau_A$ is transducer ratio of the *ARB* complex. Other symbols are as defined for Eq. (6.42). The major structural difference is that the model of Leach *et al.* does not allow the free receptor to signal so it lacks constitutive receptor activity (and strictly, therefore, should not be used in situations where constitutive activity can be measured). This was a pragmatic decision by the authors to allow the resulting equation to be fitted to experimental data investigating the effects of an allosteric modulator on responses to an orthosteric agonist. However, this formulation results in an asymmetry in the quantification of the activation cooperativity parameter (β). As defined, β is essentially equivalent to $\varepsilon_{AB}/\varepsilon_A = \delta_{AB}\varepsilon_B$ in Eq. (6.42) (constitutive activity in Eq. 6.42 means this is not a true equivalence). Hence its value differs if the roles of the orthosteric and allosteric ligands are interchanged since it implicitly contains a contribution from the intrinsic efficacy of the allosteric ligand. The value of δ_{AB} derived from Eq. (6.42) or (6.43) is not dependent on which ligand is labeled "allosteric" and is more consistent, therefore, with the definition of a cooperativity constant.

As shown in Ref. 10, the level of activity in the absence of ligands in Eq. (6.42) is

$$Basal = \frac{E_{\max}\chi^n}{1+\chi^n} \qquad [6.44]$$

Thus, the basal activity of the system is controlled by the coupling efficiency parameter, χ, and by the maximal response of the system. This allows us to derive information about the transducer function independently of the effects of a ligand by quantifying the variation in the basal activity with receptor density (or indeed by varying the midpoint of the transducer function, see Ref. 10). This then allows the absolute intrinsic efficacy of agonists to be determined in such experiments since the maximal response to a ligand (in this case ligand A) is

$$Max_A = \frac{E\max\varepsilon_A^n\chi^n}{1+\varepsilon_A^n\chi^n} \qquad [6.45]$$

The maximal response to a ligand is therefore dependent on both the coupling efficiency of the signal transduction system and the intrinsic efficacy of the ligand (as classical receptor theory tells us (e.g., Ref. 34)). In the presence of the allosteric ligand, the maximal response to the orthosteric ligand is

$$Max_{A|B} = \frac{E_{\max}\varepsilon_A^n\chi^n\left(1+\frac{\varepsilon_{AB}[B]}{\varepsilon_A\alpha K_b}\right)^n}{\left(1+\frac{[B]}{\alpha K_b}\right)^n + \varepsilon_A^n\chi^n\left(1+\frac{\varepsilon_{AB}[B]}{\varepsilon_A\alpha K_b}\right)^n}$$

In the presence of saturating concentrations of both ligands, it is, therefore,

$$Max_{AB} = \frac{E_{\max}\varepsilon_{AB}^n\chi^n}{1+\varepsilon_{AB}^n\chi^n} \qquad [6.46]$$

The potency of the orthosteric ligand from Eq. (6.42) is a rather complicated expression best given in two stages. Let

$$P=\chi\left(1+\frac{\varepsilon_B[B]}{K_b}\right),\quad Q=\chi\frac{\varepsilon_A}{K_a}\left(1+\frac{\varepsilon_{AB}[B]}{\varepsilon_A\alpha K_b}\right),\quad R=\left(1+\frac{[B]}{K_b}\right),\ \text{and}$$
$$S=\frac{1}{K_a}\left(1+\frac{[B]}{\alpha K_b}\right)$$

then Eq. (6.42) can be written

$$E = \frac{E_{\max}(P + Q[A])^n}{(R + S[A])^n + (P + Q[A])^n}$$

and the midpoint of A from Eq. (6.42) is

$$\mathrm{EC}_{50} = \frac{R(2P^nQ^n + Q^nR^n + P^nS^n)^{1/n} - P(P^nS^n + Q^nR^n + 2R^nS^n)^{1/n}}{Q(P^nS^n + Q^nR^n + 2R^nS^n)^{1/n} - S(2P^nQ^n + Q^nR^n + P^nS^n)^{1/n}}$$

This was derived by equating the abbreviated form of Eq. (6.42) with the average of its basal and its maximal value in the presence of A. In the absence of B, this becomes

$$\mathrm{EC}_{50} = \frac{K_a\left(\left(2\varepsilon_A^n\chi^n + \varepsilon_A^n + 1\right)^{1/n} - \left(\chi^n + \varepsilon_A^n\chi^n + 2\right)^{1/n}\right)}{\varepsilon_A\left(\chi^n + \varepsilon_A^n\chi^n + 2\right)^{1/n} - \left(2\varepsilon_A^n\chi^n + \varepsilon_A^n + 1\right)^{1/n}} \qquad [6.47]$$

as given in Ref. 10. In the presence of a saturating concentration of B, it gives

$$\mathrm{EC}_{50} = \frac{\alpha K_a\left(\left(2\varepsilon_{AB}^n\chi^n + \frac{\varepsilon_{AB}^n}{\varepsilon_B^n} + 1\right)^{1/n} - \left(\varepsilon_B^n\chi^n + \varepsilon_{AB}^n\chi^n + 2\right)^{1/n}\right)}{\frac{\varepsilon_{AB}}{\varepsilon_B}\left(\varepsilon_B^n\chi^n + \varepsilon_{AB}^n\chi^n + 2\right)^{1/n} - \left(2\varepsilon_{AB}^n\chi^n + \frac{\varepsilon_{AB}^n}{\varepsilon_B^n} + 1\right)^{1/n}} \qquad [6.48]$$

Comparing Eqs. (6.47) and (6.48), it can be seen that K_a in Eq. (6.47) is replaced by αK_a, in Eq. (6.48), χ is replaced by $\varepsilon_B\chi$ and ε_A is replaced by $\varepsilon_{AB}/\varepsilon_B = \delta_{AB}\varepsilon_A$ supporting the definition of activation cooperativity given earlier. For Eq. (6.43) (i.e., when $n=1$), the expression for the midpoint of the concentration–response curve to A is rather simpler:

$$\mathrm{EC}_{50} = \frac{K_a\left(1 + \chi + \frac{[B]}{K_b}(1 + \varepsilon_B\chi)\right)}{1 + \varepsilon_A\chi + \frac{[B]}{\alpha K_b}(1 + \varepsilon_{AB}\chi)}$$

It is then possible to investigate how the results of a standard pharmacological null analysis[17] relate to the parameters of the model. For this analysis, it is important to note (as shown in Figs. 6.1–6.3) that the concentration–response curves generated by this model in the presence and absence of ligand B can differ in their maximal responses to A. Therefore, equieffective concentration-ratios (DR) cannot generally be determined simply by taking the ratio of the EC_{50}s in the absence and presence of modulator. The expression for the DR caused by B is (the derivation will be presented elsewhere)

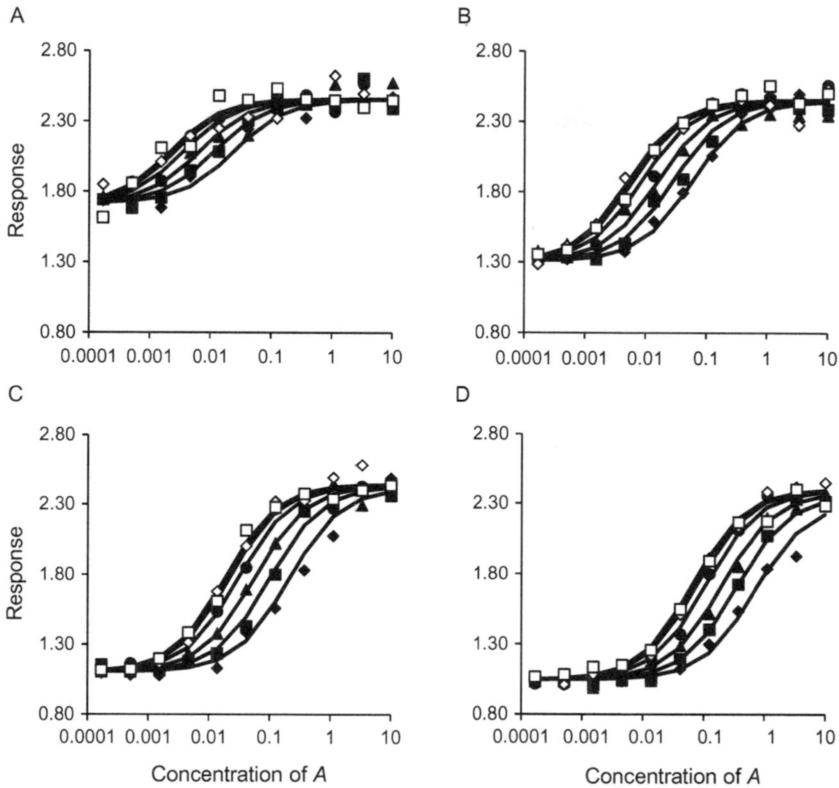

Figure 6.1 Example simulation of a positively activation cooperative interaction at a relatively high coupling efficiency ($\chi=1$ (A), 0.25 (B), 0.0625 (C), and 0.015625 (D)) using Eq. (6.43). The input parameters of the simulation are given in Table 6.1. The concentrations of *B* were 0 (filled diamonds), 1 (filled squares), 3 (triangles), 10 (circles), 30 (open diamonds), and 100 (open squares). Note then that the "control" is on the right of each set of curves. Lines show the fitted curves.

$$\mathrm{DR}=\frac{1+\dfrac{[B]}{K_b}\left(\dfrac{\varepsilon_A-\varepsilon_B}{\varepsilon_A-1}+\dfrac{K_a}{[A]}\dfrac{1-\varepsilon_B}{\varepsilon_A-1}\right)}{1+\dfrac{[B]}{\alpha K_b}\left(\dfrac{\varepsilon_{AB}-1}{\varepsilon_A-1}+\dfrac{[A]}{K_a}\dfrac{\varepsilon_{AB}-\varepsilon_A}{\varepsilon_A-1}\right)} \qquad [6.49]$$

Unlike the competitive case, it is not helpful to derive an expression for DR − 1. In this case, although the null method has eliminated the properties of the signal transduction system from the effects of *B* (since χ and n are absent from Eq. (6.49)), it has not eliminated the dependence on the properties of the orthosteric ligand (K_a and ε_A do occur). The lack of parallelism

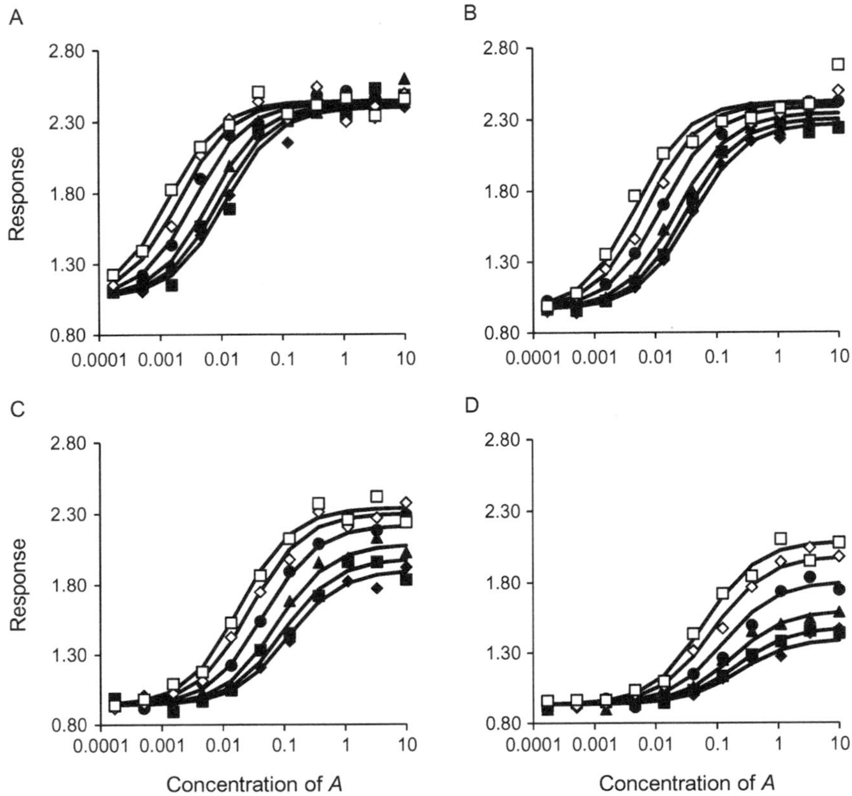

Figure 6.2 Example simulation of a positively activation cooperative interaction at a relatively low coupling efficiency (χ=0.1 (A), 0.025 (B), 0.00625 (C), and 0.0015625 (D)) using Eq. (6.43). The input parameters of the simulation are given in Table 6.1. The concentrations of *B* were 0 (filled diamonds), 1 (filled squares), 3 (triangles), 10 (circles), 30 (open diamonds), and 100 (open squares). The control is therefore on the right and at the bottom of each set of curves. Lines show the fitted curves.

between the curves in the presence and absence of *B* invalidates the standard assumption that equal responses result from equal levels of orthosteric ligand binding. It can be assumed, however, that equal responses result from equal stimuli. The latter assumption removes the contribution from the signal transduction system. The terms in $[A]$ and $[A]^{-1}$ in Eq. (6.49) describe the behavior of the DR as the concentration–response curves approach their asymptotes. When $[A]$ is small and *B* is not a neutral antagonist (i.e., when $\varepsilon_B \neq 1$), then the term in $[A]^{-1}$ in the numerator becomes significant (if $\varepsilon_B = 1$ this term cancels—the responses at $[A]=0$ are equal). Thus, this term describes the behavior of the DR at low orthosteric ligand concentrations

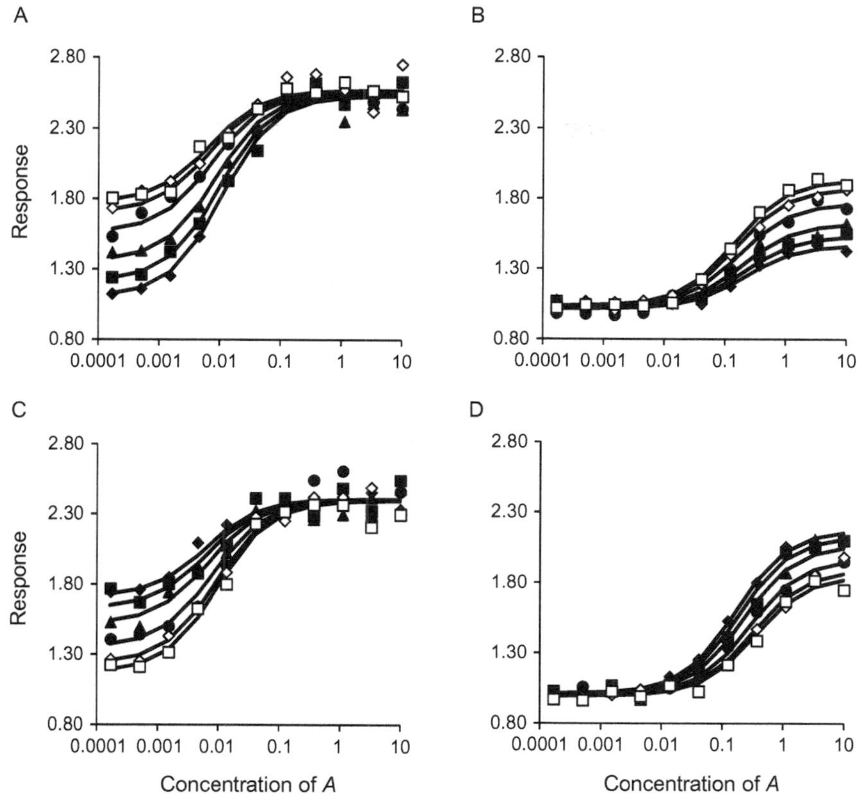

Figure 6.3 Example simulations of the effects of an allosteric agonist with weak negative activation cooperativity (A, B) and an allosteric inverse agonist with weak positive activation cooperativity (C, D) on an orthosteric agonist, using Eq. (6.43). The input parameters of the simulation are given in Table 6.1. The graphs show the curves at the highest (A, C) and lowest (B, D) coupling efficiencies simulated (these were not the same for the two simulations). The concentrations of *B* were 0 (filled diamonds), 1 (filled squares), 3 (triangles), 10 (circles), 30 (open diamonds), and 100 (open squares). Lines show the fitted curves.

when the allosteric ligand affects the receptor activity in the absence of *A*. Conversely, when [*A*] is large and the allosteric ligand changes the maximal response to *A* (i.e., when $\varepsilon_{AB} \neq \varepsilon_A$), the term in [*A*] in the denominator has a large impact (if $\varepsilon_{AB} = \varepsilon_A$, this terms cancels since the responses as $[A] \rightarrow \infty$ are equal). In both cases, it is possible for the DR to be negative (e.g., when [*A*] is small and $\varepsilon_B > 1$). This reflects the fact that, for the concentrations of *A* concerned, there is no point on the curve in the presence of *B* which gives the same response as the curve in its absence (or vice versa). Mathematically, however, Eqs. (6.42) and (6.49) are valid for negative values of [*A*] even though they are physically meaningless. This allows negative DRs to result since there may be a negative value of [A] which attains the required response level.

In between these extremes, there may be a range of concentrations of A for which the following pair of inequalities holds

$$\left|\frac{K_a(1-\varepsilon_B)}{\varepsilon_A-1}\right| \ll [A] \ll \left|\frac{K_a(\varepsilon_A-1)}{\varepsilon_{AB}-\varepsilon_A}\right|$$

Over this range of concentrations (if it is well defined), the concentration–response curves in the presence and absence of B are approximately parallel and Eq. (6.49) loses its dependence on [A]. It then simplifies to

$$\mathrm{DR} = \frac{1+\frac{[B]}{K_b}\left(\frac{\varepsilon_A-\varepsilon_B}{\varepsilon_A-1}\right)}{1+\frac{[B]}{\alpha K_b}\left(\frac{\varepsilon_{AB}-1}{\varepsilon_A-1}\right)} \qquad [6.50]$$

However, this expression does still involve the intrinsic efficacy of A and indicates that the effect of an allosteric ligand can vary systematically with the intrinsic efficacy of the orthosteric ligand as well as unpredictably due to the cooperativity. The concentration of B which causes a DR of 2 is given by

$$[B]_{\mathrm{DR}=2} = \frac{\alpha K_b(\varepsilon_A-1)}{\alpha(\varepsilon_A-\varepsilon_B)-2(\varepsilon_{AB}-1)} \qquad [6.51]$$

The largest shift in the concentration–response curve to A, $\mathrm{DR}_{\max}$, is

$$\mathrm{DR}_{\max} = \frac{\alpha(\varepsilon_A-\varepsilon_B)}{\varepsilon_{AB}-1} \qquad [6.52]$$

Note the value of $\mathrm{DR}_{\max}$ will be less than unity if there is positive allosteric modulation (since a 100-fold leftward shift gives DR = 0.01). Also, if the allosteric modulation does increase the potency of the orthosteric ligand, Eq. (6.51) is meaningless (since no concentration of B causes a rightward shift and hence a DR of 2). Apart from special cases (in particular when $\varepsilon_A \gg \varepsilon_B$ and $\varepsilon_{AB} \approx \varepsilon_A \gg 1$), it is not possible to use DRs to determine the affinity and individual allosteric constants of the allosteric ligand. This is consistent with the conclusions of Ehlert.[17] However, particularly when the effect of B on the maximal response to A is small, the maximal DR can be taken as a measure of the magnitude of the overall interaction between the two ligands: let $\varepsilon_{AB} \gg 1$ so that $\varepsilon_{AB} - 1 \approx \varepsilon_{AB}$, then

$$\mathrm{DR_{max}} \approx \alpha\left(\frac{\varepsilon_A}{\varepsilon_{AB}} - \frac{\varepsilon_B}{\varepsilon_{AB}}\right) = \alpha\left(\frac{1}{\delta_{AB}\varepsilon_B} - \frac{1}{\delta_{AB}\varepsilon_A}\right) = \frac{\alpha}{\delta_{AB}}\left(\frac{1}{\varepsilon_B} - \frac{1}{\varepsilon_A}\right)$$
$$\approx \frac{\alpha}{\delta_{AB}\varepsilon_B}$$

the last step being valid if $\varepsilon_A > \varepsilon_B$. This includes a contribution from the intrinsic efficacy of B so it measures the overall interaction not just the cooperativity. It is analogous to $1/\alpha\beta$ from the equation of Leach *et al.*[3] This can be considered an advantage as well as a disadvantage. The contribution of ε_B will be the same whether or not the allosteric ligand measurably affects the basal response in the experimental system. A neutrally cooperative allosteric inverse agonist will still inhibit the effects of an orthosteric ligand even in a system that does not show constitutive signaling due to the presence of the receptor. The mechanism of the effect will be obscured.

When α is large Eq. (6.51) simplifies to

$$[B]_{\mathrm{DR}=2} \approx \frac{\alpha K_b(\varepsilon_A - 1)}{\alpha(\varepsilon_A - \varepsilon_B)} = \frac{K_b(\varepsilon_A - 1)}{\varepsilon_A - \varepsilon_B} \approx \frac{\varepsilon_A K_b}{\varepsilon_A - \varepsilon_B} = \frac{K_b}{1 - \frac{\varepsilon_B}{\varepsilon_A}}$$

Hence, when the orthosteric ligand is sufficiently efficacious ($\varepsilon_A \gg 1$), the concentration of a strong negative allosteric modulator of binding that causes a DR of 2 should differ from its binding affinity (i.e., the affinity estimated in a binding assay) by a factor that is simply related to its relative intrinsic efficacy (compared to that of the orthosteric ligand).

3.2. Fitting the model: Monte Carlo simulations and some practical issues

The question of whether it is possible to determine the parameters of Eq. (6.42) by direct fitting will now be addressed. Monte Carlo simulations were performed using Eq. (6.43) similarly to those described in Ref. 10, although with only 25 simulated data sets per condition. The results are summarized in Table 6.1 and illustrative data sets presented in Figs. 6.1–6.3. Simulations of pure binding cooperativity were not performed since the classical null method is entirely adequate for the analysis of such data (assuming that the allosteric ligand is a neutral antagonist). Given the results of the other simulations, there is no reason to believe that good estimates of the model parameters would not be achieved in this case. Overall, it appears that it is indeed possible to derive good estimates of the parameters of Eq. (6.43), the caveat being that the data set must contain sufficient information to define the contribution of all of the parameters (lack of sufficient information is the reason for the poor estimates in the first simulation in Table 6.1). Practically this means that a sufficient range of concentrations of B must be tested to determine its maximal effect on a concentration–response curve to A at a given receptor density and that a sufficient range of receptor densities must be tested to define the pharmacological parameters of A and B alone and the background activity in the system. The latter two were explored in Ref. 10 and may require as few as two different receptor densities to be studied.

Sufficiency of information is best illustrated by comparing the simulations of the effects of ligands with positive activation cooperativity at high and low coupling efficiency (the first two simulations in Table 6.1). At high coupling efficiency (Fig. 6.1), the decrease in receptor density between

Table 6.1 Results of Monte Carlo simulations of Eq. (6.43)

Simulation	**Positive activation cooperativity, high coupling efficiency**		**Positive activation cooperativity, low coupling efficiency**		**Allosteric agonist, low coupling efficiency**	
Parameter	**Input**	**Estimate**	**Input**	**Estimate**	**Input**	**Estimate**
E_{max}	1.50	1.51 ± 0.08	1.50	1.52 ± 0.08	1.50	1.50 ± 0.08
$\log K_a$	0.48	0.13 ± 0.27	−0.52	−0.52 ± 0.05	−0.52	−0.53 ± 0.05
$\log K_b$	1.00	1.36 ± 0.20	1.48	1.49 ± 0.07	1.00	0.98 ± 0.05
$\log \varepsilon_A$	2.48	2.50 ± 0.13	2.48	2.48 ± 0.07	2.48	2.52 ± 0.10
$\log \varepsilon_B$	0.00	0.00 ± 0.05	0.00	−0.04 ± 0.14	1.00	1.04 ± 0.08
$\log \varepsilon_{AB}$	3.48	3.57 ± **0.66**	3.48	3.49 ± 0.08	3.00	3.05 ± 0.10
$\log \alpha$	0.00	0.08 ± **0.69**	0.00	0.00 ± 0.10	0.00	0.00 ± 0.07
$\log \chi_1$	0.00	0.00 ± 0.04	−1.00	−0.99 ± 0.05	−1.00	−1.04 ± 0.08
$\log \chi_2$	−0.60	−0.61 ± 0.05	−1.60	−1.60 ± 0.06	−1.60	−1.65 ± 0.09
$\log \chi_3$	−1.20	−1.20 ± 0.05	−2.20	−2.20 ± 0.05	−2.20	−2.25 ± 0.10
$\log \chi_4$	−1.81	−1.80 ± 0.06	−2.81	−2.81 ± 0.06	−2.81	−2.86 ± 0.10
Background	1.00	1.00 ± 0.06	1.00	1.01 ± 0.07	1.00	1.01 ± 0.06
$\log(\varepsilon_{AB}/\alpha\varepsilon_A\varepsilon_B)$	1.00	0.99 ± 0.08	1.00	1.04 ± 0.18	−0.52	−0.52 ± 0.10

Simulation	Allosteric inverse agonist, high coupling efficiency		Negative activation cooperativity, high coupling efficiency	
Parameter	**Input**	**Estimate**	**Input**	**Estimate**
$E_{\max}$	1.50	1.52 ± 0.13	1.50	1.51 ± 0.10
$\log K_a$	0.00	0.01 ± 0.09	−0.52	−0.53 ± 0.08
$\log K_b$	0.48	0.46 ± 0.06	0.00	0.01 ± 0.07
$\log \varepsilon_A$	2.48	2.50 ± 0.08	2.48	2.49 ± 0.09
$\log \varepsilon_B$	−1.00	−1.00 ± 0.06	0.00	0.00 ± 0.04
$\log \varepsilon_{AB}$	2.00	2.00 ± 0.06	1.48	1.47 ± 0.06
$\log \alpha$	0.00	−0.02 ± 0.10	0.00	0.00 ± 0.07
$\log \chi_1$	0.00	0.00 ± 0.03	0.00	0.01 ± 0.04
$\log \chi_2$	−0.60	−0.60 ± 0.03	−0.60	−0.59 ± 0.04
$\log \chi_3$	−1.20	−1.20 ± 0.03	−1.20	−1.20 ± 0.05
$\log \chi_4$	−1.81	−1.80 ± 0.05	−1.81	−1.80 ± 0.05
Background	1.00	1.00 ± 0.08	1.00	1.02 ± 0.05
$\log(\varepsilon_{AB}/\alpha\varepsilon_A\varepsilon_B)$	0.52	0.51 ± 0.10	−1.00	−1.01 ± 0.06

"Estimates" are the mean ± SD of the parameters derived from fitting Eq. (6.43) to 25 simulated data sets. The within-occasion standard deviation of the observations was 3% of the true mean and the between-occasion standard deviation of the parameters was 6% of the true mean. Bold to highlight the very large standard deviations associated with these estimates.

the most and least highly coupled conditions is not sufficient to make the maximal response to A in the presence of the highest concentration of B distinguishable from the maximum possible response (2.46 in this case). Thus, there is not sufficient information in the data set to estimate ε_{AB}. Indeed, even for the lowest value of χ simulated in Fig. 6.1, the effect of B is primarily to induce leftward shifts of the concentration–response curve to A, which can also be achieved with positive binding cooperativity. The random variability then results in some data sets where a sufficient depression of the maximal response to A in the presence of B is achieved and these provide accurate estimates of ε_{AB} and α and others where A remains essentially a full agonist in all conditions and the effect of B is erroneously ascribed to binding cooperativity (indeed other fits suggested very high positive activation cooperativity coupled with negative binding cooperativity). This results in ε_{AB} and α being correlated in the fits and poorly estimated individually. However, their product or, more usefully, the ratio $\varepsilon_{AB}/\alpha\varepsilon_A\varepsilon_B = \delta_{AB}/\alpha$ (which describes the overall cooperativity in this model), is accurately and precisely estimated. Thus, it is possible with these data to determine the overall effect of B on A but the precise mechanism remains obscure. Interestingly, despite this, the apparent affinity and intrinsic efficacy of the two individual ligands are well defined. In the lower coupling efficiency data set (Fig. 6.2), the reduction in receptor density from highest to lowest coupling efficiency causes a clear decrease in the maximal response to A in the presence of the highest concentration of B. The effect of B on the maximal response to A is therefore well described and the fit estimates ε_{AB} and α well, providing an unambiguous mechanism for the interaction. Indeed, in this simulation, all of the parameters of the model are estimated with a similar degree of precision.

The above comparison highlights that the key to the success of this method is to substantially reduce the intrinsic activity of the ligands (pardon the split infinitive). The intrinsic efficacy of a ligand which remains a full agonist or inverse agonist at all of the coupling efficiencies tested cannot be estimated. It also highlights the fact that it is not always possible to determine the mechanism of an allosteric interaction from a set of data at a single coupling efficiency. In the simulation illustrated in Fig. 6.1, positive activation and positive binding cooperativity were indistinguishable. It is also possible to conceive of conditions under which negative binding or activation cooperativity would be indistinguishable, or where negative cooperativity was not distinguishable from inverse agonism of the allosteric ligand. Thus, to understand the level at which two ligands are interacting, it is necessary to study them in functional assays with markedly different coupling efficiencies.

One further practical issue related to this method of analysis requires specific comment and that is the question of how the receptor density in the system is modified. This could be done chemically, for example, using irreversible alkylating agents following the classical approach of Furchgott,[34] or using a number of molecular biological approaches. The key issue is that the analysis described above is predicated on the equivalent of an actual change in the receptor density rather than on simple occlusion of the binding sites. This is automatically achieved using molecular techniques, for example, by varying multiplicities of infection when using viral methods of transient expression. However, when using chemical modification, the pharmacological properties of the modifying ligand will have an influence on the validity of the analysis. If the modifying agent is an efficacious full inverse agonist such that modified receptors no longer have any appreciable ability to signal, then this is equivalent to removing these receptors from the system. However, if the modifying agent is a neutral antagonist or only a weakly efficacious inverse agonist, the modified receptors, while unable to bind further ligands, will still contribute to the constitutive activity in the system. As presented, Eq. (6.42) cannot account for this type of intervention. Instead it would be necessary to split the response into that due to the unmodified receptors and that due to the occluded receptors (which would contribute to the basal signal but not to the effect of ligand). If a proportion, q, of the receptors is occluded, the required expression is

$$E = \frac{E_{\max}((1-q)\chi)^n\left(1+\frac{\varepsilon_B[B]}{K_b}+\frac{\varepsilon_A[A]}{K_a}\left(1+\frac{\varepsilon_{AB}[B]}{\varepsilon_A\alpha K_b}\right)\right)^n}{\left(1+\frac{[B]}{K_b}+\frac{[A]}{K_a}\left(1+\frac{[B]}{\alpha K_b}\right)\right)^n+((1-q)\chi)^n\left(1+\frac{\varepsilon_B[B]}{K_b}+\frac{\varepsilon_A[A]}{K_a}\left(1+\frac{\varepsilon_{AB}[B]}{\varepsilon_A\alpha K_b}\right)\right)^n} + \frac{E_{\max}(\varepsilon_i q\chi)^n}{1+(\varepsilon_i q\chi)^n}$$

where ε_i quantifies the effect of the irreversible modification on the constitutive activity of the receptor (NB, in this case both the allsoteric and orthosteric sites are assumed to be occluded by the chemical agent). For experimental data, a further constant term must be added to account for signaling which is not due to the receptor under consideration.

3.3. Comparison with the "transduced TSM"—Can the ATSM's parameters be measured?

Ehlert and coworkers[28] have recently suggested that it is possible to measure the affinity of a ligand for the active and inactive states of a receptor (under the TSM) allowing the relative intrinsic efficacies of a set of ligands to be determined. The method did not allow the estimation of the affinity of highly efficacious ligands for the inactive state, hence preventing a TSM intrinsic efficacy or relative efficacy being determined for these ligands. Of course, given the discussion in Section 1, these affinities should be viewed as weighted averages over a number of conformational states rather than true thermodynamic constants, particularly as it is not clear that any one conformation dominates in either state. Their model can be derived straightforwardly by letting the stimulus be $S=[R^*]+[AR^*]$ and substituting this into a Hill function as the transducer function:

$$E=\frac{E_{\max}S^n}{K_e^n+S^n}=\frac{E_{\max}([R^*]+[AR^*])^n}{K_e^n+([R^*]+[AR^*])^n}$$
$$=\frac{E_{\max}\left(\frac{L[R]_T(1+\alpha K[A])}{1+L+K[A](1+\alpha L)}\right)^n}{K_e^n+\left(\frac{L[R]_T(1+\alpha K[A])}{1+L+K[A](1+\alpha L)}\right)^n}$$
$$=\frac{E_{\max}L^n\sigma^n(1+\alpha K[A])^n}{(1+L+K[A](1+\alpha L))^n+L^n\sigma^n(1+\alpha K[A])^n} \qquad [6.53]$$

The concentrations of the active species used above can be derived from Eq. (6.13) by letting $[B]=0$. In this case, $\sigma=[R]_T/K_e$ to distinguish the parameters of this model from those of Eq. (6.42). It can be shown (see Ref. 10) that Eq. (6.53) is equivalent to that derived by Ehlert et al.[28] It is then possible to derive correspondences between the parameters of Eq. (6.42) and those of Eq. (6.53) (the derivations are presented in Ref. 10).

$$\chi=\frac{\sigma L}{1+L} \qquad [6.54]$$

$$\varepsilon_A=\frac{\alpha(1+L)}{1+\alpha L} \qquad [6.55]$$

$$K_a=\frac{1+L}{K(1+\alpha L)} \qquad [6.56]$$

Equation (6.56) is the standard definition of the apparent affinity of a ligand in the TSM and is equivalent to Eq. (6.11) with $[B]=0$. It is notable that L, the receptor isomerization constant, occurs in both Eqs. (6.54) and (6.55). In the former case it is arguable that receptor isomerization is an initial step in signal transduction and should therefore be a factor of the coupling efficiency. Its occurrence in the expression for ε_A is perhaps more concerning, but is consistent with the definition of this parameter since, in the TSM, L governs how effectively a ligand with a given value of α activates receptors and, in Eqs. (6.42) and (6.43), ε_A represents the ability of an agonist to activate receptors. Further manipulation of these expressions provides useful insight. Firstly,

$$\frac{K_a}{\varepsilon_A}=\frac{1+L}{K(1+\alpha L)}\times\frac{1+\alpha L}{\alpha(1+L)}=\frac{1}{\alpha K}$$

That is, K_a/ε_A is equal to the (weighted average) dissociation equilibrium constant of the ligand from the active state(s) of the receptor. At this point, it is instructive to note that in the absence of allosteric modulator, the basal activity, midpoint, and maximal effect of a ligand in Eq. (6.43) combine as follows to give

$$\frac{basal\mathrm{EC}_{50}}{Max_A}=\frac{E_{\max}\chi}{1+\chi}\times\frac{K_a(1+\chi)}{1+\varepsilon_A\chi}\div\frac{E_{\max}\varepsilon_A\chi}{1+\varepsilon_A\chi}=\frac{K_a}{\varepsilon_A}=\frac{1}{\alpha K}$$

In other words, when the Hill coefficient of the concentration–response curves is equal to unity and an estimate of the receptor-independent activity in the system is available (to allow accurate estimation of *basal* and Max_A), it is possible to determine the ratio of the affinity and intrinsic efficacy of the ligand directly from the parameters of fitting a linear rational function to the concentration–response curve. In the presence of a saturating concentration of allosteric ligand the same assumptions give

$$\frac{Max_A Max_B \mathrm{EC}_{50,[B]\to\infty}}{basal Max_{AB}\mathrm{EC}_{50,[B]=0}}=\frac{\alpha_{\mathrm{Eq.}(6.42)}\varepsilon_A\varepsilon_B}{\varepsilon_{AB}}$$

which, as noted above, is a measure of the overall cooperativity between the two ligands. However, from the discussion above, fitting Eq. (6.43) to the data is a superior method since it will provide separate estimates of both cooperativity constants.

Also, when $\alpha L \gg 1$,

$$\varepsilon_A \to \frac{\alpha(1+L)}{\alpha L}=\frac{1+L}{L}$$

That is, consideration of the TSM applies a theoretical upper limit to ε_A. This is again consistent with the definition of ε_A, which quantifies the ratio of the ability of free receptors and agonist:receptor complexes to interact with the signal transduction system. For a TSM, the maximal value that this ratio can take is achieved when an agonist can convert all receptors into the active state, that is, $[R]_T/[R^*]$. This limit can be used to advantage since it implies that the largest value of ε_A measured for an agonist at a given receptor can provide an upper limit on the receptor isomerization constant:

$$\varepsilon_A \le \frac{1+L}{L} \Leftrightarrow L < \frac{1}{\varepsilon_A - 1} \qquad [6.57]$$

Applying $\alpha L \gg 1$ to Eq. (6.53) gives

$$E \rightarrow \frac{E_{\max} L^n \sigma^n (1+\alpha K[A])^n}{(1+L+L\alpha K[A])^n + L^n \sigma^n (1+\alpha K[A])^n} \qquad [6.58]$$

This expression contains no term in which α or K occur independently of each other. Hence, it is impossible to determine the inactive state affinity of ligands that strongly stabilize the active state of the receptor. Thus, it becomes impossible to measure the TSM "ground state" affinity and intrinsic efficacy separately for such ligands.

Finally, if it is assumed that $L \ll 1$, that is, that the receptor is predominantly in the inactive state in the absence of ligands, then Eq. (6.55) simplifies to

$$\varepsilon_A = \frac{\alpha(1+L)}{1+\alpha L} \approx \frac{\alpha}{1+\alpha L}$$

Further assuming that $\alpha < 1/L$ so that the ligand does not stabilize the active state too strongly, then $\varepsilon_A \approx \alpha$. Under the same assumptions, Eq. (6.56) simplifies to

$$K_a = \frac{1+L}{K(1+\alpha L)} \approx \frac{1}{K(1+\alpha L)} \approx \frac{1}{K}$$

That is, for inverse agonists and agonists with relatively low intrinsic efficacy, ε_A provides a good approximation of the TSM intrinsic efficacy and K_a provides a good estimate of the (weighted average) affinity for the inactive state(s).

The foregoing discussion suggests that it may be possible to determine the parameters of the ATSM for an allosteric interaction using this approach. Let $S = [R^*] + [AR^*] + [R^*B] + [AR^*B]$, then

$$E = \frac{E_{\max} S^n}{K_e^n + S^n} = \frac{E_{\max}([R^*]+[AR^*]+[R^*\mathrm{B}]+[AR^*\mathrm{B}])^n}{K_e^n + ([R^*]+[AR^*]+[R^*\mathrm{B}]+[AR^*\mathrm{B}])^n}$$

$$= \frac{E_{\max}\left(\dfrac{L[R]_T(1+\beta M[B]+\alpha K[A](1+\beta\gamma\delta M[B]))}{1+L+M[B](1+\beta L)+K[A](1+\alpha L+\gamma M[B](1+\alpha\beta\delta L))}\right)^n}{K_e^n + \left(\dfrac{L[R]_\mathrm{T}(1+\beta M[B]+\alpha K[A](1+\beta\gamma\delta M[B]))}{1+L+M[B](1+\beta L)+K[A](1+\alpha L+\gamma M[B](1+\alpha\beta\delta L))}\right)^n}$$

$$= \frac{E_{\max}L^n\sigma^n(1+\beta M[B]+\alpha K[A](1+\beta\gamma\delta M[B]))^n}{(1+L+M[B](1+\beta L)+K[A](1+\alpha L+\gamma M[B](1+\alpha\beta\delta L)))^n + L^n\sigma^n(1+\beta M[B]+\alpha K[A](1+\beta\gamma\delta M[B]))^n} \quad [6.59]$$

It is then possible to express the parameters of Eq. (6.42) in terms of the parameters of Eq. (6.59). The expressions for χ, ε_A, and K_a are simply Eqs. (6.54)–(6.56) and by analogy

$$\varepsilon_B = \frac{\beta(1+L)}{1+\beta L} \quad \text{and} \quad K_b = \frac{1+L}{M(1+\beta L)}$$

It remains to determine expressions for the binding and activation cooperativity. Equating the expressions for the response in the presence of saturating concentrations of both ligands gives

$$\frac{E_{\max}\varepsilon_{AB}^n\chi^n}{1+\varepsilon_{AB}^n\chi^n} = \frac{E_{\max}(\alpha\beta\delta L\sigma)^n}{(1+\alpha\beta\delta L)^n+(\alpha\beta\delta L\sigma)^n}$$

This rearranges to

$$\varepsilon_{AB}\chi = \frac{\alpha\beta\delta L\sigma}{1+\alpha\beta\delta L}$$

Dividing by the expression for χ (Eq. 6.54), this gives

$$\varepsilon_{AB} = \frac{\alpha\beta\delta L\sigma}{1+\alpha\beta\delta L}\times\frac{1+L}{L\sigma} = \frac{\alpha\beta\delta(1+L)}{1+\alpha\beta\delta L}$$

To determine the expression for activation cooperativity this expression must be divided by the product $\varepsilon_A\varepsilon_B$ and gives

$$\delta_{AB} = \frac{\varepsilon_{AB}}{\varepsilon_A\varepsilon_B} = \frac{\alpha\beta\delta(1+L)}{1+\alpha\beta\delta L}\times\frac{1+\alpha L}{\alpha(1+L)}\times\frac{1+\beta L}{\beta(1+L)} = \frac{\delta(1+\alpha L)(1+\beta L)}{(1+L)(1+\alpha\beta\delta L)}$$

This has a very similar structure to the macroscopic binding cooperativity (α_{Mac}) derived in Section 1. It can also be shown, by considering the apparent affinity of A in the absence and presence of B, that α in Eq. (6.42) corresponds to

$$\alpha_{\text{Eq.}(6.42)} = \frac{(1+L)(1+\alpha\beta\delta L)}{\gamma(1+\alpha L)(1+\beta L)}$$

which is the reciprocal of the expression for α_{Mac} as, indeed, it should be since they represent (the reciprocal of) the same parameter. Then, note that

Table 6.2 Results of Monte Carlo simulations of Eq. (6.59)

Simulation	Interaction of allosteric agonists ($\alpha L < 1$)		Interaction of allosteric agonists ($\alpha L > 1$)	
Parameter	**Input**	**Estimate**	**Input**	**Estimate**
E_{max}	1.50	1.55 ± 0.08	1.50	1.53 ± 0.07
$\log K$	0.00	0.03 ± 0.10	0.00	0.68 ± 0.13
$\log M$	−1.00	-1.00 ± 0.07	−1.00	-0.98 ± 0.05
$\log \alpha$	2.04	2.02 ± 0.17	2.70	2.01 ± 0.23
$\log \beta$	1.04	1.04 ± 0.08	1.04	1.01 ± 0.12
$\log \gamma$	−0.30	-0.36 ± 0.13	0.00	0.92 ± 0.10
$\log \delta$	0.00	0.04 ± 0.10	0.00	-0.92 ± 0.12
$\log \sigma_1$	2.48	2.47 ± 0.10	1.48	2.10 ± 0.12
$\log \sigma_2$	1.88	1.87 ± 0.07	0.88	1.51 ± 0.10
$\log \sigma_3$	1.27	1.28 ± 0.06	0.27	0.91 ± 0.11
$\log \sigma_4$	0.67	0.68 ± 0.08	−0.33	0.31 ± 0.13
$\log L$	−3.00	-3.00 ± 0.14	−2.00	-2.63 ± 0.17
$\log n$	0.00	0.00 ± 0.04	0.00	0.00 ± 0.05
Background	1.00	0.99 ± 0.06	1.00	1.00 ± 0.07
$\log \alpha K$	2.04	2.04 ± 0.11	2.70	2.69 ± 0.12
$\log \beta M$	0.04	0.04 ± 0.07	0.04	0.03 ± 0.10
$\log (\gamma/\delta)$	−0.30	-0.32 ± 0.08	0.00	-0.01 ± 0.10

"Estimates" are the mean ± SD of the parameters derived from fitting Eq. (6.59) to 25 simulated data sets. The within-occasion standard deviation of the observations was 3% of the true mean and the between-occasion standard deviation of the parameters was 6% of the true mean.

$\alpha_{Mac}\delta_{AB} = \delta/\gamma$ and this product will equal unity if the ATSM binding and activation cooperativities are equal.

The results of two Monte Carlo simulations of the interaction between allosteric agonists are summarized in Table 6.2 and illustrated in Fig. 6.4. In the first simulation, the parameters were chosen such that $\alpha L < 1$ while in the second $\alpha L > 1$. The data highlight the issues raised above with attempting to recover TSM parameters from data on functional responses. When neither

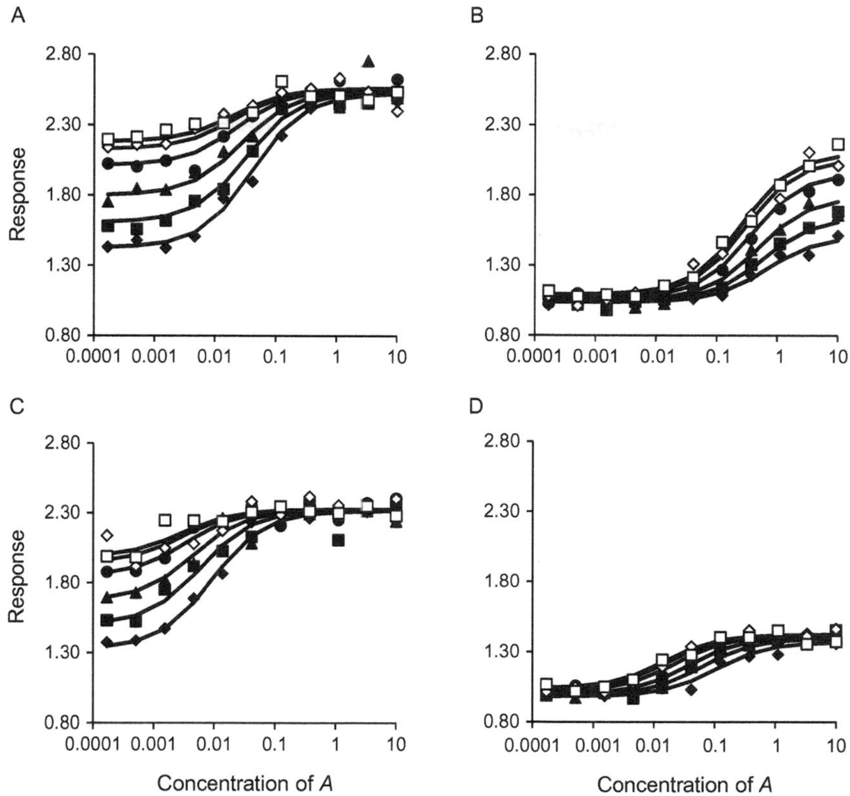

Figure 6.4 Example simulations of the interaction between two allosteric agonists, when neither agonist strongly stabilizes the active state of the receptor (A, B) and when one of the agonists is able to stabilize the active state quite strongly (C, D), using Eq. (6.59). The input parameters of the simulation are given in Table 6.2. The graphs show the curves at the highest (A, C) and lowest (B, D) coupling efficiencies simulated (these were not the same for the two simulations). The concentrations of *B* were 0 (filled diamonds), 1 (filled squares), 3 (triangles), 10 (circles), 30 (open diamonds), and 100 (open squares). Lines show the fitted curves.

ligand stabilizes the active state of the receptor particularly strongly ($\alpha L < 1$, $\beta L < 1$), it is possible to generate accurate estimates of all of the parameters in the model. However, when one of the ligands stabilizes the active state more strongly, the estimation of the model parameters is confounded and the estimates are sufficiently inaccurate to be misleading (particularly as they appear to be reasonably precise). Since L and σ are among the poorly defined parameters in the second simulation, it may be possible to identify this problem if ligands with a sufficiently wide range of intrinsic activities are studied. Interestingly, fitting Eq. (6.43) to these data did result in the "correct" values of its parameters (based on the correspondences between the parameters of Eq. (6.59) and those of Eq. (6.43) given above). Thus, the estimates from fitting the macroscopic model (Eq. 6.42) are internally consistent and the estimate of binding cooperativity should be consistent with that derived from binding experiments (under similar conditions). However, there do appear to be limitations on the ability to estimate the microscopic parameters in systems that are assumed to be at steady state.

As a final parenthetical comment to this section, following from Section 1, it is also possible to represent the pharmacological parameters of the ligands in Eq. (6.42) in terms of the multiconformational equivalents of the ATSM parameters (using the expressions Eq. 6.29). Those that cannot be deduced directly from expressions presented in Section 1 are:

$$\varepsilon_A = \frac{\alpha(1+L)}{1+\alpha L} = \frac{(\mathbf{w}_{\mathbf{a}}^T\mathbf{A}\boldsymbol{\lambda})(\mathbf{w}_{\mathbf{i}}^T\boldsymbol{\lambda})}{(\mathbf{w}_{\mathbf{a}}^T\boldsymbol{\lambda})(\mathbf{w}_{\mathbf{i}}^T\mathbf{A}\boldsymbol{\lambda})} \times \frac{\mathbf{w}^T\boldsymbol{\lambda}}{\mathbf{w}_{\mathbf{i}}^T\boldsymbol{\lambda}} \times \frac{\mathbf{w}_{\mathbf{i}}^T\mathbf{A}\boldsymbol{\lambda}}{\mathbf{w}^T\mathbf{A}\boldsymbol{\lambda}} = \frac{(\mathbf{w}_{\mathbf{a}}^T\mathbf{A}\boldsymbol{\lambda})(\mathbf{w}^T\boldsymbol{\lambda})}{(\mathbf{w}_{\mathbf{a}}^T\boldsymbol{\lambda})(\mathbf{w}^T\mathbf{A}\boldsymbol{\lambda})}$$

$$\delta_{AB} = \frac{\delta(1+\alpha L)(1+\beta L)}{(1+L)(1+\alpha\beta\delta L)} = \frac{(\mathbf{w}_{\mathbf{a}}^T\boldsymbol{\lambda})(\mathbf{w}_{\mathbf{a}}^T\mathbf{A}\mathbf{B}\boldsymbol{\lambda})(\mathbf{w}^T\mathbf{A}\boldsymbol{\lambda})(\mathbf{w}^T\mathbf{B}\boldsymbol{\lambda})}{(\mathbf{w}_{\mathbf{a}}^T\mathbf{A}\boldsymbol{\lambda})(\mathbf{w}_{\mathbf{a}}^T\mathbf{B}\boldsymbol{\lambda})(\mathbf{w}^T\boldsymbol{\lambda})(\mathbf{w}^T\mathbf{A}\mathbf{B}\boldsymbol{\lambda})}$$

$$= \frac{(\mathbf{w}_{\mathbf{a}}^T\boldsymbol{\lambda})(\mathbf{w}_{\mathbf{a}}^T\mathbf{A}\mathbf{B}\boldsymbol{\lambda})}{(\mathbf{w}_{\mathbf{a}}^T\mathbf{A}\boldsymbol{\lambda})(\mathbf{w}_{\mathbf{a}}^T\mathbf{B}\boldsymbol{\lambda})}\alpha_{Mac}^{-1}$$

The expression for ε_B can be derived by analogy with that for ε_A by replacing **A** with **B**.

4. FURTHER EXPLORATION OF THE cTCM

4.1. The cTCM when $[R]_T \gg [G]_T$

As detailed in Hall[2] and Slack and Hall,[10] consideration of the extreme case of the gTCM in which the receptor is in great excess over the G protein leads to an operational model of agonism which includes constitutive receptor activity. The elaboration of this model to include an allosteric ligand was considered in the previous section. This raises the question as to whether we may gain any further insight by considering the same extreme case of the cTCM. In Ref. 2, it was apparent that the required equation could be derived by taking the equation describing the system in which G protein is nonlimiting and

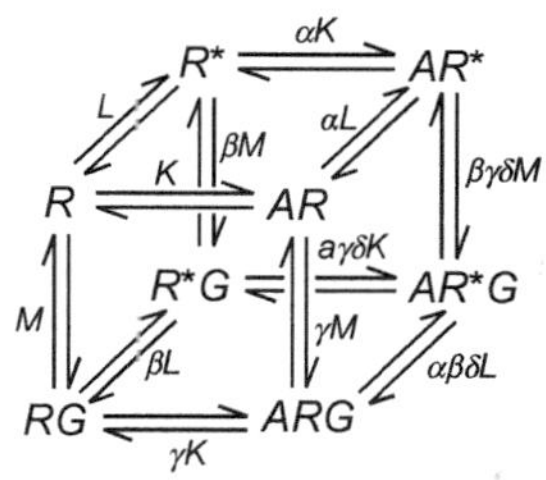

Scheme 6.6 The cubic ternary complex model. *K* and *M* are the equilibrium association constants of *A* and *G* for *R*. *L* is the receptor isomerization constant, $L=[R^*]/[R]$. α and β are the receptor state selectivity constants for *A* and *G*, respectively, so, for example, $\alpha K=[AR^*]/[A][R^*]$ and $\alpha L=[AR^*]/[AR]$. γ is the binding cooperativity between *A* and *G* and δ is the activation cooperativity between *A* and *G*.

exchanging the roles of *R* and *G*. However, for the sake of mathematical and intellectual rigor, the approach taken here will be that of Ref. 2 of deriving the equation for the system in which both *R* and *G* are limiting and specializing from this. The system under consideration is that of Scheme 6.6. Conservation of receptor and G protein mass gives the following two constraints:

$$[R]_T=[R]+[R^*]+[RG]+[R^*G]+[AR]+[AR^*]+[ARG]+[AR^*G] \quad [6.60]$$

$$[G]_T=[G]+[RG]+[R^*G]+[ARG]+[AR^*G] \quad [6.61]$$

These equations can be expressed in terms of free concentrations of *R* and *G* as

$$[R]_T=[R](1+L+M[G](1+\beta L)+K[A](1+\alpha L+\gamma M[G](1+\alpha\beta\delta L))) \quad [6.62]$$

$$[G]_T=[G]+M[G][R](1+\beta L+\gamma K[A](1+\alpha\beta\delta L)) \quad [6.63]$$

Equation (6.62) then provides an expression for the free receptor concentration that can be substituted into Eq. (6.63) to give an expression for the free concentration of *G* in terms of constants (parameters of the model) and the concentration of the ligand:

$$[G]_T=[G]+\frac{M[G][R]_T(1+\beta L+\gamma K[A](1+\alpha\beta\delta L))}{1+L+M[G](1+\beta L)+K[A](1+\alpha L+\gamma M[G](1+\alpha\beta\delta L))} \quad [6.64]$$

This can be rearranged into the following quadratic

$$M[G]^2(1+\beta L+\gamma K[A](1+\alpha\beta\delta L))-[G]_T(1+L+K[A](1+\alpha L))$$
$$+[G]\begin{pmatrix}1+L+K[A](1+\alpha L)+\\ M([R]_T-[G]_T)(1+\beta L+\gamma K[A](1+\alpha\beta\delta L))\end{pmatrix}=0 \qquad [6.65]$$

whose physically meaningful solution is

$$\frac{-b+\sqrt{b^2+4ac}}{2a} \qquad [6.66]$$

where

$$a=M(1+\beta L+\gamma K[A](1+\alpha\beta\delta L))$$

$$b=1+L+K[A](1+\alpha L)+M([R]_T-[G]_T)(1+\beta L+\gamma K[A](1+\alpha\beta\delta L))$$

$$c=[G]_T(1+L+K[A](1+\alpha L))$$

This can, of course, be used directly to give general expressions for binding isotherms, etc., and allow the exploration of the full spectrum of behaviors of the model. However, for the present discussion, the special case when $[R]_T \gg [G]_T$ will now be derived. In this case, we may write $[R]_T-[G]_T \approx [R]_T$ and then use

$$\sqrt{x^2+2y}\approx x+\frac{y}{x}\Rightarrow\sqrt{b^2+4ac}\approx b+\frac{2ac}{b}$$

[35] to approximate the square root of the determinant in Eq. (6.66) to give the following expression

$$[G]\approx\frac{[G]_T(1+L+K[A](1+\alpha L))}{1+L+K[A](1+\alpha L)+M[R]_T(1+\beta L+\gamma K[A](1+\alpha\beta\delta L))} \qquad [6.67]$$

Substituting this into the expression for the response for the cTCM then gives

$$
\begin{aligned}
E &= [R^*G] + [AR^*G] \\
&= \frac{\beta LM[G][R]_T(1+\alpha\gamma\delta K[A])}{1+L+K[A](1+\alpha L)+M[G](1+\beta L+\gamma K[A](1+\alpha\beta\delta L))} \\
&= \frac{\dfrac{\beta LM[G]_T[R]_T(1+\alpha\gamma\delta K[A])(1+L+K[A](1+\alpha L))}{1+L+K[A](1+\alpha L)+M[R]_T(1+\beta L+\gamma K[A](1+\alpha\beta\delta L))}}{1+L+K[A](1+\alpha L)+\dfrac{M[G]_T(1+L+K[A](1+\alpha L))(1+\beta L+\gamma K[A](1+\alpha\beta\delta L))}{1+L+K[A](1+\alpha L)+M[R]_T(1+\beta L+\gamma K[A](1+\alpha\beta\delta L))}} \\
&= \frac{\beta LM[G]_T[R]_T(1+\alpha\gamma\delta K[A])(1+L+K[A](1+\alpha L))}{(1+L+K[A](1+\alpha L))\left(1+L+K[A](1+\alpha L)+M\left([R]_T+[G]_T\right)(1+\beta L+\gamma K[A](1+\alpha\beta\delta L))\right)} \\
&= \frac{\beta LM[G]_T[R]_T(1+\alpha\gamma\delta K[A])}{1+L+K[A](1+\alpha L)+M[R]_T(1+\beta L+\gamma K[A](1+\alpha\beta\delta L))}
\end{aligned}
\qquad [6.68]
$$

where the approximation $[R]_T + [G]_T \approx [R]_T$ was used in the last step. This is indeed the usual form of the expression for the functional response in the cTCM (Eq. 6.69) with the role of R and G switched:

$$E = \frac{\beta LM[G]_T[R]_T(1+\alpha\gamma\delta K[A])}{1+L+K[A](1+\alpha L)+M[G]_T(1+\beta L+\gamma K[A](1+\alpha\beta\delta L))} \quad [6.69]$$

In the absence of the ligand, the level of activity in the $[R]_T \gg [G]_T$ system (i.e., that described by Eq. 6.68) is

$$Basal = \frac{\beta LM[G]_T[R]_T}{1+L+M[R]_T(1+\beta L)} \quad [6.70]$$

and the maximal response to the ligand is

$$Max_A = \frac{\alpha\beta\gamma\delta LM[G]_T[R]_T}{1+\alpha L+\gamma M[R]_T(1+\alpha\beta\delta L)} \quad [6.71]$$

A criterion for agonism can then be derived:

$$\frac{\alpha\beta\gamma\delta LM[G]_T[R]_T}{1+\alpha L+\gamma M[R]_T(1+\alpha\beta\delta L)} > \frac{\beta LM[G]_T[R]_T}{1+L+M[R]_T(1+\beta L)}$$

$$\frac{\alpha\gamma\delta(1+L+M[R]_T)}{1+\alpha L+\gamma M[R]_T} > 1$$

Comparing this inequality with inequality (6.41), the protean behavior of ligands under this assumption is governed by receptor density rather than the concentration of G protein. The limiting values of Eqs. (6.70) and (6.71) as $[R]_T$ becomes arbitrarily large are

$$Basal_\infty = \frac{\beta L[G]_T}{1+\beta L} \quad [6.72]$$

and

$$Max_{A,\infty} = \frac{\alpha\beta\delta L[G]_T}{1+\alpha\beta\delta L} \quad [6.73]$$

respectively, and we note that these two expressions are not equal (although if β is sufficiently large the values may become indistinguishable practically). Thus, in the cTCM, an agonist:receptor complex can be inherently more (or less) able to activate G protein than a free receptor,[15] something that is not possible in the gTCM (or therefore in the model described by Eq. 6.42). For example, if $[R]_T$ and hence χ becomes arbitrarily large in Eqs. (6.44) and (6.45), both approach $E_{\max}$. Indeed, this is just one of the behaviors of the cTCM that cannot be reproduced by models such as Eq. (6.42) that assume that the pharmacological stimulus is a linear function of $[AR]$. Under the assumption of a linear stimulus, the agonist:receptor complex simply acts as if it were equivalent to some multiple of the free receptor concentration (that multiple defining the intrinsic efficacy of the ligand); hence, increasing the concentration of free receptor must be able to achieve the same level of stimulus as an AR complex for some finite concentration. Thus, the derivation of an operational model with the properties of the cTCM requires a nonlinear stimulus function.

4.2. An operational model with a linear rational stimulus function

The obvious choice of stimulus function for further investigation, given that the cTCM is a linear rational function of ligand concentration, is a linear rational function. Thus, let the stimulus, S, be

$$S=\frac{[R]+\varepsilon_A[AR]}{K_s+[R]+\psi_A[AR]} \qquad [6.74]$$

In Eq. (6.74), K_s is the concentration of free receptor that gives a half-maximal stimulus, the maximum possible stimulus produced by free receptors is 1, the maximum possible stimulus produced by AR complexes is ε_A/ψ_A, and K_s/ψ_A is the concentration of AR complexes that gives half of this maximal level of stimulus. It is immediately apparent that there is now no single parameter that defines the pharmacological behavior of a ligand: the values of a pair of constants must be defined. When $[R]+\psi_A[AR]\ll K_s$, the stimulus becomes linear so the equations of Section 2 form a special case of those derived in this section. Substituting for the concentrations of the receptor species in Eq. (6.74) gives

$$S=\frac{\dfrac{[R]_T(K_a+\varepsilon_A[A])}{K_a+[A]}}{K_s+\dfrac{[R]_T(K_a+\psi_A[A])}{K_a+[A]}}=\frac{[R]_T\left(1+\dfrac{\varepsilon_A[A]}{K_a}\right)}{K_s+[R]_T+\frac{[A]}{K_a}\left(K_s+\psi_A[R]_T\right)} \qquad [6.75]$$

and taking the simplified approach of using a linear rational transducer function $\{E = E_{\max}S/(K_e + S)\}$, the response is

$$E = \frac{\dfrac{E_{\max}[R]_T\left(1+\dfrac{\varepsilon_A[A]}{K_a}\right)}{K_s+[R]_T+\dfrac{[A]}{K_a}(K_s+\psi_A[R]_T)}}{K_e+\dfrac{[R]_T\left(1+\dfrac{\varepsilon_A[A]}{K_a}\right)}{K_s+[R]_T+\dfrac{[A]}{K_a}(K_s+\psi_A[R]_T)}}$$

$$= \frac{E_{\max}[R]_T\left(1+\dfrac{\varepsilon_A[A]}{K_a}\right)}{K_eK_s+[R]_T(1+K_e)+\dfrac{[A]}{K_a}(K_eK_s+[R]_T(\varepsilon_A+\psi_A K_e))} \quad [6.76]$$

Dividing through by K_eK_s, this can be written

$$E = \frac{E_{\max}\chi\left(1+\dfrac{\varepsilon_A[A]}{K_a}\right)}{1+\chi(1+K_e)+\dfrac{[A]}{K_a}\left(1+\varepsilon_A\chi\left(1+\dfrac{\psi_A}{\varepsilon_A}K_e\right)\right)} \quad [6.77]$$

where $\chi = [R]_T/K_eK_s$. The following equivalences then show that the cTCM is indeed a model with a linear rational stimulus function:

$$E_{\max} = [G]_T, K_a = \frac{1+L}{K(1+\alpha L)},\ K_e = \frac{1}{\beta L},\ K_s = \frac{1+L}{M},\ \varepsilon_A = \frac{\alpha\gamma\delta(1+L)}{1+\alpha L},$$
$$\psi_A = \frac{\gamma(1+L)}{1+\alpha L},\ \text{and}\ \chi = \frac{[R]_T}{K_eK_s} = \frac{\beta LM[R]_T}{1+L}$$

This can be confirmed by substituting these expressions into Eq. (6.77), which will result in Eq. (6.68). Thus, for Eq. (6.68),

$$S = \frac{[R]+\varepsilon_A[AR]}{K_s+[R]+\psi_A[AR]} = \frac{[R]+\dfrac{\alpha\gamma\delta(1+L)}{1+\alpha L}[AR]}{\frac{1+L}{M}+[R]+\dfrac{\gamma(1+L)}{1+\alpha L}[AR]}$$

However, it should be stressed that these expressions apply only to the special case derived above and do not represent more general equivalences between the parameters of the cTCM and Eq. (6.77).

The response in the absence of ligand is then

$$Basal = \frac{E_{\max}\chi}{1+\chi(1+K_e)} \qquad [6.78]$$

The response in the presence of a saturating concentration of ligand is

$$Max_A = \frac{E_{\max}\varepsilon_A\chi}{1+\varepsilon_A\chi\left(1+\frac{\psi_A}{\varepsilon_A}K_e\right)} \qquad [6.79]$$

and the concentration of ligand causing half of this maximal response is

$$[A]_{50} = \frac{K_a(1+\chi(1+K_e))}{1+\varepsilon_A\chi\left(1+\frac{\psi_A}{\varepsilon_A}K_e\right)} \qquad [6.80]$$

As with the analogous expressions from Eq. (6.43), Eqs. (6.78)–(6.80) can be combined to determine the ratio of the apparent affinity and a component, in this case, of the intrinsic efficacy of the ligand:

$$\frac{Basal\mathrm{EC}_{50}}{Max_A} = \frac{E_{\max}\chi}{1+\chi(1+K_e)} \times \frac{K_a(1+\chi(1+K_e))}{1+\varepsilon_A\chi\left(1+\frac{\psi_A}{\varepsilon_A}K_e\right)} \div \frac{E_{\max}\varepsilon_A\chi}{1+\varepsilon_A\chi\left(1+\frac{\psi_A}{\varepsilon_A}K_e\right)}$$

$$= \frac{K_a}{\varepsilon_A}$$

Hence, if a suitable independent estimate of the affinity is available, this may be used to determine one component of the intrinsic efficacy of the ligand.

The limiting values of Eqs. (6.78) and (6.79) at high receptor densities are

$$Basal_{[R]_T\to\infty} = \frac{E_{\max}}{1+K_e} \qquad [6.81]$$

$$Max_{A,[R]_T\to\infty} = \frac{E_{\max}}{1+\frac{\psi_A}{\varepsilon_A}K_e} \qquad [6.82]$$

These expressions are clearly different, consistent with the behavior of the cTCM in this respect. That is, the maximal stimulatory effect of agonist: receptor complexes can differ from that of free receptors. The criterion for agonism in this model is that

$$\frac{E_{\max}\varepsilon_A\chi}{1+\varepsilon_A\chi\left(1+\frac{\psi_A}{\varepsilon_A}K_e\right)} > \frac{E_{\max}\chi}{1+\chi(1+K_e)}$$

$$\varepsilon_A(1+\chi+\chi K_e) > 1+\varepsilon_A\chi+\psi_A\chi K_e$$

$$\varepsilon_A(1+\chi K_e) > 1+\psi_A\chi K_e$$

$$\frac{\varepsilon_A(1+\chi K_e)}{1+\psi_A\chi K_e} > 1$$

$$\frac{\varepsilon_A(1+\rho)}{1+\psi_A\rho} > 1 \qquad [6.83]$$

In the last step χK_e has been replaced by $\rho=[R]_T/K_s=\chi K_e$. This is a term in the stimulus function and emphasizes that this criterion applies to the stimulus and does not include properties of the transducer function. However, it is also not a simple criterion on either of the efficacy parameters. Since the criterion applies to the pharmacological stimulus, under this model the effects of a protean ligand should be the same at each step along any subsequent linear signal transduction pathway since these only differ in the midpoint of their transducer functions. Differences in the pharmacological properties of a ligand determined for different responses or in different cell lines may indicate biased agonism rather protean behavior. It is not clear, then, that some of the examples of protean agonism that have been claimed in the literature (e.g., Refs. 36,37) are not in fact examples of biased agonism, as some authors have later noted.[38] Indeed, it is theoretically possible for a ligand that is biased to be protean for one response pathway and an agonist or inverse agonist under all conditions along another. In the standard form of the cTCM (Eq. 6.69), the criterion for agonism depends on the concentration of G protein so, again, the behavior should not change further along the signal transduction pathway at any given concentration of G protein.

When $\rho \ll 1$ in inequality (6.83) (and hence the stimulus function is linear), it approximates to $\varepsilon_A > 1$ (consistent with Ref. 10) while, when

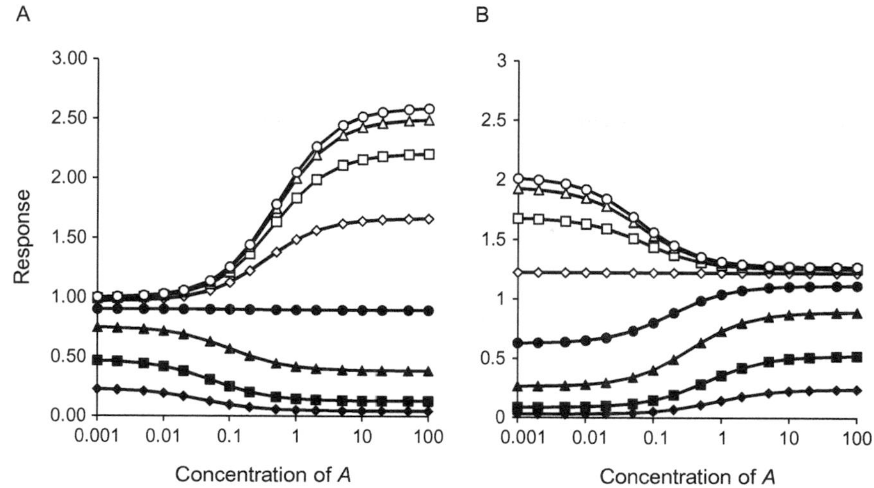

Figure 6.5 Examples of protean behavior of ligands according to Eq. (6.77). (A) Parameters that result in a ligand that transitions from an inverse agonist at low receptor densities to an agonist at high. In this case, $E_{\max}=3$, $K_a=0.03$, $\varepsilon_A=0.15$, $\psi_A=0.01$, $K_e=2$, and $\chi=0.1$ (filled diamonds), 0.3 (filled squares), 1 (filled triangles), 3 (filled circles), 10 (open diamonds), 30 (open squares), 100 (open triangles), or 300 (open circles). (B) Parameters that result in a ligand that transitions from an agonist at low receptor densities to an inverse agonist at high. In this case, $E_{\max}=3$, $K_a=1$, $\varepsilon_A=10$, $\psi_A=30$, $K_e=0.45$, and $\chi=0.01$ (filled diamonds), 0.03 (filled squares), 0.1 (filled triangles), 0.3 (filled circles), 1 (open diamonds), 3 (open squares), 10 (open triangles), or 30 (open circles).

$\rho \gg 1$, inequality (6.83) is approximately $\varepsilon_A/\psi_A > 1$. Thus, a ligand is an agonist at all receptor densities only when $\varepsilon_A > 1$ and $\varepsilon_A > \psi_A$ and is always an inverse agonist when these inequalities are reversed. However, if $1 > \varepsilon_A > \psi_A$ or $1 < \varepsilon_A < \psi_A$, then the ligand's classification will vary with receptor density and it is protean. In the former case, the ligand transitions from inverse agonism to agonism as receptor density increases while the converse is true in the latter case (this is demonstrated in Fig. 6.5). Apparent neutral antagonism is achieved when inequality (6.83) becomes an equation, which then gives

$$\varepsilon_A = \frac{1+\psi_A \rho}{1+\rho} \qquad [6.84]$$

Again, this is dependent on receptor density. Hence, variations in receptor density about the value at which a ligand behaves as a neutral antagonist

($[R]_T = K_s(\varepsilon_A - 1)/(\psi_A - \varepsilon_A)$—obtained by rearranging Eq. (6.84) and substituting $\rho = [R]_T/K_s$) would result in agonism on some occasions and inverse agonism on others. A ligand can only be a neutral antagonist at all receptor densities when $\varepsilon_A = \psi_A = 1$. It is important to note that this behavior is distinct from that exhibited in classical receptor theory.[33,34] In classical receptor theory, the intrinsic activity of an agonist increases with receptor density and that of an inverse agonist should decrease (an antagonist remains an antagonist at all receptor densities). However, in this model it is possible for a ligand to transition from agonism to an inverse agonism (or vice versa) as the receptor density increases.

4.3. The effects of a competing ligand in a system with a linear rational stimulus

Before considering allosteric modulation in a system with a linear rational stimulus function, it is sensible to consider competition between two ligands. In such a system we may define the stimulus, S, as

$$S = \frac{[R] + \varepsilon_A[AR] + \varepsilon_B[BR]}{K_s + [R] + \psi_A[AR] + \psi_B[BR]} \qquad [6.85]$$

Substituting for the concentrations of the receptor species gives

$$S = \frac{[R]_T \dfrac{\left(1 + \varepsilon_A \dfrac{[A]}{K_a} + \varepsilon_B \dfrac{[B]}{K_b}\right)}{\left(1 + \dfrac{[A]}{K_a} + \dfrac{[B]}{K_b}\right)}}{K_s + [R]_T \dfrac{\left(1 + \psi_A \dfrac{[A]}{K_a} + \psi_B \dfrac{[B]}{K_b}\right)}{\left(1 + \dfrac{[A]}{K_a} + \dfrac{[B]}{K_b}\right)}}$$

$$= \frac{[R]_T \left(1 + \varepsilon_A \dfrac{[A]}{K_a} + \varepsilon_B \dfrac{[B]}{K_b}\right)}{K_s \left(1 + \dfrac{[A]}{K_a} + \dfrac{[B]}{K_b}\right) + [R]_T \left(1 + \psi_A \dfrac{[A]}{K_a} + \psi_B \dfrac{[B]}{K_b}\right)} \qquad [6.86]$$

Taking the transducer function to be linear rational, as above, this gives

$$E = \frac{\dfrac{E_{\max}[R]_T\left(1+\varepsilon_A\dfrac{[A]}{K_a}+\varepsilon_B\dfrac{[B]}{K_b}\right)}{K_s\left(1+\dfrac{[A]}{K_a}+\dfrac{[B]}{K_b}\right)+[R]_T\left(1+\psi_A\dfrac{[A]}{K_a}+\psi_B\dfrac{[B]}{K_b}\right)}}{K_e+\dfrac{[R]_T\left(1+\varepsilon_A\dfrac{[A]}{K_a}+\varepsilon_B\dfrac{[B]}{K_b}\right)}{K_s\left(1+\dfrac{[A]}{K_a}+\dfrac{[B]}{K_b}\right)+[R]_T\left(1+\psi_A\dfrac{[A]}{K_a}+\psi_B\dfrac{[B]}{K_b}\right)}}$$

$$= \frac{E_{\max}\chi\left(1+\dfrac{\varepsilon_A[A]}{K_a}+\varepsilon_B\dfrac{[B]}{K_b}\right)}{1+\chi(1+K_e)+\dfrac{[B]}{K_b}\left(1+\varepsilon_B\chi\left(1+\dfrac{\psi_B}{\varepsilon_B}K_e\right)\right)+\dfrac{[A]}{K_a}\left(1+\varepsilon_A\chi\left(1+\dfrac{\psi_A}{\varepsilon_A}K_e\right)\right)} \quad [6.87]$$

where χ and K_e are as previously described and K_a and K_b are the equilibrium dissociation constants of A and B, respectively. It is instructive to note that the expression for the midpoint of Eq. (6.87) is

$$\mathrm{EC}_{50} = \frac{K_a\left(1+\chi(1+K_e)+\frac{[B]}{K_b}\left(1+\varepsilon_B\chi\left(1+\frac{\psi_B}{\varepsilon_B}K_e\right)\right)\right)}{1+\varepsilon_A\chi\left(1+\frac{\psi_A}{\varepsilon_A}K_e\right)}$$

This shows the expected effect of a competing ligand, that is, a progressive decrease in potency of the primary ligand with no theoretical limit as the concentration of the competitor increases. Equating equieffective concentrations of A in the absence and presence of ligand B gives

$$\frac{E_{\max}\chi\left(1+\frac{\varepsilon_A[A]}{K_a}\right)}{1+\chi(1+K_e)+\frac{[A]}{K_a}\left(1+\varepsilon_A\chi\left(1+\frac{\psi_A}{\varepsilon_A}K_e\right)\right)}=\frac{E_{\max}\chi\left(1+\frac{\varepsilon_A[A]'}{K_a}+\frac{\varepsilon_B[B]}{K_b}\right)}{1+\chi(1+K_e)+\frac{[B]}{K_b}\left(1+\varepsilon_B\chi\left(1+\frac{\psi_B}{\varepsilon_B}K_e\right)\right)+\frac{[A]'}{K_a}\left(1+\varepsilon_A\chi\left(1+\frac{\psi_A}{\varepsilon_A}K_e\right)\right)}$$

$$\left(1+\frac{\varepsilon_A[A]}{K_a}\right)\left(1+\chi(1+K_e)+\frac{[B]}{K_b}\left(1+\varepsilon_B\chi\left(1+\frac{\psi_B}{\varepsilon_B}K_e\right)\right)+\frac{[A]'}{K_a}\left(1+\varepsilon_A\chi\left(1+\frac{\psi_A}{\varepsilon_A}K_e\right)\right)\right)=\left(1+\frac{\varepsilon_A[A]'}{K_a}+\frac{\varepsilon_B[B]}{K_b}\right)\left(1+\chi(1+K_e)+\frac{[A]}{K_a}\left(1+\varepsilon_A\chi\left(1+\frac{\psi_A}{\varepsilon_A}K_e\right)\right)\right)$$

$$\frac{\varepsilon_A[A][A]'}{K_a^2}(1+\varepsilon_A\chi+\psi_A\chi K_e)+\frac{[B]}{K_b}(1+\varepsilon_B\chi+\psi_B\chi K_e)+\frac{[A]'}{K_a}(1+\varepsilon_A\chi+\psi_A\chi K_e)+\frac{\varepsilon_A[A]}{K_a}(1+\chi+\chi K_e)+\frac{\varepsilon_A[A][B]}{K_aK_b}(1+\varepsilon_B\chi+\psi_B\chi K_e)+1+\chi+\chi K_e$$

$$=\frac{\varepsilon_B[B]}{K_b}(1+\chi+\chi K_e)+\frac{[A]}{K_a}(1+\varepsilon_A\chi+\psi_A\chi K_e)+\frac{\varepsilon_A[A]'}{K_a}(1+\chi+\chi K_e)+\frac{\varepsilon_A[A][A]'}{K_a^2}(1+\varepsilon_A\chi+\psi_A\chi K_e)+1+\chi+\chi K_e+\frac{\varepsilon_B[A][B]}{K_aK_b}(1+\varepsilon_A\chi+\psi_A\chi K_e)$$

$$\frac{[B]}{K_b}(1+\psi_B\chi K_e)+\frac{[A]'}{K_a}(1+\psi_A\chi K_e)+\frac{\varepsilon_A[A]}{K_a}(1+\chi K_e)+\frac{\varepsilon_A[A][B]}{K_aK_b}(1+\psi_B\chi K_e)=\frac{[A]}{K_a}(1+\psi_A\chi K_e)+\frac{\varepsilon_A[A]'}{K_a}(1+\chi K_e)+\frac{\varepsilon_B[B]}{K_b}(1+\chi K_e)+\frac{\varepsilon_B[A][B]}{K_aK_b}(1+\psi_A\chi K_e)$$

$$\frac{[B]K_a}{K_b[A]}\frac{(1+\psi_B\chi K_e)}{(1+\psi_A\chi K_e)}+DR+\varepsilon_A\frac{(1+\chi K_e)}{(1+\psi_A\chi K_e)}+\frac{\varepsilon_A[B]}{K_b}\frac{(1+\psi_B\chi K_e)}{(1+\psi_A\chi K_e)}=1+\varepsilon_A DR\frac{(1+\chi K_e)}{(1+\psi_A\chi K_e)}+\frac{\varepsilon_B[B]K_a}{K_b[A]}\frac{(1+\chi K_e)}{(1+\psi_A\chi K_e)}+\frac{\varepsilon_B[B]}{K_b}$$

$$DR-1+\frac{[B]}{K_b}\left(\varepsilon_A\frac{(1+\psi_B\chi K_e)}{(1+\psi_A\chi K_e)}-\varepsilon_B\right)=\varepsilon_A(DR-1)\frac{(1+\chi K_e)}{(1+\psi_A\chi K_e)}+\frac{[B]K_a}{K_b[A]}\frac{(\varepsilon_B(1+\chi K_e)-(1+\psi_B\chi K_e))}{(1+\psi_A\chi K_e)}$$

$$DR-1=\frac{[B]}{K_b}\frac{(\varepsilon_A(1+\psi_B\chi K_e)-\varepsilon_B(1+\psi_A\chi K_e))+\frac{K_a}{[A]}(1+\psi_B\chi K_e-\varepsilon_B(1+\chi K_e))}{(\varepsilon_A(1+\chi K_e)-(1+\psi_A\chi K_e))}$$

$$\frac{[B]}{DR-1}=\frac{K_b(\varepsilon_A(1+\rho)-(1+\psi_A\rho))}{(\varepsilon_A(1+\psi_B\rho)-\varepsilon_B(1+\psi_A\rho))+\frac{K_a}{[A]}(1+\psi_B\rho-\varepsilon_B(1+\rho))} \qquad [6.88]$$

where ρ is defined as above. Comparing this with Eq. (6.49) there is no term in $[A]/K_a$ reflecting the fact that a competing ligand does not affect the maximal response to the primary ligand. When the concentration–response curves to A in the presence and absence of B are approximately parallel, the term in $[A]^{-1}$ becomes negligible and Eq. (6.88) simplifies to

$$\frac{[B]}{\mathrm{DR}-1}=\frac{K_b(\varepsilon_A(1+\rho)-(1+\psi_A\rho))}{\varepsilon_A(1+\psi_B\rho)-\varepsilon_B(1+\psi_A\rho)} \quad [6.89]$$

which is still a rather complicated function of the intrinsic efficacy parameters of the two ligands. It also depends on the receptor density since Eqs. (6.88) and (6.89) are dependent on ρ. When ρ is small (resulting in an approximately linear stimulus function) this simplifies to

$$\frac{[B]}{\mathrm{DR}-1}=\frac{K_b(\varepsilon_A-1)}{\varepsilon_A-\varepsilon_B} \quad [6.90]$$

If the primary ligand is an agonist of sufficiently high intrinsic efficacy ($\varepsilon_A \gg 1$) this reduces further to

$$\frac{[B]}{\mathrm{DR}-1}\approx\frac{\varepsilon_A K_b}{\varepsilon_A-\varepsilon_B}=\frac{K_b}{1-\frac{\varepsilon_B}{\varepsilon_A}} \quad [6.91]$$

This suggests that there should be a discrepancy between the estimates of the binding affinity of a partial agonist derived from a binding assay and effects on concentration–response curves to a more efficacious agonist in a constitutively active functional assay. The ratio of the two estimates can then be used to determine the ratio of one of the components of the efficacy of the two agonists. The difference for a nonprotean inverse agonist ($\varepsilon_B < 1$) would be so small that it would not be quantifiable.

When ρ is large Eq. (6.90) simplifies to

$$\frac{[B]}{\mathrm{DR}-1}=\frac{K_b(\varepsilon_A-\psi_A)}{\varepsilon_A\psi_B-\varepsilon_B\psi_A} \quad [6.92]$$

In this case, assuming that $\varepsilon_A \gg \psi_A$ (and $\varepsilon_A \gg 1$), which is equivalent to assuming that the primary agonist is of high intrinsic activity, Eq. (6.92) simplifies further to

$$\frac{[B]}{\mathrm{DR}-1}\approx\frac{\varepsilon_A K_b}{\varepsilon_A\psi_B-\varepsilon_B\psi_A}=\frac{K_b}{\psi_B-\frac{\varepsilon_B\psi_A}{\varepsilon_A}}\approx\frac{K_b}{\psi_B} \quad [6.93]$$

The apparent affinity of a partial agonist or inverse agonist would be confounded by an aspect of its intrinsic efficacy (its effect on the midpoint of the stimulus function). This could be determined if an estimate of K_b were available.

If ligand B is behaving as a neutral antagonist, Eq. (6.84) can be substituted into Eq. (6.88) to give

$$\frac{[B]}{\mathrm{DR}-1}=\frac{K_b(\varepsilon_A(1+\rho)-(1+\psi_A\rho))}{\left(\varepsilon_A(1+\psi_B\rho)-\dfrac{(1+\psi_A\rho)(1+\psi_B\rho)}{1+\rho}\right)+\dfrac{K_a}{[A]}\left(1+\psi_B\rho-\dfrac{(1+\psi_B\rho)(1+\rho)}{1+\rho}\right)}$$

$$=\frac{K_b(\varepsilon_A(1+\rho)-(1+\psi_A\rho))(1+\rho)}{(\varepsilon_A(1+\rho)-(1+\psi_A\rho))(1+\psi_B\rho)}$$

$$=\frac{K_b(1+\rho)}{(1+\psi_B\rho)}=\frac{K_b}{\varepsilon_B}$$

The final step is simply a second substitution from Eq. (6.84). Thus, unless $\varepsilon_B=1$ (in which case, from Eq. 6.84, $\psi_B=1$ and the ligand is a true neutral antagonist), the estimate of the affinity derived from a Schild analysis is again not a true estimate of the apparent affinity of an antagonist but is its ratio to the other efficacy parameter of the ligand.

4.4. The effects of an allosteric ligand with a linear rational stimulus

An allosteric interaction under this model will now be considered. In this case, the stimulus is

$$S=\frac{[R]+\varepsilon_A[AR]+\varepsilon_B[BR]+\varepsilon_{AB}[ARB]}{K_s+[R]+\psi_A[AR]+\psi_B[BR]+\psi_{AB}[ARB]} \qquad [6.94]$$

Substitution of the expression for the receptor species from the binding isotherm gives

$$S=\frac{[R]_T\dfrac{\left(1+\varepsilon_A\dfrac{[A]}{K_a}+\varepsilon_B\dfrac{[B]}{K_b}+\varepsilon_{AB}\dfrac{[A][B]}{\alpha K_aK_b}\right)}{\left(1+\dfrac{[A]}{K_a}+\dfrac{[B]}{K_b}+\dfrac{[A][B]}{\alpha K_aK_b}\right)}}{K_s+[R]_T\dfrac{\left(1+\psi_A\dfrac{[A]}{K_a}+\psi_B\dfrac{[B]}{K_b}+\psi_{AB}\dfrac{[A][B]}{\alpha K_aK_b}\right)}{\left(1+\dfrac{[A]}{K_a}+\dfrac{[B]}{K_b}+\dfrac{[A][B]}{\alpha K_aK_b}\right)}}$$

$$=\frac{[R]_T\left(1+\varepsilon_A\dfrac{[A]}{K_a}+\varepsilon_B\dfrac{[B]}{K_b}+\varepsilon_{AB}\dfrac{[A][B]}{\alpha K_aK_b}\right)}{K_s\left(1+\dfrac{[A]}{K_a}+\dfrac{[B]}{K_b}+\dfrac{[A][B]}{\alpha K_aK_b}\right)+[R]_T\left(1+\psi_A\dfrac{[A]}{K_a}+\psi_B\dfrac{[B]}{K_b}+\psi_{AB}\dfrac{[A][B]}{\alpha K_aK_b}\right)} \quad [6.95]$$

Finally, applying a linear rational transducer function gives

$$E=\frac{E_{\max}\chi\left(1+\varepsilon_A\dfrac{[A]}{K_a}+\varepsilon_B\dfrac{[B]}{K_b}+\varepsilon_{AB}\dfrac{[A][B]}{\alpha K_aK_b}\right)}{1+\chi(1+K_e)+\dfrac{[B]}{K_b}\left(1+\varepsilon_B\chi\left(1+\dfrac{\psi_B}{\varepsilon_B}K_e\right)\right)+\dfrac{[A]}{K_a}\left(1+\varepsilon_A\chi\left(1+\dfrac{\psi_A}{\varepsilon_A}K_e\right)+\dfrac{[B]}{\alpha K_b}\left(1+\varepsilon_{AB}\chi\left(1+\dfrac{\psi_{AB}}{\varepsilon_{AB}}K_e\right)\right)\right)} \quad [6.96]$$

It is immediately obvious that there can now be no single parameter that describes the activation cooperativity in this model. We must quantify the cooperativity of ligands for ε ($\delta_\varepsilon=\varepsilon_{AB}/\varepsilon_A\varepsilon_B$) and for ψ ($\delta_\psi=\psi_{AB}/\psi_A\psi_B$). This is unsurprising since there is no single parameter that describes intrinsic efficacy in this model. This is also true of the analogous extension of the cTCM (1) since two ligands acting allosterically can cause disproportionate changes in the formation of the active state of the receptor and disproportionate changes in the affinity of the receptor for G protein. Both of these factors govern the resulting level of active receptor:G protein complexes (and indeed the model contains higher order cooperativity constants governing interactions in the quaternary complex).

For completeness, the null method will be applied to this system. Equating equieffective concentrations of A in the presence and absence of B gives

$$\frac{E_{\max}\chi\left(1+\varepsilon_A\frac{[A]}{K_a}\right)}{1+\chi(1+K_e)+\frac{[A]}{K_a}\left(1+\varepsilon_A\chi\left(1+\frac{\psi_A}{\varepsilon_A}K_e\right)\right)}$$
$$=\frac{E_{\max}\chi\left(1+\varepsilon_A\frac{[A]'}{K_a}+\varepsilon_B\frac{[B]}{K_b}+\varepsilon_{AB}\frac{[A]'[B]}{\alpha K_aK_b}\right)}{1+\chi(1+K_e)+\frac{[B]}{K_b}\left(1+\varepsilon_B\chi\left(1+\frac{\psi_B}{\varepsilon_B}K_e\right)\right)+\frac{[A]'}{K_a}\left(1+\varepsilon_A\chi\left(1+\frac{\psi_A}{\varepsilon_A}K_e\right)\right.\\ \left.+\frac{[B]}{\alpha K_b}\left(1+\varepsilon_{AB}\chi\left(1+\frac{\psi_{AB}}{\varepsilon_{AB}}K_e\right)\right)\right)}$$

After some rather messy algebra that is not dissimilar to that required to derive Eq. (6.88) this gives

$$\mathrm{DR}=\frac{1+\frac{[B]}{K_b}\left(\frac{\varepsilon_A(1+\psi_B\rho)-\varepsilon_B(1+\psi_A\rho)}{\varepsilon_A(1+\rho)-(1+\psi_A\rho)}+\frac{K_a}{[A]}\frac{(1+\psi_B\rho-\varepsilon_B(1+\rho))}{(\varepsilon_A(1+\rho)-(1+\psi_A\rho))}\right)}{1+\frac{[B]}{\alpha K_b}\left(\frac{\varepsilon_{AB}(1+\rho)-(1+\psi_{AB}\rho)}{\varepsilon_A(1+\rho)-(1+\psi_A\rho)}+\frac{[A]}{K_a}\frac{(\varepsilon_{AB}(1+\psi_A\rho)-\varepsilon_A(1+\psi_{AB}\rho))}{(\varepsilon_A(1+\rho)-(1+\psi_A\rho))}\right)} \quad [6.97]$$

This has a structure which is somewhat similar to that of Eq. (6.49) but clearly the more complicated structure of the stimulus results in a much more complicated expression for DR. Over the range of concentrations of A where the concentration–response curves in the presence and absence of B are approximately parallel, if one exists, (so that the terms in $[A]$ and $[A]^{-1}$ become negligible) this simplifies to

$$\mathrm{DR}=\frac{1+\frac{[B]}{K_b}\left(\frac{\varepsilon_A(1+\psi_B\rho)-\varepsilon_B(1+\psi_A\rho)}{\varepsilon_A(1+\rho)-(1+\psi_A\rho)}\right)}{1+\frac{[B]}{\alpha K_b}\left(\frac{\varepsilon_{AB}(1+\rho)-(1+\psi_{AB}\rho)}{\varepsilon_A(1+\rho)-(1+\psi_A\rho)}\right)} \quad [6.98]$$

When ρ is small this simplifies to Eq. (6.50) as expected from the approximation to the behavior of a linear stimulus function. When ρ is large Eq. (6.98) simplifies to

$$\mathrm{DR}=\frac{1+\frac{[B]}{K_b}\left(\frac{\varepsilon_A\psi_B-\varepsilon_B\psi_A}{\varepsilon_A-\psi_A}\right)}{1+\frac{[B]}{\alpha K_b}\left(\frac{\varepsilon_{AB}-\psi_{AB}}{\varepsilon_A-\psi_A}\right)} \qquad [6.99]$$

From Eq. (6.98), the maximal DR is

$$\mathrm{DR}_{\max}=\frac{\alpha(\varepsilon_A(1+\psi_B\rho)-\varepsilon_B(1+\psi_A\rho))}{\varepsilon_{AB}(1+\rho)-(1+\psi_{AB}\rho)} \qquad [6.100]$$

which has a similar structure to Eq. (6.52) and indeed will simplify to Eq. (6.52) when ρ is small. When ρ is large

$$\mathrm{DR}_{\max}=\frac{\alpha(\varepsilon_A\psi_B-\varepsilon_B\psi_A)}{\varepsilon_{AB}-\psi_{AB}} \qquad [6.101]$$

The concentration of allosteric ligand causing a DR of 2 is given by

$$[B]_{\mathrm{DR}=2}=\frac{\alpha K_b(\varepsilon_A(1+\rho)-(1+\psi_A\rho))}{\alpha(\varepsilon_A(1+\psi_B\rho)-\varepsilon_B(1+\psi_A\rho))-2(\varepsilon_{AB}(1+\rho)-(1+\psi_{AB}\rho))} \qquad [6.102]$$

This simplifies to Eq. (6.51) for small ρ and gives

$$[B]_{\mathrm{DR}=2}=\frac{\alpha K_b(\varepsilon_A-\psi_A)}{\alpha(\varepsilon_A\psi_B-\varepsilon_B\psi_A)-2(\varepsilon_{AB}-\psi_{AB})} \qquad [6.103]$$

for large ρ. At this point it becomes largely uninformative to generate a further litany of approximations and special cases. The one final condition that will be explored is the case of Eq. (6.97) for allosteric modulators that act as neutral antagonists and that do not affect the maximal response to the orthosteric ligand. The former applies Eq. (6.84) as a constraint (with the parameters of B):

$$\varepsilon_B=\frac{1+\psi_B\rho}{1+\rho}$$

The latter is achieved by requiring that

$$\varepsilon_{AB}=\varepsilon_A\frac{1+\psi_{AB}\rho}{1+\psi_A\rho}$$

This can be shown by equating Eq. (6.79) with the analogous expression derived from Eq. (6.96). Substituting these expressions into (6.97) gives

$$\mathrm{DR}=\frac{1+\frac{[B]}{K_b}\left(\frac{\varepsilon_A(1+\psi_B\rho)-\frac{1+\psi_B\rho}{1+\rho}(1+\psi_A\rho)}{\varepsilon_A(1+\rho)-(1+\psi_A\rho)}+\frac{K_a}{[A]}\frac{\left(1+\psi_B\rho-\frac{1+\psi_B\rho}{1+\rho}(1+\rho)\right)}{(\varepsilon_A(1+\rho)-(1+\psi_A\rho))}\right)}{1+\frac{[B]}{\alpha K_b}\left(\frac{(1+\psi_{AB}\rho)\left(\varepsilon_A\frac{1+\rho}{1+\psi_A\rho}-1\right)}{\varepsilon_A(1+\rho)-(1+\psi_A\rho)}+\frac{[A]}{K_a}\frac{(\varepsilon_A(1+\psi_{AB}\rho))\left(\frac{1+\psi_A\rho}{1+\psi_A\rho}-1\right)}{(\varepsilon_A(1+\rho)-(1+\psi_A\rho))}\right)}$$

$$=\frac{1+\frac{[B]}{K_b}\left(\frac{(1+\psi_B\rho)(\varepsilon_A(1+\rho)-(1+\psi_A\rho))}{(\varepsilon_A(1+\rho)-(1+\psi_A\rho))(1+\rho)}+\frac{K_a}{[A]}\frac{(1+\psi_B\rho-(1+\psi_B\rho))}{(\varepsilon_A(1+\rho)-(1+\psi_A\rho))}\right)}{1+\frac{[B]}{\alpha K_b}\left(\frac{(1+\psi_{AB}\rho)(\varepsilon_A(1+\rho)-(1+\psi_A\rho))}{(\varepsilon_A(1+\rho)-(1+\psi_A\rho))(1+\psi_A\rho)}+\frac{[A]}{K_a}\frac{(\varepsilon_A(1+\psi_{AB}\rho)-\varepsilon_A(1+\psi_{AB}\rho))}{(\varepsilon_A(1+\rho)-(1+\psi_A\rho))}\right)}$$

$$=\frac{1+\frac{[B]}{K_b}\left(\frac{(1+\psi_B\rho)}{(1+\rho)}\right)}{1+\frac{[B]}{\alpha K_b}\left(\frac{(1+\psi_{AB}\rho)}{(1+\psi_A\rho)}\right)}$$

$$=\frac{1+\frac{\varepsilon_B[B]}{K_b}}{1+\frac{\varepsilon_{AB}[B]}{\alpha\varepsilon_A K_b}} \qquad [6.104]$$

The maximal DR caused by B is then

$$\mathrm{DR}_{\max} = \frac{\alpha\varepsilon_B\varepsilon_A}{\varepsilon_{AB}} = \frac{\alpha}{\delta_\varepsilon} \qquad [6.105]$$

which is the product of the binding cooperativity and the cooperativity at the level of ε. The concentration of B causing a DR of 2 is

$$[B]_{\mathrm{DR}=2} = \frac{\alpha K_b}{\alpha\varepsilon_B - 2(\varepsilon_{AB}/\varepsilon_A)} \qquad [6.106]$$

For large α this simplifies to

$$[B]_{\mathrm{DR}=2} = \frac{K_b}{\varepsilon_B}$$

Hence, like a competing ligand, the intercept of a Schild plot for an allosteric modulator that causes large parallel rightward shifts in the concentration–response curves to an orthosteric ligand is only a true estimate of the affinity of the modulator when it is a true neutral antagonist at the allosteric site (i.e., when $\varepsilon_B = \psi_B = 1$). Given the complexity of the expressions derived in this section, there is an argument that it may be more useful simply to reengineer the system, where possible, to give a linear stimulus function and use the models described in Section 2. Conversely, a system in which protean ligands are identified cannot be modeled by the expressions given in Section 2.

5. SUMMARY

In this chapter, I have considered three topics. In Section 1, comparison of the ATSM with explicit multiconformational models demonstrated that this model encapsulates the common behaviors of any model that has at least two active and two inactive conformations (and indeed fewer by application of certain constraints): the "states" of the model represent ensembles of active or inactive conformations. This also demonstrates that a similar interpretation of the TSM is valid. It was also demonstrated that any receptor for which allosteric modulators exist whose effects cannot be predicted on the basis of their intrinsic efficacy must have at least two active and two inactive states and cannot be represented by a strict interpretation of the TSM. A matrix representation of multiconformational models was introduced that provides a compact notation for models with arbitrarily large numbers of conformations. By considering the allosteric modulator as a transducer protein, the analysis also highlighted the possibility that the properties of allosteric modulators may differ depending on the functional response that is measured. In other words, it reiterates the ability of allosteric ligands to exhibit "permissiveness" in their effects.

In Section 2, allosteric modulation was explored in the context of a new operational model of receptor activity which includes constitutive receptor activity. It was demonstrated that, given an appropriate experimental design, it is possible to determine the apparent affinity, intrinsic efficacy, and cooperativity constants for a pair of allosteric ligands by fitting this model. Some theoretical limits on this model were derived by considering a model in which the concentrations of active species from the TSM of receptor activation were passed into a transducer function. It was also demonstrated that, within certain limits, it is possible to estimate the parameters of the ATSM that relate to an allosteric interaction. There are circumstances, however, when it is not possible to estimate the ATSM parameters accurately. Under these conditions the behavior of the macroscopic model is internally consistent.

Finally, in Section 3, a novel operational model was developed which may allow the analysis of protean ligands. This model requires a nonlinear stimulus function and two parameters to define the efficacy of a ligand. Expressions describing competitive and allosteric interactions under this model were developed and the results of applying null analyses to the data determined. The model suggests that, in general, it is not possible to determine the apparent affinity of a ligand in a functional assay. The estimates from pharmacological null analyses are confounded by a component of the ligand's intrinsic efficacy, even when protean or weakly efficacious ligands behave as neutral antagonists. This suggests that discrepancies between affinity estimates between binding and functional assays may allow efficacy parameters to be determined.

REFERENCES

1. Christopoulos A, Kenakin TP. G-protein-coupled receptor allosterism and complexing. *Pharmacol Rev* 2002;**54**:323–74.
2. Hall DA. Predicting dose–response curve behaviour: mathematical models of allosteric receptor-ligand interactions. In: Bowery NG, editor. *Allosteric receptor modulation in drug targeting*. New York: Taylor & Francis; 2006. p. 39–78.
3. Leach K, Sexton PM, Christopoulos A. Allosteric GPCR modulators: taking advantage of permissive receptor pharmacology. *Trends Pharmacol Sci* 2007;**28**:382–9.
4. Gregory KJ, Sexton PM, Christopoulos A. Overview of receptor allosterism. *Curr Protoc Pharmacol* 2010;**51**:1.21.1–1.21.34.
5. Luttrell LM, Kenakin TP. Refining efficacy: allosterism and bias in G-protein coupled receptor signalling. *Methods Mol Biol* 2011;**756**:3–35.
6. Hall DA. Modeling the functional effects of allosteric modulators at pharmacological receptors: an extension of the two-state model of receptor activation. *Mol Pharmacol* 2000;**58**:1412–23.

7. Kenakin T, Onaran O. The ligand paradox between affinity and efficacy: can you be there and not make a difference? *Trends Pharmacol Sci* 2002;**23**:275–89.
8. Onaran HO, Scheer A, Cotecchia S, Costa T. A look at receptor efficacy. From the signalling network of the cell to the intramolecular motion of the receptor. *Handb Exp Pharmacol* 2000;**148**:217–59.
9. Onaran HO, Costa T. Agonist efficacy and allosteric models of receptor action. *Ann N Y Acad Sci* 1997;**812**:98–115.
10. Slack RJ, Hall DA. Development of operational models of receptor activation including constitutive receptor activity and their use to determine the efficacy of the chemokine TARC at the CC-chemokine receptor CCR4. *Br J Pharmacol* 2012;**166**:1774–92.
11. De Léan A, Stadel JM, Lefkowitz RJ. A ternary complex model explains the agonist-specific binding properties of the adenylate cyclase-coupled beta-adrenergic receptor. *J Biol Chem* 1980;**255**:7108–17.
12. Weiss JM, Morgan PH, Lutz MW, Kenakin TP. The cubic ternary complex receptor-occupancy model I. Model description. *J Theor Biol* 1996;**178**:151–67.
13. Weiss JM, Morgan PH, Lutz MW, Kenakin TP. The cubic ternary complex receptor-occupancy model II. Understanding apparent affinity. *J Theor Biol* 1996;**178**:169–82.
14. Weiss JM, Morgan PH, Lutz MW, Kenakin TP. The cubic ternary complex receptor-occupancy model III. Resurrecting efficacy. *J Theor Biol* 1996;**181**:381–97.
15. Kenakin T. Pharmacological proteus? *Trends Pharmacol Sci* 1995;**16**:256–7.
16. Lazareno S, Birdsall NJ. Detection, quantification and verification of allosteric interactions with labelled and unlabelled ligands at G-protein-coupled receptors: interactions of strychnine and acetylcholine at muscarinic receptors. *Mol Pharmacol* 1995;**48**:362–78.
17. Ehlert FJ. Estimation of the affinities of allosteric ligands using radioligand binding and pharmacological null methods. *Mol Pharmacol* 1988;**33**:187–94.
18. Costa T, Herz A. Antagonists with negative intrinsic activity at δ opioid receptors coupled to GTP-binding proteins. *Proc Natl Acad Sci USA* 1989;**86**:7321–5.
19. Corsi M, Fina P, Trist DA. Co-agonism in drug-receptor interaction: illustrated by the NMDA receptor. *Trends Pharmacol Sci* 1996;**17**:220–2.
20. Ehlert FJ, Griffin MT. Two-state models and the analysis of the allosteric effect of gallamine at the M_2 muscarinic receptor. *J Pharmacol Exp Ther* 2008;**325**:1039–60.
21. Stockton JM, Birdsall NJM, Burgen ASV, Hulme EC. Modification of the binding properties of muscarinic receptors by gallamine. *Mol Pharmacol* 1983;**23**:551–8.
22. Karlin A. On the application of a "plausible model" of allosteric proteins to the receptor for acetylcholine. *J Theor Biol* 1967;**16**:306–20.
23. Thron CD. On the analysis of pharmacological experiments in terms of an allosteric model. *Mol Pharmacol* 1972;**9**:1–9.
24. Colquhoun D. The relation between classical and cooperative models for drug action. In: Rang HP, editor. *Drug receptors*. London: The Macmillan Company; 1973. p. 149–82.
25. Balakrishnan VK. *Schaum's outline of graph theory*. USA: McGraw-Hill Companies Inc.; 1997.
26. Lazareno S, Gharagozloo P, Kuonen D, Popham A, Birdsall NJM. Subtype-selective positive cooperative interactions between brucine analogues and acetylcholine at muscarinic receptors: radioligand binding studies. *Mol Pharmacol* 1998;**53**:573–89.
27. Jakubík J, Bačáková L, El-Fakahany EE, Tuček S. Positive cooperativity of acetylcholine and other agonists with allosteric ligands on muscarinic acetylcholine receptors. *Mol Pharmacol* 1998;**52**:172–9.
28. Ehlert FJ, Suga H, Griffin MT. Analysis of agonism and inverse agonism in functional assays with constitutive activity: estimation of orthosteric ligand affinity constants for active and inactive receptor states. *J Pharmacol Exp Ther* 2011;**338**:671–86.

29. Bindslev N. A homotropic two-state model and auto-antagonism. *BMC Pharmacol* 2004;**4**:11.
30. Kenakin T. Agonist-receptor efficacy II: agonist trafficking of receptor signals. *Trends Pharmacol Sci* 1995;**16**:232–8.
31. Leff P, Scaramellini C, Law C, McKechnie K. A three-state receptor model of agonist action. *Trends Pharmacol Sci* 1997;**18**:355–62.
32. Kenakin T. New concepts in drug discovery: collateral efficacy and permissive agonism. *Nat Rev Drug Discov* 2005;**4**:919–27.
33. Black JW, Leff P. Operational models of pharmacological agonism. *Proc R Soc Lond B Biol Sci* 1983;**220**:141–62.
34. Furchgott RF. The use of β-haloalkylamines in the differentiation of receptors and in the determination of dissociation constants of receptor-agonist complexes. In: Harper NJ, Simmonds AB, editors. *Advances in drug research*, vol. 3. New York: Academic Press; 1966. p. 21–5.
35. Riggs DS. *The mathematical approach to physiological problems*. Baltimore: The Williams and Wilkins Company; 1963.
36. Lane JR, Powney B, Wise A, Rees S, Milligan G. Protean agonism at the dopamine D_2 receptor: (s)-3-(3-hydroxyphenyl)-*N*-propylpiperidine is an agonist for activation of G_{o1} but an antagonist/inverse agonist for G_{i1}, G_{i2}, and G_{i3}. *Mol Pharmacol* 2007;**71**: 1349–59.
37. Jansson CC, Kukkonen JP, Näsman J, Huifang G, Wurster S, Virtanen R, et al. Protean agonism at α_{2A}-adrenoceptors. *Mol Pharmacol* 1998;**53**:963–8.
38. Lane JR, Powney B, Wise A, Rees S, Milligan G. G protein coupling and ligand selectivity of the D2L and D3 dopamine receptors. *J Pharmacol Exp Ther* 2008;**325**:319–30.

CHAPTER SEVEN

What Ligand-Gated Ion Channels Can Tell Us About the Allosteric Regulation of G Protein-Coupled Receptors

Frederick J. Ehlert
Department of Pharmacology, School of Medicine, University of California, Irvine, California, USA

Contents

Abstract

The $GABA_A$ receptor is the target for a number of important allosteric drugs used in medicine, including benzodiazepines and anesthetics. These modulators have variable effects on the potency and maximal response of macroscopic currents elicited by different $GABA_A$ receptor agonists, yet this modulation is consistent with a two-state

Progress in Molecular Biology and Translational Science, Volume 115
ISSN 1877-1173
http://dx.doi.org/10.1016/B978-0-12-394587-7.00007-5

model in which the allosteric ligand has invariant affinity constants for the active and inactive states. Analysis of the effects of an allosteric agonist, like etomidate, on the population current provides a means of estimating the gating constant of the unliganded $GABA_A$ receptor ($\sim 10^{-4}$). In contrast, allosteric interactions at the M_2 muscarinic receptor are often inconsistent with a two-state model. Analyzing allosterism within the constraints of a two-state model, nonetheless, provides an unbiased measure of probe dependence as well as clues to the mechanism of allosteric modulation. The rather simple allosteric effect of affinity-only modulation is difficult to explain and suggests modulation of a peripheral orthosteric ligand-docking site on the M_2 muscarinic receptor.

1. INTRODUCTION

Drugs elicit remarkable effects because they often use receptor-signaling pathways to either enhance or inhibit natural physiological responses. Drugs accomplish this feat by binding to and altering the activity of proteins in these signaling pathways, including transporters, ion channels, enzymes, and receptors for endogenous chemical messengers. Common examples include orthosteric ligands that activate receptors (agonists) or competitively antagonize (antagonists) the action of the endogenous ligand. Allosteric drugs are useful because they bind to receptors at a site distinct from that of the endogenous signaling molecule (e.g., neurotransmitter) and alter the output of the receptor. If the allosteric ligand has a moderate, but sufficient, modulatory effect, it may alter the action of the endogenous neurotransmitter without having an effect by itself. This mechanism preserves the timing of signaling through neuronal networks, and thus, the quality of the information that flows through them.[1] The limited modulation places a constraint on how much receptor activation (positive modulator) or inhibition (negative modulator) occurs at high, receptor-saturating concentrations of the allosteric modulator. This limit increases the therapeutic dose range and reduces the probability of overdose.[2] In contrast, a directly acting agonist, for example, would elicit effects continuously in time, and therefore, disrupt the normal pattern of signaling. The excessive excitation caused by overdose with an agonist could be disastrous. There are instances, of course, where continuous receptor activation is desirable, such as in the treatment of symptomatic asthma with a directly acting β_2 agonist.

The Oxford English dictionary gives two definitions for the adjective, allosteric, which I have modified to apply to receptors: (1) relating to the modification in the binding of an orthosteric ligand to its site on the receptor by another ligand that binds to a different site and (2) relating to the change in the conformation of the receptor caused by the binding of a ligand to its site on the receptor. I use both meanings in this chapter. Thus, we can say that the binding of a benzodiazepine to its allosteric site on the $GABA_A$ receptor alters the binding of GABA, and that the binding of GABA allosterically alters the transmembrane-spanning M2 helices of the receptor to enable a flux of chloride ions.

Much effort has been expended in the search for useful allosteric modulators of ligand-gated ion channels and G protein-coupled receptors (GPCRs). An obvious goal is to find an agent with the appropriate receptor selectivity and modulation. The nature of the modulation can be characterized in terms of its magnitude and sign (positive or negative). These vary depending on the orthosteric ligand whose action the allosteric drug modifies, and this phenomenon is known as probe dependence.[3] For therapeutic allosteric drugs, the relevant interacting ligand is the endogenous neurotransmitter whose actions *in vivo* are modified by administration of the modulator. Other orthosteric ligands may be relevant, particularly in pharmacological studies on the receptor (e.g., binding and functional assays) or in screening efforts to detect allosteric modulators. The goal of this chapter is to explain the origin of probe dependence and briefly outline a potential strategy for measuring probe-independent measures of allosteric regulation. I begin with a brief summary of how the interaction of a drug with a receptor relates to conformational changes in the receptor.

2. DRUG-INDUCED CONFORMATIONAL CHANGES IN RECEPTOR ACTIVITY

In describing the actions of drugs, three hierarchical levels of analysis are commonly used. Each has its own nomenclature and set of quantitative variables that describe drug action, yet nonetheless, the three levels are internally consistent.[4,5] On the surface (level 1), drug action is analyzed at a point downstream in the signaling pathway from receptor activation. The usual quantitative measures of agonist action at this level are the maximal response (E_{max}) and the concentration of agonist eliciting a half-maximal response (EC_{50}). This is the level at which drugs that bind with GPCRs are often studied. For a given agonist–receptor pair, the values of these parameters vary continuously depending on what physiological response, signaling pathway or assay system is being measured.

At a deeper level of analysis (level 2), the activation state of the receptor population is estimated. For a receptor having a single orthosteric site, the reciprocal of the concentration of ligand required for half-maximal receptor occupancy of the receptor population is defined as the observed affinity constant (K_{obs}; reciprocal of the observed dissociation constant, K_D) and the fraction of the population of ligand–receptor complexes in the active state is defined as efficacy (ε).[6,7] These are average population parameters and do not reflect the properties of single receptors.

With regard to GPCRs, it may be that only a portion of the total population of receptors in the cell is in equilibrium with the G protein that elicits the response of interest. The terms K_{obs} and ε relate specifically to this fraction of the population. Scientists usually measure drug effects at a point downstream in the signaling pathway of a GPCR because it is difficult to measure receptor activation directly. Nonetheless, techniques have been developed to infer K_{obs} and a relative value of ε using a null method (i.e., a type of response-clamp analysis)[6] or the operational model (i.e., reverse engineering).[8,9] Both methods yield the same population parameters. While the term ε is not a unique parameter of the operational model, a relative value of ε can be derived by expressing the τ value (parameter of the operational model) of one agonist relative to that of another. Like the values of the level 1 parameters EC_{50} and E_{max}, K_{obs} and ε also vary on a continuous scale for a given agonist–receptor complex, although the range of values is usually less. The concentrations of guanine nucleotides (GTP and GDP) and G protein, as well as the type of G protein, are the principal factors that modify the values of ε and K_{obs} for a given agonist–receptor pair.[4,10]

At ligand-gated ion channels, the activation state of the receptor population is measured as the whole-cell current response. This represents the summation of the unitary conductances of all of the receptors in the population (ensemble average). There are no simple parameters, like K_{obs} and ε, that describe the ensemble average, however, because ligand-gated ion channels have more than one orthosteric-binding pocket per channel. If a receptor is defined as being occupied when it has one or more ligands bound to it, then it is possible to define an empirical parameter K_{occ} that represents the reciprocal of the concentration of ligand yielding half-maximal receptor occupancy. Because the fraction of ligand–receptor complexes in the active state increases as a function of receptor occupancy, then the efficacy of the receptor complex changes with the level of receptor occupancy. Nonetheless, it is possible to define a parameter known as mean efficacy that represents the integral of the function expressing receptor activation as a function of receptor occupancy. Mean efficacy and K_{occ} have some properties that are analogous to the K_{obs} and ε parameters of a receptor having only one orthosteric-binding site as described below.

At the deepest level of analysis (level 3), drug–receptor interactions are defined in terms of the microscopic affinity constants of the ligand for active and inactive states of the receptor. These parameters have been measured directly through the analysis of single-channel recordings of the muscle-type nicotinic acetylcholine receptor using the maximum likelihood approach and a continuous Markov model.[11–13] The advantage of measuring microscopic constants is that these represent the affinity constants of real structures, and hence, they are invariant and represent the ultimate determinants of drug action. In other words, the active state is the first cause of drug action and the agonist has a specific affinity constant for its binding site on a unique active state. It is usually unfeasible to measure the activity of single GPCRs, and therefore, to estimate microscopic constants from Markov analysis. Nonetheless, there are some unique relationships between population parameters and microscopic constants that enable one to estimate relative or absolute estimates of microscopic constants at GPCRs, depending upon whether the downstream response in question exhibits constitutive activity.[14–16]

So what is a microscopic affinity constant? Consider a hypothetical receptor having a single orthosteric-binding site and very little, or essentially no, activity in the absence of ligand. When an endogenous ligand binds to the receptor, its probability of switching to the active state greatly increases. When bound with the endogenous ligand, the receptor might spend the great majority of its time (e.g., 95%) in the active state, notwithstanding rapid switching between active and inactive states and the existence of numerous evanescent transition states. If we could measure the binding of the endogenous ligand when the receptor is fixed in the active state, it might have an affinity constant of say, $10^8\ M^{-1}$. This number represents the value of the microscopic affinity constant of the agonist for its binding pocket on the active state of the receptor (K_b). In contrast, its affinity for the receptor when it is locked in the inactive state might be $10^4\ M^{-1}$, that is, the value of the microscopic affinity constant of the inactive state (K_a). The higher affinity of the endogenous ligand for the active state is what causes the induction of the active state or the turning on of the receptor.

Binding experiments are usually not done on receptors locked in hypothetical states, but rather, on natural receptors capable of isomerizing into the active state depending on the nature of the orthosteric or allosteric ligand.

If the binding of the endogenous ligand to our hypothetical receptor were measured, its observed affinity constant (K_{obs}) might be 10^5 M^{-1}, for example. This raises a dichotomy between the two different properties of the ligand–receptor affinity constant. The first is that the affinity constant is a measure of the tightness with which a ligand binds to a specified three-dimensional structure of its binding pocket, like the active state, for example. This interaction can be quantified by the change in the Gibbs-free energy of binding (ΔG_0), which is proportional to the logarithm of the affinity constant. The second meaning of affinity is related to the potency of ligand–receptor interactions, which is quantified by the K_{obs} value of the occupancy curve.

These two properties are summarized in Fig. 7.1. The solid curve represents the occupancy function of the endogenous ligand for our hypothetical receptor. The concentration of ligand that yields half-maximal receptor occupancy of both active and inactive states is known as the reciprocal of the observed affinity constant (K_{obs}). Occupancy is a measure of the potency of the ligand–receptor interaction. But there is no receptor structure that exhibits an affinity constant of K_{obs}. The dashed curve on the left illustrates how the occupancy curve of the receptor would behave if the receptor were artificially locked in the active state. The associated affinity constant is a true measure of the affinity constant of the ligand for the active state (K_b). But there is no native receptor structure (i.e., state) that exhibits half-maximal occupancy when the concentration of the endogenous ligand is 10^{-8} M,

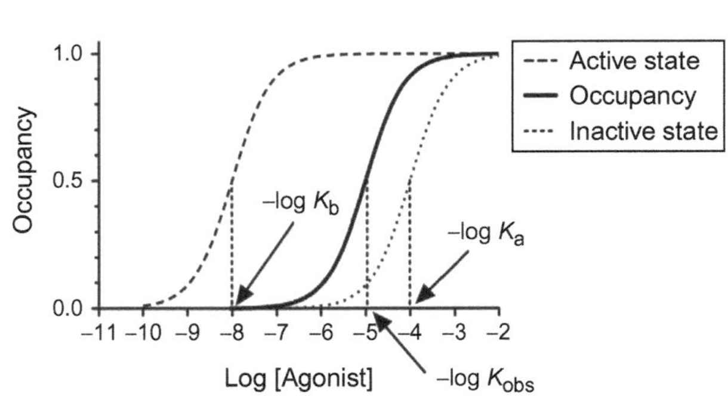

Figure 7.1 Relationship between the observed affinity constant, K_{obs}, and the microscopic affinity constants of an agonist for active, K_b, and inactive, K_a, states of a receptor. The solid curve shows the occupancy function of the agonist for a receptor existing in two states, active and inactive. The dashed lines on the left and right illustrate the imaginary occupancy curves of the agonist if the receptor were fixed in either the active state or the inactive state, respectively. The relationship between K_{obs} and the microscopic constants is, $K_{obs} = (K_bK_q + K_a)/(1 + K_q)$.

however. In this example, occupancy of the active state is at 50% when the ligand concentration is equal to the reciprocal of K_{obs} (10^{-5} M). Thus, whenever a ligand induces a conformational change in a receptor, there is always a discrepancy between its observed affinity and its microscopic affinity constant for the state that it induces.

These relationships influence the behavior of the receptor population when activated by ligands with different efficacies. Imagine that two fluorophores are attached to our hypothetical receptor such that they are either separated or close enough for Forstner resonance energy transfer to occur, depending upon whether the receptor is in the inactive or active state, respectively. Upon excitation at a specific wavelength of light, the receptor fluoresces blue in the inactive state and yellow in the active state. If an antagonist, partial agonist, or highly efficacious agonist is added to a cell expressing a population of this receptor, the cell would fluoresce blue, green, or yellow, depending upon which ligand was added. One could imagine testing a range of ligands having varying degrees of efficacy and observing a continuum of fluorescence changes from blue to blue–green to green to yellow–green to yellow. From the population perspective, it appears as though the orthosteric ligand allosterically causes a change in the shape of the receptor and that this change is greatest for the most efficacious ligand. An allosteric ligand could be introduced, and it might alter the shape of the receptor by itself or the ability of the orthosteric ligand to change the shape of the receptor. It appears that the magnitude of the observed change in shape is on a continuum, depending upon the efficacy of the orthosteric ligand and the nature of the cooperativity of the allosteric ligand.

But this interpretation is an illusion based on a low-resolution, population perspective, or rather, the average behavior of the receptor population. If we could turn up the zoom lens and see the behavior of individual receptors when saturated with an orthosteric ligand, at any given instant, something like that shown in Fig. 7.2 might be observed. In the case of the antagonist, all of the individual receptors fluoresce blue. With the highly efficacious agonist, almost all the receptors fluoresce yellow except for a few that fluoresce blue. With regard to the partial agonist, however, approximately half of the receptors fluoresce blue and half yellow. The orthosteric ligands do not really change the shape of the receptor; rather, they select, to varying degrees, natural on- and off-states of the receptor. This explains why synthetic agonists do not generate new responses in the body, but rather, signal through natural on-states of the receptor to elicit responses. There may be different types of active states as described below in the section on GPCRs.

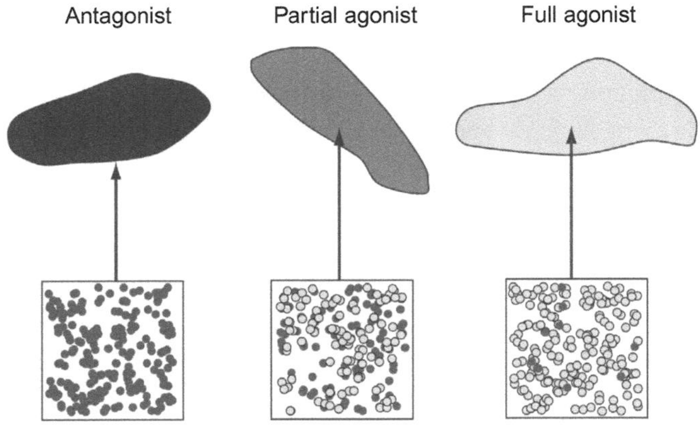

Figure 7.2 Hypothetical example of a receptor having two fluorophores that enable the receptor to fluoresce yellow or blue depending upon whether the receptor is in the active or inactive state, respectively. The upper half of the figure shows three cells, each exposed to a different type of orthosteric ligand. The receptor population on each cell fluoresces a different color on a scale from blue to yellow, depending on the efficacy of the orthosteric ligand. The lower half of the figure represents a magnification of the receptor population on each cell, which enables the resolution of single receptors. Individual receptors fluoresce only yellow or blue, depending on whether they are in the active or inactive state, respectively. (See Color Insert.)

This view of receptor activation reveals that the binding pockets and activation domains of a receptor exist in at least two quantal states and that the selectivity of ligands for these states gives rise to a continuum of efficacies and observed affinities and variation in the nature of the interaction between an allosteric ligand and various orthosteric ligands. At the level of receptor states, however, an allosteric ligand has one set of microscopic constants, regardless of the interacting orthosteric ligand. If we could discern these constants, we would have a means of defining allosteric ligands independently of the orthosteric ligand and of predicting the nature of its interaction with orthosteric ligands having known microscopic constants.

This chapter addresses this issue with regard to two classes of receptors for which abundant allosteric interactions have been described—$GABA_A$ and muscarinic receptors. Much data regarding allosteric modulation of the $GABA_A$ receptor can be rationalized by the receptor state hypothesis outlined above, so this group of receptors will be discussed first. The underlying principles of allosterism at $GABA_A$ receptors will then be applied to clarify what mechanisms might underlie allosterism at muscarinic receptors.

So that our discussion of allosterism does not rely on purely hypothetical concepts about receptor structures and their binding sites, I begin with a quick summary of the $GABA_A$ receptor structure, based on its homology with the muscle-type nicotinic acetylcholine receptor and other members of the Cys-loop receptor family.

3. LIGAND-GATED ION CHANNELS

3.1. Structure of Cys-loop receptors

The $GABA_A$ receptor is a member of the Cys-loop family of ligand-gated ion channels, which also includes nicotinic acetylcholine receptors, glycine receptors, and the $5HT_3$ receptor.[17] We know a great deal about the structure of nicotinic acetylcholine receptors because of a gift from nature in the form of two marine rays—*Torpedo marmorata* and *Torpedo californica*.[18] Their bilateral electric organs encompass a substantial portion of the flattened pectoral fins that make up the skate or ray body plan of these fishes. The electric organ consists of several columns of stacks of about 1000 flattened cells known as electroplax. Each column spans the dorsal to ventral surface of the pectoral fin, and each electroplax in the column expresses a dense matrix of nicotinic acetylcholine receptors on its ventral surface, which receives a cholinergic innervation.[19] A convenient property of membranes prepared from the *Torpedo* electric organ is that they spontaneously form elongated tubes composed of crystalline arrays of nicotinic receptors following incubation in the cold for several weeks.[20] Using helical diffraction techniques on these frozen tubes, Nigel Unwin and coworkers developed model for the nicotinic acetylcholine receptor.[21]

A huge advance in our understanding of the structure of the Cys-loop family came from the determination of the crystal structure of the acetylcholine-binding protein from the snail *Lymnaea stagnalis*.[22] This protein binds acetylcholine and regulates cholinergic transmission in molluscan ganglia. It forms a homopentamer and has 20–23% sequence identity with the corresponding heteropentameric extracellular domain of the *Torpedo* acetylcholine receptor. Similarly, the extracellular domain of the nicotinic acetylcholine receptor is homologous to that of the $GABA_A$ receptor and other Cys-loop family members.[23] The crystal structures of the snail protein bound with carbachol and other cholinergic ligands have been solved.[24–26] These structures as well as that determined by Nigel Unwin[27] have been used to develop a refined model for the nicotinic acetylcholine receptor and other members of the Cys-loop family (Fig. 7.3).

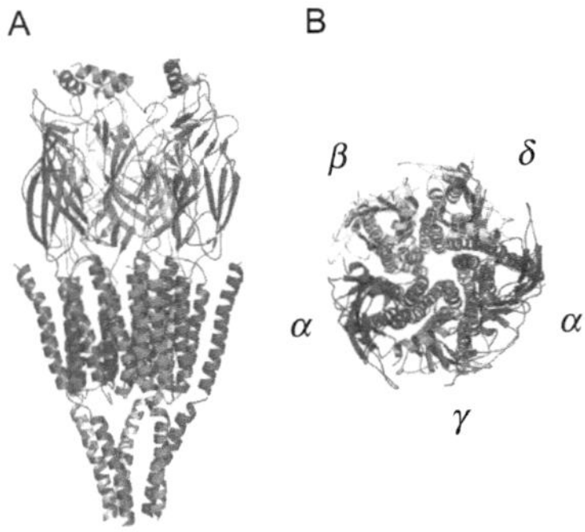

Figure 7.3 Model of the nicotinic acetylcholine receptor from *Torpedo marmorata*. (A) Side view of the secondary structure of the receptor with the extracellular domain on top. (B) View from the extracellular space looking down the channel pore. PDB Code, 2BG9. *The structure is from Unwin.*[27] (See Color Insert.)

The *Torpedo* receptor and the muscle-type nicotinic acetylcholine receptor form a pentamer of four receptor subunits, two α_1s, one β_1, one γ (ε in the mature mammalian muscle-type receptor), and one δ.[28,29] Other subtypes of mammalian nicotinic receptors are composed of different subunit combinations, usually containing two or three α subunits plus other subunits to generate a channel having five subunits.[30,31] Homomeric channels consisting of five α_7 subunits have also been described. The number of receptor subtypes and their subunit stoichiometry have yet to be completely defined for nicotinic acetylcholine receptors.

Beginning at the amino terminus, the sequences of the *Torpedo* receptor subunits encode an N-terminal α-helix, followed by 10 β-strands (β1–β10) with interconnecting loops.[27] These form inner (β1, β2, β3, β5, and β6) and outer (β4, β7, β8, β9, and β10) antiparallel β-sheets that together form a β sandwich with a one-quarter twist. The two β sheets are held together in part by a disulfide bond between cysteines located at the carboxyterminal end of β6 and the aminoterminal end of β7. These cysteines delineate a 15-residue long cys-loop that is conserved among members of the receptor family. The β sandwich forms the extracellular part of each receptor subunit, and together, the aminoterminal domains of the five receptor subunits form an outer vestibule having an internal diameter of about 20 Å.[21,32] The inner β-sheet of each subunit is dotted with about four to five acidic residues yielding an inner surface of negative charge on the outer vestibule that attracts positive cations.[27]

Proceeding in a carboxy terminal direction from the 10th β-strand, the sequence forms four membrane-spanning α helices (M1–M4) connected by two short intra- (M1–M2) and extracellular (M2–M3) loops and one long intracellular loop (M3–M4).[27] The M1, M3, and M4 helices of each subunit interact with each other and form an outer shell around the inner M2 helix, which lines the channel pore (Fig. 7.4B). The residues of M2 that line the channel pore occur every three to four turns of the α helix and define three domains of pore-lining residues—a central hydrophobic domain flanked by two hydrophilic domains. The central domains of the M2 helices of each receptor subunit are composed mostly of leucines and valines and are thought to interact with each other to form a stable channel plug. The

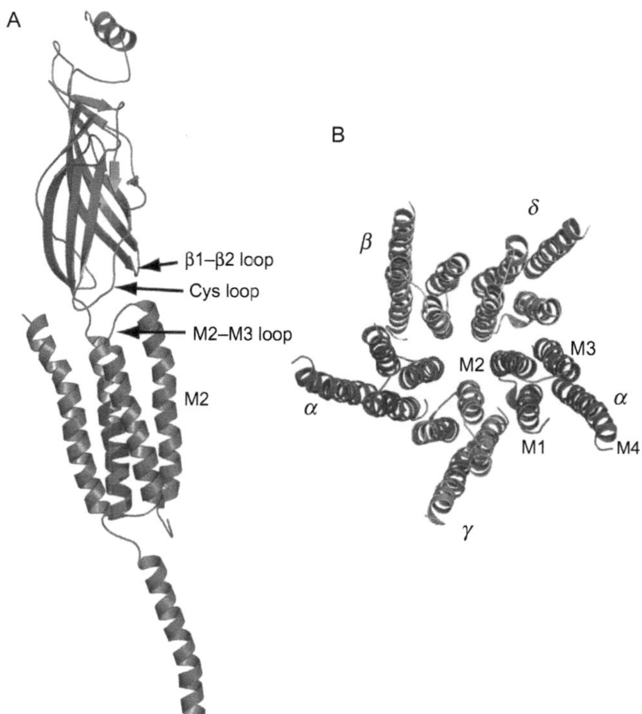

Figure 7.4 The membrane-spanning domains (M1–M4) of the nicotinic acetylcholine receptor from *Torpedo marmorata*. (A) α Subunit of the nicotinic acetylcholine receptor from *Torpedo marmorata*. The M2 helix lines the channel pore. (B) View of the membrane-spanning helical domains from the extracellular space looking down the long axis of the channel pore. The receptor subunits are indicated in Greek letters. PDB Code, 2BG9. *The structure is from Unwin.*[27] (See Color Insert.)

hydrophilic domains are composed of uncharged hydrophilic residues (serine, threonine, cysteine, and glutamine) or negatively charged residues (aspartate and glutamate). With the exception of the central channel plug, the pore-lining residues provide an ideal environment for the passage of hydrated cations.

The intracellular loop between M3 and M4 forms from each receptor subunit contributes to an inner vestibule with five orifices lined with negatively charged residues. These and a ring of negative charge on the intracellular end of M2 are thought to contribute to the cation selectivity of the nicotinic acetylcholine receptor.[27]

As described below, acetylcholine is thought to bind between the α–γ and α–δ subunit interfaces of the *Torpedo* receptor. It interacts primarily with two loops between β strands on the counterclockwise face (principal or positive face) of the α subunit. These loops are the C (between β9 and β10) and B (between β7 and β8) loops that project in a counterclockwise direction from the principal face of the α subunit.[27] Nigel Unwin and colleagues[33,34] have proposed a model to account for how the binding of acetylcholine causes and opening of the channel pore. It is thought that when acetylcholine binds, it brings the B and C closer together and causes an 11° clockwise rotation of the inner β-sheet relative to the outer β-sheet. Ultimately, it is thought that this rotation is transmitted to the M2 helices of the α subunits to cause channel opening.

This transfer of information is thought to occur through the two points where the inner and outer β sheets of the extracellular domain make contact with the M2–M3 loop of the α subunit (Fig. 7.4A). The Cys-loop has a highly conserved phenylalanine-proline-phenylalanine motif near its center that interacts with the M2–M3 loop near the top of the M3 helix. This interaction establishes a connection between the outer β sheet and the outer shell of membrane-spanning helices (M1, M3, and M4). The inner M2 helix is free of this shell, and its location is determined by the position of the M2–M3 loop. This loop makes a connection with the inner β sheet through a downward projecting loop between β1 and β2, which attaches to the M2–M3 loop near the top of the M2 helix. When acetylcholine binds and causes the inner β-sheet to rotate 11° clockwise, this rotation is transmitted to the M2–M3 loop causing its M2 end to track an 11° arch clockwise while its M3 end remains fixed as a pivot. This loop has the requisite flexibility because of a highly conserved glycine residue at its connection to the M3 helix. This clockwise rotation of the M2–M3 loop pulls the M2 helix away from the other M2 helices and breaks the hydrophobic plug formed by the central portions of the five M2 helices. This process opens the aperture of the channel and allows a flux of positive cations.

In the crystal structure of the homopentameric acetylcholine-binding protein from *L. stagnalis*, carbachol is bound between adjacent α subunits and interacts with the B and C loops on the principal face of α subunits.[24] Earlier studies that employed reactive cholinerigic ligands are consistent with the crystal structure. It has long been assumed that adjacent cysteine residues (192 and 193 in the *Torpedo californica* sequence) in α subunits are part of the acetylcholine-binding site because they react with [^{3}H] bromoacetylcholine.[35] These vicinal cysteins are located in the middle of the C loop, which projects from the principal face of the α subunit into the acetylcholine-binding pocket. Several investigators have employed a peptide-mapping strategy with reactive cholinergic ligands to identify individual amino acids that are part of the acetylcholine-binding sites. A number of residues were found to react with the photoactivated antagonist, [^{3}H] DDF (*N*,*N*-dimethylaminophenyldiazonium fluoroborate), including tryptophan 149, tyrosines 93 and 190, and cysteines 192 and 193 in the α subunit.[36,37] Similarly, photoactivated [^{3}H]nicotine reacts with tyrosine 198 and cysteine 192.[37] These residues are located between β-strands 4 and 5 (Y93), 7 and 8 (W149), and 9 and 10 (Y190, C192, C193, and Y198) in the *Torpedo* sequence (A, B, and C loops, respectively). The cognate residues in the acetylcholine-binding protein line the principal face of the acetylcholine-binding pocket. Mutagenesis has been used to identify residues in the γ and δ subunits of the nicotinic receptor essential for the binding of d-tubocurare.[38] The cognate residues on the acetylcholine-binding protein are located on β-strands 2 and 6, and in the crystal structure of the homopentameric acetylcholine-binding protein, these project into the binding pocket from the complementary face (clockwise face).

The structure of the acetylcholine-binding protein can also be used to identify binding pockets on the $GABA_A$ receptor because of the homology among members of the Cys-loop receptor family (Fig. 7.5A). The results of both mutagenesis and peptide-mapping studies indicate that phenylalanine 65 in the bovine α_1 sequence is important for GABA binding.[39,40] This residue is in the β_2 strand of the inner β-sheet and projects into the binding pocket from the clockwise face of α subunit (complementary face). Mutagenesis studies of the β subunit have identified residues essential for GABA-mediated chloride currents, including Y157, T160, T202, and Y205.[41] These are located on the B (Y157 and T160) and C (T202 and Y205) loops, which project into the binding pocket from the principal face. The binding site for GABA is located

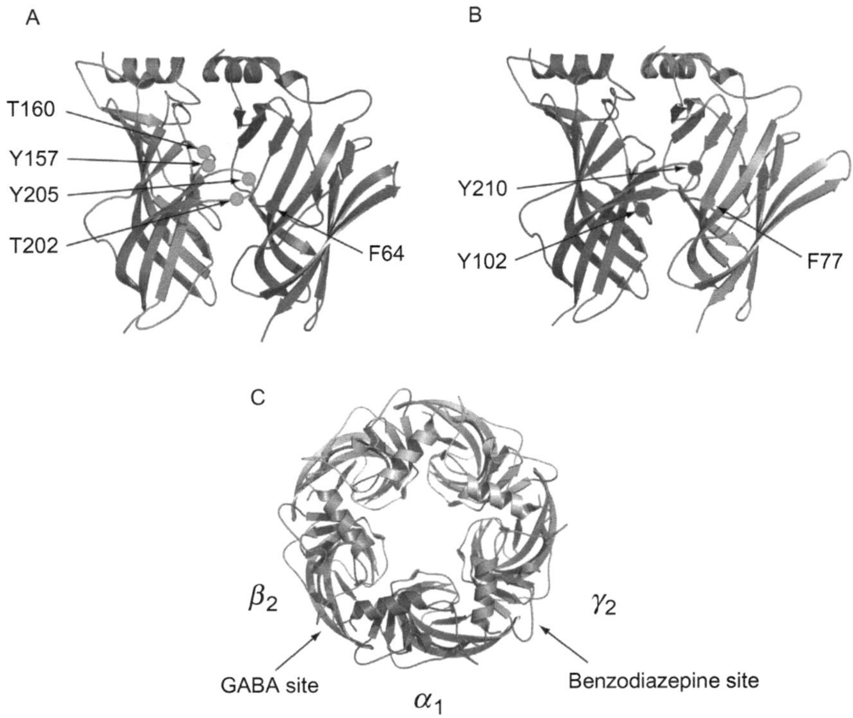

Figure 7.5 Structure of the binding sites for GABA and benzodiazepines on the $GABA_A$ receptor. The crystal structure of the acetylcholine-binding protein from *Lymnaea stagnalis*[22] is used as a template onto which critical residues involved in the binding of GABA and benzodiazepines have been mapped. The numbering and identity of the residues correspond to the sequences of the relevant $GABA_A$ receptor subunits. (A) The binding pocket for GABA. The point of view is from outside the receptor looking toward the β_2–α_1 subunit interface on the extracellular domain of the receptor. (B) The binding pocket for benzodiazepines. The point of view is from outside the receptor toward the α_1–γ_2 subunit interface. (C) Location of GABA and benzodiazepine-binding sites from an extracellular view looking down on the channel pore. PDB code, 19IB. (See Color Insert.)

between adjacent α and β subunits. The orientation is opposite to that of the nicotinic receptor, however. In the $GABA_A$ receptor, the important B and C loops of the binding pocket come from the β subunit, whereas the complementary face is formed by the α subunit.

The same approach has been used to identify the loci of the binding sites for allosteric modulators of the $GABA_A$ receptor (Fig. 7.5B). Peptide-mapping studies on the α_1 subunit have shown that H102[42] and a highly conserved residue within the Cys-loop family, P97, react covalently with photoactivated [^{3}H]flunitrazepam.[43] These are located on β4 (P97) and the A loop of the α subunit (H102). A different photoactivated benzodiazepine, [^{3}H]Ro15–4513, labels Y210 in the C loop of the α_1 subunit sequence.[44] Although the residues interacting with these different benzodiazepines are on different loops, the loops project into the subunit interface from the principal face of the α subunit. It is known that a γ subunit is necessary for benzodiazepine modulation (particularly the γ_2), and mutagenesis studies have shown that F77 in the γ_2 subunit is important for benzodiazepine binding, which is located on β strand 2. The picture that emerges is that the benzodiazepine-binding site on the $GABA_A$ receptor is formed from the principal face of the α subunit and the complementary face of the γ subunit. The orientation is the same as that of the α–γ and α–δ subunit interfaces of the acetylcholine-binding pockets on the *Torpedo* receptor. Indeed, some of the residues critical for benzodiazepine binding (H102 and Y210) in the α_1 subunit of the $GABA_A$ receptor align with residues critical for ligand binding to nicotinic receptors (Y93 and Y198 in the α_1 subunit).

Peptide-mapping studies have also been used to identify the allosteric site for anesthetic, etomidate (Fig. 7.6). The reactive derivative, [^{3}H] azietomidate, covalently reacts with residues in the α_1 (M236) and β_3 (M286) subunits.[45] Labeling in homologous residues of other α and β subunits also occurs. These residues are located in the M1 (α subunit) and M3 (β subunit) helices. Etomidate and other anesthetics are thought to bind to the M1 and M3 helices within the aqueous pocket between the M2 helix and other helices.

3.2. Why do $GABA_A$ receptors have allosteric sites?

At the outset, it was mentioned that drugs often elicit their effects by interacting with receptors for endogenous signaling molecules. It might seem odd, therefore, that allosteric sites for xenobiotics would exist on a receptor that either mediate activation (e.g., etomidate) or an enhancement of activation (i.e., benzodiazepines). There are many ways, of course, that foreign compounds might interfere with function by binding to the receptor at a site distinct from the orthosteric site. This is particularly obvious with regard to voltage-gated ion channels that lack orthosteric sites. Drugs and natural toxins that plug the channel,[46] interfere with the voltage sensor,[47] or alter inactivation kinetics[48] have been characterized. One might expect that evolutionary pressure would select for receptors that are stealth and immune

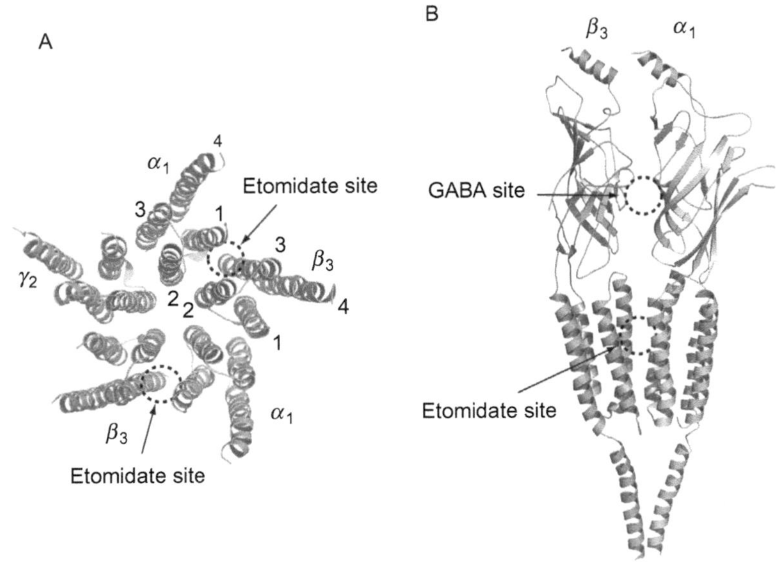

Figure 7.6 Location of the etomidate-binding site on the $GABA_A$ receptor. The structure of the nicotinic acetylcholine receptor from *Torpedo marmorata* is used as a template for mapping the binding sites for GABA and etomidate. (A) The point of view is from the extracellular space looking down the channel pore on the membrane-spanning α-helices. The numbers designate the M1–M4 helices. (B) The point of view is from outside the receptor looking on a path perpendicular to the long axis of the receptor at the interface between the α_1 and β_3 subunits. PDB Code, 2BG9. *The structure is from Unwin.*[27]

to perturbation by xenobiotics. The ability of a small molecule to act at a distance to turn on or positively modulate a ligand-gated ion channel suggests that an evolutionary adaptation might be involved.

Receptors that evolved to sense xenobiotics, like rhodopsin (photons),[49,50] nuclear receptors (toxins and drugs),[51] and taste receptors (nutrients, bitter compounds, and others)[52–54] throw some light on this question. In each case, evolution has conscripted proteins that belong to receptor families that sense endogenous signaling molecules and has adapted their orthosteric-binding sites to sense xenobiotics. Thus, sensing chemical information requires an adaptation, and once nature solves this problem, it uses the same mechanism to sense both native and foreign compounds.

This idea has interesting consequences for receptors, like ligand-gated ion channels, that are composed of distinct yet homologous receptor subunits. Acetylcholine and GABA utilize the interface between an α subunit and its adjacent subunit to activate their receptors. In the case of the nicotinic receptor, the interface incorporates the counterclockwise surface of the α subunit, whereas the opposite face is used in the $GABA_A$ receptor. Given the homology among receptor subunits, it is possible that other subunit interfaces in ligand-gated ion channels might retain the structural features to mediate receptor modulation or activation when bound by a small molecule of the appropriate structure. In a sense, these sites may represent vestigial orthosteric sites from an evolutionary past waiting for their therapeutic potential to be discovered through drug screening. Thus, ligand-gated ion channels are likely targets for allosteric drugs, and indeed, many have been identified for $GABA_A$ and nicotinic receptors.

The allosteric site for etomidate and other anesthetics on the $GABA_A$ receptor does not appear to be a vestigial orthosteric site because there may be no endogenous ligand that acts at a homologous site in the membrane-spanning helices, although endocannabinoids might.[55] The effects of anesthetics on the $GABA_A$ receptor channel are hardly allosteric, however, because they bind directly to the channel (i.e., not to a more remote allosteric site; see definition #2 of allosterism above) between the outer shell of M1, M3, and M4 helices and the inner M2 helix, and presumably, disrupt the highly ordered closed structure of the five M2 domains. It is only by default that they allosterically modulate the binding of GABA (definition #1 of allosterism) because the channel is linked to the orthosteric site. Thus, anesthetics directly bind to the channel and cause a disruptive increase in entropy that opens the channel, whereas benzodiazepines act at a distance through what appears to be a vestigial evolutionary adaptation. Both types of ligands, of course, are useful allosteric modulators of the actions of endogenous GABA.

3.3. Examples of allosteric modulation of $GABA_A$ receptor-mediated chloride currents

Numerous ligands have been reported to modify the $GABA_A$ receptor allosterically. I will not attempt to review this literature, but rather, focus on a few salient examples regarding the regulation of agonist-mediated chloride currents by benzodiazepines and etomidate. In the next section, we will analyze these examples to discern how a given modulator has differential effects on the response to orthosteric ligands.

The first example is from Maksay and coworkers[56] who investigated the tranquilizing benzodiazepine, chlordiazepoxide (30 μM), and the proconvulsant, DMCM (0.3 μM; 6,7-dimethoxy-4-ethyl-β-carboline-3-carboxylic acid methyl ester), on the whole-cell current response to GABA and the less efficacious agonist, P4S (5-(4-piperidyl)isothiazol-3-ol), in

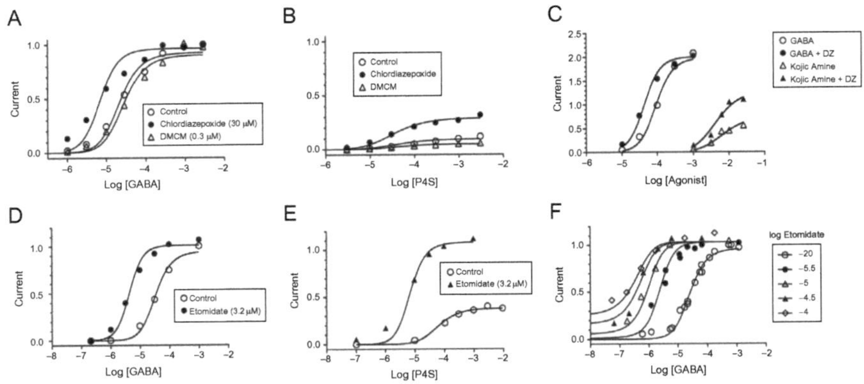

Figure 7.7 Examples of the allosteric regulation of whole-cell current responses to $GABA_A$ receptor agonists. The current responses have been estimated from published figures of the data. The data are from Maksay et al.,[56] (A and B) Downing et al.,[57] (C), and Rusch et al. (D and E) The theoretical curves represent the best fit of Eqs. (7.9) (A–C), (7.10) (D and E), and (7.11) (F) to the data.

Xenopus oocytes expressing the α_1, β_3, and γ_2 $GABA_A$ receptor subunits. In experiments with GABA, chlordiazepoxide increased and DMCM decreased the pEC_{50} value of GABA (4.52) by 0.57 and −0.36 log units, respectively, without having significant effects on the maximal current response (Fig. 7.7A). The less efficacious agonist, P4S, stimulated chloride currents with a pEC_{50} value of 3.90 and a maximal current response that was only 10% that of GABA (Fig. 7.7B). Chlordiazepoxide increased both the pEC_{50} and maximal current response to 4.62% and 25%, respectively, whereas DMCM caused a nonsignificant increase in pEC_{50} to 4.17 and a decrease in maximal current to 5%. The second example is from Downing and coworkers[57] who investigated the effects of the diazepam (0.1 μM), on the concentration-response curves of GABA and kojic amine for eliciting chloride currents in chick neurons (Fig. 7.7C). GABA elicited a robust current response characterized by a pEC_{50} value of 3.96. In the presence of diazepam, the pEC_{50} increased to 4.36, and there was no significant change in the maximal current response. The less efficacious agonist, Kojic amine, elicited chloride currents with a pEC_{50} value of 2.37 and a maximal current response of 27% that of GABA. In the presence of diazepam (0.1 μM), the pEC_{50} value was unaffected, whereas the maximal response increased to 58% that of GABA.

The third example is from Rusch and coworkers[58] who investigated the effects of the anesthetic etomidate on GABA agonist-mediated chloride currents in Xenopus oocytes expressing α_1, β_2, γ_{2L} $GABA_A$ receptors (Fig. 7.7D). GABA elicited whole-cell chloride currents with a pEC_{50} value of 4.39. Etomidate (3.2 μM) increased the potency of GABA ($pEC_{50}=5.31$) without significantly affecting the maximal current response. The less efficacious agonist P4S elicited chloride currents with a pEC_{50} value of 4.11 and a maximal current response only 39% that of GABA (Fig. 7.7E). In the presence of etomidate, the pEC_{50} value increased to 5.22 and the maximal response to 110% that of GABA.

The final example is also from the study by Rusch and coworkers.[58] In these experiments, they investigated the effects of various concentrations of edomidate on the concentration-response curve of GABA for eliciting chloride currents in *Xenopus* oocytes (Fig. 7.7F). By itself, etomidate elicits chloride currents through the $GABA_A$ receptor with a potency similar to that of GABA ($pEC_{50}=4.22$) and an E_{max} only 18% that of GABA. In the absence of etomidate, GABA elicited chloride currents with a pEC_{50} value of 4.47. Increasing concentration of etomidate increased the potency of GABA. At the highest concentration of etomidate investigated (100 μM), the pEC_{50} value of GABA increased about two log units.

3.4. Analysis of allosteric interactions at $GABA_A$ receptors

The studies just described show that when the response of a population of $GABA_A$ receptors is measured, an allosteric ligand can have differential effects on the potency and maximal current response of the GABA agonist. Do these differential effects imply that a given allosteric modulator interacts with the $GABA_A$ receptor in different ways depending on the orthosteric ligand and that different allosteric mechanisms are involved in potency and efficacy modulation? These questions can be addressed from the perspective of an allosteric model for receptor function.

3.4.1 Model for allosterism at ligand-gated ion channels

A simple and powerful way to model the interaction of agonists with ligand-gated ion channels is that used by Colquhoun and Hawkes[59] Fig. 7.8A. Their model for describing the interaction of acetylcholine with the muscle-type nicotinic receptor can be modified to account for allosterism by adding one or two allosteric-binding sites per receptor as described by Rusch *et al.*[58] Ultimately, the model is the same as that described by Monod Wyman and Changeux,[60] except for the addition of the conformational induction steps (see below). Figure 7.8B illustrates the resulting scheme for the case of two orthosteric sites and one allosteric site per receptor. This

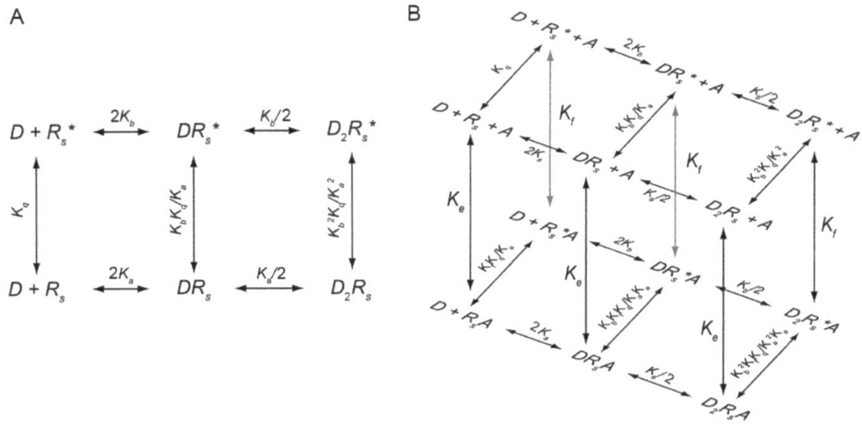

Figure 7.8 Model for the allosteric modulation of the $GABA_A$ receptor by benzodiazepines. The model is defined at the level of receptor states. (A) Model for the interaction of orthosteric ligands with a ligand-gated ion channel having two linked binding sites. (B) An allosteric model for a ligand-gated ion channel having one allosteric site and two orthosteric sites.

condition accurately describes the stoichiometry of binding sites for the major $GABA_A$ receptor subtype modulated by benzodiazepines in the mammalian brain. This model describes drug action at the level of receptors states (level 3 analysis).

In this model, K_b and K_a denote the microscopic affinity constants of the orthosteric ligand (D) for the open ($R_s{}^*$) and closed states (R_s) of the channel, K_f and K_e, the corresponding constants for the allosteric ligand (A), and K_q the isomerization constant for the unoccupied receptor. These same generic variables are used below for GPCRs for consistency. Other investigators have established more accepted variables for K_a (K_D), K_b (J_D), and K_q (E_0) in the ligand-gated ion channel field.[61] Also the isomerization constants for the mono- and bi-liganded receptors are often denoted as E_1 and E_2. I avoid the use of the variable K_D as a microscopic constant because it is so often used as a population variable (level 2 analysis) for the observed dissociation constant (i.e., reciprocal of the observed affinity constant, K_{obs}). When the variable R lacks the subscript "s", it is intended to represent the sum of the active and inactive states of a given receptor complex (e.g., R denotes $R_s{}^* + R_s$). This applies to all forms of the variable (i.e., R, DR, D_2R, RA, DRA, and D_2RA).

This model accounts for the condition where the orthosteric ligand first binds to the receptor and induces channel opening after one (very unlikely) or two ligands bind (conformational induction). It also accounts for the condition where the receptor first undergoes a rare constitutive activation and then binds one or two agonist molecules to stabilize the open state (conformational selection). Given that the probability of channel opening (P_{open}) for the unoccupied muscle-type nicotinic acetylcholine receptor is extremely low over any given time interval ($P_{open} < 10^{-6}$)[12] and that channel opening may be associated with domain closure around the orthosteric-binding pocket,[62] the conformational selection part of the scheme is unnecessary.

The orthosteric ligand *does* exhibit an affinity for the open channel (R_s^*) even though it might be impossible for it to bind to or dissociate from the open channel directly. Given appropriate values for K_a and K_q one could ask, what is the value of K_b in the isomerization expression (K_bK_q/K_a) that accurately describes the conformational induction process ($DR_s \leftrightarrow DR_s^*$)? The value of K_b in this expression is equivalent to the microscopic affinity constant of the ligand for the active state as described by the conformational selection process ($D+R_s^* \leftrightarrow DR_s^*$). If the latter process were possible, it would yield the same amount of DR_s^* at equilibrium as the conformational induction process. Thus, it is informative to include the conformational selection process in the model because it defines the affinity of the ligand for the active state even though the ligand might not directly associate with the active state of the receptor.

The ability of a ligand to open the channel depends on its selectivity for the open state. This principle is illustrated in Fig. 7.9 for the case of a ligand-gated ion channel having a gating constant ($K_q = 10^{-6}$) not much different from that estimated for the muscle-type nicotinic acetylcholine receptor (7×10^{-7}).[61] The histogram shows how the open probability of the ligand-gated ion channel changes with occupancy of the orthosteric and allosteric sites of a receptor conforming to the model in Fig. 7.8. In this example, the orthosteric and allosteric ligands have ratios of microscopic affinity constants for active and inactive states of 10^3 ($K_b/K_a = 10^3$) and 3 ($K_f/K_e = 3$), respectively. In the absence of ligands, the open probability of the channel is given by

$$P_{open} = \frac{K_q}{1+K_q}, \qquad [7.1]$$

which yields a value (10^{-6}) approximately equal to that of K_q. The term open probability (P_{open}) refers the amount of time the channel spends in

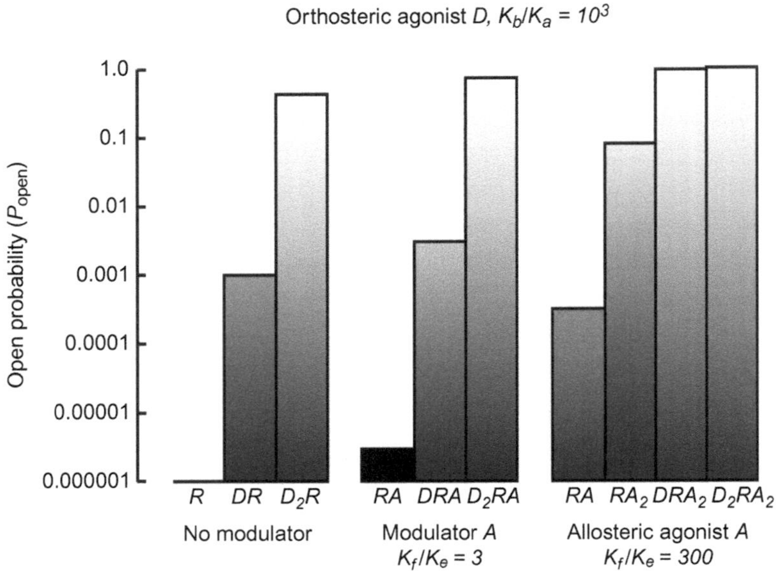

Figure 7.9 The open probability of a receptor conforming to the model shown in Fig. 7.8B when occupied by different combinations of orthosteric and allosteric ligands.

the open state divided by the observation time. When one molecule of the orthosteric ligand binds to the receptor, open probability is given by

$$P_{\text{open}} = \frac{\frac{K_b}{K_a} K_q}{1 + \frac{K_b}{K_a} K_q}. \tag{7.2}$$

This expression yields a P_{open} estimate (0.001) approximately equal to $(K_b/K_a)K_q$. Thus, when one molecule of orthosteric ligand is bound to the receptor, nearly undetectable levels of receptor activation occur. When two orthosteric ligands bind to the receptor, open probability is described by

$$P_{\text{open}} = \frac{\left(\frac{K_b}{K_a}\right)^2 K_q}{1 + \left(\frac{K_b}{K_a}\right)^2 K_q}, \tag{7.3}$$

which yields an estimate of 0.5 for this example (Fig. 7.9). Thus, substantial channel opening only occurs when two molecules of orthosteric agonist are bound to the receptor.

Equations can also be derived to describe P_{open} when only the allosteric site is occupied

$$P_{\text{open}} = \frac{\frac{K_f}{K_e} K_q}{1 + \frac{K_f}{K_e} K_q} \qquad [7.4]$$

or when the allosteric site is occupied and one,

$$P_{\text{open}} = \frac{\frac{K_f}{K_e}\frac{K_b}{K_a} K_q}{1 + \frac{K_f}{K_e}\frac{K_b}{K_a} K_q}, \qquad [7.5]$$

or two orthosteric sites are occupied,

$$P_{\text{open}} = \frac{\frac{K_f}{K_e}\left(\frac{K_b}{K_a}\right)^2 K_q}{1 + \frac{K_f}{K_e}\left(\frac{K_b}{K_a}\right)^2 K_q}. \qquad [7.6]$$

The latter three conditions yield open probabilities of 3×10^{-6}, 0.003, and 0.75, respectively (Fig. 7.9).

On the surface, this allosteric ligand behaves as a positive modulator ($K_f/K_e > 1$) because it increases the P_{open} of the receptor when both orthosteric sites are occupied (increases P_{open} from 0.5 to 0.75) but has an undetectable effect by itself (increases P_{open} from 10^{-6} to 3×10^{-6}). Nonetheless, the mechanism of the allosteric modulator is qualitatively similar to that of the orthosteric ligand. That is, it has the effect of increasing the observed gating constant by a factor equivalent to its ratio of affinity constants for the active and inactive states, (K_f/K_e). This example also illustrates that even if an allosteric modulator has selectivity for the active state equivalent to that of the orthosteric agonist (i.e., $K_f/K_e = 10^3$), it would still be unable to activate the receptor significantly because the receptor only has one allosteric site and the gating constant of the unoccupied receptor is so low (i.e., $K_q = 10^{-6}$). Although such a modulator would increase the gating constant by a factor of 1000, this would yield a nearly undetectable P_{open} of 0.001 when the modulator is present by itself. Nonetheless, this high level of selectivity for the active state would cause a colossal enhancement in P_{open} when the receptor is occupied by the agonist. In fact, it would enable the receptor to open when only one agonist molecule is bound to the receptor.

An analogous type of phenomenon occurs when constitutively activating point mutations are introduced into nicotinic acetylcholine receptors. Jha and Auerbach[63] introduced such mutations into the α subunit of the muscle-type nicotinic receptor. Two types of constitutively active α subunits were constructed, one having a wild-type ligand-binding domain and the other having a W149 mutation in the B loop preventing ligand binding. Both α subunits as well as the β_1, δ_1 and ε subunits were expressed in HEK293 cells. Patch clamp experiments showed that the hybrid receptor having a single functional binding site was activated by acetylcholine, indicating that activation of the constitutively active receptor can occur with occupancy of only one orthosteric site.

Figure 7.9 also summarizes the effects of an allosteric agonist ($K_f/K_e = 300$) on P_{open} when the receptor has two allosteric sites. This situation accurately describes the behavior of barbiturates and other anesthetics on the $GABA_A$ receptor complex. The mathematical consequences of this model are similar to those described above except that the allosteric ligand has an effect indistinguishable from causing the gating constant to increase by the square of K_f/K_e whenever both allosteric sites are occupied. This latter case yields a P_{open} of 0.083, indicating that the allosteric ligand has a small but measureable effect on P_{open}, and therefore, is empirically defined as an allosteric agonist. When both allosteric sites are occupied and either one or two orthosteric sites are occupied by the same agonist as in Fig. 7.9, the P_{open} values increase to 0.98 and essentially 1.00, respectively.

Ultimately, channel opening occurs whenever the orthosteric and allosteric ligands, working through their respective sites (one or two) is sufficient to overcome the gating constant of the unoccupied receptor (K_q). It appears that the introduction of more than one binding site for acetylcholine on the nicotinic acetylcholine receptor is an adaptation to reduce leak current in the absence of agonist. The coupling of two sites yields a Gibbs-free energy change for receptor activation proportional to the square of K_b/K_a (about 10.4 kcal/mol for acetylcholine), which is sufficient to activate a channel with a gating constant (K_q) or P_{open} of only 7×10^{-7}.[12] If there were only one site for acetylcholine per receptor, the gating constant would have to be increased to about 10^{-3}. This would increase P_{open} in the absence of acetylcholine and cause a wasteful leak current.

So far, only positive allosteric modulators and allosteric agonists have been considered. The same principles apply to negative allosteric modulators, except that negative modulators have selectivity for the inactive state ($K_b/K_a < 1.0$). Their effect is to reduce the gating constant of the receptor by a scalar equivalent to K_b/K_a or its square depending on whether there are one or two allosteric sites per receptor, respectively.

If the equilibrium constants in the model in Fig. 7.8 are replaced with bidirectional rate constants, it is possible to use the resulting model to simulate single-channel behavior. Each rate constant can be converted into a conditional probability, and the process can be modeled as a continuous Markov process in time.[13,64]

For example, let's assume that the rate constant (k_{a+}) for the binding of agonist to inactive state of the receptor is $5 \times 10^7\ M^{-1}\ sec^{-1}$. If this rate constant is multiplied by the agonist concentration (e.g., 10^{-4} M), the result is the first order rate constant for agonist binding in units of inverse time ($5 \times 10^3\ sec^{-1}$). If this constant is multiplied by a very small time increment (e.g., 10^{-7} sec), the result is the unitless probability (5×10^{-4}) that the agonist will bind to the receptor over the time increment, given that the receptor is in the R_s state at the start of the time interval. Using this approach to assign conditional probabilities to all of the reaction steps, single-channel behavior can be simulated numerically as a continuous Markov process.

Figure 7.10 shows an example of such a Monte Carlo simulation of a ligand-gated ion channel having a gating constant of 10^{-6} for the unliganded receptor. Channel openings are indicated as upward deflections in the current trace. In panel *a*, the effects of various concentrations of an agonist ($K_b/K_a = 10^4$) on single-channel currents are shown in the absence and presence of a positive allosteric modulator ($K_f/K_e = 3$) at a concentration that fully occupies the allosteric site. Panel *b* shows a plot of P_{open} for this

Figure 7.10 Monte Carlo simulation of single-channel events of a ligand-gated ion channel conforming to the model show in Fig. 7.8B. (A) Single-channel events were simulated for conditions in the absence and presence of a receptor-saturating concentration of allosteric modulator. (B) Plot of the time that the receptor spends in the open state divided by the total observation time (open probability). The rate constants used for the simulation yielded microscopic constants of log K_a, 4.0; log K_b, 8.0; log K_q, −7.0; log K_e, 8.0; and log K_f, 8.48. The concentration of allosteric ligand was 10^{-3} M.

simulation. The positive allosteric modulator has the effect of increasing the amount of time the receptor spends in the open state when the agonist is present. Empirically, these effects are manifest as a small increase in the maximal value of P_{open} (control, P_{open}, 0.89; plus modulator, P_{open}, 0.97) and a larger increase in the observed potency (pEC_{50}) of the agonist (control, pEC_{50}, 4.32; plus modulator, pEC_{50}, 4.68).

The P_{open} values of a single receptor in the presence of various concentrations of agonist are also equivalent to the fractions of the population of agonist–receptor complexes in the active state, provided that the receptor lacks constitutive activity. The model shown in Fig. 7.8 can be used to derive an equilibrium expression for the fractional amount of the receptor population in the active state ($[DR_s^* + D_2R_s^* + R_s^*]/R_T$), in which R_T denotes the total receptor concentration:

$$\frac{[R_{\text{active}}]}{[R_T]} = \frac{1}{1 + \frac{(1+AK_e)(1+DK_a)^2}{K_q(1+AK_f)(1+DK_b)^2}}. \qquad [7.7]$$

Using this equation, the equilibrium values for the fraction of the occupied receptor population in the open state can be calculated for the various concentrations of agonist and allosteric modulator that were used for the single-channel simulation shown in Fig. 7.10. These equilibrium values yield theoretical curves for receptor activation for the control condition (E_{max}, 0.89; pEC_{50}, 4.39) and in the presence of the modulator (E_{max}, 0.96; pEC_{50}, 4.66) that are nearly the same as those estimated from the single-channel Monte Carlo simulation. Thus, it follows that the whole-cell current response is proportional to the open probability of single receptors provided that desensitization is proportional to receptor activation, or more simply, that desensitization does not alter the shape of the ensemble average too drastically.

Figure 7.11 shows a simulation of the whole-cell current response of a less efficacious agonist ($K_b/K_a = 10^{3.3}$). The current responses have been normalized to the maximum possible current and were simulated in the absence and presence of a positive modulator ($K_f/K_e = 3.0$) at a concentration that fully occupied the receptor population. These fractional current responses are equivalent to open probability and have been plotted as such. In this case, the modulator increases the maximal P_{open} (Control, P_{open}, 0.27; plus modulator, P_{open}, 0.53) while having little effect on pEC_{50} (Control, pEC_{50}, 3.76; plus modulator, pEC_{50}, 3.90).

At the beginning of this section, the question was asked whether different mechanisms are involved in the allosteric modulation of the potency and

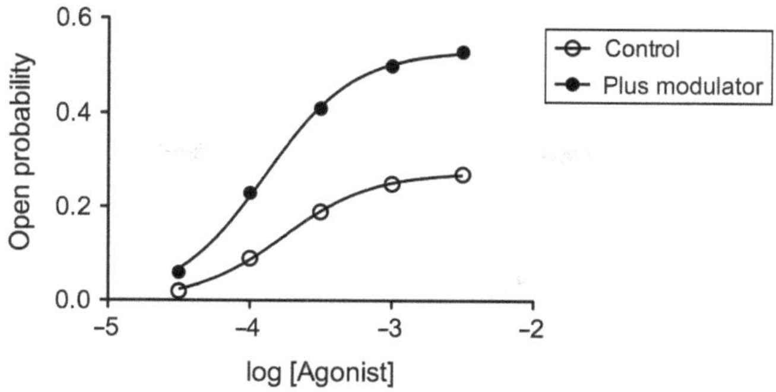

Figure 7.11 Simulation of the whole-cell current response of a partial agonist. The data have been simulated using Eq. (7.7) and the following parameter values: log K_a, 4.0; log K_b, 7.3; log K_q, −7.0; log K_e, 8.0; and log K_f, 8.48. The concentration of allosteric ligand was 10^{-3} M.

efficacy of an agonist. These empirical changes in the behavior of the receptor population occur through the same mechanism, that is, through selection of either the active or inactive state of the receptor. For a highly efficacious agonist, a moderate allosteric effect is manifest mainly as a change in potency, whereas a similar allosteric modulation of the response to a less efficacious agonist is manifest mainly as a change in efficacy. Thus, the differential effects of the allosteric modulators on the responses to GABA and the partial agonists in Fig. 7.7 do not provide evidence that the modulators affect the different orthosteric ligands through different mechanisms. This question needs to be addressed more quantitatively, however, using the approach described in the next section.

3.4.2 Analysis of data from the literature

To address whether different mechanisms are involved in the allosteric modulation of the currents elicited by GABA and the partial agonists shown in Fig. 7.7, it is necessary to analyze the data with the allosteric model shown in Fig. 7.8. If the allosteric ligand exhibits differential effects, then it must select for different conformations of the $GABA_A$ receptor. If so, then the estimates of K_f and K_e or the ratio, K_f/K_e, should vary depending on the agonist.

For most of the experiments in Fig. 7.7, only single, maximally effective concentrations of modulator were used. Under this condition (limit as A approaches infinity), Eq. (7.7) reduces to the following form:

$$\frac{[R_{\text{active}}]}{[R_T]} = \frac{1}{1 + \frac{(1+DK_a)^2}{K_q \frac{K_f}{K_e}(1+DK_b)^2}}. \qquad [7.8]$$

When there is no detectable constitutive activity ($K_q \ll 1$), an infinite number of parameter values will yield the least squares fit of Eq. (7.8) to data of the type shown in Fig. 7.7A, for example. This infinite solution set is unique, however, in that the expression $K_b^2 K_q$ is constant over the domain $K_q \ll 1$ for all of the best fitting estimates of K_q and K_b. Thus, it is easy to obtain a least-squares fit to a form of Eq. (7.8) in which the expressions $K_b^2 K_q$ and K_f/K_e occur as the single parameters, P_b and R, respectively:

$$\frac{[R_{\text{active}}]}{[R_T]} = \frac{I_{\max}}{1 + \frac{(1+DK_a)^2}{K_q R\left(1+D\sqrt{\frac{P_b}{K_q}}\right)^2}}. \quad [7.9]$$

In this equation, $I_{\max}$ denotes the maximum possible current when all of the ligand-gated ion channels in the population are open. Equation (7.9) can be fitted to the two concentration-response curves of the orthosteric ligands simultaneously by constraining K_q to an arbitrarily low value (e.g., 10^{-6}), constraining R to 1 for the control curve and sharing the estimates of the other parameters.

When this approach was used to analyze the effects of chlordiazepoxide and DMCM on GABA-mediated chloride currents (Fig. 7.7A), the following parameters were estimated: log K_a, 4.08 ± 0.24; log P_b, 9.49 ± 0.15; and I_{sys}, 0.97 ± 0.043. The estimates of log K_f/K_e (log R) when chlordiazepoxide and DMCM were present were 0.89 ± 0.19 and -0.19 ± 0.15. It is impossible to obtain a reliable estimate of $I_{\max}$ for the partial agonist P4S (Fig. 7.7B). One solution is to constrain this parameter to the value obtained with GABA (*A better approach is two analyze all of the data simultaneously sharing the estimate of $I_{\max}$ between the data obtained with the two agonists*). When this was done, the following parameters were obtained with P4S-mediated chloride currents: log K_a, 4.71 ± 0.059 and log P_b, 8.46 ± 0.12. The estimates of log K_f/K_e (R) when chlordiazepoxide and DMCM were present were 0.59 ± 0.041 and -0.30 ± 0.08. The latter estimates of K_f/K_e are not significantly different from those estimated with GABA, indicating that the modulators do not exhibit probe-dependent effects on the currents elicited by GABA and P4S.

Using the same approach to analyze the effects of diazepam on GABA- and Kojic amine-mediated chloride currents (Fig. 7.7C) yielded the following parameters: GABA, log K_a, 3.08 ± 0.77; log P_b, 8.14 ± 0.17, $I_{\max}$, 2.02 ± 0.11 and log R, 0.63 ± 0.16. For Kojic amine, the parameter estimates were: log K_a, 2.35 ± 0.095; log P_b, 4.49 ± 0.15; and log R,

0.48 ± 0.063. Again, there is no evidence for probe-dependent effects because the estimate of the log K_f/K_e value of diazepam is approximately the same for the data with both orthosteric agonists.

The anesthetic etomidate binds to the $GABA_A$ receptor between the M1 and M3 helices on adjacent α and β subunits, respectively. Thus, there are two etomidate-binding sites per $GABA_A$ receptor of the type studied by Rusch and coworkers.[58] To account for two allosteric sites, Eq. (7.9) is modified by squaring the R term:

$$\frac{[R_{\text{active}}]}{[R_T]} = \frac{I_{\max}}{1 + \dfrac{(1+DK_a)^2}{K_q R^2 \left(1 + D\sqrt{\frac{P_b}{K_q}}\right)}}. \qquad [7.10]$$

When this equation was used to analyze the effects of etomidate (3.2 μM) on GABA- and P4S-mediated currents (Fig. 7.7D and E), the following parameters were estimated: GABA, log K_a, 3.99 ± 0.24; log P_b, 9.15 ± 0.16, I_{sys}, 1.02 ± 0.035; and log R, 0.80 ± 0.09. When P4S was used as the agonist the parameter estimates were log K_a, 4.45 ± 0.25; log P_b, 8.62 ± 0.45, I_{sys}, 1.12 ± 0.059; and log R, 0.91 ± 0.19. The similarity in the estimates of log R for etomidate indicates no evidence for probe dependence. In this case, however, R does not represent an estimate of K_f/K_e, but rather, a complex function expressed in terms of A, K_e, and K_f because the concentration of etomidate (3.2 μM) did not saturate the allosteric site.

The final experiment in Fig. 7.7F investigates the effects of various concentrations of etomidate on the concentration-response curve of GABA. Because a range of concentrations of etomidate were used, it should be possible to estimate values of K_f and K_e. It should also be possible to estimate K_q because etomidate is an allosteric agonist, and estimates of the maximal effect of etomidate in the absence of GABA were made. The maximal response to etomidate is dependent solely on K_f, K_e, and K_q as shown by Eq. (7.4), and hence, it should therefore be possible to estimate all of the parameters. This is done by modifying Eq. (7.7) to account for two allosteric sites and by introducing a parameter for the maximum current of the receptor population ($I_{\max}$):

$$\frac{[R_{\text{active}}]}{[R_T]} = \frac{I_{\max}}{1 + \dfrac{(1+AK_e)^2(1+DK_a)^2}{K_q(1+AK_f)^2(1+DK_b)^2}}. \qquad [7.11]$$

This equation was fitted simultaneously to all of the data in Fig. 7.7F, and the following parameters were estimated: log K_q, -3.71 ± 0.18; K_a, 4.10 ± 0.12; log K_b, 6.54 ± 0.08; log K_e, 4.73 ± 0.09; log K_f, 6.42 ± 0.06; and $I_{max-sys}$, 1.03 ± 0.02.

There appear to be striking differences between the $GABA_A$ receptor and the muscle-type nicotinic acetylcholine receptor. The gating constant of the unoccupied nicotinic receptor is only 7×10^{-7},[12] whereas that estimated above for the $GABA_A$ receptor is 1.9×10^{-4}. The larger gating constant of the $GABA_A$ receptor reduces the energy barrier for receptor activation by about 3.4 kcal mol and causes a greater leak current in the absence of GABA. Perhaps this helps clamp the resting membrane potential to prevent excessive excitation in the same way that some constitutively active potassium channels do.[65]

The selectivity of GABA for the active state of its receptor (K_b/K_a) is only about 280, whereas that for acetylcholine at the muscle-type receptor is about 7000.[12] These translate into changes in Gibbs-free energy of 3.8 and 5.2 kcal mol^{-1}, respectively. The total available energy for inducing a conformation change is 7.6 and 10.4 kcal mol^{-1} at $GABA_A$ and nicotinic receptors, respectively, because each receptor has two orthosteric sites. Using the estimate of K_q for the $GABA_A$ receptor and that estimated by Auerbach[12] for the muscle-type nicotinic receptor, the P_{open} values for these receptors when occupied by two molecules of endogenous ligand are 0.94 and 0.97, respectively.

The P_{open} values of the allosteric modulators can be calculated using the estimate of K_q and Eq. (7.4) for the benzodiazepine ligands and the following equation for etomidate, which takes into account two allosteric sites per receptor:

$$P_{open} = \frac{K_q \left(\frac{K_f}{K_e}\right)^2}{1 + K_q \left(\frac{K_f}{K_e}\right)^2}. \qquad [7.12]$$

These estimates for chlordiazepoxide, diazepam, and etomidate are 10^{-3}, 7×10^{-4} and 0.32. These P_{open} values explain why chlordiazepoxide and diazepam are positive modulators, whereas etomidate is an allosteric agonist. At maximal concentrations, the former benzodiazepine ligands would have nearly undetectable effects, whereas etomidate would elicit a whole-cell current response of about 32% of I_{max}. The Gibbs-free energy values for these compounds to induce a conformational change in the $GABA_A$ receptor are 1.0, 0.78, and 4.7 kcal mol^{-1}.

3.5. Relation between the population parameters and the selectivity of allosteric modulators for the active state of a ligand-gated ion channel

The selectivity of an allosteric modulator for active and inactive states of a ligand-gated ion channel can also be deduced from how it modulates the population parameters of the orthosteric ligand (observed affinity, K_{occ}, and mean efficacy, ε_{mean}). The parameter, K_{occ}, represents the reciprocal of the orthosteric ligand concentration required to occupy half of the population of receptors with occupied being defined as the condition of having one or two ligands bound, and ε_{mean} represents the integral of receptor activation expressed as a function of occupancy over the domain 0–1.0. These population parameters can be calculated for the orthosteric agonist in the absence (K_{occ} and ε_{mean}) and presence (K_{occ}' and ε_{mean}') of the modulator. In this section, I summarize these relationships without going into the details of their mathematical basis.

The microscopic constants of the modulator (K_f and K_e) are related to how it appears to change the population parameters of the orthosteric ligand. The changes in the orthosteric ligand's observed affinity (ΔK_{occ}) and mean efficacy ($\Delta\varepsilon_{mean}$) caused by the allosteric modulator are defined as:

$$\Delta K_{occ} = \frac{K'_{occ}}{K_{occ}} \quad [7.13]$$

$$\Delta\varepsilon_{mean} = \frac{\varepsilon'_{mean}}{\varepsilon_{mean}}. \quad [7.14]$$

The combined effect of the allosteric ligand on the K_{occ} and ε_{mean} values of an orthosteric ligand at a ligand-gated ion channel is defined as relative activity (RA):

$$\mathrm{RA} = \Delta K_{occ}(\Delta\varepsilon_{mean})^{1/n}, \quad [7.15]$$

in which n denotes the number of orthosteric sites on the ligand-gated ion channel. For the examples of the $GABA_A$ receptors considered in this chapter, $n=2$, and hence:

$$\mathrm{RA} = \Delta K_{occ}\sqrt{\Delta\varepsilon_{mean}}. \quad [7.16]$$

The relationship between RA and the concentration of modulator (A) is given by:

$$\log \mathrm{RA} = \log\left(\frac{1 + A(\mathrm{RA}_{\max} K_{\mathrm{A}})}{1 + AK_{\mathrm{A}}}\right). \qquad [7.17]$$

In this equation, $\mathrm{RA}_{\max}$ denotes the maximum change in RA caused by the modulator and K_{A} is defined as:

$$K_A = \frac{K_{\mathrm{e}} + K_{\mathrm{f}} K_{\mathrm{q}}}{1 + K_{\mathrm{q}}}. \qquad [7.18]$$

For a ligand-gated ion channel having two orthosteric sites and one allosteric site, the ratio, $K_{\mathrm{f}}/K_{\mathrm{e}}$, for the allosteric modulator is given by:

$$\mathrm{RA}_{\max} \approx \sqrt{\frac{K_{\mathrm{f}}}{K_{\mathrm{e}}}} \qquad [7.19]$$

provided that $K_{\mathrm{f}}/K_{\mathrm{e}} < 1/K_{\mathrm{q}}$. When there are two orthosteric and two allosteric sites on the ligand-gated ion channel, then $\mathrm{RA}_{\max}$ is equivalent to $K_{\mathrm{f}}/K_{\mathrm{e}}$, provided that $(K_{\mathrm{f}}/K_{\mathrm{e}})^2 < 1/K_{\mathrm{q}}$:

$$\mathrm{RA}_{\max} \approx \frac{K_{\mathrm{f}}}{K_{\mathrm{e}}}. \qquad [7.20]$$

Thus, in the prior examples of published studies on allosteric effects, the $\mathrm{RA}_{\max}$ values of chloridazepoxide and diazepam are equivalent to the square root of $K_{\mathrm{f}}/K_{\mathrm{e}}$, whereas in the case of etomidate, $\mathrm{RA}_{\max}$ is equivalent to $K_{\mathrm{f}}/K_{\mathrm{e}}$.

Finally, when the number of allosteric sites is equivalent to the number of orthosteric sites (e.g., etomidate modulation of the $\mathrm{GABA_A}$ receptor), there is a simple relationship between occupancy of the receptor by the allosteric ligand and its effect on the normalized RA value ($\mathrm{RA}_{\mathrm{norm}}$), which is calculated as:

$$\mathrm{RA}_{\mathrm{norm}} = \frac{\mathrm{RA} - 1}{\mathrm{RA}_{\max} - 1}. \qquad [7.21]$$

In the absence of modulator, the RA value of the orthosteric ligand is equal to one, and hence, from Eq. (7.21), $\mathrm{RA}_{\mathrm{norm}}$ is equivalent to 0. When the concentration of allosteric ligand is maximally effective, the RA value is equivalent to $\mathrm{RA}_{\max}$, and hence, $\mathrm{RA}_{\mathrm{norm}}$ is equivalent to 1.0. It can be shown that the relationship between $\mathrm{RA}_{\mathrm{norm}}$ and the modulator concentration is consistent with a simple one-site, occupancy function provided that $K_f/K_e \ll 1/K_q$:

$$\mathrm{RA}_{\mathrm{norm}} = \frac{A}{A - \frac{1}{K_A}}. \qquad [7.22]$$

Under this condition, K_A is essentially equal to K_e, and $\mathrm{RA}_{\mathrm{norm}}$ is a measure of receptor occupancy by the allosteric ligand. Thus, by investigating the influence of the modulator on the changes in observed affinity and mean efficacy of the orthosteric ligand, it is possible to estimate receptor occupancy by A, K_e and K_f/K_e, and therefore, K_f.

Obviously, whenever the requisite data for this type of analysis is measured (e.g., Fig. 7.7F), it is much easier to estimate K_f and K_e using nonlinear regression analysis with the appropriate forms of Eqs. (7.7) (one allosteric site) and (7.11) (two allosteric sites). Nonetheless, it is insightful to understand how the population parameters relate to more fundamental state parameters. We will use these relationships in analyzing allosterism at GPCRs.

4. G PROTEIN-COUPLED RECEPTORS

4.1. Structure of the amine cluster of GPCRs within the α-group of the rhodopsin family

The remainder of this chapter focuses on the origin of allosterism at the M_2 muscarinic receptor. I begin with a brief summary of the structures of the M_2 muscarinic and β_2 adrenergic receptors so that we will have a better idea of how selection of different receptor conformations generates allosteric effects.

These two receptors belong to the amine cluster within the α-group of the rhodopsin family, which is the largest of the five families of GPCRs.[54,66,67] Like other GPCRs, members of the amine cluster have seven membrane-spanning α-helices (TM1–7), three extracellular loops (E1–3), three intracellular loops (I1–3), an extracellular amino terminus, and an intracellular carboxyl terminus. The second extracellular loop is the longest, but it contains a conserved cysteine near its center that forms a disulfide linkage with a conserved cysteine in helix three near its extracellular end, resulting in two short loops connecting the top of helix three with those of helices four and five. On the intracellular side, the third loop is by far the longest particularly among subtypes of the muscarinic receptor. The membrane proximal ends of this loop have a role in G protein interactions.

We know much about the structures of the β-adrenergic and muscarinic receptors because the crystal structures of the inactive states of M_2, M_3, and β_2 receptors have been solved in addition to the active state of the β_2 receptor in a co-crystal with agonist and G_s.[67–70] From an extracellular point of view down their long axes, the helices (TM1–7) of the M_2 and β_2-adrenergic receptors circumscribe in a counterclockwise direction a comma-like structure with the tail formed by helices one and two and the central ring formed by helices three, five, six, and seven. Helix four lies peripherally to this ring between helix three and five. In the crystal structures of the M_2 and M_3 receptors, the respective antagonist ligands, 3-quinuclidinyl benzilate and tiotropium, are oriented so that their ionized amines coordinate with a highly conserved aspartic acid (D3.32) (numbering system of Ballesteros and Weinstein[71]) about three helical turns in from the extracellular end of helix three. The analogous interaction also occurs in the crystal structure of the β_2-adrenergic receptor in a complex with the inverse agonist, carazolol. These structures are consistent with studies on how mutagenesis of D3.32 affects the binding and responses of orthosteric ligands including the endogenous neurotransmitters.

The largest differences between the helical domains of the active and inactive states of the β_2 receptor are the outward projection of helix six and the downward displacement of helix five in the active structure.[70] These movements open a pocket on the underside of the receptor that enables the docking of the carboxy terminal α5 helix of the α subunit of G_s with the receptor. The points of contact include the interaction of Y391 near the carboxy terminis of $G_{\alpha s}$ with the R131 and Y326 of the highly conserved DRY and NPxxY sequences on the cytosolic ends of helices three and seven of the receptor, respectively. Several interactions between other residues on the α5 helix of $G_{\alpha s}$ and the cytosolic ends of helix three and five also occur. This structure is consistent with the known role that the α5 helix of G proteins plays in determining the specificity of receptor–G protein interactions.[72–74]

In the crystal structure of the holoprotein form of G_i bound with GDP, the ras and helical domains of $G_{\alpha s}$ form a closed structure around the binding pocket for GDP.[75] The $G_{\beta\gamma}$ subunits bury a substantial surface area on the ras domain of $G_{\alpha s}$ in the α subunit including the α2 helix. In the co-crystal of the β_2-adrenerigc receptor and G_s, the ras and helical domains of $G_{\alpha s}$ are splayed apart, which presumably facilitates the exchange of GTP for GDP.[70] Part of the binding pocket of guanine nucleotides on $G_{\alpha s}$ includes a highly conserved TCAV sequence on a loop between β-strand 6 and the α5 helix.[76] This position is ideally suited so that the interaction of the α5 helix with the receptor would alter the guanine nucleotide-binding pocket.

Numerous ligands have been shown to interact allosterically with the M_2 muscarinic receptor to modify the binding of the orthosteric radioligands, $[^3H]$*N*-methylscopolamine ($[^3H]$NMS) and $[^3H]$3-quinuclidinyl benzilate ($[^3H]$QNB), and the functional responses to acetylcholine and other orthosteric agonists.[77–79] Studies on mutant and chimeric receptors indicate that there is an allosteric site near the extracellular face of the receptor involving residues on the extracellular loops, or just above the orthosteric site, depending on the nature of the modulator.[80–83] This location is consistent with the observation that occupancy of the allosteric site often prevents the access and egress of orthosteric ligands to and from their binding pocket. In addition to the M_2, other subtypes of the muscarinic receptor are differentially regulated by allosteric modulators.[77–79,84]

4.2. Model for the interaction of orthosteric ligands with GPCRs

One of the first population models to described receptor–G protein interactions is the ternary complex model of DeLean and coworkers.[85] It illustrates how the positive cooperativity between the binding of agonist (small ligand) and G protein (large ligand) to the receptor generates the ternary complex (agonist–receptor–G protein complex). Ligands that lack positive cooperativity behave as antagonists. Weiss *et al.*,[86] extended this model by dividing the receptor population into two populations of active and inactive states governed by a series of sub-cooperativities. To model how the receptor functions physiologically to catalyze guanine nucleotide exchange, a guanine nucleotide (X)-binding step was added to the ternary complex model[10] (Fig. 7.12A). The resulting model was expanded to include one inactive and one or two active states (Fig 7.12B).[4,14,87] Ultimately, this model predicts that functional receptor activation is proportional to the amount of receptor in the active state in a quaternary complex consisting of the agonist–receptor–G protein–GDP complex ($DR_s{}^*GX$). In the remainder of this chapter, this model is designated as the quaternary complex model.

At the population level (Fig. 7.12A), the affinity constants of the orthosteric ligand and G protein for the receptor are denoted by K_1 and K_2, respectively. When the receptor is bound with agonist, the affinity of the G protein for the receptor increases by the scalar, α. Similarly, when the receptor is bound with G protein, the affinity of the agonist increases

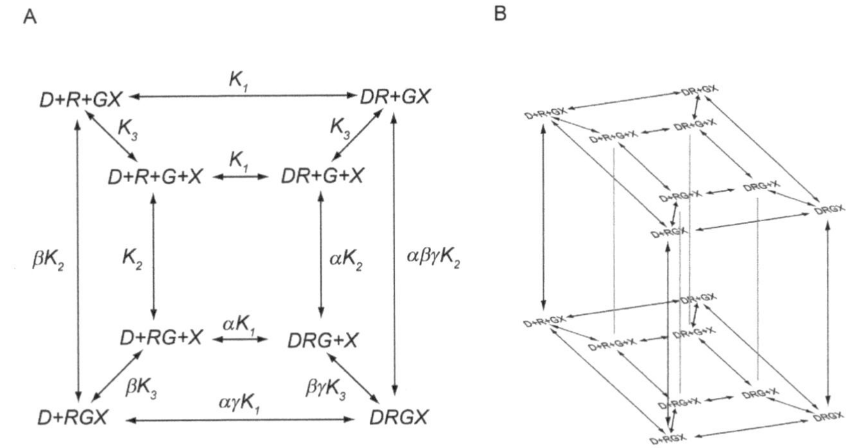

Figure 7.12 The quaternary complex model for G protein-coupled receptors. (A) Population model for the interaction of an orthosteric ligand (D) with a GPCR (R) in equilibrium with a G protein (G) and guanine nucleotide (X). At this level of analysis, the model is defined in terms of the observed affinity constants of ligand–receptor complexes (K_1, K_2, and K_3) and cooperativity constants (α, β, and γ). (B) Receptor-state model for the population model shown in panel A. The top side of the square represents the active state of the receptor and the bottom side the inactive state.

by the same amount (α). Thus, α is a population parameter that defines the cooperativity between the binding of orthosteric ligands and the G protein with the receptor.[88]

The cooperativity constant, γ, plays a pivotal role in receptor function. It represents the allosteric interaction between the binding of agonist and guanine nucleotide (GDP or GTP) to the receptor–G protein complex (RG). The affinity constants of the orthosteric ligand and GDP for RG are denoted by αK_1 and βK_3, respectively. When an agonist is bound to RG, the affinity of GDP for the G protein is reduced by the scalar, γ. Similarly, when GDP is bound to RG, the affinity of the agonist is also reduced by γ. Thus, γ represents the negative cooperativity between the binding of the agonist and GDP with the RG complex. This cooperativity represents the mechanism by which the endogenous neurotransmitter reduces the affinity of GDP for the G protein, causing an increase in its rate of dissociation and allowing intracellular GTP to rapidly displace it from the G protein. Once bound, GTP causes the dissociation of the G protein subunits and an activation of downstream effectors.

From the scheme in Fig. 7.12A, it might appear that the cooperativity constants, α and γ, are independent and can assume any values. By expanding the model into active and inactive states (Fig. 7.12B), it can be shown that the values of α and γ are inversely related.[4,87] This is because the agonist has higher affinity for the receptor when it is in the active state, and the active state binds to the G protein with high affinity when its ras and helical domains are open.[70] In contrast, GDP has higher affinity for the G protein when the ras and helical domains are closed around the guanine nucleotide-binding pocket with $G_{\beta\gamma}$ bound to the α subunit.[75] Thus, the agonist–receptor complex pushes the G protein into an open state, whereas GDP pushes it into a closed state. This antagonism is manifest as negative cooperativity (γ). At M_2 muscarinic receptors, this negative cooperativity is inversely correlated with efficacy and αK_2.[4,88]

Again, the mutual negative cooperativity, γ, reduces the affinity of the interacting ligands (agonist and GDP) and enables GTP to compete GDP off the G protein. In the absence of agonist, the GTPase activity of the G protein locks G proteins in an inactive state from which GDP dissociates only very slowly. The purpose of the agonist–receptor complex is to drive this steady state closer to a thermodynamic equilibrium in which GTP can readily compete GDP off the G protein.[4,89] For in the absence of agonist, GTP resembles a suicide substrate in the sense that its hydrolysis product (GDP) does not readily dissociate from the receptor. The agonist also reduces the affinity of GTP, but this does not prevent GTP from binding to the G protein because (1) the resting steady state nearly prevents any binding of GTP regardless of its concentration, and (2) the cytosolic concentration of GTP is usually high. If the agonist actually increased the affinity of GTP for the G protein, this would reduce signaling by stabilizing the receptor–G protein complex and hindering the liberation of the G protein from the receptor and preventing its interaction with downstream effectors.

This interpretation is consistent with the well-known observations that both GTP and GDP reduce the affinity of agonists for GPCRs, including the M_2 muscarinic.[90–92] The magnitude of this negative cooperativity is proportional to efficacy and the value of αK_2, and hence, α.[88,93] These results imply that agonists must reduce the affinity of GDP and GTP for the receptor–G protein complex because allosteric interactions are reciprocal.

The consequences of the quaternary complex model predict that orthosteric ligand binding is consistent with a simple one-site model with regard to the fraction of the receptor population in equilibrium with the G protein.[4,88] The reciprocal of the concentration of agonist required for half-maximal occupancy is denoted by K_{obs}, and this constant accurately describes occupancy of all components of the model including the *DR*,

DRG, and *DRG-GDP* components. In the absence of GTP, the agonist exhibits high affinity and the receptor lacks activity because it does not generate GTP-bound G_α subunits. As the concentration of GTP increases, observed affinity decreases, yet the maximal amount of the quaternary complex (DR_sG-GDP) increases and that of the ternary complex (DR_sG) decreases.[4,88] At saturating concentrations of GTP, there is no ternary complex (DRG), only a mixture of quaternary (*DRGX*) and binary (*DR*) receptor complexes. Agonist efficacy is proportional to the maximal amount of DR_sG-GDP. For a highly efficacious agonist having a large receptor reserve, changing the concentration of GTP over a wide range has no effect on the concentration-response curve of a downstream response because GTP-induced changes in observed affinity are opposed by reciprocal changes in observed efficacy.[4,88] Thus, changing the concentration of GTP and G proteins has an influence on the population parameters of affinity and efficacy, but not the fundamental microscopic constants (e.g., K_a and K_b) that define the system.

4.3. Two-state models for GPCRs

Given the widespread appreciation that GPCRs can adopt more than one active state,[94,95] it may seem exclusive to limit the number of active states of the receptor to one or two. I use such models in this chapter as educational devices and also as a realistic model to account for how endogenous ligands interact with their receptors. If more than one state is involved in the function of a particular receptor, one can apply the concepts learned from simple models to predict how a more complex model behaves.

Some arguments against two-state models are rhetorical, based on the language of the name, Two-State Model, and the logical argument that evidence for multiple transition states excludes the possibility of only two states. The two-state model was designed to track stable active and inactive states, however, and not the multitude of evanescent transitions states. This use is very reasonable because a ligand-gated ion channel, for example, may make thousands of transitions over a time interval of seconds, yet it will spend more than 99.99% of its time in one of two states—open and closed. A two-state model is useful because it is a type of accounting device that provides a means of estimating the activity of stable receptor states and the affinity of ligands for them. As shown in Fig. 7.13, the existence of two transitions states (i.e., R_{s1} and R_{s2}) does not influence the estimation of the microscopic constants (K_b, K_a and K_q) for the stable active (R_s*)

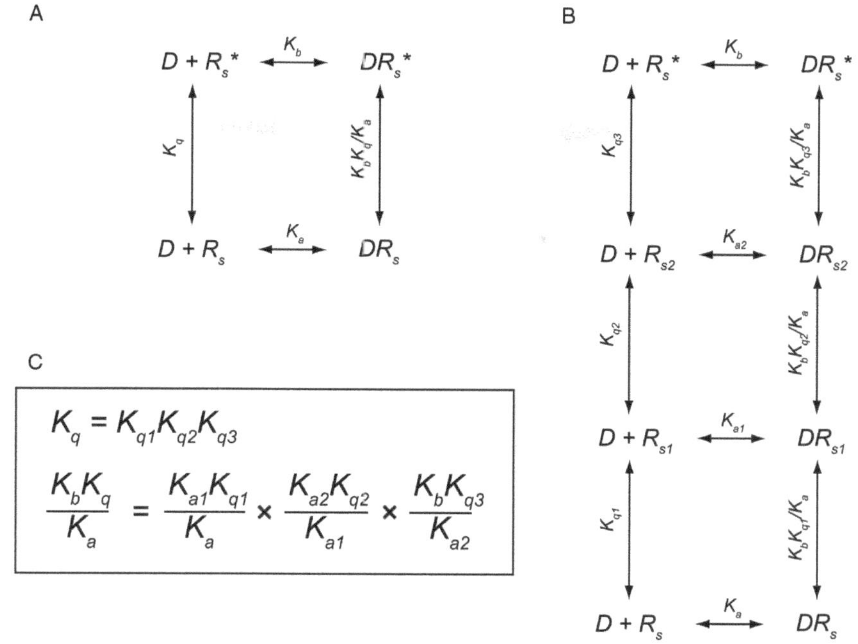

Figure 7.13 Relationship between the microscopic constants of a simple two-state model and a scheme showing two transition states between the stable active R_s^* and inactive states R_s. (A) Simple two-state model involving stable active (R_s^*) and inactive (R_s) states. (B) Simple two-state model with two transition states, R_{s1} and R_{s2}. (C) Relationships between the microscopic constants of the stable states (K_a, K_b, and K_q) and those of the transition states.

and inactive (R_s) states. In the case of multiple active states, the two-state model provides a weighted average for the estimates of the stable active states.[14,16] The same applies to inactive states.

The use of the word "model" in "two-state model" does not imply that the structures of the two states are invariant and completely defined. Rather, a two-state model implies that within the global structure of the receptor, there is a quantal switch that couples an activation domain with the orthosteric-binding pocket. There may be other disordered domains on the receptor, but this does not detract from the two-state nature of the quantal switch. Simply put, two-state models account for the affinity and activity of the R_s and R_s^* states. No other implications are intended.

For the switch to be highly efficient, it must have a very low probability of turning on in the absence of agonist.[96] This condition implies a stable inactive domain linked to a low affinity form of the orthosteric ligand's binding pocket. For the endogenous ligand to cause a robust activation of the switch, it must have much higher affinity for its binding site when the receptor is in the active state. If there were multiple conformations of the active state many would have low activity because of the structural requirements for high activity, and hence, the efficiency of the switch would be low.[96] If there were an infinite number of states, the switch would cease to function because the ligand would always induce a high affinity inactive state. Thus a striking property of molecular switches is that despite their ability to undergo ligand-induced conformational changes, the stable states of the quantum switch are they are conformationally restricted when bound with endogenous ligand.

Some of the strongest evidence for a two-state model is the output from single-channel recordings of ligand-gated ion channels.[46,97] These display an abrupt change in activation with no evidence of variable activation states. Another recent line of evidence comes from a comparison of the microscopic affinity constants of cholinergic ligands for nicotinic and muscarinic receptors. Acetylcholine has a log affinity constant of about 7.8 for the open state of the muscle-type nicotinic receptor ($1/K_b = 25$ nM). The close structural analog of acetylcholine, carbachol, has log affinity constants of about 7.2 and 6.8 for the active states of M_2 and M_3 muscarinic receptors.[16] Given that the affinity of carbachol for muscarinic receptors is about one-third to one-fifth that of acetylcholine, the predicted log estimate of acetylcholine for the active states of M_2 and M_3 receptors would be about 7.3–7.9, very similar to that for the open state of the muscle-type nicotinic acetylcholine receptor.

The binding pockets on muscarinic and nicotinic receptors both consist of aromatic cages with a residue that coordinates the positive charge of acetylcholine even though the primary sequences and secondary structures are completely different.[70,98] This convergence of evolution on the same optimal binding pocket for acetylcholine suggests a unique active state for the acetylcholine-binding pocket. The high electron density at the interface between receptor and G protein in the crystal structure of the β_2 adrenergic receptor–G_s complex suggests that the activation domains of GPCRs are also optimal in the active state.[70]

While the endogenous ligand is likely to interact with the quantal switch of receptors in an efficient manner, xenobiotics may deviate from this pattern. Many ligands, particularly allosteric modulators of muscarinic receptors, are much larger than acetylcholine, and thus, interact with the binding pocket in regions that are not evolutionary adaptations.[78,99] It is possible that this extra binding energy is sufficient to stabilize conformations of the receptor foreign to the endogenous ligand. Presumably, such interactions underlie the phenomenon of agonist bias.[94,95] One might expect that these unnatural altered states are not as active as the natural active state. Nonetheless, such ligands may be useful therapeutic agents because high efficacy in the Stephensonian sense is usually not a desirable property of drugs. Benzodiazepines are useful because they gently push the receptor in the active direction while having nearly immeasurable efficacy ($\sim 10^{-3}$ or less).

4.4. Allosteric model for GPCRs

An additional binding step for an allosteric ligand can be added to the quaternary complex model to describe allosteric ligand interactions with GPCRs. A much simpler model is valid, however, because of the consistency of the quaternary complex binding component with one-site behavior. A useful minimalist approach is the allosteric ternary complex model shown in Fig. 7.14A.[100] The affinity constants, K_1 and K_2, denote those of the orthosteric (K_1) and allosteric (K_2) ligands, and α denotes the cooperativity constant for their allosteric interaction. Hence, the affinity constants of the orthosteric and allosteric ligands for the ternary complex (*DRA*) are denoted as αK_1 and αK_3. The model is essentially the same as the ternary complex model of DeLean *et al.*,[85] but with the allosteric ligand taking the place of the G protein. Once again, this model represents an explanation of receptor behavior at the population level of analysis.

A quantum improvement in the analysis of allosterism came with the introduction of an altered efficacy component for the ternary complex (ε^*) as compared to that of the orthosteric ligand–receptor complex (ε), thus enabling a characterization of efficacy modulation.[100] In the original description of this model, the ratio $\varepsilon^*/\varepsilon$ was defined as C and the product, αC, as B. In the model in Fig. 7.14A, the original variables, $1/\alpha$, $1/C$, and B are replaced with α, β, and γ, respectively. Efficacy values can also be introduced for the allosteric ligand (ε_A) and the free receptor complex (ε_{sys}, i.e., constitutive activity). Using these mathematics expressions for the receptor activation function (stimulus) describing affinity and efficacy modulation can be derived for an allosteric modulator at a receptor lacking constitutive activity[100,101] as well as conditions involving a modulator with efficacy[102] and a constitutively active receptor.

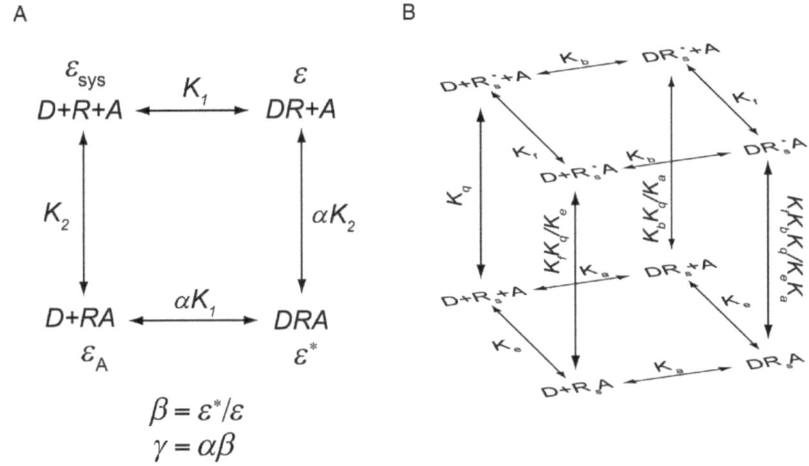

Figure 7.14 Allosteric model for ligand interactions at a receptor having one orthosteric and one allosteric site. (A) Population analysis of the allosteric model. At this level of analysis, the model is defined in terms of the observed affinity constants of ligand–receptor complexes (K_1 and K_2), cooperativity constants (α and β), and the fraction of the population of receptor complexes (*DR, DRA, RA,* and *R*) in the active state (ε, ε^*, ε_A, and ε_{sys}). (B) Analysis of the allosteric model at the level of receptor states. Each receptor complex in panel A has been divided into two states, R_s and R_s^*, representing the inactive (bottom side) and active (top side) states of the receptor. Each state has a unique affinity constant for the orthosteric and allosteric ligand.

It is difficult to measure receptor activation directly at a GPCR, but not the response elicited by the receptor at a point downstream in the signaling pathway where the signal has been amplified. Innumerable types of measurements can be made including entirely artificial ones in heterologous cells expressing the receptor of interest and other components of the signaling pathway. Using classic pharmacological methods of response-clamp analysis (null method[6]) or reverse engineering (operational model[8]), it is possible to infer some of the population parameters of the allosteric ternary complex model (level 2 analysis) from measurements made downstream in the signaling pathway (level 1 analysis). The typical experimental data involve measuring the concentration-response curve of an orthosteric ligand in the absence and presence of various concentrations of the allosteric ligand. The range of concentrations of the allosteric ligand needs to include maximally effective concentrations so that its limiting effect is apparent. With these measurements, K_2 and γ can be estimated for allosteric modulators,[100,101] as well as more complicated interactions involving constitutive activity and allosteric agonists. It is usually impossible to resolve γ into its affinity and efficacy components unless additional information is available or experimental conditions are explored.

The allosteric ternary complex model has been expanded into two receptor populations corresponding to the active and inactive states of the receptor, each defined by a series of sub-cooperativities.[103] It is also possible to expand the allosteric ternary complex model into active and inactive states as shown in Fig. 7.14B (level 3 analysis).[1,104] Each of the population parameters (K_1, K_3, α, β, γ, and ε) can be expressed in terms of the microscopic constants (K_q, K_a, K_b, K_e, and K_f), the latter being defined as described above for ligand-gated ion channels. For allosteric ligands having a K_f/K_e ratio less than the reciprocal of the gating constant for the free receptor (K_q), the ratio K_f/K_e is essentially equal to γ. Thus, while knowledge of the population parameters affinity (α) and efficacy (β) may seem important, these are soft population parameters, whereas γ is a quantal constant equivalent to K_f/K_e. This relationship explains why it requires less information to estimate γ compared to α or β.

The foregoing two-state model is highly constrained, and it behaves similarly to that of the model for ligand-gated ion channels (Fig. 7.8), notwithstanding differences related to the number of orthosteric and allosteric sites per receptor, which are both equal to one for the allosteric ternary complex model in Fig. 7.14. It predicts that a negative allosteric modulator would inhibit the effects of all orthosteric agonists, with the affinity and efficacy components of the modulation being greater for agonists and partial agonists, respectively.[1,104,105] Analogous behavior applies to positive allosteric modulators. This statement can be extended further to define the γ value of a modulator as constant for the two-state model, regardless of whether the orthosteric ligand is a full agonist, partial agonist, neutral antagonist, or inverse agonist.[104]

This may seem erroneous because, at the population level of analysis, γ is equivalent to the product of affinity (α) and efficacy (β) modulation. Within the constraints of the two-state model, would not a ligand lacking efficacy, like a neutral antagonist, be unaffected by an allosteric modulator? In this context, efficacy is used consistently as a population term as originally defined by Stephenson.[7] Many assume that Stephenson's explanation of efficacy was an exclusive hypothesis to describe receptor activation. It is not; it explained the activity of the ligand–receptor complex. The fraction of the total population of ligand–receptor complexes activated was defined as efficacy. Any receptor or ligand-activated molecular switch must have some small level of activity ($K_q > 0$); otherwise, ligand-induced receptor activation could not occur through conformational induction,

$$D + R_s \xrightarrow{K_a} DR_s \xrightarrow{K_b K_q / K_a} DR_s^* \quad [7.23]$$

or selection:

$$D + R_s \xrightarrow{K_q} D + R_s^* \xrightarrow{K_b} DR_s^*. \quad [7.24]$$

In the absence of ligands, the efficacy of the receptor (ε_{sys}) is approximately equal to K_q, and hence, that of the neutral antagonist is also equivalent to K_q because it does not influence the fraction of the receptor population in the active state. From Eq. (7.2), it follows that an allosteric modulator increases P_{open} or ε_{sys} of the unoccupied receptor by the scalar, K_f/K_e. Thus, the allosteric modulator increases the efficacy of neutral antagonists, and the product (γ) of its affinity ($\alpha = 1.0$) and efficacy ($\beta = K_f/K_e$) modulation is equivalent to K_f/K_e. A related argument can be advanced for allosteric modulation of inverse agonists. Clearly, the allosteric modulation of the M_2 muscarinic receptors deviates from this simple relationship as described below. Nonetheless, the behavior of allosteric models having one or two each of active and inactive states provides clues to the mechanism of allosteric regulation of the M_2 muscarinic receptor.[104]

The operational model and null approaches described above can be used to analyze the effects of allosteric ligands on the downstream response of agonists to provide unbiased and quantitative estimates of allosteric effects. Here, I briefly summarize how occupancy of the allosteric site affects a combined estimate of affinity and efficacy modulation (RA). The analysis is related to that described above for the more complicated situation at ligand-gated ion channels, involving two orthosteric sites and one or two allosteric sites.

For the allosteric ternary complex model, the changes in the orthosteric ligand's observed affinity (ΔK_{obs}) and efficacy ($\Delta\varepsilon$) caused by the allosteric modulator are defined as:

$$\Delta K_{obs} = \frac{K'_{obs}}{K_{obs}} \quad [7.25]$$

$$\Delta\varepsilon = \frac{\varepsilon'}{\varepsilon} \quad [7.26]$$

in which, K_{obs} and ε denote the observed affinity and efficacy of the orthosteric ligand and K_{obs}' and ε' denote those measured in the presence of the modulator. The combined effect of the allosteric ligand on the K_{obs} and ε values of an orthosteric ligand is defined as RA[101]:

$$RA = \Delta K_{obs} \Delta \varepsilon. \quad [7.27]$$

The RA values measured in the presence of different concentrations of the modulator can be estimated using the operational model, regardless of the Hill slopes the concentration-response curves. When the Hill slopes of these are equivalent to one, the transducer slope factor in the operational model is also equal to one, and the RA values can be estimated from the changes in E_{max} and EC_{50} of the concentration-response curve of the orthosteric ligand.[101] Equation (7.27) is analogous to the Eq. (7.15) for ligand-gated ion channels for the condition of $n=1$. Under these conditions, K_{occ} is equivalent K_{obs}, and ε_{mean} is equivalent to ε. When there is only one orthosteric site per receptor complex, efficacy is a simple scalar that is constant for all levels of receptor occupancy.

The relationships among log RA, the concentration of modulator and its observed affinity constant (K_A) are given by Eq. (7.17) described above. For the allosteric ternary complex model, RA_{max} is equivalent to γ. Finally, receptor occupancy by the allosteric ligand is proportional to the normalized RA value (RA_{norm}) as defined by Eqs. (7.21) and (7.22).

In the next section, these relationships are used to attempt to understand what types of mechanisms might be involved in the allosteric regulation of muscarinic receptors using the M_2 receptor as the primary example.

4.5. Allosteric modulation of M_2 muscarinic receptors

A variety of ligands have been shown to modify the functional responses of muscarinic receptors allosterically, and several excellent reviews summarize these effects.[78,106] As in the prior section on $GABA_A$ receptors, my goal here is simply to highlight a few prototypical examples and attempt to rationalize allosteric effects in terms of receptor states.

One of the first demonstrations of allosteric interactions at M_2 muscarinic receptors was described by Clark and Mitchelson,[107] who investigated the antagonistic effects of the neuromuscular-blocking agent, gallamine, on the negative inotropic effect of carbachol on the electrically paced guinea pig atrium. These investigators noted that gallamine had a competitive-like effect in the sense that it caused a parallel shift to the right in the concentration-response curve of carbachol. In contrast to a competitive mechanism, however, there was a limit to the shift caused by high concentrations of gallamine (60-fold). Competitive antagonists should shift the curve to the right without a limit. Clark and Mitchelson[107] suggested that the antagonistic effect of gallamine could be explained by an allosteric mechanism. This was later confirmed by Stockton *et al.*,[108] in binding experiments with [^{3}H]NMS on cardiac homogenates. Gallamine was shown to cause a maximal reduction in the affinity of carbachol to one-eightieth of control ($\alpha = 0.0125$), similar to the maximal gallamine-induced shift in the concentration-response curve of carbachol (i.e., $RA_{max} = 0.017$) observed by Clark and Mitchelson.

Surprisingly, gallamine also reduced the affinity of [^{3}H]NMS to 1/14 of control ($\alpha = 0.071$) at the cardiac M_2 muscarinic receptor.[108] At the time, allosteric regulation of the $GABA_A$ receptor was known to be consistent with the consequences of a two-state model described above.[1,105,109] The latter includes a modification of the Monod Wyman and Changeux[60] model to account for conformational induction. The essential feature of the model is that allosterism arises from selection of the natural on and off states of the receptor. This model predicts that allosteric modulators that reduce the affinity of agonists should have no effect on the affinity of neutral antagonists and increase the affinity of antagonists with selectivity for the inactive state (i.e., inverse agonists). It is now clear that NMS behaves as an inverse agonist.[110] Thus, it would appear that the states recognized by gallamine do not correspond to the natural active and inactive states of the M_2 receptor. This is not surprising because gallamine is a large molecule and appears to interact with an area of the muscarinic receptor above the orthosteric ligand-binding pocket[83], and therefore, may be able to select for unnatural states of the receptor. Nonetheless, it is interesting that this site is analogous to the orthosteric site of other members of the GPCR family.

Another puzzle is that gallamine appears to affect some of the actions of the M_2 receptor solely through a mechanism of affinity modulation.[104] This is hard to explain from the perspective of a peripheral allosteric site linked to an orthosteric site because selection for an altered state of the orthosteric site would be expected to alter its ability to interact with G proteins. This effect would therefore have the capacity to alter both affinity and efficacy. While highly efficacious agonists might be immune to efficacy modulation by a negatively cooperative modulator because of high selectivity for the active state,[104] this condition seems unlikely, firstly because gallamine has highly negatively cooperative effects, and secondly, because the M_2 receptor is a GPCR. As mentioned above, high intracellular GTP limits the maximal amount of quaternary complex that can form (i.e., limits efficacy), even for highly efficacious agonist. An allosteric change in the state of the orthosteric site would therefore be expected to alter both the affinity and efficacy of agonist. A potential explanation for affinity-only modulation is given in the next section.

Given this unusual behavior, it is difficult to relate the value of γ for the allosteric effects of gallamine with its ratio of affinity constants for active and inactive states of the receptor (K_f/K_e). It raises caution in attempting to do the same with regard to the effects other allosteric modulators of muscarinic receptors.

Additional allosteric ligands having unusual effects have also been described. Alcuronium, for example, was found to increase the affinity of one antagonist [^{3}H]NMS, but decrease that of another, QNB, as well as that of agonists.[111] In functional assays, however, alcuronium has been found to have a powerful inhibitory effect on the responses to agonists, and in the case of the agonist pilocarpine, this has been shown to be caused by a reduction in efficacy.[112]

Various derivatives of strychnine have been shown to have allosteric effects on muscarinic receptor subtypes.[113] In some respects, these follow the behavior predicted by an allosteric two-state model of the type described above. That is, strychnine was shown to cause a modest enhancement in the affinity of NMS for the M_2 receptor, but a greater reduction in the affinity of ACh. In functional assays, the effects of strychnine involved a reduction in the affinity and efficacy of acetylcholine.[113]

Analogs of brucine have subtype selective effects on muscarinic receptors.[114,115] For example, *N*-chloromethylbrunce allosterically reduces the affinity of acetylcholine at the M_2 receptor, but exhibits positive cooperativity with acetylcholine at the M_3 receptor. In both cases, *N*-chloromethylbrucine has small, but opposite, effects on the affinity of NMS, which is consistent with an allosteric two-state model.

Perhaps the most striking consistency with the two-state model is seen in the allosteric actions of benzyl quinolone carboxylic acid (BQCA).[116] This compound exhibits negative cooperativity with [^{3}H]NMS and behaves as an allosteric agonist. It causes a substantial increase in the potency of carbachol for eliciting responses through the M_1 receptor. The estimates of log γ for enhancing carbachol-stimulated cyclic AMP accumulation, ERK 1/2 phosphorylation, and Ca^{2+} mobilization were 3.6, 4.1, and 4.3, respectively. Assuming that the data are consistent with a two-state model and that the reported K_{obs} value of BQCA is not much different from K_e, these data yield log K_e and K_f values of approximately 10^4 and 10^8 M^{-1}, respectively. This high selectivity for the active state rivals that of acetylcholine, although it is difficult to estimate the K_a values of highly efficacious agonists at GPCRs.[15]

4.6. Difficulties in explaining affinity-only modulation

As mentioned above, the allosteric mechanism of gallamine is difficult to reconcile at the level of receptor states because only the affinity, and not the efficacy, of several agonists is modified by gallamine. This conclusion is based on the observations that gallamine shifted the concentration-response curves of agonists, including the partial agonist BM5 (*N*-methyl-*N*-(1-methyl-4-pyrrolidino-2-butynyl) acetimide), to the right without affecting E_{max} and these shifts were equivalent to the change in binding affinity caused by gallamine[104] . In addition, no evidence of depression in E_{max} was observed when functional experiments were performed on cells in which part of the receptor population was inactivated following treatment with a covalent antagonist.[104] Under these conditions, one might have expected that a change in efficacy would be more readily detected as a reduction in E_{max}.

While the data are inconsistent with a simple two-state model, adding multiple inactive and active states does not solve the problem.[104] It could be argued that there are two pairs of active ($R_{1s}{}^*$ and $R_{2s}{}^*$) and inactive (R_{1s} and R_{2s}) states. In this example, let's assume that in the absence of gallamine, receptor activation by orthosteric ligands proceeds through an R_{1s} to $R_{1s}{}^*$ transition because the abundances of the R_{2s} and $R_{2s}{}^*$ states (0.001 and 10^{-6}) are much less than those of R_{1s} and $R_{1s}{}^*$ (0.998 and 0.001). Let's also assume that the microscopic affinity constants of carbachol for the R_{1s} (K_a) and $R_{1s}{}^*$ (K_b) states are 10^5 and 10^8 M^{-1}, and those of the R_{2s} (K_c) and $R_{2s}{}^*$ (K_d) states are 10^3 and 10^6 M^{-1}, respectively. Thus the selectivity of carbachol for R_{1s} relative to $R_{1s}{}^*$ ($K_b/K_a = 10^3$) is the same as its selectivity for $R_{2s}{}^*$ relative to R_{2s} ($K_d/K_c = 10^3$). Finally, let's assume that gallamine has equal affinity for the R_1 states ($K_e = K_f = 10^5$ M^{-1}) and also equal affinity for the R_2 states ($K_g = K_h = 10^9$ M^{-1}), but that its affinity for the R_2 states is much greater than that of the R_1 states. In the presence of gallamine, agonist–receptor activation would proceed mainly through the R_2 states because gallamine has much higher affinity for these. This would reduce the affinity of carbachol because it has lower affinity for the R_2 states, but its efficacy would be unaffected because its selectivity for the active state of $R_{s2}{}^*$ is the same as that of $R_{s1}{}^*$.

While this combination of microscopic constants seems unlikely, it could be argued that perhaps, in the case of carbachol, something like this occurs. The problem is that once an allosteric–orthosteric ligand pair is found that follows this relationship among microscopic constants, other orthosteric ligands must exhibit the same profile. This seems highly unlikely, yet several muscarinic agonists only exhibit affinity modulation by gallamine.

It is possible that the activity of the R_{1s}^* state is different from that of R_{2s}^*. If so, affinity-only modulation could still occur provided that there are appropriate differences in K_b/K_a, K_d/K_c, K_f/K_e, and K_h/K_q to offset the difference. If an agonist is found for this condition, all of the other agonists must have the same K_b/K_a and K_d/K_c ratios as the first agonist in order for them to display affinity-only modulation with gallamine. One would expect that the ratio of affinity constants of a group of agonists for four different receptor structures would vary randomly and not satisfy the unique relationship of $K_b/K_a = K_d/K_c$, however. Thus, it is impossible to rationalize affinity-only modulation with an allosteric model that is based on the assumption that a single orthosteric site is coupled to an allosteric site.[104]

A clue to solving this conundrum comes from Jakubic *et al.*,[117] who suggested that the interaction of ligands with the M_1 receptor first involves the attachment of the ligand to a peripheral relay site followed by shuttling to the orthosteric activation site. These investigators postulated that D99 in the M_1 sequence might be part of the relay site based on the kinetic behavior of [^{3}H]NMS binding following a D99N mutation. Hulme *et al.*[98] have postulated that W157 in TM4 of the M_1 receptor (W155 in the M_2 receptor) may be part of a relay site because its mutation to alanine greatly reduces the affinity of acetylcholine. W155 in too remote to be considered part of the orthosteric site. Finally, Matsui *et al.*[83] have suggested that tryptophan residues on TM3 lining the water pocket above the orthosteric site (W91 and W101 in the M_1 sequence) may line this aqueous vestibule and provide an efficient pathway for the shuttling of acetylcholine to the orthosteric site in the same way that comparable residue do on acetylcholinesterase.[118]

If the putative relay site is coupled to the allosteric site for gallamine, this could explain why gallamine only alters the affinity of allosteric ligands.[104] An allosteric reduction in the affinity of the relay site has the effect of reducing the affinity of the agonist for active and inactive states of the receptor to the same extent. This would cause a reduction in affinity with no change in efficacy.

Recently modeling studies on the docking of the antagonist, tiotropium, to the crystal structure of the M_2 receptor are consistent with a peripheral relay site.[67] In modeling the dissociation of tiotropium from the receptor, tiotropium was found to associate with a binding pocket near W155 in the M_2 receptor, the same site postulated by Hulme and coworkers[98] to be a relay site. At later times, tiotropium associated with aromatic residues lining the lip of the outer vestibule of the receptor in regions that are thought to be part of the allosteric site. Thus, it may be that allosteric modulation in the M_2 receptor involves an alteration in a peripheral relay site.

While this model incorporates two sites for the orthosteric ligand at the level of receptor states, it still behaves according to the consequences of the allosteric ternary complex model at the population level of analysis.[104] It provides an explanation for changes in α without corresponding change in β that is otherwise an essential feature of two-state models, regardless of the number of active and inactive receptor conformations. It is likely that, depending on the allosteric ligand, allosteric modulation involves differential effects on both relay and orthosteric sites.

4.7. Steric effects on receptor function

So far, only allosteric mechanisms that involve ligand-mediated changes in receptor conformation have been discussed. It is possible that a ligand might bind to a site on the receptor distinct from the orthosteric site and inhibit the activity of the receptor sterically, without causing a change in the observed affinity of the orthosteric ligand.[100] This mechanism is known noncompetitive inhibition. For example, a ligand might bind to a GPCR and prevent the binding of G protein. This mode of inhibition is consistent with the allosteric ternary complex model and involves a selective reduction in β without affecting α. An example like this may be the inhibitory effects of 4-[1-(2-fluoropyridin-3-yl)-5-methyl-1H11,2,3-triazol-4-yl]-*N*-isopropyl-*N*-methyl-3,6-dihydropyridine-1(2H)-carboxamide (FTIDC) on L-glutamate-stimulation of Ca^{2+} mobilization through the mGluR1a in CHO cells.[119]

Another mechanism involves the selective binding of an allosteric inhibitor to the active state. This mechanism is known as uncompetitive inhibition and is equivalent to the open-channel block at ligand-gated ion channels.[46,120] It is possible that a membrane-diffusible ligand could bind selectively to the active state of a GPCR, perhaps by binding in the space opened by movement of the TM5 and 6 during receptor activation. Occupancy of an allosteric site in this location might prevent receptor–G protein interactions and cause inhibition of signaling. A unique property of this mode of inhibition is that it causes opposing effects on affinity (increase in α) and efficacy (decrease in β).[104] The net effect is that substantial occupancy of the allosteric site has no effect on the concentration-response curve of a highly efficacious agonist for a signaling pathway that elicits a maximal response at a low level of receptor occupancy.[101] It is only at maximal levels of occupancy of the allosteric site that inhibition of the response occurs. At ligand-gated ion channels, open-channel blockers lack strong inhibitory effect on population current, and remarkably, do not reduce the mean open time per burst, but rather, increase the mean closed time within a burst.[13]

5. CONCLUSIONS

Models incorporating unique microscopic affinity constants for receptor states are useful for quantifying allosteric interactions and probe-dependent effects. At $GABA_A$ receptors, the behavior of many allosteric modulators is consistent with a simple two-state model. At GPCRs, however, probe-dependent effects are not uncommon. In cases where multiple active states are involved, a two-state model provides an estimate of the weighted average value of the microscopic affinity constants of a modulator for active states of the receptor mediating the same response. With a sufficient number of interacting ligands, it should be possible to deconvolute probe-dependent interactions and estimate unique microscopic affinity constants for different active states.

REFERENCES

1. Ehlert FJ, Roeske WR, Gee KW, Yamamura HI. An allosteric model for benzodiazepine receptor function. *Biochem Pharmacol* 1983;**32**:2375–83.
2. Burgen AS. Conformational changes and drug action. *Fed Proc* 1981;**40**:2723–8.
3. Valant C, Felder CC, Sexton PM, Christopoulos A. Probe dependence in the allosteric modulation of a G protein-coupled receptor: implications for detection and validation of allosteric ligand effects. *Mol Pharmacol* 2012;**81**:41–52.
4. Ehlert FJ. Ternary complex model. In: Christopoulos A, editor. *Biomedical applications of computer modeling*. Boca Raton: CRC Press; 2000. p. 21–85.
5. Figueroa KW, Griffin MT, Ehlert FJ. Selectivity of agonists for the active state of M1–M4 muscarinic receptor subtypes. *J Pharmacol Exp Ther* 2008;**328**:331–42.
6. Furchgott RF. The use of β-haloalkylamines in the differentiation of receptors and in the determination of dissociation constants of receptor-agonist complexes. *Adv Drug Res* 1966;**3**:21–55.
7. Stephenson RP. A modification of receptor theory. *Br J Pharmacol* 1956;**11**:379–93.
8. Black JW, Leff P. Operational models of pharmacological agonism. *Proc R Soc Lond B Biol Sci* 1983;**220**:141–62.
9. Leff P. Analysis of agonist action using the operational model. *Trends Pharmacol Sci* 1988;**9**:395–8.
10. Ehlert FJ, Rathbun BE. Signaling through the muscarinic receptor-adenylate cyclase system of the heart is buffered against GTP over a range of concentrations. *Mol Pharmacol* 1990;**38**:148–58.
11. Qin F, Auerbach A, Sachs F. Maximum likelihood estimation of aggregated Markov processes. *Proc Biol Sci* 1997;**264**:375–83.

12. Auerbach A. The gating isomerization of neuromuscular acetylcholine receptors. *J Physiol* 2010;**588**:573–86.
13. Colquhoun D, Hawkes AG. The principles of the stochastic interpretation of ion-channel mechanisms. In: Sakmann B, Neher E, editors. *Single-channel recording*. 2nd ed. New York: Plenum Press; 1995.
14. Tran JA, Chang A, Matsui M, Ehlert FJ. Estimation of relative microscopic affinity constants of agonists for the active state of the receptor in functional studies on M_2 and M_3 muscarinic receptors. *Mol Pharmacol* 2009;**75**:381–96.
15. Ehlert FJ, Griffin MT, Suga H. Analysis of functional responses at G protein coupled receptors: estimation of relative affinity constants for the inactive receptor state. *J Pharmacol Exp Ther* 2011;**338**:658–70.
16. Ehlert FJ, Suga H, Griffin MT. Analysis of agonism and inverse agonism in functional assays with constitutive activity: estimation of orthosteric ligand affinity constants for active and inactive receptor states. *J Pharmacol Exp Ther* 2011;**338**:671–86.
17. Ortells MO, Lunt GG. Evolutionary history of the ligand-gated ion-channel superfamily of receptors. *Trends Neurosci* 1995;**18**:121–7.
18. Keesey J. How electric fish became sources of acetylcholine receptor. *J Hist Neurosci* 2005;**14**:149–64.
19. Sheridan MN. The fine structure of the electric organ of torpedo marmorata. *J Cell Biol* 1965;**24**:129–41.
20. Brisson A, Unwin PN. Tubular crystals of acetylcholine receptor. *J Cell Biol* 1984;**99**:1202–11.
21. Unwin N. Nicotinic acetylcholine receptor at 9 A resolution. *J Mol Biol* 1993;**229**:1101–24.
22. Brejc K, van Dijk WJ, Klaassen RV, Schuurmans M, van Der Oost J, Smit AB, et al. Crystal structure of an ACh-binding protein reveals the ligand-binding domain of nicotinic receptors. *Nature* 2001;**411**:269–76.
23. Thompson AJ, Lester HA, Lummis SC. The structural basis of function in Cys-loop receptors. *Q Rev Biophys* 2010;**43**:449–99.
24. Celie PH, van Rossum-Fikkert SE, van Dijk WJ, Brejc K, Smit AB, Sixma TK. Nicotine and carbamylcholine binding to nicotinic acetylcholine receptors as studied in AChBP crystal structures. *Neuron* 2004;**41**:907–14.
25. Bourne Y, Talley TT, Hansen SB, Taylor P, Marchot P. Crystal structure of a Cbtx-AChBP complex reveals essential interactions between snake alpha-neurotoxins and nicotinic receptors. *EMBO J* 2005;**24**:1512–22.
26. Dellisanti CD, Yao Y, Stroud JC, Wang ZZ, Chen L. Crystal structure of the extracellular domain of nAChR alpha1 bound to alpha-bungarotoxin at 1.94 A resolution. *Nat Neurosci* 2007;**10**:953–62.
27. Unwin N. Refined structure of the nicotinic acetylcholine receptor at 4A resolution. *J Mol Biol* 2005;**346**:967–89.
28. Raftery MA, Hunkapiller MW, Strader CD, Hood LE. Acetylcholine receptor: complex of homologous subunits. *Science* 1980;**208**:1454–6.
29. Mishina M, Takai T, Imoto K, Noda M, Takahashi T, Numa S, et al. Molecular distinction between fetal and adult forms of muscle acetylcholine receptor. *Nature* 1986;**321**:406–11.
30. Albuquerque EX, Pereira EF, Alkondon M, Rogers SW. Mammalian nicotinic acetylcholine receptors: from structure to function. *Physiol Rev* 2009;**89**:73–120.
31. Millar NS, Gotti C. Diversity of vertebrate nicotinic acetylcholine receptors. *Neuropharmacology* 2009;**56**:237–46.
32. Toyoshima C, Unwin N. Ion channel of acetylcholine receptor reconstructed from images of postsynaptic membranes. *Nature* 1988;**336**:247–50.

33. Unwin N, Miyazawa A, Li J, Fujiyoshi Y. Activation of the nicotinic acetylcholine receptor involves a switch in conformation of the alpha subunits. *J Mol Biol* 2002;**319**:1165–76.
34. Miyazawa A, Fujiyoshi Y, Unwin N. Structure and gating mechanism of the acetylcholine receptor pore. *Nature* 2003;**423**:949–55.
35. Moore HP, Raftery MA. Studies of reversible and irreversible interactions of an alkylating agonist with Torpedo californica acetylcholine receptor in membrane-bound and purified states. *Biochemistry* 1979;**18**:1862–7.
36. Dennis M, Giraudat J, Kotzyba-Hibert F, Goeldner M, Hirth C, Chang JY, et al. Amino acids of the Torpedo marmorata acetylcholine receptor alpha subunit labeled by a photoaffinity ligand for the acetylcholine binding site. *Biochemistry* 1988;**27**:2346–57.
37. Middleton RE, Cohen JB. Mapping of the acetylcholine binding site of the nicotinic acetylcholine receptor: [3H]nicotine as an agonist photoaffinity label. *Biochemistry* 1991;**30**:6987–97.
38. O'Leary ME, Filatov GN, White MM. Characterization of d-tubocurarine binding site of Torpedo acetylcholine receptor. *Am J Physiol* 1994;**266**:C648–C653.
39. Sigel E, Baur R, Kellenberger S, Malherbe P. Point mutations affecting antagonist affinity and agonist dependent gating of GABAA receptor channels. *EMBO J* 1992;**11**:2017–23.
40. Smith GB, Olsen RW. Identification of a [3H]muscimol photoaffinity substrate in the bovine gamma-aminobutyric acidA receptor alpha subunit. *J Biol Chem* 1994;**269**:20380–7.
41. Amin J, Weiss DS. GABAA receptor needs two homologous domains of the beta-subunit for activation by GABA but not by pentobarbital. *Nature* 1993;**366**:565–9.
42. Duncalfe LL, Carpenter MR, Smillie LB, Martin IL, Dunn SM. The major site of photoaffinity labeling of the gamma-aminobutyric acid type A receptor by [3H] flunitrazepam is histidine 102 of the alpha subunit. *J Biol Chem* 1996;**271**:9209–14.
43. Smith GB, Olsen RW. Deduction of amino acid residues in the GABA(A) receptor alpha subunits photoaffinity labeled with the benzodiazepine flunitrazepam. *Neuropharmacology* 2000;**39**:55–64.
44. Sawyer GW, Chiara DC, Olsen RW, Cohen JB. Identification of the bovine gamma-aminobutyric acid type A receptor alpha subunit residues photolabeled by the imidazobenzodiazepine [3H]Ro15-4513. *J Biol Chem* 2002;**277**:50036–45.
45. Li GD, Chiara DC, Sawyer GW, Husain SS, Olsen RW, Cohen JB. Identification of a GABAA receptor anesthetic binding site at subunit interfaces by photolabeling with an etomidate analog. *J Neurosci* 2006;**26**:11599–605.
46. Neher E, Steinbach JH. Local anaesthetics transiently block currents through single acetylcholine-receptor channels. *J Physiol* 1978;**277**:153–76.
47. Leipold E, Borges A, Heinemann SH. Scorpion beta-toxin interference with NaV channel voltage sensor gives rise to excitatory and depressant modes. *J Gen Physiol* 2012;**139**:305–19.
48. Jeong I, Choi BH, Hahn SJ. Rosiglitazone inhibits Kv4.3 potassium channels by open-channel block and acceleration of closed-state inactivation. *Br J Pharmacol* 2011;**163**:510–20.
49. Filipek S, Stenkamp RE, Teller DC, Palczewski K. G protein-coupled receptor rhodopsin: a prospectus. *Annu Rev Physiol* 2003;**65**:851–79.
50. Stryer L. Cyclic GMP, cascade of vision. *Annu Rev Neurosci* 1986;**9**:87–119.
51. Laudet V. Evolution of the nuclear receptor superfamily: early diversification from an ancestral orphan receptor. *J Mol Endocrinol* 1997;**19**:207–26.
52. Sternini C. Taste receptors in the gastrointestinal tract IV. Functional implications of bitter taste receptors in gastrointestinal chemosensing. *Am J Physiol Gastrointest Liver Physiol* 2007;**292**:G457–G461.

53. Wu SV, Rozengurt N, Yang M, Young SH, Sinnett-Smith J, Rozengurt E. Expression of bitter taste receptors of the T2R family in the gastrointestinal tract and enteroendocrine STC-1 cells. *Proc Natl Acad Sci USA* 2002;**99**:2392–7.
54. Lagerstrom MC, Schioth HB. Structural diversity of G protein-coupled receptors and significance for drug discovery. *Nat Rev Drug Discov* 2008;**7**:339–57.
55. Baur R, Gertsch J, Sigel E. The cannabinoid CB1 receptor antagonists rimonabant (SR141716) and AM251 directly potentiate GABA(A) receptors. *Br J Pharmacol* 2012;**165**:2479–84.
56. Maksay G, Thompson SA, Wafford KA. Allosteric modulators affect the efficacy of partial agonists for recombinant GABA(A) receptors. *Br J Pharmacol* 2000;**129**:1794–800.
57. Downing SS, Lee YT, Farb DH, Gibbs TT. Benzodiazepine modulation of partial agonist efficacy and spontaneously active GABA(A) receptors supports an allosteric model of modulation. *Br J Pharmacol* 2005;**145**:894–906.
58. Rusch D, Zhong H, Forman SA. Gating allosterism at a single class of etomidate sites on alpha1beta2gamma2L GABA A receptors accounts for both direct activation and agonist modulation. *J Biol Chem* 2004;**279**:20982–92.
59. Colquhoun D, Hawkes AG. On the stochastic properties of bursts of single ion channel openings and of clusters of bursts. *Philos Trans R Soc Lond B Biol Sci* 1982;**300**:1–59.
60. Monod J, Wyman J, Changeux JP. On the nature of allosteric transitions: a Plausible model. *J Mol Biol* 1965;**12**:88–118.
61. Auerbach A. Thinking in cycles: MWC is a good model for acetylcholine receptor-channels. *J Physiol* 2012;**590**:93–8.
62. Lape R, Colquhoun D, Sivilotti LG. On the nature of partial agonism in the nicotinic receptor superfamily. *Nature* 2008;**454**:722–7.
63. Jha A, Auerbach A. Acetylcholine receptor channels activated by a single agonist molecule. *Biophys J* 2010;**98**:1840–6.
64. Smith GD. Modeling the stochastic gating of ion channels. In: Fall CP, Marland ES, Wagner JM, Tyson JJ, editors. *Computational cell biology*. New York: Springer; 2002. p. 285–319.
65. Roepke TK, King EC, Reyna-Neyra A, Paroder M, Purtell K, Koba W, et al. Kcne2 deletion uncovers its crucial role in thyroid hormone biosynthesis. *Nat Med* 2009;**15**:1186–94.
66. Fredriksson R, Lagerstrom MC, Lundin LG, Schioth HB. The G-protein-coupled receptors in the human genome form five main families. Phylogenetic analysis, paralogon groups, and fingerprints. *Mol Pharmacol* 2003;**63**:1256–72.
67. Kruse AC, Hu J, Pan AC, Arlow DH, Rosenbaum DM, Rosemond E, et al. Structure and dynamics of the M3 muscarinic acetylcholine receptor. *Nature* 2012;**482**:552–6.
68. Haga K, Kruse AC, Asada H, Yurugi-Kobayashi T, Shiroishi M, Zhang C, et al. Structure of the human M2 muscarinic acetylcholine receptor bound to an antagonist. *Nature* 2012;**482**:547–51.
69. Rasmussen SG, Choi HJ, Rosenbaum DM, Kobilka TS, Thian FS, Edwards PC, et al. Crystal structure of the human beta2 adrenergic G-protein-coupled receptor. *Nature* 2007;**450**:383–7.
70. Rasmussen SG, DeVree BT, Zou Y, Kruse AC, Chung KY, Kobilka TS, et al. Crystal structure of the beta2 adrenergic receptor-Gs protein complex. *Nature* 2011;**477**:549–55.
71. Ballesteros JA, Weinstein H. Integrated methods for modeling G-protein-coupled receptors: implications of the high-resolution structure of rhodopsin for structure-function analysis of rhodopsin-like receptors. *Methods Neurosci* 1995;**25**:366–428.

72. Conklin BR, Farfel Z, Lustig KD, Julius D, Bourne HR. Substitution of three amino acids switches receptor specificity of Gq alpha to that of Gi alpha. *Nature* 1993;**363**:274–6.
73. Conklin BR, Herzmark P, Ishida S, Voyno-Yasenetskaya TA, Sun Y, Farfel Z, et al. Carboxyl-terminal mutations of Gq alpha and Gs alpha that alter the fidelity of receptor activation. *Mol Pharmacol* 1996;**50**:885–90.
74. Hamm HE, Deretic D, Arendt A, Hargrave PA, Koenig B, Hofmann KP. Site of G protein binding to rhodopsin mapped with synthetic peptides from the alpha subunit. *Science* 1988;**241**:832–5.
75. Wall MA, Coleman DE, Lee E, Iniguez-Lluhi JA, Posner BA, Gilman AG, et al. The structure of the G protein heterotrimer Gi alpha 1 beta 1 gamma 2. *Cell* 1995;**83**:1047–58.
76. Johnston CA, Siderovski DP. Receptor-mediated activation of heterotrimeric G-proteins: current structural insights. *Mol Pharmacol* 2007;**72**:219–30.
77. Tucek S, Proska J. Allosteric modulation of muscarinic acetylcholine receptors. *Trends Pharmacol Sci* 1995;**16**:205–12.
78. Birdsall NJ, Lazareno S. Allosterism at muscarinic receptors: ligands and mechanisms. *Mini Rev Med Chem* 2005;**5**:523–43.
79. Christopoulos A, Lanzafame A, Mitchelson F. Allosteric interactions at muscarinic cholinoceptors. *Clin Exp Pharmacol Physiol* 1998;**25**:185–94.
80. Gregory KJ, Hall NE, Tobin AB, Sexton PM, Christopoulos A. Identification of orthosteric and allosteric site mutations in M2 muscarinic acetylcholine receptors that contribute to ligand-selective signaling bias. *J Biol Chem* 2010;**285**:7459–74.
81. Avlani VA, Gregory KJ, Morton CJ, Parker MW, Sexton PM, Christopoulos A. Critical role for the second extracellular loop in the binding of both orthosteric and allosteric G protein-coupled receptor ligands. *J Biol Chem* 2007;**282**:25677–86.
82. Jager D, Schmalenbach C, Prilla S, Schrobang J, Kebig A, Sennwitz M, et al. Allosteric small molecules unveil a role of an extracellular E2/transmembrane helix 7 junction for G protein-coupled receptor activation. *J Biol Chem* 2007;**282**:34968–76.
83. Matsui H, Lazareno S, Birdsall NJ. Probing of the location of the allosteric site on m1 muscarinic receptors by site-directed mutagenesis. *Mol Pharmacol* 1995;**47**:88–98.
84. Melancon BJ, Hopkins CR, Wood MR, Emmitte KA, Niswender CM, Christopoulos A, et al. Allosteric modulation of seven transmembrane spanning receptors: theory, practice, and opportunities for central nervous system drug discovery. *J Med Chem* 2012;**55**:1445–64.
85. De Lean A, Stadel JM, Lefkowitz RJ. A ternary complex model explains the agonist-specific binding properties of the adenylate cyclase-coupled beta-adrenergic receptor. *J Biol Chem* 1980;**255**:7108–17.
86. Weiss JM, Morgan PH, Lutz MW, Kenakin TP. The cubic ternary complex receptor-occupancy model. 1. Model description. *J Theor Biol* 1996;**178**:151–67.
87. Ehlert FJ. On the analysis of ligand directed signaling at G protein coupled receptors. *Naunyn Schmiedebergs Arch Pharmacol* 2008;**377**:549–77.
88. Ehlert FJ. The relationship between muscarinic receptor occupancy and adenylate cyclase inhibition in the rabbit myocardium. *Mol Pharmacol* 1985;**28**:410–21.
89. Ehlert FJ. Correlation between the binding parameters of muscarinic agonists and their inhibition of adenylate cyclase activity. *Adv Exp Med Biol* 1988;**236**:265–76.
90. Berrie CP, Birdsall NJ, Burgen AS, Hulme EC. Guanine nucleotides modulate muscarinic receptor binding in the heart. *Biochem Biophys Res Commun* 1979;**87**:1000–5.
91. Rosenberger LB, Yamamura HI, Roeske WR. Cardiac muscarinic cholinergic receptor binding is regulated by Na+ and guanyl nucleotides. *J Biol Chem* 1980;**255**:820–3.

92. Hayashi MK, Haga T. Purification and functional reconstitution with GTP-binding regulatory proteins of hexahistidine-tagged muscarinic acetylcholine receptors (m2 subtype). *J Biochem* 1996;**120**:1232–8.
93. Ehlert FJ. Coupling of muscarinic receptors to adenylate cyclase in the rabbit myocardium: effects of receptor inactivation. *J Pharmacol Exp Ther* 1987;**240**:23–30.
94. Kenakin T. Functional selectivity and biased receptor signaling. *J Pharmacol Exp Ther* 2011;**336**:296–302.
95. Kenakin T, Miller LJ. Seven transmembrane receptors as shapeshifting proteins: the impact of allosteric modulation and functional selectivity on new drug discovery. *Pharmacol Rev* 2010;**62**:265–304.
96. Jackson MB. Perfection of a synaptic receptor: kinetics and energetics of the acetylcholine receptor. *Proc Natl Acad Sci U.S.A* 1989;**86**:2199–203.
97. Neher E, Sakmann B. Single-channel currents recorded from membrane of denervated frog muscle fibres. *Nature* 1976;**260**:799–802.
98. Hulme EC, Lu ZL, Saldanha JW, Bee MS. Structure and activation of muscarinic acetylcholine receptors. *Biochem Soc Trans* 2003;**31**:29–34.
99. Valant C, Gregory KJ, Hall NE, Scammells PJ, Lew MJ, Sexton PM, et al. A novel mechanism of G protein-coupled receptor functional selectivity. Muscarinic partial agonist McN-A-343 as a bitopic orthosteric/allosteric ligand. *J Biol Chem* 2008;**283**:29312–21.
100. Ehlert FJ. Estimation of the affinities of allosteric ligands using radioligand binding and pharmacological null methods. *Mol. Pharmacol* 1988;**33**:187–94.
101. Ehlert FJ. Analysis of allosterism in functional assays. *J Pharmacol Exp Ther* 2005;**315**:740–54.
102. Stewart GD, Sexton PM, Christopoulos A. Detection of novel functional selectivity at M3 muscarinic acetylcholine receptors using a Saccharomyces cerevisiae platform. *ACS Chem Biol* 2010;**5**:365–75.
103. Hall DA. Modeling the functional effects of allosteric modulators at pharmacological receptors: an extension of the two-state model of receptor activation. *Mol Pharmacol* 2000;**58**:1412–23.
104. Ehlert FJ, Griffin MT. Two-state models and the analysis of the allosteric effect of gallamine at the M2 muscarinic receptor. *J Pharmacol Exp Ther* 2008;**325**:1039–60.
105. Ehlert FJ. 'Inverse agonists', cooperativity and drug action at benzodiazepine receptors. *Trends Pharmacol Sci* 1986;**7**:28–32.
106. Mohr K, Trankle C, Holzgrabe U. Structure/activity relationships of M2 muscarinic allosteric modulators. *Receptors Channels* 2003;**9**:229–40.
107. Clark AL, Mitchelson F. The inhibitory effect of gallamine on muscarinic receptors. *Br J Pharmacol* 1976;**58**:323–31.
108. Stockton JM, Birdsall NJ, Burgen AS, Hulme EC. Modification of the binding properties of muscarinic receptors by gallamine. *Mol Pharmacol* 1983;**23**:551–7.
109. Ehlert FJ, Ragan P, Chen A, Roeske WR, Yamamura HI. Modulation of benzodiazepine receptor binding: insight into pharmacological efficacy. *Eur J Pharmacol* 1982;**78**:249–53.
110. Jakubik J, Bacakova L, el-Fakahany EE, Tucek S. Constitutive activity of the M1-M4 subtypes of muscarinic receptors in transfected CHO cells and of muscarinic receptors in the heart cells revealed by negative antagonists. *FEBS Lett* 1995;**377**:275–9.
111. Hejnova L, Tucek S, el-Fakahany EE. Positive and negative allosteric interactions on muscarinic receptors. *Eur J Pharmacol* 1995;**291**:427–30.
112. Zahn K, Eckstein N, Trankle C, Sadee W, Mohr K. Allosteric modulation of muscarinic receptor signaling: alcuronium-induced conversion of pilocarpine from an agonist into an antagonist. *J Pharmacol Exp Ther* 2002;**301**:720–8.

113. Lazareno S, Birdsall NJ. Detection, quantitation, and verification of allosteric interactions of agents with labeled and unlabeled ligands at G protein-coupled receptors: interactions of strychnine and acetylcholine at muscarinic receptors. *Mol Pharmacol* 1995;**48**:362–78.
114. Birdsall NJ, Farries T, Gharagozloo P, Kobayashi S, Lazareno S, Sugimoto M. Subtype-selective positive cooperative interactions between brucine analogs and acetylcholine at muscarinic receptors: functional studies. *Mol Pharmacol* 1999;**55**:778–86.
115. Gharagozloo P, Lazareno S, Popham A, Birdsall NJ. Allosteric interactions of quaternary strychnine and brucine derivatives with muscarinic acetylcholine receptors. *J Med Chem* 1999;**42**:438–45.
116. Canals M, Lane JR, Wen A, Scammells PJ, Sexton PM, Christopoulos A. A Monod-Wyman-Changeux mechanism can explain G protein-coupled receptor (GPCR) allosteric modulation. *J Biol Chem* 2012;**287**:650–9.
117. Jakubik J, El-Fakahany EE, Tucek S. Evidence for a tandem two-site model of ligand binding to muscarinic acetylcholine receptors. *J Biol Chem* 2000;**275**:18836–44.
118. Harel M, Schalk I, Ehret-Sabatier L, Bouet F, Goeldner M, Hirth C, et al. Quaternary ligand binding to aromatic residues in the active-site gorge of acetylcholinesterase. *Proc Natl Acad Sci USA* 1993;**90**:9031–5.
119. Suzuki G, Kimura T, Satow A, Kaneko N, Fukuda J, Hikichi H, et al. Pharmacological characterization of a new, orally active and potent allosteric metabotropic glutamate receptor 1 antagonist, 4-[1-(2-fluoropyridin-3-yl)-5-methyl-1H-1,2,3-triazol-4-yl]-N-isopropyl-N-methyl-3,6-dihydropyridine-1(2H)-carboxamide (FTIDC). *J Pharmacol Exp Ther* 2007;**321**:1144–53.
120. Ogden DC, Colquhoun D. Ion channel block by acetylcholine, carbachol and suberyldicholine at the frog neuromuscular junction. *Proc R Soc Lond B* 1985; **225**:329–55.

CHAPTER EIGHT

Modeling Cooperativity Effects in Dimeric G Protein-Coupled Receptors

Jesús Giraldo
Laboratory of Systems Pharmacology and Bioinformatics, Institut de Neurociències and Unitat de Bioestadística, Universitat Autònoma de Barcelona, Bellaterra, Spain

Contents

Abstract

G protein-coupled receptors organize into oligomeric arrangements to exert their function. In this chapter, three models of dimeric receptors, the two-state dimer receptor model, the metabotropic glutamate receptor model, and the asymmetric/symmetric three-state dimer receptor model are revisited focusing on the cooperative effects between their binding sites and the subunits they are composed of. The mathematical analysis reveals the complexity of the intra-receptor interactions providing insights on the mechanistic aspects of receptor function.

Progress in Molecular Biology and Translational Science, Volume 115
ISSN 1877-1173
http://dx.doi.org/10.1016/B978-0-12-394587-7.00008-7

1. INTRODUCTION

The term allostery derives from the Greek "allos" meaning "other," and it was first used by Monod *et al.* to explain cooperativity effects in enzymes.[1,2] The model was originally formulated to account for oligomeric proteins, by assuming that the protomers of which they are composed occupy equivalent positions and there is one, and only one, site on each protomer.[3] Accordingly, the translation of the original model to the G protein-coupled receptor (GPCR) world has involved many structural and functional differences.[4]

GPCRs exert their function by the combination of three elements: the ligand, the receptor, and the transduction protein (the G protein, the β-arrestin, or others; let's say the T protein, in general). Here, as in the novel "The third man" by Graham Green, allostery involves the disturbing presence of a third element. The T protein modifies the binding of the ligand to the receptor in the same way that the ligand modifies the binding of the receptor to the T protein. If we assume that protein allosterism is the change in protein reactivity at one site arising from a molecule binding on the protein at another site,[5] then GPCRs are intrinsic allosteric systems. Yet GPCR allosterism involves other molecular interactions apart from the ternary agonist–receptor–T protein complex, and with different degrees of complexity depending on whether GPCRs signal or they are considered to signal as monomers or oligomers.

In the case of monomeric GPCRs, an allosteric site is defined as a site other than that to which the endogenous hormone or neurotransmitter binds (the orthosteric site). Thus we distinguish between agonists and antagonists, which bind to the orthosteric site and compete directly with the endogenous ligand, and allosteric modulators, which bind to the allosteric site and modulate indirectly the binding and the function of orthosteric ligands. In the case of oligomeric GPCRs, apart from the allosteric sites that can be present in each of the protomers, the cooperativity effects between the orthosteric sites need to be considered (this is similar to the original model of Monod *et al.*[3]). Moreover, GPCRs oligomerization may involve also the possibility of receptor hetero-oligomerization, providing more complexity to the system.[5–7]

The aim of this chapter is to present an overview of some of the current approaches for the analysis of cooperativity effects in dimeric GPCRs. To this end, a model for allosterism in monomeric receptors, the allosteric two-state model[8], will be taken as a reference for comparison.

2. MODELING COOPERATIVITY EFFECTS IN MONOMERIC GPCRs: THE ALLOSTERIC TWO-STATE MODEL

The allosteric two-state model[8] is a combination of the ternary complex model of allosteric modulation[9–14] and the two-state model of agonism.[15–18] Basically, the receptor is considered as a monomeric entity with two binding sites, one (the orthosteric site) specific for molecule A, the agonist, and the other, (the allosteric site) specific for molecule B, the modulator (Fig. 8.1). Thus, the receptor can be either free, singly bound to A, singly bound to B or bound to A and B, simultaneously. The model assumes that either of these states can be an active state. Thus, the constitutive activity of the receptor is possible (via R^*) and the allosteric modulator can be active by itself (via R^*B). It is worth noting that the model does not make a pharmacological distinction between A and B compounds nor between A and B binding sites. The model accounts for the constitutive activity of the receptor (L), the affinity of A and B to the free receptors (K and M, respectively), the intrinsic efficacy of A and B (α and β, respectively), the binding cooperativity between A and B (γ), and the activation cooperativity between A and B (δ). Receptor functionality is quantified as the fraction of active receptors (Appendix A).

Depending on the sign and magnitude of the intrinsic efficacies and the binding and activation cooperativities, a diverse collection of pharmacologic concentration-effect curves can be obtained.[8] It is worth examining the effect of B in the bottom and top curve geometric descriptors of ligand A.

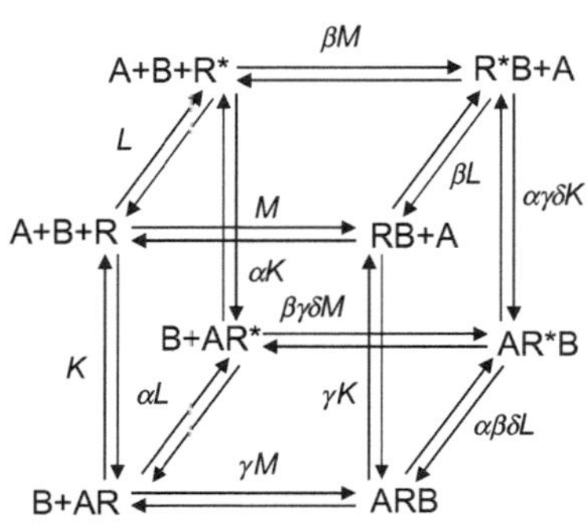

Figure 8.1 The allosteric two-state model. The receptor contains two sites, one is recognized by the agonist A and the other by the allosteric modulator B. The receptor can be either in an inactive state or in an active state. The definition of the constants and functional response are included in Appendix A. *Adapted from Ref. 8.*

We see (Appendix A) that the bottom (left asymptote of ligand A) depends on the constitutive activity of the receptor (L), the affinity of B for the free receptor (M), the intrinsic efficacy of B (β), and [B]. Logically, because A is absent, there is no constant associated to it neither directly nor indirectly through cooperativity effects. These constants appear, actually, for the top descriptor (right asymptote as [A] increases). For the latter property, and as expected, the intrinsic efficacy of A (α) and the cooperativity constants for binding and function (γ and δ, respectively) are included. Note that because the system is at maximum [A] concentration (infinitum in mathematical words), the affinity constant of A (K) is not present.

It is worth commenting on some of the parameters of the model. For instance, the binding cooperativity constant γ. In the presence of modulator B, values of γ greater than one favor the binding of A whereas the contrary happens for values lower than one. Interestingly, in the limit of $\gamma \to 0$ the receptor cannot be occupied by the two ligands simultaneously ($[ARB] = [AR^*B] = 0$) and the top of the curve as $[A] \to \infty$ is equal to that from A in the absence of B (two-state receptor model of agonism).[18] In addition, values of β and/or δ greater than one increase the top of the A curve whereas the contrary occurs for values lower than one. Also if $\beta = 1$ (B is a neutral modulator, no intrinsic activity) and, in addition, $\delta = 1$ (there is no functional cooperativity) then the top of the A curve is the same as that of the two-state receptor model (equivalent to the case in which B were absent).

It is worth noticing a limitation of the approach we are using. The property we are considering as the functional quantity is the fraction of active receptors (f). The maximum value of f is one and a full agonist can asymptotically yield this value without the help of any modulator (if the intrinsic efficacy of A (α) is sufficiently high). The question arises as to how we can model the activity of those allosteric modulators that increase the maximum response of full agonists. To do this we should use a function connecting the concentration of active receptors with response (this is related to the operational model of allosterism[19]) but this is out of scope of the present study.

3. MODELING COOPERATIVITY EFFECTS IN DIMERIC GPCRs

3.1. The two-state dimer receptor model

In the allosteric two-state model, we have three elements, the agonist, A, the receptor, R, and the modulator, B ("the third man"). In the recent modeling framework proposed by Kenakin for allosterism,[5] A would be the guest and

R the conduit, whereas B would be the modulator. B modulates the activity of the guest A through the conduit R. Generally speaking, we could consider also that a compound could perform the functions of both the guest and the modulator. This happens in dimeric receptor where each protomer has an orthosteric binding site and the molecule that binds first to the receptor modulates the binding of the molecule that binds the receptor secondly. We illustrate the case with the two-state dimer receptor model.[20,21]

The two-state dimer receptor model considers the dimer receptor as a global functional entity and makes no distinction between the protomers (Fig. 8.2). As in the previous model, cooperativity is present both for binding and for activation.

In comparison with the allosteric two-state model, there are fewer parameters because there is only one compound class, A, though the receptor can be doubly occupied. As in the former model there is a constant for the constitutive activity of the receptor, L. The parameter α measures the intrinsic activity of the first-bound ligand either by the active-state induction (αL) or by differential selection between inactive and active states (αK). As before there are cooperativity constants both for binding and function. The binding cooperativity μ is equivalent in meaning to the previous γ whereas the functional cooperativity β is equivalent to the former δ. As in the case of α, the effect of β can be seen either by the induction ($\alpha\beta L$) or the selection ($\mu\beta K$) approaches. It is worth noting that the expression for the top is

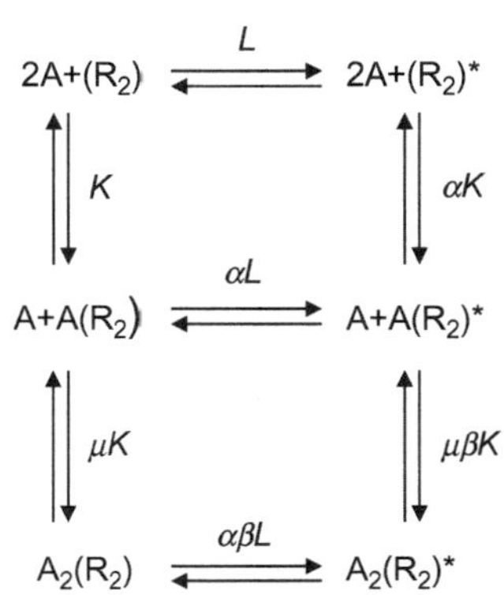

Figure 8.2 The two-state dimer receptor model. The receptor contains two sites. The receptor can be either in an inactive state or in an active state. The dimer receptor is considered as a global entity. The protomers that constitute the dimer are not distinguishable. The definition of the constants and functional response are included in Appendix B. *Adapted from Ref. 21.*

simpler in the two-state dimer model than in the allosteric two-state model (Appendices A and B). Interestingly, it yields the same expression after removing the concentration of the modulator, [B], the binding constant of the modulator, M, the binding cooperativity, γ, and the intrinsic activity of the modulator, β. The two-state dimer model can be made more complex by including a binding site for a modulator in each of the protomers but this is beyond of the scope of this study.

The differences in the models should reflect differences in the shapes of the resulting curves they produce both for binding and function. One particular characteristic of the two-state dimer receptor model, which is not present in the monomeric allosteric two-state model, is the possibility of biphasic curves. As it has been shown,[22] the existence of a biphasic curve is an indication though not a proof of receptor oligomerization. Moreover, the Hill coefficient at the mid-point is a parameter measuring the steepness of the curve and Hill coefficients greater than one are indicative of receptor oligomerization. As expected, the Hill coefficient applied to a dimeric receptor should reflect the cooperativity between their sites as it embodies the mechanistic features of the receptor. A detailed analysis of the Hill coefficient for the binding to dimeric receptors is included in Section 3.4.

3.2. The metabotropic glutamate receptors: A paradigmatic system of receptor dimerization and receptor inter-domain modulation

Metabotropic glutamate (mGlu) receptors belong to the class C of the GPCR family,[23,24] and play essential roles in the central nervous system by regulating fast excitatory and inhibitory transmission. mGlu receptors form constitutive dimers that are stabilized by a disulfide bridge. Each protomer of a mGlu dimer is composed of three structural domains: an extracellular Venus flytrap (VFT) domain where agonists bind, a transmembrane heptahelical domain (HD) responsible for G protein activation, and a cystein-rich domain (CRD) that interconnects the VFT and the HD both structurally and functionally.[25] Allosteric modulators bind to the HD but the modeling of the allosteric interactions between allosteric modulators at the HD and orthosteric agonists at the VFT will not be included in the present study. Rather, I will focus on the binding of agonists to the dimeric VFT and the inter-domain allosteric interactions between VFT and HD linked to the functional response.

The modeling framework for allosterism proposed recently by Kenakin[5] allows to visualize the different allosteric models in a general picture. Under this approach, the agonist could be considered as a modulator that through the binary conduit VFT+CRD is able to modulate the activation of the transmembrane domain (the guest). The VFT dimeric domain can be present in three different states open–open (OO), open–closed (OC), and closed–closed (CC) states, either free, singly bound, or doubly bound (Fig. 8.3). The functional response is consequence of the activation of the HD. It has been suggested both from experimental[26,27] and theoretical[28] data that only one HD per dimer is in the active conformation at a time. This is consistent with recent results for the rhodopsin dimer.[29] Furthermore, asymmetrical functioning of the purified NTS1 receptor dimer has also reported recently.[30] To account for these findings, a model was proposed[31]

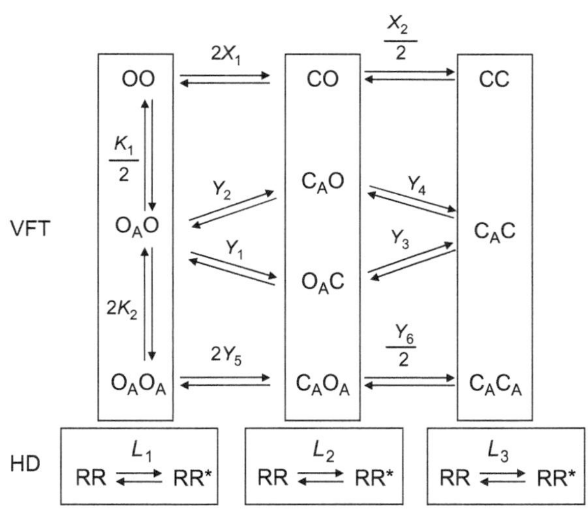

Figure 8.3 The metabotropic glutamate receptor model. The model includes two domains: the extracellular VFT domain to which agonists bind and the transmembrane heptahelical domain (HD) to which the signaling G protein binds. The receptor is dimeric. For the VFT domain, open (O) and closed (C) states of the subunits, leading to OO, OC, and CC dimeric states, are considered. The VFT domain can be free (top row), singly bound (middle row), or doubly bound (bottom row). The model follows a conformational induction approach in which closed states are induced from the ligand binding to OO states. For the HD dimer, a symmetric disposition of the two protomers (RR) for the inactive state and an asymmetric one (RR*) for the active state is proposed. The relative populations of active and inactive HD dimers are determined by an equilibrium constant (L). The binding to the extracellular domain modulates the activity of the transmembrane domain by making the constant L to vary according to the VFT state to which the HD is linked. In the simulations, it is assumed that the proportion of active HD (RR*) increases with the number of closed subunits of the VFT domain. The definition of the constants and functional response are included in Appendix C. *Adapted from Ref. 31.*

consisting of two identical protomers (RR) for the inactive and two different protomers (RR^*) for the active dimer state, whose relative populations are governed by the equilibrium constant L (Appendix C).

3.2.1 The allosteric modulation of the HD by the VFT

In mGlu receptors, the activation of the HD is modulated by the VFT through the inter-connecting CRD.[25] The ligand-bound state of the VFT regulates HD activation: while agonist-induced closure of one of the VFTs is required to activate the HD, the closure of the two VFTs is necessary for full activation. In the proposed model,[31] the allosteric modulation of HD by VFT was quantified by assuming that the equilibrium constant L governing HD activation depends on the state of the VFT dimer, with L_1, L_2, and L_3 for OO, OC, and CC, respectively, and supposing the $L_1 < L_2 < L_3$ ranking order (see Fig. 8.3 and Appendix C). Thus, the active RR^* HD was associated to each of the VFT states but with a higher propensity of formation as more VFT subunits are closed. In addition, and in agreement with experimental findings,[32] it was contemplated that activation of a particular HD protomer could result from agonist stimulation of the CRD-connected VFT (cis-activation) or from the agonist stimulation of the VFT associated to the other HD protomer (trans-activation).

The complexity of both structure and mechanism [two states (open and closed), dimerization, and two inter-connected binding and functional domains] makes mGlu receptors a unique machine especially designed along evolution for a fine and versatile regulation of function. Cooperativity is in the very nature of these proteins. In their study,[31] the authors distinguished two kinds of cooperativity for the VFT domain: one related to binding and the other to function. Figure 8.3 shows the induction approach that was followed in which the ligands bind to the inactive OO state and induce its activation by facilitating the closure of its subunits. Both cis- and trans-activation of the VFT subunits were supposed to exist. In the first case, the agonist closes the subunit to which it is bound (governed by Y_2) whereas in the latter it closes the partner one (governed by Y_1). By positive binding cooperativity we mean that the binding of the first ligand facilitates the binding of the second or, in other words, the affinity of the second ligand for the binding site is greater than the affinity of the first (the contrary would happen for negative binding cooperativity). Binding cooperativity is determined by the relative values between K_1 and K_2. By positive functional cooperativity, we mean that the closure of the first VFT subunit facilitates the closure of the second (the contrary would happen for negative functional cooperativity).

Different combinations of induction constants control functional cooperativity, for instance, the relative values between Y_4 and Y_1 ($Y_4 > Y_1$ is an indication of positive functional cooperativity). The complex network of binding and functional cooperativities generates a wide pharmacologic space for ligand diversity. If we accept that there can be three categories, absent (0), positive (+), and negative (−), for both binding and induction cooperativities, nine types of ligands could, in principle, exist from the (binding, induction) combination of cooperativity values. If we are able to link cooperativity pharmacologic concepts to structural chemical features of both the ligands and the receptor-binding sites, we could potentially find a common language between pharmacologists and medicinal chemists; a common language that could be of help in focusing drug discovery within a mechanistic pharmacological framework.

In the model constructed for mGlu receptor function,[31] an asymmetric dimeric RR^* was proposed for the active receptor species. However, the particular features of this structural arrangement were not examined because the binding of ligands (allosteric modulators in the case of mGlu receptors) to the HD was not included in the study. Yet, the crosstalk between the protomers of a dimeric receptor in both asymmetric and symmetric active states was taken into account by some of the same authors in a latter study.[33] We analyze the cooperative effects of this model in the following section.

3.3. The asymmetric/symmetric three-state dimer receptor model

The asymmetric/symmetric three-state dimer receptor model was proposed[33] as a theoretical construction aiming at providing a mathematical tool for describing functional selectivity in a receptor dimeric context (Fig. 8.4). In contrast to previous dimer receptor models,[20,21,34] the protomers within the dimer were allowed to display distinct receptor conformations. Thus, whereas one single (RR) state was assigned to represent the overall class of receptor inactive conformations, two states, one asymmetric (R^*R) and the other symmetric (R^*R^*) were proposed for functionally distinguishable active receptor states. In the model,[33] the two active receptor states were assumed to correspond to different signaling pathways: The asymmetric active receptor state was proposed to be associated to a G protein-dependent pathway whereas the symmetric receptor state was proposed to correspond to a G protein-independent pathway. The proposal was consistent with some experimental data. Thus, recent data on the BLT_1 receptor dimer suggested an asymmetric arrangement of the two-subunits in the G

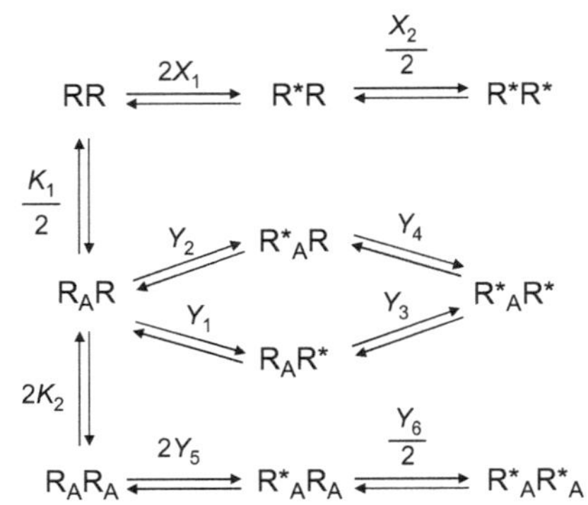

Figure 8.4 The asymmetric/symmetric three-state dimer receptor model. The model proposes two structurally determined signaling pathways: one G protein-mediated associated to the asymmetric R*R state and the other, the G protein-independent pathway, associated to the symmetric R*R* state. *Adapted from Ref. 33.*

protein-coupled state.[27] Interestingly, the authors speculated with the possibility that a symmetric dimer might be associated to arrestin.[27] Furthermore, an asymmetric interaction with the G protein has been obtained from other receptors as for instance from nanodiscs having two rhodopsins[29] or from a dimeric 5-HT_{2C} receptor.[35] Asymmetric disposition of the protomers within the active receptor state has also been found for class C GPCRs both the homodimeric mGlu receptors and for the heterodimeric $GABA_B$[36] and T1R taste[37] receptors. mGlu receptors are characterized by an activation mechanism in which only one HD of the homodimer changes its conformation upon receptor activation.[26] Consistently with an asymmetric dimeric structure, it was found that only one positive allosteric modulator (PAM) per dimer was sufficient to increase receptor activity.[38]

It is worth noting that the asymmetric/symmetric three-state dimer receptor model is formally equivalent to the proposed mechanistic model for the VFT domain in mGlu receptors, with the inactive R and active R^* of the HD corresponding to the inactive O and active C of the VFT domain, respectively. The difference resides in that whereas in the functional model proposed for mGlu receptors all the states of the VFT domain are associated to a single signaling pathway, two distinctive pathways were proposed for R^*R and R^*R^* active HD states.

3.3.1 What use has protomer distinction within a dimer model?

The proposal of two functionally distinguishable states one associated to an asymmetric protomer arrangement and the other to a symmetric disposition within the HD dimer may explain different experimental data. For instance, the apparently paradoxical results obtained on dopamine class A dimers.[39] In this study, the authors found that whereas maximal functional response was achieved by agonist binding to a single protomer of the dopamine D_2 receptor, in agreement with the hypothesis of an asymmetrically activated dimer, the functional response associated to the binding to the second protomer was dependent on the pharmacologic profile of the ligand. Thus, either decreasing or increasing of the signaling resulted depending on whether the secondly-bound ligand was an agonist or an inverse agonist, respectively.[39] This result could be explained by the asymmetric/symmetric three-state dimer receptor model.[33] If we accept that the G protein-dependent R^*R state is induced by an agonist (A) binding to the R^* protomer (R^*_AR), the addition of an inverse agonist (B) may stabilize this dimeric state ($R^*_AR_B$). On the contrary, an excess of agonist may lead to the formation of the G protein-independent R^*R^* state (because the agonist A induces the formation of the active protomer R^*). If the symmetric R^*R^* state signals through a G protein-independent pathway, the formation of this receptor species has a detrimental effect on the G protein-dependent pathway yielded by the asymmetric R^*R state.

Figure 8.4 shows all the receptor species present in the system in the absence and in the presence of an agonist A. The top row includes the free receptor species, (RR, R^*R, and R^*R^*); the middle row, the singly bounded receptor species (R_AR, R^*_AR, R_AR^*, $R^*_AR^*$); and the bottom row, the doubly bounded receptor species (R_AR_A, $R^*_AR_A$, $R^*_AR^*_A$). Increasing [A] leads the receptor from top row receptor species to bottom row by passing through middle row. As in the case of the VFT model, we can distinguish between binding and functional cooperativity and associate the latter to the higher or lower propensity of formation of the second R^* when the first protomer is already active (positive and negative functional cooperativity, respectively). Comparison between Y_6 and Y_5 is an indication of the functional cooperativity of a particular agonist–receptor system. Y_6 greater than Y_5 favors the formation of the doubly activated $R^*_AR^*_A$ whereas Y_6 lower than Y_5 would stabilize the doubly bounded receptor in an asymmetric $R^*_AR_A$ state. Furthermore, comparison of the constants in the middle and bottom rows with those in the top shows the capacity of the ligand to activate one or both protomers with respect to the free receptor. In particular, Y_6, and Y_4 and Y_3, in comparison with X_2 indicate how the ligand changes the intrinsic functional cooperativity of the free receptor to activate the second protomer when the first is already active, either when the receptor is doubly or singly occupied. Thus, we note that a complex relationship involving ligand receptor concentration and the populations of singly and doubly bound asymmetric and symmetric active receptor species is determined by the network of equilibrium constants that is included in the model.

The structural differentiation between the protomers within the dimer allowed the authors of the asymmetric/symmetric three-state dimer receptor model to explain the dosage-dependent switch from G protein-coupled to G protein-independent signaling found for β_2-adrenoceptors (β_2-ARs).[40] In their experimental study,[40] the authors observed two signaling pathways, one which is mediated by $G\alpha_s$ at low agonist concentration to activate the mitogen-activated protein kinase pathway and the other at high agonist concentration, which is not G protein—but tyrosine kinase Src dependent. The authors suggested that SCR activation could be facilitated by β_2-AR dimerization, which would allow the intermolecular autophosphorylation and activation of two molecules of Src.[40] Mathematical models are useful tools for quantitative analysis of mechanistic hypotheses derived from experimental studies. Simulations performed with the asymmetric/symmetric three-state dimer receptor[33] model allowed the authors to obtain concentration-effect curves displaying a dosage-dependent switch from G protein-coupled to G protein-independent signaling. The network of equilibrium constants included among others the relationship $Y_6 > Y_5$, which implies a positive cooperative effect of protomer activation linked to agonist occupation or in other words that the major species at high [A] is the doubly activated $R^*{}_AR^*{}_A$ (a proposal similar to that commonly accepted for the VFT domain of mGluRs, that is the closure of a subunit facilitates the closure of the second subunit if they are occupied by agonists[41]). The resulting theoretical plots with a bell-shaped curve for the R^*R-mediated G protein-dependent pathway followed at higher [A] by a sigmoid curve with an asymptotic maximum value for the R^*R^*-mediated G protein-independent pathway are compatible with the observed dosage-dependent signaling pathway switch.[40]

3.4. Quantifying cooperativity in a dimeric receptor context

In the previous sections, we have seen that cooperativity is intrinsically linked to receptor oligomerization. Obviously, the complexity of cooperativity effects depends on the complexity of the model used (as an example, see an analysis of binding data for a dimer-tetramer oligomeric receptor[42]). Yet, we should not forget that models are mere conceptual representations of reality and reality can be much more complex than their representations.

Nevertheless, the prudent principle of parsimony advises us not to use a complex model when a simpler one can sufficiently explain experimental results.

The cooperativity effects resulting from the allosteric interactions between the protomers composing the dimeric receptors (structure) are reflected in the measured response curves they generate (function). Typically, absence of cooperativity yields curves correctly fitted by Hill equations with a Hill coefficient of one whereas positive and negative cooperativity lead to Hill coefficients greater and lower than one, respectively. It is worth noting that the opposite is not true and we cannot always infer the presence of receptor oligomerization from the shape of the curves.[22]

Quantification of cooperativity by the calculation of the Hill coefficient can be done by making use of the definition of the Hill coefficient at the mid-point ($n_{H_{50}}$) for a given $y(x)$ function.[43]

$$n_{H_{50}} = \frac{4\left(\frac{dy}{dx}\right)_{x_{50}}}{a \ln 10} \qquad [8.1]$$

with $x = \log[A]$; x_{50}, the mid-point; a, the maximum asymptote; ln, the natural logarithm; and d/dx, the derivative operator.

To illustrate the use of $n_{H_{50}}$ we will apply its expression to a particular situation: the binding of a ligand to a dimeric receptor. To this end, the model in Fig. 8.2 and equations in Appendix B were used.

$$y = \frac{[A_{bound}]}{2R_T} = \frac{1}{2}\left(\frac{c_1[A] + 2[A]^2}{c_2 + c_1[A] + [A]^2}\right) \qquad [8.2]$$

with

$$c_1 = \frac{1+\alpha L}{K\mu(1+\alpha\beta L)} \quad \text{and} \quad c_2 = \frac{1+L}{K^2\mu(1+\alpha\beta L)} \qquad [8.3]$$

Applying Eq. (8.1) to Eq. (8.2) yields Eq. (8.4).

$$n_{H_{50}} = \frac{4}{2 + \left(c_1/\sqrt{c_2}\right)} \qquad [8.4]$$

The $n_{H_{50}}$ parameter ranges between 0 and 2. Traditionally, it is considered that there is no cooperativity when $n_{H_{50}} = 1$ (if $c_1/\sqrt{c_2} = 2$), positive cooperativity when $1 < n_{H_{50}} \leq 2$ (if $0 \leq c_1/\sqrt{c_2} < 2$), and negative cooperativity when $0 \leq n_{H_{50}} < 1$ (if $c_1/\sqrt{c_2} > 2$). Yet, what information about the drug–receptor interactions is hidden inside the Hill coefficient. c_1 and c_2 are empirical parameters embodying the mechanistic equilibrium constants included in the model displayed in Fig. 8.2. It is worth examining the effect on the Hill coefficient of the changes made in the different mechanistic parameters. We considered a situation in which the association constant for the binding of the first ligand to the receptor (K) is equal to 10^6 and the conformational constant for the equilibrium between inactive and active states of the free receptor (L) is 10^{-3}. Ten cases were tested; nine for the pairwise combinations between α (1, 10^3, 10^{-3}) and β (1, 10^3, 10^{-3}) and an additional 10th case for $\alpha = 10^{-3}$ and $\beta = 10^6$. In all the cases, μ was fixed to 1/4. This is the expected value for null cooperativity in a two-site receptor if no other factors are involved. The 1/4 value results from the fact that the first ligand has two receptor sites for binding and that, once bound, the second ligand can dissociate from either one binding site or the other. The objective of these simulations is to measure the effect of the agonist/inverse agonist features on the observed Hill coefficient. We will conclude that absence, positive, or negative cooperativity is present if values of $n_{H_{50}}$ equal, greater, or lower than one, respectively, are obtained. Table 8.1 displays the results. We see that positive cooperativity is obtained only in those cases in which $\beta > 1$ and, in addition, $\alpha\beta > 1$. These conditions assure an extra affinity for doubly occupied receptors arisen from agonist capacity. Negative cooperativity results from the joint condition of $\beta \leq 1$ and $\alpha > 1$.

Table 8.1 Calculation of the Hill coefficient at the mid-point ($n_{H_{50}}$) for the binding of a ligand to a two-state dimer receptor

Case	α	β	$n_{H_{50}}$
1	1	1	1
2	1	10^{-3}	0.9998
3	1	10^3	1.1713
4	10^3	1	0.8287
5	10^3	10^{-3}	0.6671
6	10^3	10^3	1.8811
7	10^{-3}	1	1.0002
8	10^{-3}	10^{-3}	1.0002
9	10^{-3}	10^3	1.0005
10	10^{-3}	10^6	1.1718

Fixed values in the model (Fig. 8.2 and Appendix A.2) $K = 10^6$, $L = 10^{-3}$, and $\mu = 1/4$.

This corresponds to a situation in which there is an agonist-induced propensity to form active singly bound—but not doubly bound-receptors. Null cooperativity is observed in the other cases. It is worth noting that (phenotypic) observed positive or negative cooperative may result from different combinations of (genotypic) mechanistic parameters. Thus, caution should be taken when proposing mechanistic hypotheses from the shape of experimental concentration-effect curves.[22]

4. CONCLUSIONS

Cooperativity effects have been examined in various dimer receptor models: The two-state dimer receptor model, the mGlu receptor model, and the asymmetric/symmetric three-state dimer receptor model. To this end, the allosteric two-state model corresponding to monomeric receptors was taken first as a reference for comparison. Structural differences are reflected in the shape of the curves that receptors yield, with steep curves (Hill coefficients at the mid-point greater than one) allowed for dimer receptors. Analysis of the shape of the curves by mechanistic models may provide information concerning the mode the receptors act and justify the assembling of receptors in oligomeric arrangements. Apart from other properties, such as trafficking and internalization,[7] which are beyond of the scope of the present study, oligomerization through cooperativity effects can be a practical way for proteins to regulate their function by a complex network of equilibrium constants. The asymmetric/symmetric three-state dimer receptor model is of particular interest inasmuch as the relative weight of two signaling pathways depends on the activated state of each of the protomers. This model suggests another issue: receptor hetero-oligomerization. Many receptors regulate themselves by grouping in hetero-oligomeric constructs. This can be a useful way for nature to use proteins and their binding in complex structures as a chemist would use reactants in the laboratory. Analysis of cooperativity effects in hetero-oligomeric receptors is an important issue that can be of interest in drug design, but this is out the scope of the present chapter.

ACKNOWLEDGMENTS

This work was supported in part by grants from Ministerio de Economía y Competitividad (SAF2010-19257), Fundació La Marató de TV3 (110230), and RecerCaixa (2010ACUP 00378).

APPENDIX A. THE ALLOSTERIC TWO-STATE MODEL

The allosteric two-state model is shown in Fig. 8.1.

A.1. The equilibrium constants of the model

$$R \overset{L}{\leftrightarrow} R^*, \quad L = \frac{[R^*]}{[R]}$$

$$A + R \overset{K}{\leftrightarrow} AR, \quad K = \frac{[AR]}{[A][R]}$$

$$B + R \overset{M}{\leftrightarrow} RB, \quad M = \frac{[RB]}{[B][R]}$$

$$AR + B \xleftrightarrow{\gamma M} ARB, \quad \gamma = \frac{[R][ARB]}{[AR][RB]}$$

$$AR \xleftrightarrow{\alpha L} AR^*, \quad \alpha = \frac{[R][AR^*]}{[R^*][AR]}$$

$$RB \xleftrightarrow{\beta L} R^*B, \quad \beta = \frac{[R][R^*B]}{[R^*][RB]}$$

$$ARB \xleftrightarrow{\alpha\beta\delta L} AR^*B, \quad \delta = \frac{[R^*][AR][RB][AR^*B]}{[R][AR^*][R^*B][ARB]}$$

A.2. Functional response: Fraction of active receptors

$$f = \frac{[R]_{Active}}{[R]_T} = \frac{L(1 + \alpha K[A] + \beta M[B](1 + \alpha\gamma\delta K[A]))}{1 + L + M[B](1 + \beta L) + K[A](1 + \alpha L + \gamma M[B](1 + \alpha\beta\delta L))}$$

where $[R]_{Active} = [R^*] + [AR^*] + [R^*B] + [AR^*B]$
and $[R]_T = [R] + [AR] + [RB] + [ARB] + [R^*] + [AR^*] + [R^*B] + [AR^*B]$

I will take f as the effect of the agonist A in the presence of fixed concentrations of the modulator B.

A.3. The asymptotes of the functional curves: Extracting pharmacological information from geometric descriptors

- Left asymptote in the absence of A and B (Basal response: f for $[A] = 0$ and $[B] = 0$)

$$\text{Basal} = \frac{1}{1+\frac{1}{L}}$$

- Left asymptote in the absence of A (Bottom: f for [A] = 0)

$$\text{Bottom} = \frac{1}{1+\frac{1+M[\text{B}]}{L(1+\beta M[\text{B}])}}$$

- Right asymptote, the asymptotic f value as [A] increases (Top: $\lim f_{[\text{A}]\to\infty}$)

$$\text{Top} = \frac{1}{1+\frac{1+\gamma M[\text{B}]}{L\alpha(1+\beta M\gamma\delta[\text{B}])}}$$

It is worth noting that if $\gamma=0$, that is, $[\text{ARB}]=[\text{AR}^*\text{B}]=0$, then $\text{Top}=1/(1+1/L\alpha)$

APPENDIX B. THE TWO-STATE DIMER RECEPTOR MODEL

The two-state dimer receptor model is shown in Fig. 8.2.

B.1. The equilibrium constants of the model

$$(\text{RR}) \overset{L}{\leftrightarrow} (\text{RR})^*, \quad L = \frac{[(RR)^*]}{[(RR)]}$$

$$\text{A} + (\text{RR}) \overset{K}{\leftrightarrow} (\text{RR})_\text{A}, \quad K = \frac{[(\text{RR})\text{A}]}{[\text{A}][(\text{RR})]}$$

$$(\text{RR})_\text{A} \overset{\alpha L}{\longleftrightarrow} (\text{RR}^*{}_\text{A}), \; \text{A} + (\text{RR})^* \overset{\alpha K}{\longleftrightarrow} (RR)^*{}_\text{A}, \; \alpha = \frac{[\text{RR}^*{}_\text{A}][(\text{RR})]}{[(\text{RR})^*][(\text{RR})_\text{A}]}$$

$$\text{A} + (\text{RR})_\text{A} \overset{\mu K}{\longleftrightarrow} (\text{RR})_{\text{AA}}, \quad \mu = \frac{[(\text{RR})_{\text{AA}}][(\text{RR})]}{[(\text{RR})_\text{A}][(\text{RR})_\text{A}]}$$

$$(\text{RR})_{\text{AA}} \overset{\alpha\beta L}{\longleftrightarrow} (\text{RR})^*{}_{\text{AA}}, \; \text{A} + (\text{RR})^*{}_\text{A} \overset{\mu\beta K}{\longleftrightarrow} (\text{RR})^*{}_{\text{AA}}, \; \beta = \frac{[(\text{RR})^*{}_{\text{AA}}][(\text{RR})_\text{A}]}{[(\text{RR})_{\text{AA}}][(\text{RR})^*{}_\text{A}]}$$

B.2. Functional response: Fraction of active receptors

$$f = \frac{[(\text{RR})]_{\text{Active}}}{[(\text{RR})]_\text{T}} = \frac{a_1 + a_2[\text{A}] + a_3[\text{A}]^2}{a_4 + a_5[\text{A}] + [\text{A}]^2}$$

where

$$[(\mathrm{RR})]_{\mathrm{Active}} = \left[(\mathrm{RR})^*\right] + \left[(\mathrm{RR})^*{}_{\mathrm{A}}\right] + \left[(\mathrm{RR})^*{}_{\mathrm{AA}}\right]$$

$$[(\mathrm{RR})]_{\mathrm{T}} = [(\mathrm{RR})] + \left[(\mathrm{RR})_{\mathrm{A}}\right] + \left[(\mathrm{RR})_{\mathrm{AA}}\right] + \left[(\mathrm{RR})^*\right] + \left[(\mathrm{RR})^*{}_{\mathrm{A}}\right] + \left[(\mathrm{RR})^*{}_{\mathrm{AA}}\right]$$

and

$$a_1 = \frac{L}{K^2\mu(1+\alpha\beta L)}$$

$$a_2 = \frac{\alpha L}{K\mu(1+\alpha\beta L)}$$

$$a_3 = \frac{\alpha\beta L}{1+\alpha\beta L}$$

$$a_4 = \frac{1+L}{K^2\mu(1+\alpha\beta L)}$$

$$a_5 = \frac{1+\alpha L}{K\mu(1+\alpha\beta L)}$$

B.3. The asymptotes of the functional curves: Extracting pharmacological information from geometric descriptors

- Left asymptote: Basal activity

$$\underset{\text{for}\,[\mathrm{A}]=0}{f} = \frac{a_1}{a_4} = \frac{1}{1+\frac{1}{L}}$$

- Right asymptote: Efficacy

$$\lim f_{[\mathrm{A}]\to\infty} = a_3 = \frac{1}{1+\frac{1}{\alpha\beta L}}$$

APPENDIX C. THE METABOTROPIC GLUTAMATE RECEPTOR MODEL

The metabotropic glutamate receptor model is shown in Fig. 8.3.

C.1. The equilibrium constants of the model for the VFT domain

$$\mathrm{OO} \xleftrightarrow{2X_1} \mathrm{OC}, \quad 2X_1 = \frac{[\mathrm{OC}]}{[\mathrm{OO}]}$$

$$\mathrm{OC} \overset{\frac{X_2}{2}}{\longleftrightarrow} \mathrm{CC}, \quad \frac{X_2}{2} = \frac{[\mathrm{CC}]}{[\mathrm{OC}]}$$

$$\mathrm{A} + \mathrm{OO} \overset{\frac{K_1}{2}}{\longleftrightarrow} \mathrm{O_AO}, \quad \frac{K_1}{2} = \frac{[\mathrm{OO}][\mathrm{A}]}{[\mathrm{O_AO}]}$$

$$\mathrm{O_AO} \overset{Y_1}{\leftrightarrow} \mathrm{O_AC}, \quad Y_1 = \frac{[\mathrm{O_AC}]}{[\mathrm{O_AO}]}$$

$$\mathrm{O_AO} \overset{Y_2}{\leftrightarrow} \mathrm{C_AO}, \quad Y_2 = \frac{[\mathrm{C_AO}]}{[\mathrm{O_AO}]}$$

$$\mathrm{O_AC} \overset{Y_3}{\leftrightarrow} \mathrm{C_AC}, \quad Y_3 = \frac{[\mathrm{C_AC}]}{[\mathrm{O_AC}]}$$

$$\mathrm{C_AO} \overset{Y_4}{\leftrightarrow} \mathrm{C_AC}, \quad Y_4 = \frac{[\mathrm{C_AC}]}{[\mathrm{C_AO}]}$$

$$\mathrm{A} + \mathrm{O_AO} \overset{2K_2}{\longleftrightarrow} \mathrm{O_AO_A}, \quad 2K_2 = \frac{[\mathrm{O_AO}][\mathrm{A}]}{[\mathrm{O_AO_A}]}$$

$$\mathrm{O_AO_A} \overset{2Y_5}{\longleftrightarrow} \mathrm{O_AC_A}, \quad 2Y_5 = \frac{[\mathrm{O_AC_A}]}{[\mathrm{O_AO_A}]}$$

$$\mathrm{O_AC_A} \overset{\frac{Y_6}{2}}{\longleftrightarrow} \mathrm{C_AC_A}, \quad \frac{Y_6}{2} = \frac{[\mathrm{C_AC_A}]}{[\mathrm{O_AC_A}]}$$

C.2. The equilibrium constants of the model for the HD

The equilibrium between inactive (RR) and active (RR*) HDs

$$\mathrm{RR} \overset{2L}{\longleftrightarrow} \mathrm{RR}^*$$

where R and R* stand for the inactive and active HDs within the dimer, respectively, and $2L = [\mathrm{RR}^*]/[\mathrm{RR}]$ is the macroscopic equilibrium constant for the equilibrium between HD dimer states.

The equilibrium constant L depends on the state of the VFT domain. Thus, three apparent constants for the equilibrium between inactive (RR) and active (RR*) HDs are defined:

$$2L_1 = \frac{\left[\mathrm{RR}^*_{\mathrm{OO_T}}\right]}{\left[\mathrm{RR_{OO_T}}\right]}, \quad 2L_2 = \frac{\left[\mathrm{RR}^*_{\mathrm{OC_T}}\right]}{\left[\mathrm{RR_{OC_T}}\right]}, \quad 2L_3 = \frac{\left[\mathrm{RR}^*_{\mathrm{CC_T}}\right]}{\left[\mathrm{RR_{CC_T}}\right]}$$

where $\left\{\left[\mathrm{RR}^*_{\mathrm{OO_T}}\right], \left[\mathrm{RR}^*_{\mathrm{OC_T}}\right], \text{and} \left[\mathrm{RR}^*_{\mathrm{CC_T}}\right]\right\}$ and $\{[\mathrm{RR_{OO_T}}], [\mathrm{RR_{OC_T}}],$ and $[\mathrm{RR_{CC_T}}]\}$ stand for total active and inactive HD connected to open–open (OO), open–closed (OC), and closed–closed (CC) VFT, respectively.

C.3. Functional response: Fraction of active receptors

$$f = \frac{[(\mathrm{RR})]_{\mathrm{Active}}}{[(\mathrm{RR})]_{\mathrm{T}}} = 2\frac{c_1 + c_2[\mathrm{A}] + c_3[\mathrm{A}]^2}{c_4 + c_5[\mathrm{A}] + c_6[\mathrm{A}]^2} = 2\frac{a_1 + a_2 + [\mathrm{A}] + a_3[\mathrm{A}]^2}{a_4 + a_5[\mathrm{A}] + [\mathrm{A}]^2}$$

where

$$[(\mathrm{RR})]_{\mathrm{Active}} = \left[\mathrm{RR}^*_{\mathrm{OO_T}}\right] + \left[\mathrm{RR}^*_{\mathrm{OC_T}}\right] + \left[\mathrm{RR}^*_{\mathrm{CC_T}}\right]$$

$$[(\mathrm{RR})]_{\mathrm{T}} = \left[\mathrm{RR}^*_{\mathrm{OO_T}}\right] + [\mathrm{RR}_{\mathrm{OO_T}}] + \left[\mathrm{RR}^*_{\mathrm{OC_T}}\right] + [\mathrm{RR}_{\mathrm{OC_T}}] + \left[\mathrm{RR}^*_{\mathrm{CC_T}}\right] + [\mathrm{RR}_{\mathrm{CC_T}}]$$

$$c_1 = K_1K_2(L_1 + 2L_2X_1 + L_3X_1X_2)$$
$$c_2 = 2K_2(L_1 + L_2(Y_1 + Y_2) + L_3Y_2Y_4)$$
$$c_3 = L_1 + 2L_2Y_5 + L_3Y_5Y_6$$
$$c_4 = K_1K_2(2L_1 + 1 + 2(2L_2 + 1)X_1 + (2L_3 + 1)X_1X_2)$$
$$c_5 = 2K_2(2L_1 + 1 + (2L_2 + 1)(Y_1 + Y_2) + (2L_3 + 1)Y_2Y_4)$$
$$c_6 = 2L_1 + 1 + 2(2L_2 + 1)Y_5 + (2L_3 + 1)Y_5Y_6; \text{ and } a_i = \frac{c_i}{c_6} \quad \text{for } i = 1 - 6$$

C.4. The asymptotes of the functional curves: Extracting pharmacological information from geometric descriptors

- Left asymptote: Basal activity

$$\underset{\text{for }[\mathrm{A}]=0}{f} = 2\frac{a_1}{a_4} = \frac{1}{1 + \frac{1}{2}\frac{1+2X_1+X_1X_2}{L_1+2L_2X_1+L_3X_1X_2}}$$

- Right asymptote: Efficacy

$$\lim f_{\mathrm{R}^*[\mathrm{A}]\to\infty} = 2a_3 = \frac{1}{1 + \frac{1}{2}\frac{1+2Y_5+Y_5Y_6}{L_1+2L_2Y_5+L_3Y_5Y_6}}$$

APPENDIX D. THE ASYMMETRIC/SYMMETRIC THREE-STATE RECEPTOR MODEL

The asymmetric/symmetric three-state receptor model is shown in Fig. 8.4.

D.1. The equilibrium constants of the model

$$\mathrm{RR} \xleftrightarrow{2X_1} \mathrm{R^*R},\quad 2X_1 = \frac{[\mathrm{R^*R}]}{[\mathrm{RR}]}$$

$$\mathrm{R^*R} \xleftrightarrow{\frac{X_2}{2}} \mathrm{R^*R^*},\quad \frac{X_2}{2} = \frac{[\mathrm{R^*R^*}]}{[\mathrm{R^*R}]}$$

$$\mathrm{A} + \mathrm{RR} \xleftrightarrow{\frac{K_1}{2}} \mathrm{R_AR},\quad \frac{K_1}{2} = \frac{[\mathrm{RR}][\mathrm{A}]}{[\mathrm{R_AR}]}$$

$$\mathrm{R_AR} \xleftrightarrow{Y_1} \mathrm{R_AR^*},\quad Y_1 = \frac{[\mathrm{R_AR^*}]}{[\mathrm{R_AR}]}$$

$$\mathrm{R_AR} \xleftrightarrow{Y_2} \mathrm{R^*{}_AR},\quad Y_2 = \frac{[\mathrm{R^*{}_AR}]}{[\mathrm{R_AR}]}$$

$$\mathrm{R_AR^*} \xleftrightarrow{Y_3} \mathrm{R^*{}_AR^*},\quad Y_3 = \frac{[\mathrm{R^*{}_AR^*}]}{[\mathrm{R_AR^*}]}$$

$$\mathrm{R^*{}_AR} \xleftrightarrow{Y_4} \mathrm{R^*{}_AR^*};\quad Y_4 = \frac{[\mathrm{R^*{}_AR^*}]}{[\mathrm{R^*{}_AR}]}$$

$$\mathrm{A} + \mathrm{R_AR} \xleftrightarrow{2K_2} \mathrm{R_AR_A},\quad 2K_2 = \frac{[\mathrm{R_AR}][\mathrm{A}]}{[\mathrm{R_AR_A}]}$$

$$\mathrm{R_AR_A} \xleftrightarrow{2Y_5} \mathrm{R^*{}_AR_A},\quad 2Y_5 = \frac{[\mathrm{R^*{}_AR_A}]}{[\mathrm{R_AR_A}]}$$

$$\mathrm{R^*{}_AR_A} \xleftrightarrow{\frac{Y_6}{2}} \mathrm{R^*{}_AR^*{}_A},\quad \frac{Y_6}{2} = \frac{[\mathrm{R^*{}_AR^*{}_A}]}{[\mathrm{R^*{}_AR_A}]}$$

D.2. Functional response through the G protein-mediated signaling pathway: Fraction of active receptors

$$f_{\mathrm{R^*R}} = \frac{[\mathrm{R^*R}] + [\mathrm{R^*{}_AR}] + [\mathrm{R_AR^*}] + [\mathrm{R^*{}_AR_A}]}{[\mathrm{R_t}]} = 2\frac{c_1 + c_2[\mathrm{A}] + c_3[\mathrm{A}]^2}{c_4 + c_5[\mathrm{A}] + c_6[\mathrm{A}]^2}$$

$$= 2\frac{a_1 + a_2[\mathrm{A}] + a_3[\mathrm{A}]^2}{a_4 + a_5[\mathrm{A}] + [\mathrm{A}]^2}$$

where

$$[\mathrm{R}_t] = [\mathrm{RR}] + [\mathrm{R^*R}] + [\mathrm{R^*R^*}] + [\mathrm{R_AR}] + [\mathrm{R^*{}_AR}] + [\mathrm{R_AR^*}] + [\mathrm{R^*{}_AR^*}] + [\mathrm{R_AR_A}] + [\mathrm{R^*{}_AR_A}] + [\mathrm{R^*{}_AR^*{}_A}]$$

$$
\begin{aligned}
c_1 &= K_1K_2X_1\\
c_2 &= K_2(Y_1+Y_2)\\
c_3 &= Y_5\\
c_4 &= K_1K_2(1+2X_1+X_1X_2)\\
c_5 &= 2K_2(1+Y_1+Y_2+Y_1Y_3)\\
c_6 &= 1+2Y_5+Y_5Y_6 \quad \text{and} \quad a_i=\frac{c_i}{c_6} \quad \text{for} \quad i=1-6
\end{aligned}
$$

D.2.1 The asymptotes of the functional curves: Extracting pharmacological information from geometric descriptors

- Left asymptote: Basal activity

$$
\underset{[\mathrm{A}]=0}{f_{\mathrm{R^*R}}} = 2\frac{a_1}{a_4} = \frac{1}{1+\frac{1}{2}\left(\frac{1}{X_1}+X_2\right)}
$$

- Right asymptote: Efficacy

$$
\underset{[\mathrm{A}]\to+\infty}{f_{\mathrm{R^*R}}} = 2a_3 = \frac{1}{1+\frac{1}{2}\left(\frac{1}{Y_5}+Y_6\right)}
$$

D.3. Functional response through the G protein-independent signaling pathway

$$
\begin{aligned}
f_{\mathrm{R^*R^*}} &= \frac{[\mathrm{R^*R^*}]+[\mathrm{R^*{}_AR^*}]+[\mathrm{R^*{}_AR^*{}_A}]}{[\mathrm{R_t}]} = 2\frac{d_1+d_2[\mathrm{A}]+d_3[\mathrm{A}]^2}{d_4+d_5[\mathrm{A}]+d_6[\mathrm{A}]^2}\\
&= 2\frac{b_1+b_2[\mathrm{A}]+b_3[\mathrm{A}]^2}{b_4+b_5[\mathrm{A}]+[\mathrm{A}]^2}
\end{aligned}
$$

where

$$
\begin{aligned}
[R_t] = {} & [RR]+[R^*R]+[R^*R^*]+[R_AR]+[R^*{}_AR]+[R_AR^*]\\
& +[R^*{}_AR^*]+[R_AR_A]+[R_AR^*{}_A]+[R^*{}_AR^*{}_A]
\end{aligned}
$$

$$
\begin{aligned}
d_1 &= K_1K_2X_1X_2\\
d_2 &= 2K_2Y_1Y_3\\
d_3 &= Y_5Y_6\\
d_4 &= 2K_1K_2(1+2X_1+X_1X_2)\\
d_5 &= 4K_2(1+Y_1+Y_2+Y_1Y_3)\\
d_6 &= 2(1+2Y_5+Y_5Y_6) \quad \text{and} \quad b_i=\frac{d_i}{d_6} \quad \text{for} \quad i=1-6
\end{aligned}
$$

- Left asymptote: Basal activity

$$f_{\substack{R^*R^*\\ [A]=0}} = 2\frac{b_1}{b_4} = \frac{1}{1 + \frac{2}{X_2} + \frac{1}{X_1 X_2}}$$

- Right asymptote: Efficacy

$$f_{\substack{R^*R^*\\ [A]\to+\infty}} = 2b_3 = \frac{1}{1 + \frac{2}{Y_6} + \frac{1}{Y_5 Y_6}}$$

REFERENCES

1. Monod J, Jacob F. General conclusions: teleonomic mechanisms in cellular metabolism, growth and differentiation. *Cold Spring Harb Symp Quant Biol* 1961;**26**:389–401.
2. Monod J, Changeux J, Jacob F. Allosteric proteins and cellular control systems. *J Mol Biol* 1963;**6**:306–29.
3. Monod J, Wyman J, Changeux JP. On the nature of allosteric transitions: a plausible model. *J Mol Biol* 1965;**12**:88–118.
4. Gregory KJ, Sexton PM, Christopoulos A. Overview of receptor allosterism. *Curr Protoc Pharmacol* 2010;**1**:1.21.1–1.21.34.
5. Kenakin TP. '7TM receptor allostery: putting numbers to shapeshifting proteins. *Trends Pharmacol Sci* 2009;**30**:460–9.
6. Christopoulos A, Kenakin T. G protein-coupled receptor allosterism and complexing. *Pharmacol Rev* 2002;**54**:323–74.
7. Smith NJ, Milligan G. Allostery at G protein-coupled receptor homo- and heteromers: uncharted pharmacological landscapes. *Pharmacol Rev* 2010;**62**:701–25.
8. Hall DA. Modeling the functional effects of allosteric modulators at pharmacological receptors: an extension of the two-state model of receptor activation. *Mol Pharmacol* 2000;**58**:1412–23.
9. Stockton JM, Birdsall NJ, Burgen AS, Hulme EC. Modification of the binding properties of muscarinic receptors by gallamine. *Mol Pharmacol* 1983;**23**:551–7.
10. Tucek S, Proska J. Allosteric modulation of muscarinic acetylcholine receptors. *Trends Pharmacol Sci* 1995;**16**:205–12.
11. Waelbroeck M. Identification of drugs competing with d-tubocurarine for an allosteric site on cardiac muscarinic receptors. *Mol Pharmacol* 1994;**46**:685–92.
12. Lazareno S, Birdsall NJ. Detection, quantitation, and verification of allosteric interactions of agents with labeled and unlabeled ligands at G protein-coupled receptors: interactions of strychnine and acetylcholine at muscarinic receptors. *Mol Pharmacol* 1995;**48**:362–78.
13. Proska J, Tucek S. Competition between positive and negative allosteric effectors on muscarinic receptors. *Mol Pharmacol* 1995;**48**:696–702.
14. Hoare SR, Strange PG. Regulation of D2 dopamine receptors by amiloride and amiloride analogs. *Mol Pharmacol* 1996;**50**:1295–308.
15. Karlin A. On the application of "a plausible model" of allosteric proteins to the receptor for acetylcholine. *J Theor Biol* 1967;**16**:306–20.
16. Thron CD. On the analysis of pharmacological experiments in terms of an allosteric receptor model. *Mol Pharmacol* 1973;**9**:1–9.
17. Colquhoun D. The relationship between classical and cooperative models for drug action. In: Rang HP, editor. *A symposium on drug receptors*. Baltimore: University Park Press; 1973. p. 149–82.

18. Leff P. The two-state model of receptor activation. *Trends Pharmacol Sci* 1995;**16**:89–97.
19. Leach K, Sexton PM, Christopoulos A. Allosteric GPCR modulators: taking advantage of permissive receptor pharmacology. *Trends Pharmacol Sci* 2007;**28**:382–9.
20. Franco R, Casadó V, Mallol J, Ferré S, Fuxe K, Cortés A, et al. Dimer-based model for heptaspanning membrane receptors. *Trends Biochem Sci* 2005;**30**:360–6.
21. Franco R, Casadó V, Mallol J, Ferrada C, Ferré S, Fuxe K, et al. The two-state dimer receptor model: a general model for receptor dimers. *Mol Pharmacol* 2006;**69**:1905–12.
22. Giraldo J. On the fitting of binding data when receptor dimerization is suspected. *Br J Pharmacol* 2008;**155**:17–23.
23. Pin JP, Galvez T, Prezeau L. Evolution, structure, and activation mechanism of family 3/C G-protein-coupled receptors. *Pharmacol Ther* 2003;**98**:325–54.
24. Pin JP, Kniazeff J, Goudet C, Durroux T, Rondard P, Prezeau L. Metabotropic glutamate receptors: a paradigm of structural and functional receptor complexity. In: Giraldo J, Pin JP, editors. *G protein-coupled receptors: from structure to function*. Cambridge: Royal Society of Chemistry; 2011. p. 232–54.
25. Rondard P, Liu J, Huang S, Malhaire F, Vol C, Pinault A, et al. Coupling of agonist binding to effector domain activation in metabotropic glutamate-like receptors. *J Biol Chem* 2006;**281**:24653–61.
26. Hlavackova V, Goudet C, Kniazeff J, Zikova A, Maurel D, Vol C, et al. Evidence for a single heptahelical domain being turned on upon activation of a dimeric GPCR. *EMBO J* 2005;**24**:499–509.
27. Damian M, Martin A, Mesnier D, Pin JP, Baneres JL. Asymmetric conformational changes in a GPCR dimer controlled by G-proteins. *EMBO J* 2006;**25**:5693–702.
28. Filizola M, Wang SX, Weinstein H. Dynamic models of G-protein coupled receptor dimers: indications of asymmetry in the rhodopsin dimer from molecular dynamics simulations in a POPC bilayer. *J Comput Aided Mol Des* 2006;**20**:405–16.
29. Bayburt TH, Leitz AJ, Xie G, Oprian DD, Sligar SG. Transducin activation by nanoscale lipid bilayers containing one and two rhodopsins. *J Biol Chem* 2007;**282**:14875–81.
30. White JF, Grodnitzky J, Louis JM, Trinh LB, Shiloach J, Gutierrez J, et al. Dimerization of the class A G protein-coupled neurotensin receptor NTS1 alters G protein interaction. *Proc Natl Acad Sci USA* 2007;**104**:12199–204.
31. Rovira X, Roche D, Serra J, Kniazeff J, Pin JP, Giraldo J. Modeling the binding and function of metabotropic glutamate receptors. *J Pharmacol Exp Ther* 2008;**325**:443–56.
32. Brock C, Oueslati N, Soler S, Boudier L, Rondard P, Pin JP. Activation of a dimeric metabotropic glutamate receptor by inter-subunit rearrangement. *J Biol Chem* 2007; **282**:33000–8.
33. Rovira X, Pin JP, Giraldo J. The asymmetric/symmetric activation of GPCR dimers as a possible mechanistic rationale for multiple signalling pathways. *Trends Pharmacol Sci* 2010;**31**:15–21.
34. Brea J, Castro M, Giraldo J, Lopez-Gimenez JF, Padin JF, Quintian F, et al. Evidence for distinct antagonist-revealed functional states of 5-hydroxytryptamine(2A) receptor homodimers. *Mol Pharmacol* 2009;**75**:1380–91.
35. Mancia F, Assur Z, Herman AG, Siegel R, Hendrickson WA. Ligand sensitivity in dimeric associations of the serotonin 5HT2c receptor. *EMBO Rep* 2008;**9**:363–9.
36. Pin JP, Kniazeff J, Binet V, Liu J, Maurel D, Galvez T, et al. Activation mechanism of the heterodimeric GABA(B) receptor. *Biochem Pharmacol* 2004;**68**:1565–72.
37. Xu H, Staszewski L, Tang H, Adler E, Zoller M, Li X. Different functional roles of T1R subunits in the heteromeric taste receptors. *Proc Natl Acad Sci USA* 2004;**101**:14258–63.
38. Goudet C, Kniazeff J, Hlavackova V, Malhaire F, Maurel D, Acher F, et al. Asymmetric functioning of dimeric metabotropic glutamate receptors disclosed by positive allosteric modulators. *J Biol Chem* 2005;**280**:24380–5.

39. Han Y, Moreira IS, Urizar E, Weinstein H, Javitch JA. Allosteric communication between protomers of dopamine class A GPCR dimers modulates activation. *Nat Chem Biol* 2009;**5**:688–95.
40. Sun Y, Huang J, Xiang Y, Bastepe M, Juppner H, Kobilka BK, et al. Dosage-dependent switch from G protein-coupled to G protein-independent signaling by a GPCR. *EMBO J* 2007;**26**:53–64.
41. Kniazeff J, Bessis AS, Maurel D, Ansanay H, Prezeau L, Pin JP. Closed state of both binding domains of homodimeric mGlu receptors is required for full activity. *Nat Struct Mol Biol* 2004;**11**:706–13.
42. Rovira X, Vivo M, Serra J, Roche D, Strange PG, Giraldo J. Modelling the interdependence between the stoichiometry of receptor oligomerization and ligand binding for a coexisting dimer/tetramer receptor system. *Br J Pharmacol* 2009;**156**:28–35.
43. Giraldo J. Empirical models and Hill coefficients. *Trends Pharmacol Sci* 2003;**24**:63–5.

CHAPTER NINE

Chemokine Receptor Oligomerization and Allostery

Bryan Stephens, Tracy M. Handel
Skaggs School of Pharmacy and Pharmaceutical Science, University of California, San Diego, La Jolla, California, USA

Contents

Abstract

Oligomerization of chemokine receptors has been reported to influence many aspects of receptor function through allosteric communication between receptor protomers. Allosteric interactions within chemokine receptor hetero-oligomers have been shown to cause negative cooperativity in the binding of chemokines and to inhibit receptor activation in the case of some receptor pairs. Other receptor pairs can cause enhanced

Progress in Molecular Biology and Translational Science, Volume 115
ISSN 1877-1173
http://dx.doi.org/10.1016/B978-0-12-394587-7.00009-9

signaling and even activate entirely new, hetero-oligomer-specific signaling complexes and responses downstream of receptor activation. Many mechanisms contribute to these effects including direct allosteric coupling between the receptors, G protein-mediated allostery, G protein stealing, ligand sequestration, and recruitment of new intracellular proteins by exposing unique binding interfaces on the oligomerized receptors. These effects present both challenges as well as exciting opportunities for drug discovery. One of the most difficult challenges will involve determining if and when hetero-oligomers versus homomeric receptors are involved in specific disease states.

1. INTRODUCTION

The chemokine family of G protein-coupled receptors (GPCRs) and their protein ligands control the migration, activation, differentiation, and survival of leukocytes in many normal physiological contexts including development, hematopoiesis, immune surveillance, and inflammation. However, inappropriate expression, regulation or exploitation of these proteins contributes to a wide spectrum of inflammatory and autoimmune diseases, cancer, heart disease, and HIV, making chemokine receptors prime targets for therapeutic intervention.[1–4] Understanding the molecular details that control chemokine receptor interactions with ligands and their signaling responses should contribute to drug discovery efforts, and add to the list of approved therapeutics, which now include the CCR5 HIV entry inhibitor, Maraviroc, and the CXCR4-targeted stem cell mobilizer, Mozobil.

Approximately 45 chemokines have been identified in humans and are classified into four families (CC, CXC, XC, and CX3C) on the basis of the pattern of conserved cysteine residues.[4] The majority of the ligands are secreted in response to inflammatory signals while others are constitutively produced and involved in homeostatic processes such as lymphopoiesis and immune cell patrol of abnormal physiology. There are 22 known human receptors, most of which couple to heterotrimeric Gαi protein complexes. Four of the receptors (D6, DARC, CCX-CCKR1, and CXCR7) are classified as "atypical receptors" that lack canonical DRY boxes and consequently do not signal through Gαi. Instead they have scavenging, decoy, transport, presentation, and other accessory functions.[5,6] CXCR7 has also been reported to be a β-arrestin-biased signaling receptor[7,8] although a recent report suggests some signaling through Gαi in astrocytes and glioma cells.[9]

A subset of the receptors (CXCR4, CXCR6, CCR6, CCR8, and CCR9) have only one known ligand while most have multiple ligands

(~11 in the case of CCR1, ~12 for CCR3). The atypical receptors (DARC, D6) and virally encoded receptors (e.g., US28) tend to be particularly promiscuous with respect to ligand recognition. Similarly, many of the ligands bind multiple receptors making for a complex network of interactions, just considering the receptors and ligands alone (see Refs. 4,10 for an up-to-date matrix of the chemokine receptors and the ligands that they bind). This promiscuous pairing of ligands and receptors initially gave rise to the notion that there is significant redundancy built into the chemokine system for robustness of the immune response.[11,12] Furthermore, redundancy has been used as a potential explanation for the failure of drug candidates targeting a given receptor for the treatment of specific diseases.[10,13] However, there are reasons to believe that the system is not as redundant as initially believed,[14] and mechanisms for regulation and fine-tuning of signaling responses are beginning to emerge.

Initially, different spatial and temporal patterns of expression of chemokines and receptors were hypothesized to impose some level of functional non-redundancy.[14] However, there is now extensive evidence for homo- and hetero-oligomerization of chemokine receptors, as well as oligomerization of chemokine receptors with GPCRs outside the chemokine family and with non-GPCR receptors, which can modulate aspects of signaling and cause diverse functional responses, even with the same ligand. In this chapter, we provide examples of the pharmacological effects that these oligomeric interactions have been reported to have on the function of chemokine receptors compared to the receptors in (apparent) isolation. Note that we primarily use the term oligomer rather than dimer to refer to these complexes since it is not known whether they are predominantly dimers or higher-order assemblies.

2. BACKGROUND: CHEMOKINE STRUCTURE AND INTERACTIONS WITH RECEPTORS

Before delving into oligomerization and allostery, it is useful to review concepts regarding chemokine:receptor structure and interactions that prevailed prior to knowledge that they form homo- and hetero-oligomers. Much is known about chemokine structure and function from a wealth of NMR, X-ray, and mutagenesis studies, and recently, the first structure of a chemokine receptor was solved.

2.1. Chemokines have conserved tertiary structures but diverse oligomerization states

Despite their functional diversity, chemokines are small 8–12 kDa proteins with remarkably conserved tertiary structures stabilized by one to three disulfide bonds.[15] The basic ~70 residue chemokine module generally consists of a disordered N-terminus, which is a critical signaling domain tethered to a folded α/β core domain (Fig. 9.1). Some chemokines (e.g., SDF-1γ/CXCL12γ and SLC/CCL21) also have extended C-terminal domains that are thought to function in binding to glycosaminoglycans (GAGs). The two most unique chemokines (CXCL16 and fractalkine/CX3CL1) are fused to the N-terminus of a large mucin-like stalk that tethers them to the cell membrane and allows them to function as adhesion molecules when membrane bound, and as canonical chemokines after proteolytic release from the transmembrane (TM) domain.

In solution, different chemokines adopt a broad range of oligomerization states with some forming stable monomers (e.g., MCP-3/CCL7, SLC/CCL21), while others form reversible dimers (MCP-1/CCL2, IL-8/CXCL8, SDF-1/CXCL12), tetramers (PF-4/CXCL4), and polymers (MIP-1α/CCL3,

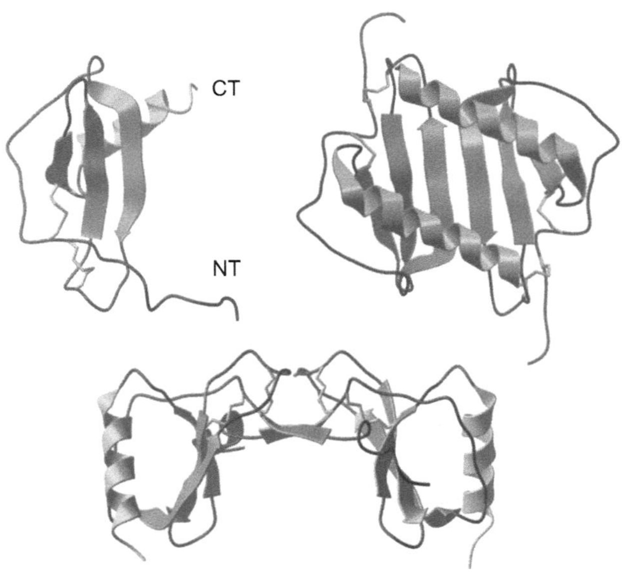

Figure 9.1 Ribbon diagrams of: (top left) a typical chemokine monomer (MCP-1/CCL2, PDB ID 1dol); (top right) a CXC chemokine dimer (IL-8/CXCL8, PDB ID 1il8); and (bottom) a CC chemokine dimer (MCP-1/CCL2, PDB ID 1dok). (See Color Insert.)

MIP-1β/CCL4, RANTES/CCL5).[2,16] There are basically two types of dimer structures—CC dimers that are characteristic of the CC chemokine family and CXC-type dimers that are formed predominantly by CXC chemokines (Fig. 9.1). These dimers form the basic substructure of the higher-order oligomers and polymers.[17,18] Furthermore, interaction of chemokines with GAGs promotes or stabilizes further oligomerization of many if not all chemokines.[19,20] Oligomerization and interactions with GAGs are important for locally sequestering chemokines on cell surfaces to prevent diffusion and facilitate the formation of chemokine gradients that help guide cell movement. GAG interactions have also been shown to facilitate transcytosis of chemokines across cells, chemokine-mediated signaling, and they can act as cofactors in promoting receptor interactions.[21–23] Nevertheless, as demonstrated using monomeric variants in bare filter transwell migration assays, the reversibility of chemokine oligomerization is necessary because it is the monomeric form that binds to the receptor with highest affinity and promotes cell migration.[24,25]

2.2. Evidence for the two-site model of chemokine:receptor binding and activation

Early mutagenesis studies of IL-8/CXCL8 from Clark–Lewis revealed that the chemokine N-terminus is a critical signaling domain, and more specifically, the prominent signaling role of the N-terminal ELR motif in a subset of CXC chemokines.[26] Subsequent mutagenesis studies of many chemokines revealed that if the N-terminus is mutated, deleted, or extended, the signaling properties of a given chemokine can be dramatically altered without significantly affecting receptor-binding affinity. For example, deletion of seven residues from MCP-1/CCL2 converts it from an agonist into a high-affinity antagonist,[27] as does addition of methionine to RANTES/CCL5 or CCL2,[28,29] or the introduction of a Pro2Gly or Lys1Arg mutation into CXCL12.[30] By contrast, deletion of the first eight residues of HCC-1/CCL14 produces a more potent agonist,[31] and deletion of 4 and 15 residues from the precursor CTAP-IIII generates the chemokines β-thromboglobulin and NAP-II/CXCL7, respectively, which have distinct biological activities.[32] Chemical and genetic modification of the N-terminus of RANTES/CCL5 has yielded superagonists and antagonists with more potent abilities than the WT chemokine to internalize the receptor CCR5, making these modified chemokines more effective in inhibiting HIV entry into cells.[33,34] Thus, the N-termini of chemokines are thought to interact with receptor-binding pockets formed primarily by the receptor TM domains (referred to as chemokine recognition site 2, CRS2 in

Ref. 4), perhaps mimicking the binding and activation of other GPCRs by small molecules ligands. In other words, this small domain seems to have the largest influence on the conformational state of the receptor and thus the signaling response.

By contrast, mutations of the chemokine core domain (everything beyond the first cysteine) generally modulate binding affinity and signaling to a proportional extent but without producing dramatic switches in pharmacology such as the conversion of agonists into antagonists. Numerous studies including the structure of a sulfated N-terminal peptide from CXCR4 in complex with SDF-1/CXCL12[35] have demonstrated that the core domain interacts with the N-terminus of the receptor (referred to as chemokine recognition site 1, CRS1 in Ref. 4), which is largely unstructured in the absence of ligand engagement. Together, these data along with evidence that the monomeric forms of chemokines promote cell migration have given rise to the concept of a two-site model of receptor activation.[36] In this model, the CRS1 binds to the chemokine core domain in an initial docking interaction. This interaction then orients the chemokine N-terminal signaling domain into the receptor CRS2, which triggers the requisite conformational change (Fig. 9.2).

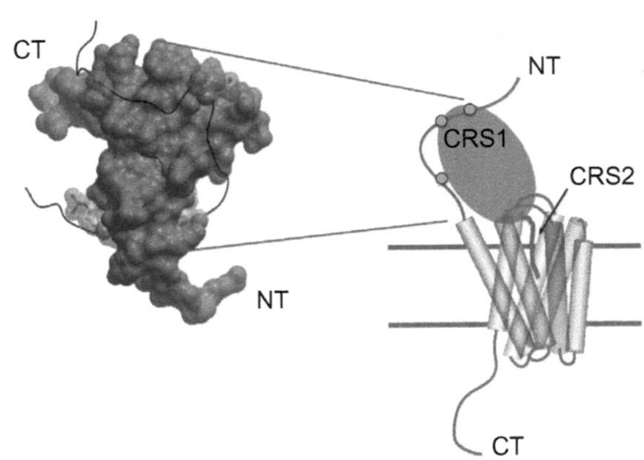

Figure 9.2 Two-site model of receptor activation. On the left is a surface topology model of CXCL12 bound to the N-terminal CRS1 of CXCR4 (black string with sulfated tyrosines side chains shown). The right illustrates the binding of the chemokine core domain to the N-terminus of the receptor (CRS1, circles represent sulfated tyrosines) and the N-terminus of the chemokine binding into the receptor helical bundle (CRS2). (See Color Insert.)

The two-site model has been supported by recent NMR studies of an *in vitro* reconstituted CXCL12:CXCR4 complex.[37] In this study, NMR signals from isotopically labeled CXCL12 were broadened beyond detection when in complex with CXCR4; however, when the small molecule antagonist AMD3100 was added, signals from the chemokine N-terminus but not the core domain became visible, presumably because the N-terminus became mobile after being displaced from the receptor by AMD3100, while the core remained bound to CRS1 (Fig. 9.3A). The fact that AMD3100 binds in the TM region of the receptor is consistent with the N-terminus of CXCL12 also binding in this region, although allosteric mechanisms of displacement cannot be ruled out.

In the interpretation of the above studies, it was assumed that the receptor would be monomeric and thus that the functionally relevant form of the complex is 1:1 chemokine:receptor. This hypothesis may well be valid but the accumulating evidence that chemokine receptors homo- and hetero-oligomerize raises the possibility of alternative stoichiometries and modes of binding of monomeric chemokines to oligomerized receptors.

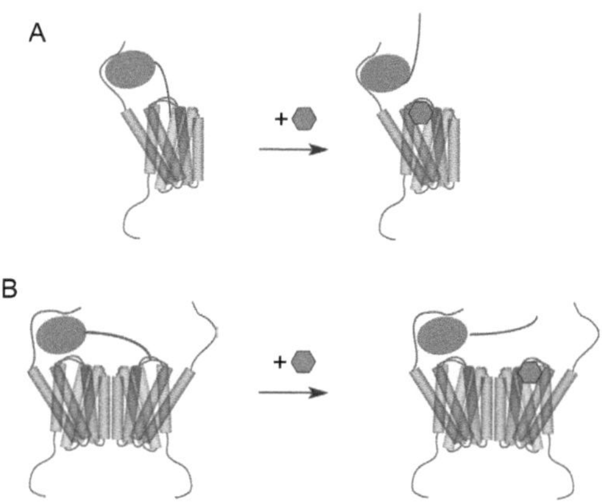

Figure 9.3 Cartoon of the two-site model where the small molecule antagonist AMD3100 (hexagon) binds CXCR4 in the TM domain CRS2 and displaces the N-terminus of SDF-1/CXCL12. Figure A illustrates the hypothetical displacement of the chemokine N-terminus in the context of a 1:1 chemokine-monomer:receptor-monomer interaction. Figure B illustrates the displacement in the context of a 1:2 chemokine-monomer:receptor-dimer interaction. Other stoichiometries are also possible such as 2:2 interactions where two chemokine monomers bind a receptor dimer or one chemokine dimer binds to a receptor dimer. (See Color Insert.)

3. EVIDENCE FOR HETERO- AND HOMO-OLIGOMERIZATION OF CHEMOKINE RECEPTORS

As implied above, for quite some time, GPCRs were assumed to function as monomeric units. Moreover, by reconstituting the β2-adrenergic receptor, the μ-opioid receptor, and rhodopsin into nanodiscs, it was demonstrated that they can function as monomers with respect to G protein coupling,[38–40] and in the case of rhodopsin, the monomer is also sufficient for rhodopsin kinase phosphorylation and arrestin binding.[41] However, a large body of evidence suggests that many GPCRs form dimers and higher-order homo- and hetero-oligomers, and chemokine receptors are no exception (Table 9.1). These oligomers may be required for assembling large functional signaling complexes and for allosteric communication within the complexes. For example, natively expressed CXCR4, CCR5, and CD4 have been identified in homogeneous microclusters, predominantly on microvilli, in many cells types.[105] Furthermore, these microclusters were identified in small trans golgi vesicles, suggesting their assembly shortly after synthesis and prior to transport to the cell membrane. The authors proposed that this localization and clustering might facilitate more precise sensing of the microenvironment during cell migration and noted that selectins and integrins, which are also important for cell migration, are located on microvilli. Whether these microclusters contain stable contact-mediated oligomerized receptors was not determined, but it seems likely given the number of studies that have demonstrated homo- and hetero-oligomerization of CXCR4 and CCR5 (Table 9.1).

While the validation of oligomerization in native tissues has yet to be convincingly demonstrated for most chemokine receptors, and the functional relevance of oligomerization on chemokine receptor activity/signaling and dynamics/trafficking is far from well understood, the majority of the chemokine receptors have been reported to homo and/or hetero-oligomerize (Table 9.1). Questions that have been probed in these studies include: (1) What is the affect of ligand binding, the nature of the ligand (agonist vs. antagonist) and the activation state of the receptor on oligomerization? (2) What is the functional significance of homo- and hetero-oligomerization? (3) Is oligomerization required for transport to the membrane surface? (4) What factors regulate receptor oligomerization? (5) Is there allosteric communication between receptors and what are the mechanisms? (6) Are G proteins involved? Answering these questions has

Table 9.1 Chemokine receptor oligomers

Receptors involved	Interesting observations	Methods used	References
Chemokine receptor homomers			
CCR2/CCR2	Homodimerization induced by CCL2	Co-IP with chemical cross-linking, divalent antibody cross-linking	42–44
	Constitutive homodimerization, conformational change caused by CCL2 stimulation	BRET	45,46
	Constitutive, CCL2 stimulation had no effect	BRET	47
	Constitutive, negative cooperativity in agonist binding	BRET	48
	Simultaneous higher-order heteromerization	BiLC-BRET	49
CCR5/CCR5	Trafficking-defective CCR5Δ32 dimerized with WT CCR5 to reduce surface expression	^{35}S pulse labeling gel analysis IMF Co-IP	50
	Homodimerization induced by CCL5	Co-IP with chemical cross-linking	44,51,52
	CCR5Δ32 mutant defect not related to normal trafficking	Flow cytometry	53
	Divalent antibodies stabilized dimers and promoted internalization	BRET	54
	Constitutive, unaffected by CCL5, BRET signal increased by divalent dimer-stabilizing antibody	BRET	55
	Constitutive	Co-IP	47,56,57

Continued

Table 9.1 Chemokine receptor oligomers—cont'd

Receptors involved	Interesting observations	Methods used	References
	TM1 and TM4 implicated in homodimer interface	FRET-based mutational analysis	58
	Homodimer interface mutations in TM1 and TM4 called into question	BRET	59
	GRK-mediated "cross-phosphorylation" across the homodimer interface	BRET	60
	Constitutive, negative cooperativity in agonist binding, evidence for G protein involvement in negative cooperativity	BRET	48
	Constitutive, homodimer-specific adaptor protein	BiFC	61
	Constitutive, homodimers depend on specific Rabs for cell surface delivery	BiFC	62
	Constitutive, homodimer-specific chaperone	BiFC	63
CXCR1/CXCR1	Constitutive	Co-IP, tr-FRET, single cell FRET, BRET, ER trapping	64
	Homodimers stabilized by CXCL8	FRET	65
CXCR2/CXCR2	Constitutive, TM3 and ICL2 implicated in homodimerization	Co-IP	66
	Disulfide involvement in homodimerization unclear	WB	67
	Constitutive	Co-IP, tr-FRET, single cell FRET, BRET, ER trapping	64
	Homodimers stabilized by CXCL8	FRET	65
CXCR3/CXCR3	Constitutive	Single cell FRET	128

CXCR4/CXCR4	Homodimerization induced by CXCL12	Co-IP with chemical cross-linking	68
	Constitutive	Co-IP, BRET, sucrose gradient centrifugation, BiFC, Bivalent ligand synthesis	61,69–73
	Constitutive, CXCL12 altered FRET signal	FRET	74
	Constitutive, CXCL12 caused conformational change in pre-formed homodimers	BRET	45
	Constitutive, homodimerization reduced by cholesterol depletion and a TM4 synthetic peptide	Single cell FRET, pbFRET	75
	Constitutive, higher-order homo-oligomerization	BRET, BiFC-BRET	76
	Constitutive, signal altered after CXCL12 incubation	PCA	77
	Constitutive, signal modified by ligand stimulation	BRET	7
	Constitutive, signal modified by ligand stimulation	BRET	78
	Simultaneous higher-order heteromerization	BiLC-BRET	49
	Homodimers reconstituted into proteoliposomes	Thermal inactivation	79
	TM 5 and 6 comprised homodimer interface	X-ray crystallography	80
	Constitutive, homodimers depend on specific Rabs for cell surface delivery	BiFC	62
	TM 5 and 6 interface supported with minor adjustments	MD simulation	81
	Sphingomyelin deficiency increased dimerization and signaling, presumably by causing accumulation of receptor in lipid rafts	FRET	82

Continued

Table 9.1 Chemokine receptor oligomers—cont'd

Receptors involved	Interesting observations	Methods used	References
CXCR7/CXCR7	Constitutive, CXCL12-modulated reporter signals	BRET, PCA	7,77,78
DARC/DARC	Constitutive	BRET	83
Chemokine receptor heteromers			
CCR2/CCR5	Required stimulation with both CCL2 and CCL5	Co-IP with chemical cross-linking	43
	CCR2/CCR5 heterodimerization proposed to result from relatively recent gene duplication and resultant high sequence similarity	Correlated mutation analysis	84
	Required co-stimulation with CCL2 and CCR5, cooperative stimulation at lowered chemokine concentrations, distinct signaling proteins recruited and cellular responses elicited	Co-IP with chemical cross-linking	44
	Stimulated by chemokines and divalent antibodies	Co-IP with chemical cross-linking, FRET	85
	No cooperative signaling, negative cooperativity in chemokine binding	BRET	47
	Constitutive, negative cooperativity in agonist binding, evidence for G protein involvement in negative cooperativity	BRET	48
	Gene conversion proposed to be the cause of heterodimerization	Comparative and phylogenetic analysis	86
	Simultaneous higher-order heteromerization and homomerization of CCR2, negative cooperativity in both chemokine and antagonist binding	BiLC-BRET	49
	Heterodimer recruited β-arrestin	GPCR-HIT	87

CCR2/CXCR4	Required stimulation with both CCL2 and CXCL12	Co-IP with chemical cross-linking	43
	Stimulated by chemokines and divalent antibodies	Co-IP with chemical cross-linking, FRET	85
	Constitutive, chemokines caused conformational change	BRET	45
	Constitutive, negative cooperativity in both chemokine and antagonist binding	BRET	88
	Simultaneous higher-order heteromerization and homomerization of CCR2, negative cooperativity in both chemokine and antagonist binding	BiLC-BRET	49
	Heterodimer recruited β-arrestin	GPCR-HIT	87
CCR5/CXCR4	Transinhibition by both agonists and antagonists	Co-IP, BRET, FRET, BiFC	43,49,56,61,89,90
	Heterodimerization seemed to be CD4 expression-dependent	IMF, Co-IP	90
	Constitutive; co-recruited into the Immunological Synapse (IS) of T cells; elicited heterodimer-specific signaling pathways when together in the IS	BRET, BiFC	56,63
	Constitutive, FRET signal modulated by chemokine ligands	FRET	89
	Simultaneous higher-order heteromerization, negative cooperativity in both chemokine and antagonist binding	BRET, BiLC-BRET	49
	Constitutive, distinct adaptor protein from CCR5 homodimer	BiFC	61
	Constitutive, heterodimers depend on specific Rabs for cell surface delivery	BiFC	62

Continued

Table 9.1 Chemokine receptor oligomers—cont'd

Receptors involved	Interesting observations	Methods used	References
CXCR1/CXCR2	Constitutive, signal disrupted by CXCL8	Co-IP, tr-FRET, single cell FRET, BRET, ER trapping	64
	Constitutive, signal altered by CXCL8	FRET	65
CXCR4/CXCR7	Constitutive, CXCR4 signaling was enhanced	Single cell FRET, pbFRET	91
	Constitutive	PCA	77
	CXCR7 impairs CXCR4-mediated G protein signaling	BRET	78
	Increased β-arrestin recruitment, enhanced CXCR4-mediated migration	Co-IP	92
CCR2/CCR5/CXCR4 Higher-order oligomer	Transinhibition by both agonists and antagonists	BiLC-BRET	49
DARC/CCR5	Constitutive, DARC inhibits CCR5 activation	BRET	83
CCX-CKR/CXCR3	Inhibits CXCR3 signaling, transinhibition of ligand binding	Single cell FRET	128
CXCR3/CXCR4	Constitutive, transinhibition by agonists in isolated membranes	Co-IP, tr-FRET, BRET, GPCR-HIT	129
Chemokine receptor heteromers with nonchemokine GPCRs			
CXCR2/α1A-adrenoceptor	α1A-AR activation by norepinephrine inhibited by the CXCR2 inverse agonist SB265610; dimerization itself unaffected by ligands	BRET	93
CCR5/μ-OR CCR5/δ-OR CCR5/κ-OR	Cooperative ligand effects	Co-IP with chemical cross-linking	94

CCR5/μ-OR	Constitutive, negative cooperativity between CCR5 and μ-OR agonists; cross-phosphorylation observed in both directions	Co-IP; bivalent ligand synthesis	95,96
CCR5/C5aR	GRK-mediated "cross-phosphorylation" across the heterodimer interface	BRET	60
CXCR2/AMPA Glu 1	Constitutive	Co-IP	66
	Dimerization reduces activation by CXCL8; CXCL8 modulates AMPA Glu 1 phosphorylation	BRET	97
CXCR2/δ-OR	CXCR2 antagonists increased δ-OR activation	Co-IP, BRET, FRET	98
CXCR4/μ-OR CXCR4/κ-OR	Competed for CXCR4 homodimer formation	FRET	74
CXCR4/δ-OR	Heterodimerization proposed to silence receptor functions	FRET	99
CCR6/BILF1 CCR7/BILF1 CCR9/BILF1 CCR10/BILF1 CXCR3/BILF1 CXCR4/BILF1 CXCR5/BILF1 CXCR7/BILF1	Constitutive	Co-IP, BRET, tr-FRET	100
CXCR4/BILF1	BILF1 inhibits CXCR4 activation	BiFC, BiLC	101

Continued

Table 9.1 Chemokine receptor oligomers—cont'd

Receptors involved	Interesting observations	Methods used	References
CXCR5/EBI2	EBI2 inhibits CXCR5 activation	FRET	102
Chemokine receptor heteromers with receptors outside the GPCR family			
CXCR4/CD4	CD4 expression was required for changes in CXCR4 homodimer FRET caused by HIV-1 coat protein gp120IIIB	FRET	74
CXCR4/TCR	Required CXCL12 stimulation; CXCL12 activated downstream signaling pathways through TCR	FRET, Co-IP	103
CCR5/CD4	Constitutive; CCR5 reported to Co-IP with CD4 to a greater extent than CXCR4	Co-IP	104

motivated the development of many methods for investigating receptor oligomerization in living cells as described in the next section.

3.1. Methods used for studying GPCR oligomerization

Many biochemical and biophysical methods have been used to investigate receptor oligomerization, and are summarized in Table 9.2, along with their pros and cons. These methods include chemical cross-linking followed by co-immunoprecipitation (Co-IP), protein fragment complementation (PFC, PCA), and many variants of fluorescence energy transfer (FRET) and bioluminescence energy transfer (BRET) techniques, including time-resolved FRET (tr-FRET, HTRF), bimolecular fluorescence/luminescence complementation BRET (BiFC/BiLC-BRET), and FRET after photobleaching (pbFRET). However, one must be cautious in the interpretation of the data and aware of the artifacts that can arise as a consequence of all of these methods. The biggest criticism is that most methods require heterologous expression of modified receptors, for example, with fluorescent or other tags for detection, or for resonance energy transfer (RET)-based experiments, which can lead to apparent oligomerization due to the unnatural high density of the expressed receptors. The tags may also inhibit interactions with intracellular proteins and alter receptor trafficking. Aggregation of receptors during Co-IP experiments can result in apparent but artificial oligomerization, and chemical cross-linking and protein fragment complementation can stabilize otherwise transient interactions between receptors. These issues have been extensively reviewed,[1,3,107] and it is now broadly appreciated that methods that identify oligomerization of receptors in native tissues are critically needed. To this end, RET assays based on fluorescent labeling of GPCR ligands rather than the receptors have been developed. However, these ligand-based methods also have limitations because the ligand can modulate the basal state of the receptor. Ligands may alter the oligomerization state, agonists will often cause receptor internalization or modulate receptor trafficking, and allostery between oligomerized receptors can result in transinhibition or cooperative binding of ligands (see Section 4), leading to a lack of correlation between results from RET experiments and receptor oligomerization. In the case of chemokine receptors, the use of labeled chemokine ligands is also complicated by their propensity to bind to and oligomerize on cell-surface GAGs, although this issue is not relevant to synthetic small molecule ligands.

All of the methods are also fraught with difficulties in quantitative interpretation. For example, although the half maximal BRET signal ($BRET_{50}$) in saturation experiments has been interpreted as a measure of receptor

Table 9.2 Methods used in studying chemokine receptor oligomerization

Method	Description	Pros and cons	Other notes	Reference(s)
Co-IP	Immunoprecipitation, electrophoresis, and immunoblotting of one receptor followed by immunoblotting of a candidate dimer partner receptor	The most well-established experimental technique used to study GPCR dimerization; requires the least technologically advanced equipment	Sometimes carried out after chemical and/or antibody cross-linking of dimerized receptors; Co-IP methods were the earliest used to establish chemokine receptor dimerization and tended to suggest that dimerization was agonist-induced	3,42
FRET	Dimerization indicated by fluorescence resonance energy transfer between fluorophores (either fluorescent proteins such as CFP and YFP or small organic fluorophores such as Cy3 and Cy5) coupled to candidate receptors	More sensitive than Co-IP methods; requires equipment capable of FRET detection	This method tends to show constitutive dimerization, but the FRET signal is often increased or reduced upon agonist stimulation, which could indicate either a change in dimerization equilibrium or conformational changes within pre-formed dimers	3,65
Single cell FRET	Specific type of FRET in which microscopy is used to collect fluorescent signal from a specific region of a single cell chosen by the experimenter	Can be used in combination with microscopy to analyze specific cellular regions (e.g., plasma membrane, ER); requires microscope capable of FRET detection	In one interesting case, this was used to demonstrate an absence of dimerization where a cell population-average method, BRET, failed to reach the same conclusions	58
tr-FRET	Time-resolved FRET. Specific type of FRET in which a donor with a long fluorescent half-life is used to detect FRET after a time delay	Increased sensitivity over simple FRET due to reduced autofluorescence; often used to detect cell-surface dimerization specifically; and requires	This method is promising for the future, as studies with nonchemokine receptors using long-lived fluorophores coupled to agonists and antagonists have	64,106

	(i.e., after the autofluorescence of the cells being assayed has subsided)	equipment capable of detecting FRET	produced interesting results with respect to ligand:GPCR oligomer stoichiometry. This method can also be used to investigate endogenous receptor oligomers on native cell types of interest	
pbFRET	FRET after photobleaching. This method relies on deducing FRET from recovered donor fluorescence signal after photobleaching the acceptor fluorophore	Can be used in combination with microscopy to analyze specific cellular regions	Usually performed in the context of confocal microscopy-based single cell FRET	75,91
BRET	Similar to FRET, except that a bioluminescent enzyme is used as the donor for resonance energy transfer, so that a chemical substrate is added to produce the observed signal	Increased signal-to-noise ratio over FRET due to the use of a bioluminescent enzyme as a donor rather than a fluorophore that must be excited with light; requires both luminescence and fluorescence detection capabilities	Tends to show constitutive dimerization; this method allows distinction between agonist-mediated disruption/formation of dimers and comformational changes within pre-formed dimers, and agonists are almost always found to cause conformational changes within dimers without affecting the dimer equilibrium	3,45
BiFC-BRET/ BiLC-BRET	Derivative of BRET in which the fluorescent protein (BiFC) and/or bioluminescent enzyme (BiLC) is split, with part of the protein placed on each candidate receptor	Allows the detection of higher-order multimerization; stabilization of split YFP derivatives upon fusion, which will lead to increase in signal unrelated to oligomerization of the actual	These methods have been used to show higher-order homomerization and heteromerization of chemokine receptors	49,76

Continued

Table 9.2 Methods used in studying chemokine receptor oligomerization—cont'd

Method	Description	Pros and cons	Other notes	Reference(s)
		fused GPCRs, may complicate interpretation of results		
GPCR-HIT	BRET-based method in which the fluorescent protein is coupled to one of the candidate GPCRs and the bioluminescent enzyme is fused to β-arrestin	Allows the identification of active, functional heteromers; use is restricted to heterodimer identification	Chemokine receptors were among those used to establish the initial validation of this method	87
PFC (BiFC/BiLC)	Either a bioluminescent enzyme or fluorescent protein is split, with part fused to each candidate receptor, and a functional fluorescent or bioluminescent protein is interpreted to result from dimerization of the candidate receptors	Can be used with microscopy to zoom in on single cells/cellular regions; stabilization of split YFP derivatives upon fusion, which will lead to increase in signal unrelated to oligomerization of the actual fused GPCRs, may complicate interpretation of results	Again, this method tends to show constitutive dimerization, with agonist stimulation often affecting the signal, which could indicate either a change in dimerization equilibrium or conformational changes within pre-formed dimers	77
Radio-ligand displacement	Ligand affinity measured by displacement of radio-labeled ligand	Can be used on live cells; can be used to obtain evidence of allosteric functional effects of receptor oligomerization, such as negative cooperativity; and does not directly establish dimerization	This method has been used several times in recent studies to demonstrate negative cooperativity/transinhibition resulting from chemokine receptor oligomerization	47–49,88

stability, in reality it is not possible to compare different receptors. This is due to the fact that many factors influence the results, including receptor expression levels and cellular localization, which are difficult to control and properly quantify between experiments. For example, the stabilizing or destabilizing effect of mutations on the oligomerization of a given receptor may be difficult to address if they alter stability to an extent that is outside the detection range of BRET, and if receptor expression levels differ but the amount of transfected DNA (the standard protocol) is used as a measure of receptor density.[108] Most methods generally reflect steady state average views of receptor oligomerization over the whole cell although some studies have focused on the cell membrane and subcellular organelles.[55,58,105] Dynamic, reversible association of receptors is also not captured by the most commonly used approaches, although exciting efforts in this direction have been reported.[109] Nevertheless, these approaches have at least provided leads that oligomerization may have functional relevance for specific receptors, allowing for follow-up studies; and despite the caveats and numerous conflicting reports, much has been learned or at least brought on radar about this fundamentally important feature of GPCRs.

3.2. Chemokine receptor homo- and hetero-oligomerization: Evidence for constitutive ligand-independent oligomer formation early after biosynthesis

Table 9.1 summarizes at least the majority of the reports related to chemokine receptor homo- and hetero-oligomerization along with some of the key observations in these publications. CCR2b was the first chemokine receptor that was shown to oligomerize.[42] In these studies, chemical cross-linking coupled with Co-IP and western blotting was initially used to demonstrate oligomerization that was induced by binding of its ligand MCP-1/CCL2. Similarly, SDF-1/CXCL12 and RANTES/CCL5 were shown to induce oligomerization of CXCR4 and CCR5, respectively.[51,68] Subsequently, many reports using RET-based methods showed constitutive association of many receptors without the requirement of ligand binding, and the current consensus is that receptors probably form in the absence of ligand binding.[4,45,55] Whether the ligand has a significant effect on stabilizing receptor oligomers remains to be seen, as the commonly used RET approaches may not be sufficiently sensitive or quantitative to detect relevant changes. However, BRET studies of CCR2 and CXCR4 homo- and hetero-oligomers suggest that chemokine ligands and small molecule inhibitors affect the conformation but not the basal number of associated receptors based on observed ligand-induced changes in the $BRET_{max}$ but not $BRET_{50}$.[45] FRET studies of CCR5/CXCR4 also suggest that the heteromers are preformed in the absence of ligand; however in these studies, stabilization of the hetero-oligomer by CCR5 ligands MIP-1α/CCL3 and RANTES/CCL5, but destabilization by SDF-1/CXCL12 was reported.[89] These data are consistent with the early chemical cross-linking/Co-IP results which suggested that chemokines can stabilize receptor dimers.[42,43,51,68,91] However, it is difficult to judge whether such changes observed in single-expression point FRET studies result from an actual change in the dimerization equilibrium or from conformational changes within constitutive dimers, illustrating a potential advantage of using BRET saturation titration curves.

The presence of basal, ligand-independent formation of chemokine receptor oligomers is consistent with the idea that oligomers form early along the biosynthetic pathway and can be detected during transport through the ER and golgi.[64,110] The first and most convincing example of this concept was demonstrated with the class C gamma-aminobutyric acid $(GABA)_B$ receptors that form obligate heterodimers: $GABA_B$-R1 requires dimerization with $GABA_B$-R2 in the ER in order to traffick to the cell surface, and although $GABA_B$-R2 can be transported to the cell surface in the absence of $GABA_B$-R1, it is not functional unless oligomerized with $GABA_B$-R1.[111,112] Along these lines, constitutive homo- and hetero-oligomerization of CXCR1 and CXCR2 was demonstrated by a combination of BRET and Co-IP, and a novel-trapping strategy showed interactions between CXCR1 homomers and CXCR1:CXCR2 heteromers in the endoplasmic reticulum (ER). In these experiments, an ER retention signal was added to the C-terminus of CXCR1 and resulted in a significant reduction in the amount of CXCR1 and CXCR2 that was translocated to the cell surface.[64] Similarly, CCR5 has been shown to oligomerize in a ligand-independent fashion, both at the plasma membrane and in ER subfractions.[55] More recent studies using both BRET and FRET show constitutive association of virtually all chemokine receptors studied,[1,3,4,45,47,58,59,90,107,110] and the assumption is, that if these hetero-oligomers are relevant in native cells, they probably form prior to reaching the cell membrane.

3.3. Crystal structures of CXCR4 reveal homodimers

In keeping with earlier biochemical studies that showed that CXCR4 sedimented as a dimer when purified in nondenaturing detergent,[69] recent crystal structures of CXCR4 have provided structural validation to the relevance of contact-mediated receptor dimerization rather than simple clustering within the range detectable by RET. In 2010, five structures were reported and in all cases showed a dimer with subunit interactions primarily between TM helices V and VI.[80] While one cannot exclude that the dimer was due to crystal contacts, the fact that all five structures showed the same dimer interface despite being in different crystal forms, suggests that they are probably not artifacts of crystallization. This data supports numerous cell-based studies that suggest that CXCR4 forms homo- and heterodimers (Table 9.1).

The structures of CXCR4 were in complex with a small molecule antagonist, It1t, and a 16-residue cyclic peptide inhibitor, CVX15. They revealed a rather large acidic binding pocket and showed that It1t bound in the minor pocket involving TM helices I, II, III, and VII while CVX15 bound in the major pocket (TM helices III–VII). These compounds interact with several acidic residues that line the pocket and are known to be involved in binding to SDF-1/CXCL12. Consistent with the two-site model, the current thinking is that the N-terminus of CXCL12 interacts with the pocket CRS2, formed by the TM helices and ECL2, a β-hairpin structure that helps shape the entry to the pocket, and has been highly implicated in chemokine interactions. No density was observed for the receptor N-terminus up to the first cysteine consistent with the hypothesis that in the absence of chemokine, the CRS1 domain is unstructured.

The observed dimers raise questions about the stoichiometry of chemokine:receptor binding in cells. Originally 1:1 chemokine-monomer: receptor-monomer complexes were assumed, but given the CXCR4 dimer structures, it is possible to envision a 1:2 chemokine-monomer: receptor-dimer complex that still conforms to the two-site model (Fig. 9.3B). This issue has yet to be resolved, but it is noteworthy that some biochemical studies suggest that only one chemokine can bind to a receptor dimer at a time (see Section 4 and Refs. 47–49,88). Furthermore, as described in Section 7, CXCL12 dimers can interact with CXCR4 to produce different downstream signals than those stimulated by CXCL12 monomers,[113] suggesting 2:2 chemokine-dimer:receptor-dimer complexes may also be functionally relevant.

Other questions related to the dimer structures include the following: (1) What is the variability and plasticity in chemokine receptor dimer interfaces? Assuming that the CXCR4 structures show a relevant dimer interface, is the TM V/VI interface always the interaction surface in CXCR4 dimers? Or are other interfaces used in higher-order oligomers or with other heteromeric interactions? (2) Is it possible to engineer non-oligomerizing receptors to answer the functional relevance of chemokine receptor oligomerization? (3) Is there allosteric coupling across the interfaces, what is the functional consequence, and what is the mechanism? (4) Do dimer interfaces make good therapeutic targets?

Answers to these questions are only beginning to emerge. With regard to questions 1, it is worth noting that one of the CXCR4 structures showed an additional interface involving TMS I and II.[80,108] Furthermore, structures of other GPCRs suggest a great deal of variability in dimer interfaces.[114] Regarding question 2, in order to engineer non-oligomerizing receptors, structural knowledge of the interface is obviously useful but even then, identifying appropriate mutations that destabilize oligomeric receptors without affecting receptor stability, folding and ligand binding may be difficult. Quantifying the effect of mutations is also nontrivial as described above and in Ref. 108. Furthermore, the extent to which the stability of oligomerized receptors is affected by intracellular G or other proteins as well as the lipid environment, is unclear. If the receptors exit in assemblies larger than dimers, as has been suggested,[49,76,101] then more than one interface may need to be simultaneously disrupted to achieve a monomeric status.

3.4. Attempts to disrupt receptor dimerization

In an early attempt to engineer a non-oligomerizing variant of CCR5, a TM I/TM IV dimer interface was predicted in a bioinformatic analysis.[58] Subsequent mutation of I52V and V150A in TMS I and IV, respectively, were reported to prevent dimerization according to both FRET and cross-linking studies, and TM peptides encompassing Ile52 and Val150 were shown to block dimerization. However, this data was subsequently contested based on BRET analysis by a different group and remains unresolved.[59] Nevertheless, regardless of the affect on dimerization, in contrast to the WT protein, the mutant CCR5 was unable to signal in calcium flux, chemotaxis, and JAK-STAT activation assays despite retention of binding affinity for RANTES/CCL5.[58] Similarly, the peptide blocked WT CCR5 receptor function. It therefore remains an open question as to whether the inhibitory effects of the TM peptide and the CCR5 mutant are through the controversial dimer disruption mechanism or through allosteric induction of nonsignaling receptor conformations.

More recent BRET studies as well as the report by Lemay[59] tend to argue for allosteric induction of nonsignaling receptor conformations. In a 1999 report from Tarasova and coworkers, multiple TM peptides from CXCR4 and CCR5 were shown to block signaling of the receptors and to inhibit HIV replication.[115] Follow-up studies using BRET suggested that the effects were due to inhibition of ligand-induced conformational changes rather than disruption of receptor dimers, since none of the peptides affected the basal BRET signal but did produce changes in the ligand-induced BRET signals.[45] The authors reasoned that the inhibitory effect of the TM peptide could best be explained by blockade of the allosteric communication between dimerized receptors. Similarly, in 2006, Wang and coworkers reported that a peptide corresponding to TM IV of CXCR4 blocked the migration of monocytes and cancer cells to CXCL12.[75] In this case, FRET between CXCR4-CFP and CXCR4-YFP was reduced by the peptide, but again changes in FRET efficiency at a single-expression level of FRET pairs are difficult to assign to reductions in actual numbers of dimers/oligomers versus conformational changes within stable complexes. Whatever the mechanism, the data overall suggest that targeting TM helices (dimer interfaces or otherwise) can be an effective strategy for chemokine receptor inhibition; the question is whether they can be specifically targeted with small molecules.

4. FUNCTIONAL EFFECTS OF CHEMOKINE RECEPTOR HETERO-OLIGOMERIZATION ON LIGAND BINDING

In a series of detailed studies, transinhibition of ligand binding between hetero-oligomerized receptors (CCR2/CCR5, CCR2/CXCR4, and CCR2/CCR5/CXCR4) was demonstrated by BRET and ligand-binding experiments. These studies made a compelling case for the importance of allostery in controlling the function of chemokine receptors, the potential impact that heterodimerization can have on drug efficacy, and provided insight into the mechanism for allosteric communication between receptor subunits. The first of these reports on the heterodimerization of CCR2 and CCR5 showed that these receptors were basally associated in the absence of chemokine agonists.[47] Furthermore, $BRET_{50}$ values, which are considered a measure of affinity (with the caveats described in Section 3.1), were similar for the CCR2 and CCR5 homomers and for the CCR2/CCR5 heteromers suggesting similar propensities for the homo- and hetero-oligomers to form. This is not surprising given the high-sequence conservation between CCR2 and CCR5, especially in their TM domains (78.2% identity, 89.4% similarity). Binding of the respective chemokine ligands had no effect on the $BRET_{50}$ but affected the $BRET_{max}$ signal, suggesting that the ligands induce conformational changes rather than changes in the number of associated receptors, similar to related studies of CCR2/CXCR4.[45] Binding of ^{125}I-MCP-1/CCL2 tracer to CCR2 was unaffected by unlabeled CCR5 ligands (MIP-1α/CCL3, MIP-1β/CCL4, and RANTES/CCL5) when CCR2 was expressed alone. Similarly, no competitive binding between ^{125}I-MIP-1β/CCL4 tracer and CCR2 ligands (MCP-1/CCL2 and MCP-2/CCL8) to CCR5 was observed when CCR5 was expressed alone. The surprising finding was that coexpression of the two receptors made the reporter ligands of one receptor susceptible to binding inhibition by the ligands of the other receptor: CCR5 ligands inhibited binding of ^{125}MCP-1/CCL2 to cells coexpressing both CCR2 and CCR5, and CCR2 ligands inhibited the binding of ^{125}MIP-1β/CCL4 to the coexpressing cells. Furthermore, the extent of the transinhibition corresponded with the approximate proportion of expressed heterodimers, suggesting some sort of negative binding allostery between the coexpressed receptors. Similar results were observed in both transfected cells as well as T lymphoblasts that naturally express both receptors. These data suggested the potential for allosteric communication between coexpressed and apparently oligomerized CCR2 and CCR5.

To further investigate the mechanism and demonstrate the allosteric nature of the binding inhibition observed for the CCR2/CCR5 hetero-oligomers, a subsequent study used "infinite dilution tracer" experiments.[48] These experiments showed that dissociation of the CCR2-specific ligand ^{125}I-MCP-1/CCL2 from cells coexpressing CCR2 and CCR5 was accelerated significantly by the CCR5-specific ligand MIP-1β/CCL4 compared to the dissociation rate from cells expressing CCR2 alone. Likewise, dissociation of ^{125}MIP-1β/CCL4 from CCR5 was accelerated by CCL2 when the cells expressed both receptors compared to CCR5 alone. Again, the results were demonstrated in both transfected cells and in T lymphoblasts suggesting the physiological relevance of the observations. Transinhibition of ligand binding was also demonstrated for CCR2/CXCR4 hetero-oligomers in transfected and primary leukocytes[88] and for CCR2/CCR5/CXCR4 multimers in transfected cells as well as primary T cells and monocytes.[49]

It has been argued that "G protein stealing" can be the source of trans-inhibition of ligand binding without the need to invoke heterodimerization.[107,116,117] The idea here is that many agonists require G protein coupling for high-affinity receptor binding. Thus, independent of receptor hetero-oligomerization, depletion of G protein due to ligation of one receptor with its agonist can result in the apparent lowering of the affinity of the other receptor for its ligand, especially if the pool of G proteins is limiting. However, the partial agonist [10–68]RANTES/CCL5, and the antagonist MET-RANTES/CCL5 were also effective in promoting dissociation of MCP-1/CCL2 from CCR2 in CCR2/CCR5 coexpressing cells. Similarly, the small molecule CXCR4 antagonist AMD3100 and the CCR2 inverse agonist TAK-779 were able to compete off the binding of chemokine from CCR2 and CXCR4, respectively, but only when the two receptors were coexpressed.[88] These data are not consistent with a G protein steal since high-affinity binding of antagonists and inverse agonists typically does not require G protein coupling. Along with the results from the infinite dilution tracer experiments, the data suggest that there is direct allosteric communication between hetero-oligomerized receptors (Fig. 9.4).

That is not to say that G proteins do not play a role in the observed trans-inhibition. Springael and coworkers showed that the addition of pertussis toxin (PTx) or Gpp(NH)p, a nonhydrolyzable analog of GTP, strongly reduced the binding of MIP-1β/CCL4 to CCR5.[48] Similarly binding of MIP-1β/CCL4 was reduced on cells expressing an R126N mutant

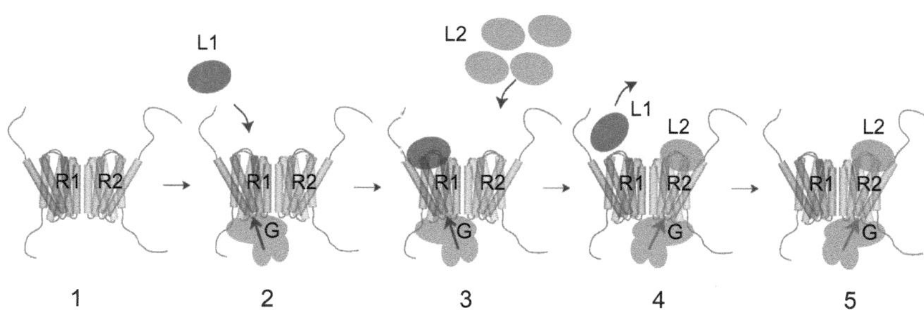

Figure 9.4 A model for the allosteric ligand-binding transinhibition of agonists from chemokine heterodimers as suggested by Springael and coworkers.[48] (1) Heterodimerization of receptor 1 (R1) and receptor 2 (R2). (2 and 3) Ligand 1 (L1) binds with high affinity to R1 when it is coupled to heterotrimeric Gαi proteins (G). The arrow illustrates allosteric coupling between the G protein and R1, which allows R1 to adopt a conformation that leads to high-affinity binding of L1. (3 and 4) Subsequent binding of ligand 2 (L2) to R2 induces a conformational change in R2 that results in the G protein interacting with R2 rather than R1, and subsequently (4 and 5) dissociation of L1 from R1 since G protein is required for high-affinity interaction of L1/R1. Note that the mechanism for transinhibition of agonist by binding of small molecule antagonists does not need to involve changes in G protein coupling but could be explained simply by allosteric communication between the two receptors.[88] (See Color Insert.)

of CCR5 that does not couple to G proteins, indicating the requirement of G proteins for high-affinity binding of CCL4 to CCR5. Although the CCR5-R126N mutant retained the ability to hetero-oligomerize as efficiently with CCR2 as WT CCR5, and CCR5-R126N did not show a dominant negative effect on signaling of coexpressed CCR5 and CCR2 as assessed by calcium flux, MIP-1β/CCL4 was no longer able to increase the dissociation rate of MCP-1/CCL2 from CCR2 in cells coexpressing both CCR2 and CCR5-R126N. Together the data suggest that the G protein is needed for allosteric communication through the dimer, at least when agonists are involved, and the authors proposed a mechanism for G protein-dependent binding transinhibition (Fig. 9.4). However, since antagonists also promote transinhibition, the requirement for G protein is likely to be first and foremost necessary for high-affinity binding of agonist, regardless of whether the G protein directly contributes to allostery across the heterodimer.

Other conclusions reached from these studies include the fact that (1) ligand binding does not need to induce the activated state of a receptor in order to inhibit ligand binding to the other receptor in the hetero complex[48,49,88]; (2) that homo and heterodimers likely interact, suggesting larger allosterically coupled arrays of chemokine receptors[49]; (3) that binding transinhibition is not likely due to steric blockade by the competing ligand since small molecules are as effective as chemokines in causing ligand dissociation from the partner receptor; (4) that a receptor heterodimer, and most probably a homodimer, can only bind a single chemokine with high affinity.[48,49,88] The latter hypothesis is interesting in light of the CXCR4 dimer structure, but further studies are obviously needed to determine chemokine:receptor stoichiometries. Likewise, the stoichiometry between chemokine receptors and downstream signaling partners including G proteins and β-arrestins, may provide key insights that reconcile some of the data.

What is particularly surprising about the results is the fact that ligands with different efficacies and sizes (inverse agonists, antagonists, and agonist variants of small both molecules and chemokines) were all capable of ligand-binding transinhibition. Whether the effect is common or specific to the receptors and ligands reported in these studies remains to be seen. However, it is worth noting that different ligands have been shown to produce different conformational changes in receptor homo- and heterodimers[45,89]; this suggests that many receptor conformations may produce cross-competition whether it be due to distortions of the binding pocket or perturbations of

G protein coupling that affect the ligand affinity of the opposing oligomeric partner. It will be interesting to see if similar findings are observed with other chemokine receptors that have been shown to heterodimerize (Table 9.1) and in what contexts.

Transinhibition of ligand binding makes sense when there is inhibition of signaling of one receptor by the ligand of the partner receptor as described in the next section. However, there are also examples where the signaling is amplified or completely altered, and how this correlates with ligand binding is an open question. As discussed in Section 5, DARC heterodimerizes with CCR5 and blocks its downstream signaling without affecting ligand binding or CCR5 internalization. The observed combination of effects has functional implications consistent with other behaviors of this atypical chemokine receptor. For example, can chemokine receptor heteromers bind different chemokines than chemokine homomers? Given the known ligand:receptor promiscuity, and the fact that proteolytic processing of chemokines can cause receptor-specificity changes, this scenario would not be difficult to imagine.

5. EFFECTS OF CHEMOKINE HETERO- AND HOMO-OLIGOMERIZATION ON SIGNALING

5.1. Transinhibition of signaling by ligands in hetero-oligomeric complexes

Early reports suggested synergy in calcium flux due to CCR2/CCR5 heterodimers[44]; however, subsequent studies failed to show cooperative calcium signaling with CCR2/CCR5 and with CCR2/CXCR4 and thus this finding remains controversial.[49,88] Cooperative signaling in cell migration was also not observed with agonists of the coexpressed receptors in transfected as well as native cells. Instead, in both functional assays, small molecule antagonists of one receptor caused inhibition of functional responses of the other receptor to which it does not bind.[49,88] Thus the CXCR4 antagonist AMD3100 inhibited not only signaling of CXCR4 in response to its ligand SDF-1/CXCL12, but it also blocked signaling of MCP-1/CCL2 to CCR2. Similarly, the CCR2 inverse agonist TAK-779 blocked signaling of SDF-1/CXCL12 in primary CD4+ lymphoblasts coexpressing both CCR2 and CXCR4.[88] These data are in line with the ligand-binding transinhibition; however, particularly interesting was the fact that the inhibition of functional responses was stronger than the binding cross-competition. The authors proposed that this might reflect allosteric functional effects across larger arrays of receptors than heterodimers.[88] Along these lines, subsequent studies showed cooperative interactions and hetero-oligomerization between CCR2, CCR5, and CXCR4 in T lymphoblasts, and similar inhibition of calcium signaling and migration of receptors by antagonists of the orthogonal receptors. This cross-competition also translated into an *in vivo* air pouch migration model in which the small molecule TAK-779 (which antagonizes both CCR5 and CCR2), blocked migration of cells to CXCL12/CXCR4.[49] The implications for drug discovery here are quite striking; in principal, it may be possible to inhibit the activity of one chemokine receptor indirectly by targeting another receptor with which it oligomerizes.

5.2. Activation of alternative signaling pathways by hetero-oligomers

In addition to inhibiting signaling, Mellado and coworkers subsequently demonstrated that heterodimerzation of CCR2 and CCR5 can produce unique signaling responses compared to the classical Gαi signals characteristic of chemokine receptors expressed in isolation.[44] Coexpression of both receptors and stimulation with their respective ligands (RANTES/CCL5 and MCP-1/CCL2) resulted in a PTx-insensitive calcium flux through Gq/11, in contrast to the normal inhibitory effect that PTx has on Gαi-mediated calcium flux when the receptors are expressed alone. Furthermore, simultaneous stimulation of the presumed hetero-oligomers by both ligands failed to cause receptor downregulation and produced a delayed and sustained activation of phosphatidyl inositol 3-kinase (PI3K). The consequence of the altered signaling was linked to more efficient adhesion instead of cell migration. A plausible interpretation was presented in which the homomers and heteromers cooperate to augment the versatility of the signaling responses, with the ligands and their concentrations controlling the formation/stability of the homo- or hetero-complexes. In the Mellado study, it was suggested that hetero-oligomers might contribute to cell adhesion and "parking," once the cells reach their destination in tissues, while the homo-oligomers promoted migration.

Subsequent studies showed the relevance of the above findings to the recruitment of CXCR4 and CCR5 into the immunological synapse of T cells, and their role in co-stimulation of the T cell receptor during activation through Gq/G11 mediated responses.[118] Similar to the above studies, rather than promoting normal Gαi-mediated migration, the recruitment of the receptors resulted in an insensitivity to chemokine gradients, enhanced adhesion to antigen-presenting cells, and promoted increased proliferation and cytokine production. These results were shown later to be due to the physical association of CXCR4 and CCR5 by BRET and Co-IP; furthermore, it was shown that CXCR4 requires CCR5 for recruitment to the immunological synapse.[56] Together this series of studies demonstrates the ability of chemokine receptor hetero-complexes to differentially signal compared to the homomeric counterparts, in this case due to specificity changes in coupling with G proteins. These studies represent particularly good examples of signaling versatility bestowed by hetero-oligomerization.

5.3. Modulation of signaling by atypical and virally encoded chemokine receptors

5.3.1 DARC/CCR5

The Duffy antigen for chemokines (DARC), an atypical receptor that does not signal through G proteins, has also been show to homo-oligomerize and to hetero-oligomerize with CCR5.[83] Coexpression of CCR5 and DARC showed a marked attenuation of chemotaxis and calcium flux in response to the CCR5 ligand RANTES/CCL5 (which also binds DARC with equal affinity) as well as to the CCR5-specific CCL3 isoform ligand, LD78β/CCL3L1. On the other hand, ligand-stimulated CCR5 was internalized to the same extent whether it was coexpressed with DARC or not, even though DARC itself does not internalize upon ligand stimulation or interact with β-arrestin in the cells used in this study. On the basis of these and other data, it was proposed that the DARC/CCR5 interaction inhibits ligand-induced CCR5 signaling by altering the affinity of CCR5 for G proteins or the responsiveness of CCR5 to its ligands, but not by altering its affinity for ligands. One could imagine, for example, that DARC induces a conformation in CCR5 that remains competent for ligand binding and internalization but does not allow CCR5 to adopt conformations required for calcium signaling and chemotaxis. DARC has been suggested to function as a "chemokine rheostat" on endothelial cells by supporting the transport, presentation, and concentration of chemokines to balance the inflammatory response.[6,119] This study suggests a mechanism by which DARC "rheostats" the function of CCR5, turning it down by blocking signaling through hetero-oligomerization.[83] Secondly, the fact that heterodimerization with DARC does not affect CCR5's high affinity for CCL5 or its ability to internalize (presumably with ligand), adds a second stage to the "dial-down" switch.

5.3.2 CXCR7/CXCR4

It was recently demonstrated that CXCR7 forms hetero-oligomers with CXCR4.[77,78,91,92] In the original report,[91] hetero-oligomerization coincided with an increased Ca^{2+} flux response to CXCL12 stimulation. Furthermore, it was shown that the time course of ERK activation was altered when CXCR7 was coexpressed in CXCR4-expressing cells. Specifically, when only CXCR4 was present, ERK was activated in a biphasic fashion, whereas when the two receptors were coexpressed, only the second delayed peak in ERK activation was observed.

In a later report, however, coexpression of CXCR7 along with CXCR4 decreased the potency of CXCL12-induced Ca^{2+} flux, though the maximal efficacy was unchanged.[78] G protein activation as measured by an ^{35}S-GTP-γS binding assay was similarly reduced in potency when the two receptors were coexpressed, and BRET between CXCR4-YFP and Gαi1-Rluc demonstrated that CXCR7 causes a conformational rearrangement within pre-coupled CXCR4- and Gαi-containing complexes. Finally, this report demonstrated that CXCR7 knockdown in T lymphocytes, which endogenously express both CXCR4 and CXCR7, led to an increased migratory response to a lower concentration (0.3 nM) of CXCL12, which was attributed to the propensity of CXCR7 to scavenge CXCL12 (discussed below). Overall, it was suggested that the effects of CXCR7 on CXCL12:CXCR4-mediated signaling cell migration was due both to allosteric modulation of CXCR4:Gαi interactions and hoarding of CXCL12 by CXCR7.

One recent report, which did not directly demonstrate dimerization, nevertheless obtained intriguing results that could involve CXCR4/CXCR7 hetero-oligomerization.[120] In this study, it was shown that both CXCL11, a chemokine ligand for CXCR7 but not CXCR4, and CCX771, a small molecule inhibitor of CXCR7, were able to inhibit CXCL12-mediate transendothelial migration (TEM), specifically in the case of migrating cells that endogenously expressed both CXCR4 and CXCR7. Since TEM is driven entirely by CXCR4, the most striking observation was that the CXCR7 ligand CCX771, was substantially more potent than the CXCR4 specific ligand AMD3100 in blocking TEM. These results have tremendous ramifications for drug discovery; for example, the authors note that CCX771 might be a particularly potent CXCR4 inhibitor in cells that express both receptors (usually cancer cells) and could provide greater selectivity than blocking CXCR4 indiscriminately with CXCR4 antagonists like AMD3100.

Finally, a recent report demonstrated that CXCR7 coexpression along with CXCR4 decreased the Gαi-mediated inhibition of cAMP production resulting from CXCL12 stimulation.[92] At the same time, coexpression of the receptors greatly increased the resting and CXCL12-induced β-arrestin recruitment to CXCR7. Interestingly, CXCL11 both reversed the decrease in Gαi activity and slightly attenuated the increased arrestin recruitment. This study also showed that CXCR7 coexpression increased ERK, p38 MAPK, and SAPK activation upon CXCL12 stimulation, suggesting a broad change in the signaling response elicited in the case of the heteromer. The authors also found that the CXCL12 response of cells in a transwell migration assay was increased when CXCR7 was coexpressed along with CXCR4. The increase in the activation of downstream signaling proteins as well as the increased chemotactic response were dependent upon β-arrestin expression.

As alluded to above, the allosterically regulated functional effects of CXCR4/CXCR7 hetero-oligomerization may be difficult to dissect from other cooperative interactions between the two receptors. In the interpretation of all of the above studies, for example, it should be noted that CXCR7 has significantly higher affinity than CXCR4 for SDF-1/CXCL12. Thus some of the effects, for instance, any increase in CXCR4's responsiveness to CXCL12 when CXCR7 is blocked, could be attributed to the inhibition of CXCR7 from binding and effectively sequestering (from CXCR4) a large proportion of the SDF-1/CXCL12 present. In this regard, it was recently demonstrated that CXCR7 can aid in the pro-migratory, pro-metastatic effects of CXCR4 by scavenging CXCL12 rather than (or in addition to) allosteric modulation of CXCR4 as suggested in the study by Levoye.[121] In this intriguing study, CXCR7 was even shown to affect the function of CXCR4 when it was expressed primarily on different populations of malignant cells compared to CXCR4. *In vivo* imaging demonstrated that CXCR7 reduced SDF-1/CXCL12 levels in the primary tumor microenvironment, which in turn reduced CXCR4 internalization and downregulation. While this scavenging function of CXCR7 would potentially limit the effects of CXCL12:CXCR4 on tumor growth, it was proposed that it allows CXCR4 to maintain responsiveness to external CXCL12 gradients that would draw metastatic cells to other tissues.[121]

Several aspects of the above observations are in line with emerging principles of chemokine receptor oligomerization. The ability of CXCR7-specific agonists as well as antagonists to alter signaling responses mediated by CXCR4 is especially compelling evidence of allosteric communication within receptor oligomers, as CXCR7 does not activate G proteins and as shown by Levoye, likely does not interfere with CXCR4 signaling simply by stealing G proteins.[78] Furthermore, the change in CXCR4-Gαi interaction as well as the increased activation of β-arrestin-mediated signaling proteins (MAPK, ERK, p38, SAPK) suggests that the CXCR4/CXCR7 heteromer is a functionally unique signaling complex. However, not all of the above data can be reconciled with a consistent story, illustrating the dependence on methods and the context dependence of the studies (e.g., the effect of the relative densities of the two receptors as well as cell background).

5.3.3 BILF1/CXCR4

Many viruses, including herpesviruses, encode GPCRs with considerable homology to chemokine receptors.[122,123] Most of these vGPCRs show significant constitutive activity although they also tend to bind numerous ligands. One of the most famous of these receptors is ORF74 from the Karposi's sarcoma-associated herpesvirus (HHV-8), which was initially identified as being the cause of the highly vascularized Karposi's sarcoma lesions in AIDS patients, and other proliferative disorders. It binds to at least 12 chemokine ligands whose activities range from inverse agonists to full agonists. On the other end of the spectrum of known ligands and relationship to chemokines, BILF1 is a GPCR encoded by the Epstein-Barr virus (EBV or HHV-4) that persists in B cells following primary infection and contributes to Burkitt's lymphoma and Hodgkin's lymphoma among other oncogenic disorders.[124] It has limited homology to chemokine receptors and currently is considered an orphan GPCR with no known ligands. This receptor seems to be involved in immune evasion by a number of mechanisms including downregulation of MHC class I receptors and inhibition of RNA-dependent protein kinase activity that would otherwise put a stop to cellular translation and therefore viral replication.

One of the more recent mechanisms discovered for BILF1 is that it heterodimerizes with a number of chemokine receptors including CXCR4.[100,101] In fact, by combining bimolecular luminescence complementation and bimolecular fluorescence complementation with BRET measurements, it was shown that heteromeric complexes between BILF1 and CXCR4 consist of the concurrent interaction of at least four GPCR subunits. BILF1 was shown to inhibit binding of SDF-1/CXCL12 to CXCR4 with the consequence of blocking chemokine-mediated signaling. Since BILF1 is a constitutively active receptor and CXCL12 requires G protein coupling for high affinity, it was hypothesized that this

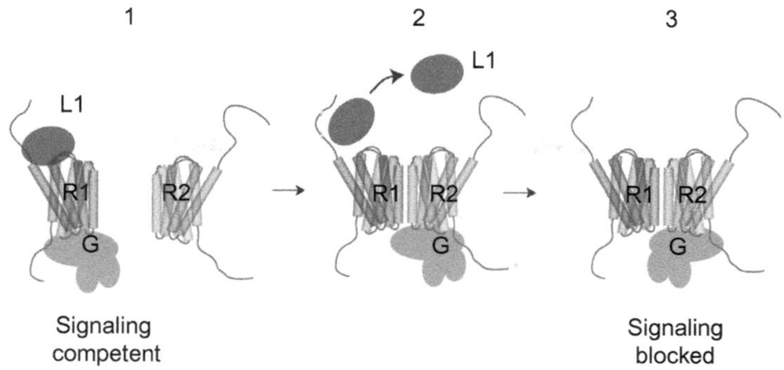

Figure 9.5 A model for inhibition of CXCR4 signaling by the EBV viral GPCR, BILF1.[101] (1) CXCR4 (R1), shown here as monomer for simplicity but probably exists as homo-oligomer, is coupled to heterotrimeric Gαi proteins (G) and therefore competent for high affinity binding and signaling in response to SDF-1/CXCL12 (L1). BILF1 is represented as R2. (2) Hetero-oligomerization of CXCR4 (R1) with BILF1 (R2) scavenges the G protein heterotrimer from CXCR4 due to its constitutive activity. The lack of G protein shifts CXCR4 into a low affinity state and CXCL12 dissociates. (3) Uncoupled, ligand-free CXCR4 R1 does not signal. (See Color Insert.)

receptor might scavenge Gαi protein from CXCR4, forcing it into a low affinity state for its ligand (Fig. 9.5). Indeed, overexpression of Gαi1 restored the ability of CXCL12 to bind and signal through CXCR4. Furthermore, a G protein uncoupled mutant of BILF1 was much less effective in inhibiting CXCL12-mediated signaling. Together these data suggest that inhibition of CXCR4 by BILF1 is a consequence of its constitutive activity, and contrasts with the allosteric mechanisms described above for the ligand-binding transinhibition between CXCR4, CCR2, and CCR5. In principle, this G protein-scavenging mechanism could represent a general mechanism of viral GPCRs for inhibiting the function of chemokine receptors that require G protein coupling for high-affinity binding.

6. HETERODIMERIZATION OF CHEMOKINE RECEPTORS WITH NONCHEMOKINE RECEPTORS

In addition to BILF1, other receptors outside the chemokine receptor family have been shown to hetero-oligomerize with chemokine receptors and impact signaling (Table 9.1). In particular, members of the opioid family of GPCRs have been shown to form heterodimers with CCR5, CXCR4, and CXCR2.[94,95,98,99] These studies were motivated by the fact that opioid receptors and chemokine receptors are often coexpressed on immune system cells as well as neurons and glial cells in the brain,[99] and the fact that opioids have been shown to inhibit migration of leukocytes.[125] In a study of the μ-opioid receptor (MOR) and CCR5, for example, transinhibition of functional responses was observed.[95] Whereas cells coexpressing both receptors were responsive to the MOR agonist (DAMGO) and to the CCR5 agonist (RANTES/CCL5) when they were added individually, pre-treatment of the cells with CCL5 inhibited migration to DAMGO, and DAMGO but not the antagonist Naloxone, inhibited migration to CCL5. Similarly DAMGO caused increased phosphorylation of CCR5 and inhibited binding of GTPγS, while CCL5 treatment caused enhanced phosphorylation and decreased GTPγS binding to MOR. In contrast to the ligand-binding transinhibition described previously for CCR5, CCR2, and CXCR4, there was no significant bi-directional inhibitory effect on the binding affinity of the agonists. Overall, the data suggest that agonist stimulation of each receptor promoted cross-desensitization of the other receptor.

CXCR4 and the δ-opioid receptor (DOR) have also been shown to hetero-oligomerize. In these studies, hetero-oligomerization and simultaneous stimulation with their respective agonist ligands, SDF-1/CXCL12 and [D-Pen2, D-Pen5]enkephalin (DPDPE), inhibited migration of CXCL12-mediated cell migration and adhesion of primary monocytes and monocytic cell lines. These results were also validated with *in vivo* studies of cell migration into the peritoneal cavity of mice.[99] In contrast to the above studies of CCR5 and MOR, the inhibitory effect was not due to heterologous desensitization, nor did DPDPE affect the affinity of CXCL12 for CXCR4. Instead the silencing of CXCR4 was shown to be due to the inability of the ligand-engaged heterodimers to activate JAK2, which is a pre-requisite for Gαi coupling to CXCR4. While FRET studies showed that the receptors formed oligomers independent of ligand binding, additional studies suggested that the receptors function in a ligand-regulated dynamic equilibrium between homo and hetero-oligomers. Specifically, FRET signals from CXCR4 homomers were disrupted with increasing expression of DOR. Furthermore, the heterodimer formation was reversed by DPDPE but not by simultaneous addition of CXCL12 and DPDPE. Overall the results suggest that CXCL12 or DPDPE alone allows signaling by their respective homomeric receptors, but treatment with both ligands stabilizes the hetero-oligomers and blocks signaling presumably by stabilizing the receptors in inactive conformations. This ligand-dependent regulation of the heteromers was proposed to provide a mechanism that might have consequences on physiological processes involving pain and inflammation; for example, by increasing sensitivity to pain while simultaneously curtailing migration of cells to sites of inflammation.

Entirely different functional effects were observed in a study of hetero-oligomers between CXCR2 and DOR.[98] In this case, several complementary methods (BRET, FRET, tr-FRET, and Co-IP) were used to demonstrate heterodimers in transfected cells. Saturation BRET studies suggested that the heterodimers have a greater tendency to associate than the homodimers. Most importantly, the small molecule CXCR2 antagonist resulted in enhanced signaling responsiveness of DOR to agonists, due to allosteric communication between the receptors while the CXCR2 agonist IL-8/CXCL8 did not. That the nature of the ligand matters makes intuitive sense since agonists and antagonists would likely promote different conformations of the partner receptor in the heterodimer.

In contrast, the Epstein-Barr virus-induced receptor 2 (EBI2) exerts a negative allosteric effect on the function of CXCR5.[102] In these studies coexpression of EBI2 lead to decreased responsiveness to CXCL13, as measured by Ca^{+2} flux, chemotaxis, and ERK1/2 phosphorylation. These results were attributed to the reduced affinity of CXCR5 for its ligand CXCL13 when the two receptors were coexpressed. Whether the ligand affinity was reduced due to a direct allosteric modulation of EBI2 on the CXCR5-binding pocket, the coupling with G protein, or a combination of both mechanisms was not determined. The cooperation between these two receptors is thought to regulate B cell movement into lymphoid follicles.

CXCR2 has also been shown to associate with the α1A-andrenoreceptor.[93] In this case, heterodimerization with CXCR2 changes the pharmacology of α1A such that it strongly recruits β-arrestin upon stimulation with norephinephrine. This effect was inhibited not only by the α1A antagonist Terazosin, but also by a CXCR2-specific small molecule inverse agonist, SB265610.

7. OTHER SOURCES OF ALLOSTERY IN CHEMOKINE RECEPTOR SIGNALING: CHEMOKINE OLIGOMERIZATION

The bulk of this review has been focused on chemokine receptor oligomerization and its functional consequences. However, there are many other sources of allostery related to ligands that are worth noting. Similar to small molecule agonists of GPCRs, functional selectivity of different chemokine ligands of the same receptor have been reported. One of the best examples involves the ligands of CCR7, SLC/CCL21, and ELC/CCL19, which work together to allow for temporally and spatially distinct responses of CCR7-expressing T cells.[126] Both of these ligands show a similar binding affinity for CCR7 and equipotent ability to activate G proteins and cause calcium flux. However, only CCL19 promotes robust desensitization and phosphorylation of the receptor, as well as β-arrestin recruitment and ERK1/2 activation. This finding and the fact that these ligands are differentially expressed *in vivo* has been used to explain how CCR7 can produce the diverse responses of T cells as they migrate into the T cell zones of peripheral lymph nodes. CCL21 is expressed on the high endothelial venules, and thus it makes sense that it would not cause desensitization, allowing subsequent migration of the cells into T cell zones where CCL19 is expressed.

Another important example is that of synthetic N-terminally modified variants of RANTES/CCL5 that can cause internalization of CCR5 and inhibit it from being recycled back to the cell surface. These studies have provided proof of concept that engineering this type of allosteric functional selectivity can be a powerful approach to inhibiting HIV. There are also many reports of allosteric small molecules of chemokine receptors (for an excellent review see Ref. 4).

All of the above examples of functionally selective ligands fit into the classic view of chemically different variants of related ligands. However, recently in the chemokine field, it has been shown that different oligomerization states of the same ligand can also show biased signaling. Early investigations of RANTES/CCL5 showed that although oligomerization-deficient mutants were equally capable as WT CCL5 in promoting transendothelial cell migration, only the oligomerizing WT chemokine could promote monocyte arrest.[127] More recent studies of an obligate disulfide-locked dimer of SDF-1/CXCL12 have shown that it is capable of binding CXCR4 but it has a different signaling profile than WT CXCL12 (which is effectively monomeric below millimolar concentrations). Whereas WT CXCL12 stimulated calcium flux, inhibited cAMP, and promoted cell migration and β-arrestin association, dimeric CXCL12 was impaired in its ability to stimulate cell migration and recruit β-arrestin.[113] Whether dimeric CXCL12 binds and stabilizes monomeric or dimeric CXCR4 remains to be seen, but one can imagine an effect of the oligomerization state of the CXCL12 on the homo or hetero-oligomerization state of CXCR4 and vice versa—that the state of CXCR4 affects whether monomeric or dimeric CXCL12 binds to CXCR4.

8. CONCLUSIONS AND FUTURE PERSPECTIVES

In the last decade, there has been an explosion in the number of studies focused on demonstrating the presence and functional relevance of chemokine receptor homo- and hetero-oligomerization. Notwithstanding acknowledgement of the fact that there are conflicting reports and perhaps erroneous conclusions because of limitations in methods used, it appears that receptor oligomerization, particularly hetero-oligomerization, can result in many different types of pharmacological responses compared to chemokine receptors in isolation, and as a consequence of many different types of cellular mechanisms. These functional effects and mechanisms include (1) transinhibition of ligand binding whether it be through a G protein steal when the pool of G proteins is limiting, or purely from allosteric effects because of changes in the conformation of a given receptor (R1) due to the presence of the second receptor (R2) and/or the presence of the agonist or antagonist ligand of R1 or R2; (2) activation of receptor functional responses whether it be through ligand-dependent or independent effects; (3) inhibition of functional responses that are ligand independent, or dependent on ligands of one or both of the interacting receptors; (4) ligand-regulated formation of receptor heteromers or destabilization of receptor heteromers.

In addition to these functional effects, the next level of complexity and a key issue is whether there are cell- and tissue-dependent effects given differences in receptor expression levels, intracellular signaling partners, and intra- and inter-cellular microenvironments. The idea that these effects can impact drug discovery seems indisputable and while it complicates high-throughput screening campaigns, in principle there are major opportunities for drug discovery if the biology of the disease and detailed pharmacology of the receptors in question are understood. Technology development to validate the presence of interacting receptors, their functional consequences and mechanisms of action, and determining their role in disease will be required to capitalize on these new insights into the complex function and regulation of chemokine receptors and other GPCRs.

ACKNOWLEDGMENTS

The authors acknowledge support from the National Institutes of Health/National Institute of General Medical Sciences U01GM094612 and R01GM081763 to T. M. H., and the Graduate Training Program in Cellular and Molecular Pharmacology T32GM007752, which supports B. S. We also thank Catherina L. Salanga for assistance with figure preparation.

REFERENCES

1. Munoz LM, Lucas P, Holgado BL, Barroso R, Vega B, Rodriguez-Frade JM, et al. Receptor oligomerization: a pivotal mechanism for regulating chemokine function. *Pharmacol Ther* 2011;**131**:351–8.
2. Salanga CL, Handel TM. Chemokine oligomerization and interactions with receptors and glycosaminoglycans: the role of structural dynamics in function. *Exp Cell Res* 2011;**317**:590–601.
3. Munoz LM, Holgado BL, Martinez AC, Rodriguez-Frade JM, Mellado M. Chemokine receptor oligomerization: a further step toward chemokine function. *Immunol Lett* 2012;**145**:23–9.
4. Scholten DJ, Canals M, Maussang D, Roumen L, Smit MJ, Wijtmans M, et al. Pharmacological modulation of chemokine receptor function. *Br J Pharmacol* 2012;**165**:1617–43.
5. Graham GJ. D6 and the atypical chemokine receptor family: novel regulators of immune and inflammatory processes. *Eur J Immunol* 2009;**39**:342–51.
6. Graham GJ, Locati M, Mantovani A, Rot A, Thelen M. The biochemistry and biology of the atypical chemokine receptors. *Immunol Lett* 2012;**145**:30–8.
7. Kalatskaya I, Berchiche YA, Gravel S, Limberg BJ, Rosenbaum JS, Heveker N. AMD3100 is a CXCR7 ligand with allosteric agonist properties. *Mol Pharmacol* 2009;**75**:1240–7.
8. Zabel BA, Lewen S, Berahovich RD, Jaen JC, Schall TJ. The novel chemokine receptor CXCR7 regulates trans-endothelial migration of cancer cells. *Mol Cancer* 2011;**10**:73.
9. Odemis V, Lipfert J, Kraft R, Hajek P, Abraham G, Hattermann K, et al. The presumed atypical chemokine receptor CXCR7 signals through G(i/o) proteins in primary rodent astrocytes and human glioma cells. *Glia* 2012;**60**:372–81.
10. Schall TJ, Proudfoot AE. Overcoming hurdles in developing successful drugs targeting chemokine receptors. *Nat Rev Immunol* 2011;**11**:355–63.
11. Lukacs NW, Oliveira SH, Hogaboam CM. Chemokines and asthma: redundancy of function or a coordinated effort? *J Clin Invest* 1999;**104**:995–9.
12. Mantovani A. The chemokine system: redundancy for robust outputs. *Immunol Today* 1999;**20**:254–7.
13. Proudfoot AE, Power CA, Schwarz MK. Anti-chemokine small molecule drugs: a promising future? *Expert Opin Investig Drugs* 2010;**19**:345–55.
14. Devalaraja MN, Richmond A. Multiple chemotactic factors: fine control or redundancy? *Trends Pharmacol Sci* 1999;**20**:151–6.
15. Handel TM, Lau EK. Chemokine structure and receptor interactions. *Ernst Scher Res Found Workshop* 2004;**45**:101–24.
16. Ray P, Lewin SA, Mihalko LA, Lesher-Perez SC, Takayama S, Luker KE, et al. Secreted CXCL12 (SDF-1) forms dimers under physiological conditions. *Biochem J* 2012;**442**:433–42.
17. Ren M, Guo Q, Guo L, Lenz M, Qian F, Koenen RR, et al. Polymerization of MIP-1 chemokine (CCL3 and CCL4) and clearance of MIP-1 by insulin-degrading enzyme. *EMBO J* 2010;**29**:3952–66.
18. Wang X, Watson C, Sharp JS, Handel TM, Prestegard JH. Oligomeric structure of the chemokine CCL5/RANTES from NMR, MS, and SAXS data. *Structure* 2011;**19**: 1138–48.
19. Hoogewerf AJ, Kuschert GS, Proudfoot AE, Borlat F, Clark-Lewis I, Power CA, et al. Glycosaminoglycans mediate cell surface oligomerization of chemokines. *Biochemistry* 1997;**36**:13570–8.

20. Lau EK, Paavola CD, Johnson Z, Gaudry JP, Geretti E, Borlat F, et al. Identification of the glycosaminoglycan binding site of the CC chemokine, MCP-1: implications for structure and function in vivo. *J Biol Chem* 2004;**279**:22294–305.
21. Vega B, Munoz LM, Holgado BL, Lucas P, Rodriguez-Frade JM, Calle A, et al. Technical advance: surface plasmon resonance-based analysis of CXCL12 binding using immobilized lentiviral particles. *J Leukoc Biol* 2011;**90**:399–408.
22. Wang L, Fuster M, Sriramarao P, Esko JD. Endothelial heparan sulfate deficiency impairs L-selectin- and chemokine-mediated neutrophil trafficking during inflammatory responses. *Nat Immunol* 2005;**6**:902–10.
23. Yin X, Truty J, Lawrence R, Johns SC, Srinivasan RS, Handel TM, et al. A critical role for lymphatic endothelial heparan sulfate in lymph node metastasis. *Mol Cancer* 2010;**9**:316.
24. Proudfoot AE, Handel TM, Johnson Z, Lau EK, LiWang P, Clark-Lewis I, et al. Glycosaminoglycan binding and oligomerization are essential for the in vivo activity of certain chemokines. *Proc Natl Acad Sci USA* 2003;**100**:1885–90.
25. Rajarathnam K, Sykes BD, Kay CM, Dewald B, Geiser T, Baggiolini M, et al. Neutrophil activation by monomeric interleukin-8. *Science* 1994;**264**:90–2.
26. Clark-Lewis I, Kim KS, Rajarathnam K, Gong JH, Dewald B, Moser B, et al. Structure-activity relationships of chemokines. *J Leukoc Biol* 1995;**57**:703–11.
27. Zhang Y, Rollins BJ. A dominant negative inhibitor indicates that monocyte chemoattractant protein 1 functions as a dimer. *Mol Cell Biol* 1995;**15**:4851–5.
28. Proudfoot AE, Power CA, Hoogewerf AJ, Montjovent MO, Borlat F, Offord RE, et al. Extension of recombinant human RANTES by the retention of the initiating methionine produces a potent antagonist. *J Biol Chem* 1996;**271**:2599–603.
29. Jarnagin K, Grunberger D, Mulkins M, Wong B, Hemmerich S, Paavola C, et al. Identification of surface residues of the monocyte chemotactic protein 1 that affect signaling through the receptor CCR2. *Biochemistry* 1999;**38**:16167–77.
30. Crump MP, Gong JH, Loetscher P, Rajarathnam K, Amara A, Arenzana-Seisdedos F, et al. Solution structure and basis for functional activity of stromal cell-derived factor-1; dissociation of CXCR4 activation from binding and inhibition of HIV-1. *EMBO J* 1997;**16**:6996–7007.
31. Blain KY, Kwiatkowski W, Zhao Q, La Fleur D, Naik C, Chun TW, et al. Structural and functional characterization of CC chemokine CCL14. *Biochemistry* 2007;**46**:10008–15.
32. Brandt E, Van Damme J, Flad HD. Neutrophils can generate their activator neutrophil-activating peptide 2 by proteolytic cleavage of platelet-derived connective tissue-activating peptide III. *Cytokine* 1991;**3**:311–21.
33. Mack M, Luckow B, Nelson PJ, Cihak J, Simmons G, Clapham PR, et al. Aminooxypentane-RANTES induces CCR5 internalization but inhibits recycling: a novel inhibitory mechanism of HIV infectivity. *J Exp Med* 1998;**187**:1215–24.
34. Gaertner H, Cerini F, Escola JM, Kuenzi G, Melotti A, Offord R, et al. Highly potent, fully recombinant anti-HIV chemokines: reengineering a low-cost microbicide. *Proc Natl Acad Sci USA* 2008;**105**:17706–11.
35. Veldkamp CT, Seibert C, Peterson FC, De la Cruz NB, Haugner 3rd JC, Basnet H, et al. Structural basis of CXCR4 sulfotyrosine recognition by the chemokine SDF-1/CXCL12. *Sci Signal* 2008;**1**:ra4.
36. Blanpain C, Doranz BJ, Bondue A, Govaerts C, De Leener A, Vassart G, et al. The core domain of chemokines binds CCR5 extracellular domains while their amino terminus interacts with the transmembrane helix bundle. *J Biol Chem* 2003;**278**:5179–87.
37. Kofuku Y, Yoshiura C, Ueda T, Terasawa H, Hirai T, Tominaga S, et al. Structural basis of the interaction between chemokine stromal cell-derived factor-1/CXCL12 and its G-protein-coupled receptor CXCR4. *J Biol Chem* 2009;**284**:35240–50.

38. Kuszak AJ, Pitchiaya S, Anand JP, Mosberg HI, Walter NG, Sunahara RK. Purification and functional reconstitution of monomeric mu-opioid receptors: allosteric modulation of agonist binding by Gi2. *J Biol Chem* 2009;**284**:26732–41.
39. Whorton MR, Bokoch MP, Rasmussen SG, Huang B, Zare RN, Kobilka B, et al. A monomeric G protein-coupled receptor isolated in a high-density lipoprotein particle efficiently activates its G protein. *Proc Natl Acad Sci USA* 2007;**104**:7682–7.
40. Whorton MR, Jastrzebska B, Park PS, Fotiadis D, Engel A, Palczewski K, et al. Efficient coupling of transducin to monomeric rhodopsin in a phospholipid bilayer. *J Biol Chem* 2008;**283**:4387–94.
41. Bayburt TH, Vishnivetskiy SA, McLean MA, Morizumi T, Huang CC, Tesmer JJ, et al. Monomeric rhodopsin is sufficient for normal rhodopsin kinase (GRK1) phosphorylation and arrestin-1 binding. *J Biol Chem* 2011;**286**:1420–8.
42. Rodriguez-Frade JM, Vila-Coro AJ, de Ana AM, Albar JP, Martinez AC, Mellado M. The chemokine monocyte chemoattractant protein-1 induces functional responses through dimerization of its receptor CCR2. *Proc Natl Acad Sci USA* 1999;**96**:3628–33.
43. Mellado M, Rodriguez-Frade JM, Vila-Coro AJ, de Ana AM, Martinez AC. Chemokine control of HIV-1 infection. *Nature* 1999;**400**:723–4.
44. Mellado M, Rodriguez-Frade JM, Vila-Coro AJ, Fernandez S, Martin de Ana A, Jones DR, et al. Chemokine receptor homo- or heterodimerization activates distinct signaling pathways. *EMBO J* 2001;**20**:2497–507.
45. Percherancier Y, Berchiche YA, Slight I, Volkmer-Engert R, Tamamura H, Fujii N, et al. Bioluminescence resonance energy transfer reveals ligand-induced conformational changes in CXCR4 homo- and heterodimers. *J Biol Chem* 2005;**280**:9895–903.
46. Berchiche YA, Gravel S, Pelletier ME, St-Onge G, Heveker N. Different effects of the different natural CC chemokine receptor 2b ligands on beta-arrestin recruitment, Galphai signaling, and receptor internalization. *Mol Pharmacol* 2011;**79**:488–98.
47. El-Asmar L, Springael JY, Ballet S, Andrieu EU, Vassart G, Parmentier M. Evidence for negative binding cooperativity within CCR5-CCR2b heterodimers. *Mol Pharmacol* 2005;**67**:460–9.
48. Springael JY, Le Minh PN, Urizar E, Costagliola S, Vassart G, Parmentier M. Allosteric modulation of binding properties between units of chemokine receptor homo- and hetero-oligomers. *Mol Pharmacol* 2006;**69**:1652–61.
49. Sohy D, Yano H, de Nadai P, Urizar E, Guillabert A, Javitch JA, et al. Hetero-oligomerization of CCR2, CCR5, and CXCR4 and the protean effects of "selective" antagonists. *J Biol Chem* 2009;**284**:31270–9.
50. Benkirane M, Jin DY, Chun RF, Koup RA, Jeang KT. Mechanism of transdominant inhibition of CCR5-mediated HIV-1 infection by ccr5delta32. *J Biol Chem* 1997;**272**:30603–6.
51. Vila-Coro AJ, Mellado M, Martin de Ana A, Lucas P, del Real G, Martinez AC, et al. HIV-1 infection through the CCR5 receptor is blocked by receptor dimerization. *Proc Natl Acad Sci USA* 2000;**97**:3388–93.
52. Rodriguez-Frade JM, Vila-Coro AJ, Martin A, Nieto M, Sanchez-Madrid F, Proudfoot AE, et al. Similarities and differences in RANTES- and (AOP)-RANTES-triggered signals: implications for chemotaxis. *J Cell Biol* 1999;**144**:755–65.
53. Chelli M, Alizon M. Determinants of the trans-dominant negative effect of truncated forms of the CCR5 chemokine receptor. *J Biol Chem* 2001;**276**:46975–82.
54. Blanpain C, Vanderwinden JM, Cihak J, Wittamer V, Le Poul E, Issafras H, et al. Multiple active states and oligomerization of CCR5 revealed by functional properties of monoclonal antibodies. *Mol Biol Cell* 2002;**13**:723–37.
55. Issafras H, Angers S, Bulenger S, Blanpain C, Parmentier M, Labbe-Jullie C, et al. Constitutive agonist-independent CCR5 oligomerization and antibody-mediated clustering occurring at physiological levels of receptors. *J Biol Chem* 2002;**277**:34666–73.

56. Contento RL, Molon B, Boularan C, Pozzan T, Manes S, Marullo S, et al. CXCR4-CCR5: a couple modulating T cell functions. *Proc Natl Acad Sci USA* 2008;**105**:10101–6.
57. Chelli M, Alizon M. Rescue of HIV-1 receptor function through cooperation between different forms of the CCR5 chemokine receptor. *J Biol Chem* 2002;**277**:39388–96.
58. Hernanz-Falcon P, Rodriguez-Frade JM, Serrano A, Juan D, del Sol A, Soriano SF, et al. Identification of amino acid residues crucial for chemokine receptor dimerization. *Nat Immunol* 2004;**5**:216–23.
59. Lemay J, Marullo S, Jockers R, Alizon M, Brelot A. On the dimerization of CCR5. *Nat Immunol* 2005;**6**:535 Author reply 535–536.
60. Huttenrauch F, Pollok-Kopp B, Oppermann M. G protein-coupled receptor kinases promote phosphorylation and beta-arrestin-mediated internalization of CCR5 homo- and hetero-oligomers. *J Biol Chem* 2005;**280**:37503–15.
61. Hammad MM, Kuang YQ, Yan R, Allen H, Dupre DJ. Na+/H+ exchanger regulatory factor-1 is involved in chemokine receptor homodimer CCR5 internalization and signal transduction but does not affect CXCR4 homodimer or CXCR4-CCR5 heterodimer. *J Biol Chem* 2010;**285**:34653–64.
62. Charette N, Holland P, Frazer J, Allen H, Dupre DJ. Dependence on different Rab GTPases for the trafficking of CXCR4 and CCR5 homo or heterodimers between the endoplasmic reticulum and plasma membrane in Jurkat cells. *Cell Signal* 2011;**23**:1738–49.
63. Kuang YQ, Charette N, Frazer J, Holland PJ, Attwood KM, Dellaire G, et al. Dopamine receptor-interacting protein 78 acts as a molecular chaperone for CCR5 chemokine receptor signaling complex organization. *PLoS One* 2012;**7**:e40522.
64. Wilson S, Wilkinson G, Milligan G. The CXCR1 and CXCR2 receptors form constitutive homo- and heterodimers selectively and with equal apparent affinities. *J Biol Chem* 2005;**280**:28663–74.
65. Martinez Munoz L, Lucas P, Navarro G, Checa AI, Franco R, Martinez AC, et al. Dynamic regulation of CXCR1 and CXCR2 homo- and heterodimers. *J Immunol* 2009;**183**:7337–46.
66. Trettel F, Di Bartolomeo S, Lauro C, Catalano M, Ciotti MT, Limatola C. Ligand-independent CXCR2 dimerization. *J Biol Chem* 2003;**278**:40980–8.
67. Limatola C, Di Bartolomeo S, Catalano M, Trettel F, Fucile S, Castellani L, et al. Cysteine residues are critical for chemokine receptor CXCR2 functional properties. *Exp Cell Res* 2005;**307**:65–75.
68. Vila-Coro AJ, Rodriguez-Frade JM, Martin De Ana A, Moreno-Ortiz MC, Martinez AC, Mellado M. The chemokine SDF-1alpha triggers CXCR4 receptor dimerization and activates the JAK/STAT pathway. *FASEB J* 1999;**13**:1699–710.
69. Babcock GJ, Farzan M, Sodroski J. Ligand-independent dimerization of CXCR4, a principal HIV-1 coreceptor. *J Biol Chem* 2003;**278**:3378–85.
70. Hamdan FF, Percherancier Y, Breton B, Bouvier M. Monitoring protein-protein interactions in living cells by bioluminescence resonance energy transfer (BRET). *Curr Protoc Neurosci* 2006; Chapter 5, Unit 5 23.
71. Lagane B, Chow KY, Balabanian K, Levoye A, Harriague J, Planchenault T, et al. CXCR4 dimerization and beta-arrestin-mediated signaling account for the enhanced chemotaxis to CXCL12 in WHIM syndrome. *Blood* 2008;**112**:34–44.
72. Tanaka T, Nomura W, Narumi T, Masuda A, Tamamura H. Bivalent ligands of CXCR4 with rigid linkers for elucidation of the dimerization state in cells. *J Am Chem Soc* 2010;**132**:15899–901.
73. Choi WT, Kumar S, Madani N, Han X, Tian S, Dong CZ, et al. A novel synthetic bivalent ligand to probe chemokine receptor CXCR4 dimerization and inhibit HIV-1 entry. *Biochemistry* 2012;**51**:7078–86.

74. Toth PT, Ren D, Miller RJ. Regulation of CXCR4 receptor dimerization by the chemokine SDF-1alpha and the HIV-1 coat protein gp120: a fluorescence resonance energy transfer (FRET) study. *J Pharmacol Exp Ther* 2004;**310**:8–17.
75. Wang J, He L, Combs CA, Roderiquez G, Norcross MA. Dimerization of CXCR4 in living malignant cells: control of cell migration by a synthetic peptide that reduces homologous CXCR4 interactions. *Mol Cancer Ther* 2006;**5**:2474–83.
76. Hamatake M, Aoki T, Futahashi Y, Urano E, Yamamoto N, Komano J. Ligand-independent higher-order multimerization of CXCR4, a G-protein-coupled chemokine receptor involved in targeted metastasis. *Cancer Sci* 2009;**100**:95–102.
77. Luker KE, Gupta M, Luker GD. Imaging chemokine receptor dimerization with firefly luciferase complementation. *FASEB J* 2009;**23**:823–34.
78. Levoye A, Balabanian K, Baleux F, Bachelerie F, Lagane B. CXCR7 heterodimerizes with CXCR4 and regulates CXCL12-mediated G protein signaling. *Blood* 2009;**113**:6085–93.
79. Zhukovsky MA, Basmaciogullari S, Pacheco B, Wang L, Madani N, Haim H, et al. Thermal stability of the human immunodeficiency virus type 1 (HIV-1) receptors, CD4 and CXCR4, reconstituted in proteoliposomes. *PLoS One* 2010;**5**:e13249.
80. Wu B, Chien EY, Mol CD, Fenalti G, Liu W, Katritch V, et al. Structures of the CXCR4 chemokine GPCR with small-molecule and cyclic peptide antagonists. *Science* 2010;**330**:1066–71.
81. Rodriguez D, Gutierrez-de-Teran H. Characterization of the homodimerization interface and functional hotspots of the CXCR4 chemokine receptor. *Proteins* 2012;**80**:1919–28.
82. Asano S, Kitatani K, Taniguchi M, Hashimoto M, Zama K, Mitsutake S, et al. Regulation of cell migration by sphingomyelin synthases: sphingomyelin in lipid rafts decreases responsiveness to signaling by the CXCL12/CXCR4 pathway. *Mol Cell Biol* 2012;**32**:3242–52.
83. Chakera A, Seeber RM, John AE, Eidne KA, Greaves DR. The duffy antigen/receptor for chemokines exists in an oligomeric form in living cells and functionally antagonizes CCR5 signaling through hetero-oligomerization. *Mol Pharmacol* 2008;**73**:1362–70.
84. Gouldson PR, Dean MK, Snell CR, Bywater RP, Gkoutos G, Reynolds CA. Lipid-facing correlated mutations and dimerization in G-protein coupled receptors. *Protein Eng* 2001;**14**:759–67.
85. Rodriguez-Frade JM, del Real G, Serrano A, Hernanz-Falcon P, Soriano SF, Vila-Coro AJ, et al. Blocking HIV-1 infection via CCR5 and CXCR4 receptors by acting in trans on the CCR2 chemokine receptor. *EMBO J* 2004;**23**:66–76.
86. Vazquez-Salat N, Yuhki N, Beck T, O'Brien SJ, Murphy WJ. Gene conversion between mammalian CCR2 and CCR5 chemokine receptor genes: a potential mechanism for receptor dimerization. *Genomics* 2007;**90**:213–24.
87. See HB, Seeber RM, Kocan M, Eidne KA, Pfleger KD. Application of G protein-coupled receptor-heteromer identification technology to monitor beta-arrestin recruitment to G protein-coupled receptor heteromers. *Assay Drug Dev Technol* 2011;**9**:21–30.
88. Sohy D, Parmentier M, Springael JY. Allosteric transinhibition by specific antagonists in CCR2/CXCR4 heterodimers. *J Biol Chem* 2007;**282**:30062–9.
89. Isik N, Hereld D, Jin T. Fluorescence resonance energy transfer imaging reveals that chemokine-binding modulates heterodimers of CXCR4 and CCR5 receptors. *PLoS One* 2008;**3**:e3424.
90. Wang J, Alvarez R, Roderiquez G, Guan E, Norcross MA. Constitutive association of cell surface CCR5 and CXCR4 in the presence of CD4. *J Cell Biochem* 2004;**93**:753–60.
91. Sierro F, Biben C, Martinez-Munoz L, Mellado M, Ransohoff RM, Li M, et al. Disrupted cardiac development but normal hematopoiesis in mice deficient in the second CXCL12/SDF-1 receptor, CXCR7. *Proc Natl Acad Sci USA* 2007;**104**: 14759–64.

92. Decaillot FM, Kazmi MA, Lin Y, Ray-Saha S, Sakmar TP, Sachdev P. CXCR7/CXCR4 heterodimer constitutively recruits beta-arrestin to enhance cell migration. *J Biol Chem* 2011;**286**:32188–97.
93. Mustafa S, See HB, Seeber RM, Armstrong SP, White CW, Ventura S, et al. Identification and profiling of novel alpha1A-adrenoceptor-CXC chemokine receptor 2 heteromer. *J Biol Chem* 2012;**287**:12952–65.
94. Suzuki S, Chuang LF, Yau P, Doi RH, Chuang RY. Interactions of opioid and chemokine receptors: oligomerization of mu, kappa, and delta with CCR5 on immune cells. *Exp Cell Res* 2002;**280**:192–200.
95. Chen C, Li J, Bot G, Szabo I, Rogers TJ, Liu-Chen LY. Heterodimerization and cross-desensitization between the mu-opioid receptor and the chemokine CCR5 receptor. *Eur J Pharmacol* 2004;**483**:175–86.
96. Yuan Y, Arnatt CK, Li G, Haney KM, Ding D, Jacob JC, et al. Design and synthesis of a bivalent ligand to explore the putative heterodimerization of the mu opioid receptor and the chemokine receptor CCR5. *Org Biomol Chem* 2012;**10**:2633–46.
97. Catalano M, Trettel F, Cipriani R, Lauro C, Sobrero F, Eusebi F, et al. Chemokine CXCL8 modulates GluR1 phosphorylation. *J Neuroimmunol* 2008;**198**:75–81.
98. Parenty G, Appelbe S, Milligan G. CXCR2 chemokine receptor antagonism enhances DOP opioid receptor function via allosteric regulation of the CXCR2-DOP receptor heterodimer. *Biochem J* 2008;**412**:245–56.
99. Pello OM, Martinez-Munoz L, Parrillas V, Serrano A, Rodriguez-Frade JM, Toro MJ, et al. Ligand stabilization of CXCR4/delta-opioid receptor heterodimers reveals a mechanism for immune response regulation. *Eur J Immunol* 2008;**38**:537–49.
100. Vischer HF, Nijmeijer S, Smit MJ, Leurs R. Viral hijacking of human receptors through heterodimerization. *Biochem Biophys Res Commun* 2008;**377**:93–7.
101. Nijmeijer S, Leurs R, Smit MJ, Vischer HF. The Epstein-Barr virus-encoded G protein-coupled receptor BILF1 hetero-oligomerizes with human CXCR4, scavenges Galphai proteins, and constitutively impairs CXCR4 functioning. *J Biol Chem* 2010;**285**:29632–41.
102. Barroso R, Martinez Munoz L, Barrondo S, Vega B, Holgado BL, Lucas P, et al. EBI2 regulates CXCL13-mediated responses by heterodimerization with CXCR5. *FASEB J* 2012;**26**:4841–54.
103. Kumar A, Humphreys TD, Kremer KN, Bramati PS, Bradfield L, Edgar CE, et al. CXCR4 physically associates with the T cell receptor to signal in T cells. *Immunity* 2006;**25**:213–24.
104. Xiao X, Wu L, Stantchev TS, Feng YR, Ugolini S, Chen H, et al. Constitutive cell surface association between CD4 and CCR5. *Proc Natl Acad Sci USA* 1999;**96**:7496–501.
105. Singer II, Scott S, Kawka DW, Chin J, Daugherty BL, DeMartino JA, et al. CCR5, CXCR4, and CD4 are clustered and closely apposed on microvilli of human macrophages and T cells. *J Virol* 2001;**75**:3779–90.
106. Doumazane E, Scholler P, Zwier JM, Trinquet E, Rondard P, Pin JP. A new approach to analyze cell surface protein complexes reveals specific heterodimeric metabotropic glutamate receptors. *FASEB J* 2011;**25**:66–77.
107. Vischer HF, Watts AO, Nijmeijer S, Leurs R. G protein-coupled receptors: walking hand-in-hand, talking hand-in-hand? *Br J Pharmacol* 2011;**163**:246–60.
108. Kufareva I, Stephens B, Gilliland CT, Wu B, Fenalti G, Hamel D, et al. Novel approach to quantify GPCR dimerization equilibrium using bioluminescence resonance energy transfer. *Methods Mol Biol* 2012. In press.
109. Hern JA, Baig AH, Mashanov GI, Birdsall B, Corrie JE, Lazareno S, et al. Formation and dissociation of M1 muscarinic receptor dimers seen by total internal reflection fluorescence imaging of single molecules. *Proc Natl Acad Sci USA* 2010;**107**:2693–8.

110. Milligan G. G protein-coupled receptor hetero-dimerization: contribution to pharmacology and function. *Br J Pharmacol* 2009;**158**:5–14.
111. Jones KA, Borowsky B, Tamm JA, Craig DA, Durkin MM, Dai M, et al. GABA(B) receptors function as a heteromeric assembly of the subunits GABA(B)R1 and GABA (B)R2. *Nature* 1998;**396**:674–9.
112. White JH, Wise A, Main MJ, Green A, Fraser NJ, Disney GH, et al. Heterodimerization is required for the formation of a functional GABA(B) receptor. *Nature* 1998;**396**:679–82.
113. Drury LJ, Ziarek JJ, Gravel S, Veldkamp CT, Takekoshi T, Hwang ST, et al. Monomeric and dimeric CXCL12 inhibit metastasis through distinct CXCR4 interactions and signaling pathways. *Proc Natl Acad Sci USA* 2011;**108**:17655–60.
114. Milligan G, Wilson S, Lopez-Gimenez JF. The specificity and molecular basis of alpha1-adrenoceptor and CXCR chemokine receptor dimerization. *J Mol Neurosci* 2005;**26**:161–8.
115. Tarasova NI, Rice WG, Michejda CJ. Inhibition of G-protein-coupled receptor function by disruption of transmembrane domain interactions. *J Biol Chem* 1999;**274**:34911–5.
116. Chabre M, Deterre P, Antonny B. The apparent cooperativity of some GPCRs does not necessarily imply dimerization. *Trends Pharmacol Sci* 2009;**30**:182–7.
117. Birdsall NJ. Class A GPCR heterodimers: evidence from binding studies. *Trends Pharmacol Sci* 2010;**31**:499–508.
118. Molon B, Gri G, Bettella M, Gomez-Mouton C, Lanzavecchia A, Martinez AC, et al. T cell costimulation by chemokine receptors. *Nat Immunol* 2005;**6**:465–71.
119. Pruenster M, Rot A. Throwing light on DARC. *Biochem Soc Trans* 2006;**34**:1005–8.
120. Zabel BA, Wang Y, Lewen S, Berahovich RD, Penfold ME, Zhang P, et al. Elucidation of CXCR7-mediated signaling events and inhibition of CXCR4-mediated tumor cell transendothelial migration by CXCR7 ligands. *J Immunol* 2009;**183**:3204–11.
121. Luker KE, Lewin SA, Mihalko LA, Schmidt BT, Winkler JS, Coggins NL, et al. Scavenging of CXCL12 by CXCR7 promotes tumor growth and metastasis of CXCR4-positive breast cancer cells. *Oncogene* 2012;**45**:4750–8.
122. Maussang D, Vischer HF, Schreiber A, Michel D, Smit MJ. Pharmacological and biochemical characterization of human cytomegalovirus-encoded G protein-coupled receptors. *Methods Enzymol* 2009;**460**:151–71.
123. Slinger E, Langemeijer E, Siderius M, Vischer HF, Smit MJ. Herpesvirus-encoded GPCRs rewire cellular signaling. *Mol Cell Endocrinol* 2011;**331**:179–84.
124. Hsu JL, Glaser SL. Epstein-barr virus-associated malignancies: epidemiologic patterns and etiologic implications. *Crit Rev Oncol Hematol* 2000;**34**:27–53.
125. Miyagi T, Chuang LF, Lam KM, Kung H, Wang JM, Osburn BI, et al. Opioids suppress chemokine-mediated migration of monkey neutrophils and monocytes - an instant response. *Immunopharmacology* 2000;**47**:53–62.
126. Kohout TA, Nicholas SL, Perry SJ, Reinhart G, Junger S, Struthers RS. Differential desensitization, receptor phosphorylation, beta-arrestin recruitment, and ERK1/2 activation by the two endogenous ligands for the CC chemokine receptor 7. *J Biol Chem* 2004;**279**:23214–22.
127. Baltus T, Weber KS, Johnson Z, Proudfoot AE, Weber C. Oligomerization of RANTES is required for CCR1-mediated arrest but not CCR5-mediated transmigration of leukocytes on inflamed endothelium. *Blood* 2003;**102**:1985–8.
128. Vinet J, van Zwam M, Dijkstra IM, Brouwer N, van Weering HR, Watts A, et al. Inhibition of CXCR3-mediated chemotaxis by the human chemokine receptor-like protein CCX-CKR. *Br J Pharmacol* 2012. Epub ahead of print.
129. Watts AO, van Lipzig MM, Jaeger WC, Seeber RM, van Zwam M, Vinet J, et al. Identification and profiling of CXCR3-CXCR4 chemokine receptor heteromer complexes. *Br J Pharmacol* 2012. Epub ahead of print.

CHAPTER TEN

Fine-Tuning of GPCR Signals by Intracellular G Protein Modulators

Peishen Zhao[*,1], **Wendy Cladman**[*], **Hubert H.M. Van Tol**[†,‡,§], **Peter Chidiac**[*]

[*]Department of Physiology and Pharmacology, Western University, London, Ontario, Canada
[†]Department of Psychiatry, University of Toronto, Toronto, Canada
[‡]Department of Pharmacology, University of Toronto, Toronto, Canada
[§]Laboratory of Molecular Neurobiology, Centre for Addiction and Mental Health, Toronto, Ontario, Canada
[1]Current affiliation: Monash Institute of Pharmaceutical Sciences, Parkville, VIC, Australia

Contents

Abstract

Heterotrimeric G proteins convey receptor signals to intracellular effectors. Superimposed over the basic GPCR–G protein–effector scheme are three types of auxiliary proteins that also modulate Gα. Regulator of G protein signaling proteins and G protein signaling modifier proteins respectively promote GTPase activity and hinder GTP/GDP exchange to limit Gα activation. There are also diverse proteins that, like GPCRs, can promote nucleotide exchange and thus activation. Here we review the impact of these auxiliary proteins on GPCR signaling. Although their precise physiological functions are not yet clear, all of them can produce significant effects in experimental systems. These signaling changes are generally consistent with established effects on isolated Gα; however, the activation state of Gα is seldom verified and many such changes appear also to reflect the physical disruption of or indirect effects on interactions between Gα and its associated GPCR, Gβγ, and/or effector.

Progress in Molecular Biology and Translational Science, Volume 115
ISSN 1877-1173
http://dx.doi.org/10.1016/B978-0-12-394587-7.00010-5

1. INTRODUCTION: HETEROTRIMERIC G PROTEINS AND THEIR BINDING PARTNERS

G protein-coupled receptors (GPCRs) play a major role in physiological functions and also serve as essential therapeutic targets. In spite of their fundamental importance, some key aspects of how GPCRs convey extracellular signals into cells remain poorly understood. The basic unit of G protein-mediated signaling comprises five gene products: a seven transmembrane-spanning receptor that recognizes extracellular signals, an effector that produces a change in cellular homeostasis, and the Gα, Gβ, and Gγ subunits of the heterotrimeric G protein that work together to carry signals from the receptor to the effector. Over the years numerous other proteins have been identified that interact with one or more of these components and thereby have an impact on signaling. This review will deal with auxiliary proteins that bind to Gα and alter its ability to bind or hydrolyze the activating nucleotide GTP. Specifically we will focus on how such proteins affect G protein-mediated signaling, and in some cases also interface with GPCRs themselves.

G protein Gα subunits belong to a superfamily of GTPases that function as molecular switches to control a wide array of cellular processes.[1,2] Other such GTPases include the smaller Ras-like signaling proteins, initiation and elongation factors such as eIF2 and eEF1a, and specialized proteins such as tubulin and dynamin. Generally speaking, GTPases are activated by the binding of GTP, and self-deactivate by hydrolyzing this to GDP plus inorganic phosphate. The rates at which these biochemical steps take place can be altered by modulatory proteins: guanine nucleotide exchange factors (GEFs) promote GDP dissociation and thus facilitate the binding of GTP, guanine nucleotide dissociation inhibitors (GDIs) impede GDP dissociation and thus delay GTP binding, and GTPase-accelerating proteins (GAPs) facilitate GTP hydrolysis.

Gα proteins have multiple-binding partners that can act as GEFs, GDIs, and GAPs. Of primary importance among these are the other members of the basic receptor–G protein–effector (R–G–E) signaling unit. GPCRs activate Gα proteins by stimulating GDP dissociation and consequently facilitate GTP binding, thus they act as GEFs. Also, the rate of GDP dissociation from heterotrimeric G proteins is slower than from free Gα, which indicates that the stable Gβγ heterodimer (Gβ and Gγ tend to come apart only under denaturing conditions) actually functions as a GDI.[3] In spite of this inhibitory effect, Gβγ is still required for a GPCR to productively couple to and activate Gα.[4] Whereas receptors and Gβγ have effects on nucleotide exchange, some effector proteins have been found to increase the rate at which their Gα-binding partners hydrolyze GTP. Specifically, phospholipase Cβ acts as a GAP for Gαq,[5] while the Rho-activating factors p115RhoGEF and leukemia-associated RhoGEF (LARG) are GAPs for Gα12/13.[6]

Besides their immediate signaling partners, Gα proteins are also governed by a variety of other proteins. Perhaps the best known among these are the regulator of G protein signaling (RGS) proteins, which act as GAPs for members of the Gαi/o and Gαq subfamilies of Gα proteins. By promoting G protein deactivation, RGS proteins can limit the duration of GPCR signals, and also serve to decrease agonist efficacy and/or potency.[7] The RGS protein family is encoded by 20 different genes, some of which yield multiple splice variants. The RGS proteins ranges in size from about 200 to 1400 amino acid residues and can be subdivided into four distinct subfamilies (A/RZ, B/R4, C/R7, and D/R12).[7] Common to all of these is a conserved 120-amino acid-residue RGS domain that binds preferentially to activated Gα and facilitates GTP hydrolysis.[8] Additionally there are a number of structurally diverse RGS-like (RL) proteins that contain an RGS homology domain with either minimal (e.g., G protein-coupled receptor kinases)[9] or atypical GAP activity (e.g., p115RhoGEF, LARG).[6]

In contrast to RGS proteins, which affect GTP hydrolysis, there are also auxiliary Gα-binding proteins that can alter the rate of nucleotide exchange on Gα. G protein signaling modifier (GPSM) proteins contain one to four conserved 20–15 amino acid residue motifs that bind to and stabilize the inactive GDP-bound forms of their target Gαi/o proteins.[10] These interactions produce a GDI effect on Gα and also appear to promote Gβγ dissociation from the heterotrimer, both of which are changes that may impact GPCR signaling. In addition, a handful of proteins with GEF activity have been identified that, like GPCRs, can promote Gα activation. Most of these novel GEFs bear little structural resemblance to one another.

Mounting evidence indicates that proteins that alter Gα nucleotide exchange rates can affect G protein-mediated receptor signals. However unlike RGS proteins, which clearly act as negative regulators of GPCR signals, how novel GEFs and GDIs contribute to signaling has been difficult to ascertain, as their biochemical properties overlap respectively with those of Gβγ and GPCRs. Both GPSM-type proteins and the nonreceptor GEF Ric-8a are also known to be important in intracellular Gα-dependent processes such as asymmetric cell division, a topic that has been covered in several recent reviews.[11–13]

2. ORGANIZATION OF G PROTEIN-MEDIATED SIGNALS

Most or all of the proteins involved in G protein-mediated signaling presumably have been identified, but the interplay among these is complex and not fully understood. Early studies showed that the basic components (namely the GPCR, Gα, Gβγ, and the effector) could be physically separated from one another and recombined experimentally, while other functional data seemed to suggest that the depletion of cellular effectors and G proteins respectively could have limiting effects on receptor signaling and high-affinity agonist binding.[14] Such findings led to the widespread acceptance of the idea that Gα and Gβγ could dissociate from each other and from their activating GPCR, after which they could either converge on or interact independently with their effector targets.[14]

A mechanism based on random collisions between G proteins and their receptors and effectors would imply that all compatible signaling partners can interact within the plasma membrane of a cell; however, signaling among the dozens of GPCRs within a typical cell tends to be discrete and many biochemically possible pathways evidently are not utilized.[15–18] An alternate view is that G proteins do not shuttle freely between GPCR and effector proteins,[19] which in turn suggests that signaling complexes containing receptor, G protein and effector can exist.[14] The latter idea is supported by a considerable body of physical and functional evidence,[20] and in some cases GPCRs, G proteins, and effectors are found to form stable complexes through which signaling can occur.[20,21] In contrast, other findings show that in some cases G proteins can be dissociated from their GPCRs or effectors, and also that trimeric G proteins can dissociate into their constituent Gα and Gβγ subunits in living cells. Thus it remains a topic of discussion which signaling systems dissociate, and whether this occurs as part of the process of signal transduction.[21,22] On a longer time scale, it is clear that molecular rearrangements do occur, such as the substitution of arrestins for G proteins that occurs during receptor desensitization or activation of some MAP kinase-signaling pathways.[23]

There are a number of cellular-organizing factors, such as scaffolding proteins and lipid microdomains, that can facilitate the existence of signaling complexes and thereby enhance signaling specificity. Scaffolding (or anchoring) proteins are multidomain structures that bind to two or more proteins to bring them into proximity with one another, for example, a receptor and its effector.[24–26] Scaffolding proteins may have additional binding partners that turn off signaling or localize them to particular cellular structures, and thus respectively can temporally or spatially focus signaling events within the cell.[23,27] Furthermore, scaffolds can also connect to downstream signaling components or other signaling systems such as receptor tyrosine kinases or ligand-gated ion channels.[25] Scaffolded signaling complexes are not necessarily stable entities, as these may be assembled or disassembled as part of the signaling process.[23]

A further layer of complexity comes from the fact that some proteins may be present in multiple copies within signaling complexes. It is well documented that many GPCRs can form homo- and hetero-oligomers,[28] and cooperative interactions within such complexes may profoundly affect observed dose–response relationships[29] and agonist-binding profiles.[30] GPCRs are typically thought to function as dimers, but emerging data suggest that at least some exist in nature as larger oligomers.[31] In some cases, more than one receptor protomer is required for the attainment of proper function, particularly with hetero-oligomeric GPCRs[32]; however, as has been reported with receptor tyrosine kinases,[33,34] at least some GPCRs when in monomeric form still exhibit the ability to signal.[35]

In addition to GPCRs, at least some effector proteins appear to oligomerize, including phospholipase Cβ[36] and adenylyl cyclase.[37] G protein heterotrimers are generally assumed to act as single units, although it is clear that some distantly related GTPases function as requisite dimers,[2] and also there is some evidence to suggest that individual Gα subunits may be able to associate[19] and/or communicate with one another,[38] perhaps via GPCR oligomers. Gα monomers may also be brought into proximity with one another by proteins with multiple Gα-binding domains such as RGS14 and GPSM3. In addition, scaffolding proteins that bind to GPCRs and/or effectors, such as the PDZ-containing protein INAD and the protein kinase A anchoring protein AKAP79,[39] can also self-associate, possibly leading to the formation of extended signaling complexes.

The arrangement and stability of signaling complexes is important from the point of view of understanding the effects of auxiliary GAPs, GDIs, and GEFs, as such proteins are likely to target these as well as free G proteins. In some cases, the presence of these proteins may impede G protein-mediated signaling due, for example, to steric effects,[40] but conversely in other cases it is clear that they are required for normal signaling to take place (e.g., GRK channel regulation in multiple tissues and rapid Gαt deactivation in the retina[7]). RGS and GPSM proteins can interact with free Gα proteins,[41,42] which in the context of the shuttling model would be consistent with the view that such proteins "intercept" free G proteins in transit between GPCRs and effectors. In contrast to this idea, there is ample evidence to show that G protein modulators additionally can interact with GPCRs, effectors, Gβγ, and scaffolding proteins.[7,43] While it is clear that such arrangements can contribute to the specificity and temporal focusing of GPCR signals, the overall impact of RGS proteins, GPSM proteins, and nonreceptor GEFs on G protein-mediated signals remains an area of active study.

3. RGS PROTEINS

RGS proteins were first discovered to be Gα GAPs in 1995,[41] and since then they have been shown to play an integral role in GPCR signaling. Early studies focused on the effects of RGS proteins on isolated Gα proteins in solution, and revealed that an RGS protein molecule could act catalytically to deactivate multiple equivalents of Gα–GTP.[7] In addition to their ability to deactivate G proteins by promoting GTP hydrolysis, RGS proteins may also prevent activated Gα proteins from productively interacting with their effectors through steric or competitive mechanisms.[40] This is particularly true in the case of Gαs-mediated signaling, where clear inhibitory effects of RGS proteins have been observed[44,45] in spite of the inability of RGS proteins to promote GTP hydrolysis by this Gα protein.[46–48]

While bimolecular RGS–Gα interactions presumably do occur in cells, it may be overly simplistic to view free Gα proteins as the sole or perhaps even the primary target of RGS proteins *in vivo*. For reasons noted above, one might not expect Gα–GTP to routinely venture from its site of activation into the cytosol, and correspondingly the ability of RGS proteins to limit GPCR signaling to some extent correlates with their ability to associate with the plasma membrane.[49,50] Whereas RGS–Gα interactions in solution-based assays would occur randomly in three-dimensional space, interactions with other proteins and with the plasma membrane would tend to restrict the movement, orientation, and available surface area of receptor-activated G proteins. Depending on the exact situation, this might either impede or enhance RGS–Gα interactions.

Gβγ subunits, which are required for GPCRs to activate Gα proteins[4], strongly inhibit RGS GAP effects on isolated Gα–GTP in solution,[5,51,52]. This effect does not preclude the ability of RGS proteins to promote GTPase activity over multiple cycles of nucleotide binding and hydrolysis in the presence of an activated receptor,[53–55] but it again becomes manifest with superstoichiometric Gβγ, which apparently competes with RGS proteins for Gα.[52] The targeting of RGS proteins to receptor-activated Gα in contrast may be facilitated by other membrane proteins. As previously noted, the inhibitory effects of RGS proteins on signaling can be promoted or otherwise regulated by associating with affiliated GPCRs, effectors, or scaffolding proteins.[7] Such protein–protein interactions may go beyond mere scaffolding effects. For example, the GAP effects of PLCβ on isolated Gαq[5] appear to be weaker than with M1 muscarinic receptor-activated G protein.[56,57] Similarly the maximal GTPase-accelerating effects of PLCβ and RGS4 on Gq co-reconstituted with M1 receptor can exceed three orders of magnitude[53] whereas maximal GAP effects on free Gα proteins are typically found to be about two orders of magnitude.[5,41,51] Thus the ability of RGS proteins (or other GAPs) to stimulate G protein hydrolytic activity may be sensitive to interactions between the GPCR and the G protein and/or RGS protein.

By targeting RGS proteins to G proteins that they have activated, receptors seem to play a role in assisting RGS proteins to achieve selectivity. The underlying basis of this selectivity toward certain receptors but not others, however, remains unclear but may involve particular structural elements. For example, the interaction between the N-terminus of RGS4 and the D2 and D3 Dopaminergic receptors leads to a potent negative effect on receptor-mediated inhibition of cAMP production. This activity is abolished upon deletion of the RGS4-N-terminus. Interestingly, other RGS proteins coexpressed with these receptors such as RGS2 and RGS9 have little or no effect on cAMP production.[58] In an earlier paper, Zeng *et al.* reported a similar observation where deletion of the RGS4-N-terminus led to a 10-fold decrease in RGS4 potency, and this also eliminated RGS4 receptor selectivity.[59] It thus seems like association between receptor and RGS4 may be critical in regulating RGS4 activity. The importance of RGS–receptor interactions is also demonstrated by the relationship between RGS2 and the AT1 angiotensin II receptor. A recent study has shown that the inhibitory effect of RGS2 on AT1 receptor stimulated Ca^{2+} activity is regulated via its N-terminus, possibly by directly interacting with the receptor. Deletion of this region or replacement with the N-terminus from other RGS proteins greatly decreased the ability of RGS2 to inhibit AT1 receptor signaling.[60] Similarly, the N-terminal domain of RGS2 has also been shown to affect M1 muscarinic receptor signaling.[50] Besides regulating RGS protein activity by changing their binding selectivity, the interaction between RGS protein and receptor may also alter RGS protein conformation and thus influence its activity. For example, the association between M3 receptor and RGS7 can induce the DEP domain of RGS7 to separate from Gβ and bind to the C-tail of the receptor, thereby inhibiting M3 receptor-mediated signaling.[61]

Besides direct association between RGS proteins and the receptor, another potential mechanism for RGS protein selectivity in regulating receptor signaling is that other factors such as scaffolding proteins may direct the RGS protein to the receptor and its associated G protein. Neurabin, a multidomain-scaffolding protein expressed in neural tissues, has been shown to assemble a complex between AT1R and RGS4, which allows the latter to attenuate AT1R signaling.[62] In addition, Spinophilin, a scaffolding protein known to regulate many GPCR signaling, is able to interact with multiple RGS proteins, such as RGS1, RGS2, RGS4, RGS16, and GAIP.[63] Recently, this protein has also been reported to inhibit α1-adrenergic receptor signaling and M3 receptor signaling by recruiting either RGS2 or RGS4 to the receptor in a ligand-dependent fashion.[63,64]

Reported effects of GPCRs and scaffolding proteins on RGS proteins have typically been ascertained indirectly, for example, through second messenger or electrophysiological assays.[65–69] To be sure that receptor-dependent differences in RGS protein inhibitory effects on signaling are indeed due to G protein deactivation, one would need to compare actual GAP activities; however, such experiments are technically challenging due to high background GTPase activities in cell-derived systems (multiple purified, reconstituted receptor–G protein preparations would not be practical for most labs). Membranes from baculovirus-infected insect cells, where high levels of protein expression are obtainable, do allow for the reliable measurement of RGS GAP activities on GPCR-activated G proteins,[54,70] and we have used this system to coexpress various GPCRs together with heterotrimeric G proteins. Membranes derived from Sf9 cells were used to compare different purified RGS proteins with different receptors (Fig. 10.1), anticipating that some GPCR–RGS combinations might work better than others. Contrary to this expectation we found no obvious selectivity between RGS proteins and GPCRs. Somewhat surprisingly a number of receptors tested did not appear to support RGS protein GAP activity at all. For some of these (δ-opioid, D2 dopaminergic, and 5HT1B receptors) there was little measurable agonist-stimulated GTPase activity (data not shown) so we could not be completely certain that there was no RGS GAP activity; however, the D4.4 dopaminergic receptor yielded an agonist signal comparable to those seen with the RGS protein-sensitive M2 muscarinic and α2a-adrenergic receptors coexpressed with

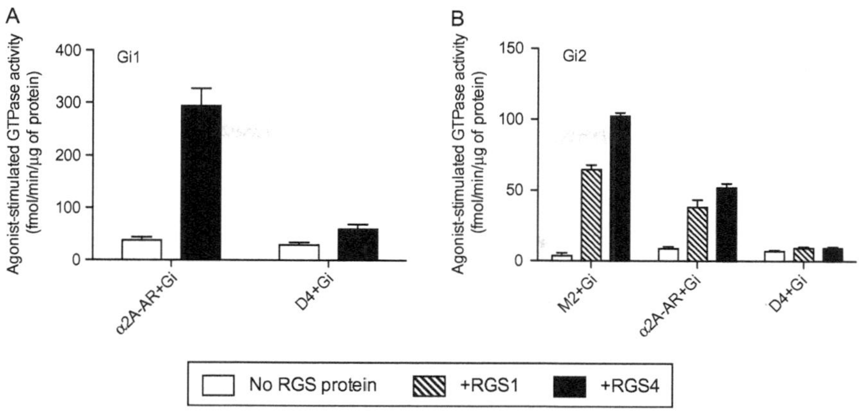

Figure 10.1 Sf9 insect cells were infected with baculoviruses encoding Gβ1, Gγ2, Gαi1 (A) or Gαi2 (B) and GPCRs as indicated and membranes from these were prepared as described by Cladman and Chidiac.[54] Agonist-stimulated GTPase activities were assayed essentially as described previously.[54] Agonists used were Epinephrine (α2a-adrenergic receptor), carbachol (M2 muscarinic receptor), and Dopamine (D4.4 dopaminergic receptor). The steady-state hydrolysis of [γ^{32}P]GTP by Sf9 membranes was measured in the absence and presence of purified RGS proteins. 50 μl reaction mixtures containing 20 mM HEPES, pH 7.5, 1 mM EDTA, 1 mM DTT, 0.1 mM PMSF, 10 μg/ml leupeptin, 1 μg/ml aprotinin, plus 10–50 mM NaCl, and 10 mM $MgCl_2$ were incubated for 5 min at 30 °C. The assay was stopped by adding 90 μl of ice-cold 5% Norit in 0.05 M NaH_2PO_4, the mix was centrifuged and the level of ^{32}Pi in the supernatant was determined by liquid-scintillation counting. Agonist-stimulated GTP hydrolysis was determined by subtracting the value from un-stimulated or antagonist-treated samples. Graphs are mean ± SEM values of 2 (A) or 3 (B) experiments with triplicate samples.

either Gi1 or Gi2. Notwithstanding this, essentially no RGS protein GAP activity could be detected with G proteins activated by the D4.4 dopaminergic receptor (Fig. 10.1).

The findings shown in Fig. 10.1 suggest that some GPCRs may be generally less supportive than others of RGS protein GAP activity. Although the reasons for this are unclear, it is consistent with the requirement for scaffolding proteins to enable certain RGS proteins to target particular signaling pathways.[71] Alternatively, another possible (and not mutually exclusive) explanation is that observed inhibitory effects of RGS proteins on signaling do not always reflect just their ability to promote GTP hydrolysis but may at least partly stem from other inhibitory effects, as noted above. To gain insight into the possible contributions of GAP-independent inhibitory mechanisms of RGS proteins on GPCR signaling, Anger and coworkers compared the effects on PLCβ activity of various RGS proteins coexpressed with M3 muscarinic receptor and either wild-type or GTPase-deficient Gαq.[72] Based on the different rank orders observed among RGS proteins with the two different Gαq proteins, it was concluded that some RGS proteins were more reliant than others on GAP-independent mechanisms for their inhibitory effects on signaling.[72] While that is a reasonable interpretation, a more stringent test would be to compare the GAP effects of RGS proteins to their inhibitory effects on second messenger regulation. While that apparently has never been done systematically, another study did compare the effects of purified RGS16 and full length and N-terminally truncated RGS2 in membrane-based PLCβ and GTPase assays with activated M1 muscarinic receptor.[73] In the second messenger-based assay, full length RGS2 was the most potent of the three, RGS16 was intermediate, and truncated RGS2 was essentially without effect; in stark contrast to this, all three produced similar maximal GAP effects, with RGS16 the most potent, and full length and truncated RSG2 characterized by EC_{50} values respectively about three- and tenfold greater than RGS16.[73] Based on this finding, it would seem premature to assume that an RGS protein-induced decrease in GPCR signaling could generally be taken as an accurate readout of its GAP activity. Rather, G protein deactivation due to increased GTP hydrolysis may only be a partial contributing factor to such inhibitory effects.

4. GPSM PROTEINS

Similar to RGS proteins that regulate the deactivation of Gα proteins, GEFs and GDI proteins are able to modulate the activation rate of G protein-mediated cell signaling. GDI proteins directly interact with and stabilize the inactive (GDP-bound) Gα proteins and inhibit GDP dissociation, which will in turn slow down the activation of Gα. Being recognized as a GDI, the Gβγ dimer is considered to stabilize Gα in its inactive state, suppressing spontaneous Gα activation while also facilitating Gα coupling to receptors.[52,74]

In recent years, proteins other than Gβγ have been identified as GDIs. Oxidized human neuroglobin, a heme protein that is expressed in the brain binds to Gαi–GDP, competes with Gβγ binding, and inhibits GDP release.[75] Interestingly, a 25–35% homology was found between neuroglobin and the RGS domain of GPCR kinases (GRKs).[76]

Among the GDIs identified for heterotrimeric G proteins, GPSM proteins have been studied extensively over the last decade or so. These proteins all share a highly conserved 15–20 amino acid Gα-binding motif that impedes GDP dissociation and also promotes Gβγ dissociation.[77–79] Although there is broad agreement regarding the biochemical nature of this domain, there is no consensus on what to call it. The first protein found to contain the domain was loco, an RGS12 homologue, found in Drosophila melanogaster.[80] The observation that RGS12 shared a similar Gα protein-interacting region distinct from the RGS domain led to the name Gαi/o-Loco motif, or GoLoco for short.[81] The same conserved region was named G protein-regulatory (GPR) domain at around the same time by Cismowski and colleagues, who identified it as a receptor-independent activator of Gβγ signaling.[77,78] The domain also was found in the Leu-Gly-Asn-enriched protein LGN and thus has also been referred to as the LGN motif,[82,83] and proteins bearing this domain are sometimes referred to collectively as Group II Activator of G Protein Signaling (AGS) proteins.[10] The term GPR while adequately descriptive is also used for GPCRs (particularly orphans) (www.genenames.org), whereas the remaining terms do not convey functional information. Therefore we prefer the term *GPSM domain*, in accordance with the names proposed by the HUGO Gene Nomenclature Committee for several of the proteins that contain it (GPSM1/AGS3; GPSM2/AGS5/LGN; GPSM3/AGS4/G18; GPSM4/Pcp-2/L7) (www.genenames.org). Although to date only four proteins have been named in this manner, we also suggest the collective term *GPSM proteins*.

Among the Gα subfamilies, most GPSM domains favor binding to Gαi/o proteins with typically higher affinity toward Gαi than Gαo.[84] Some GPSM proteins also interact with G proteins other than Gi or Go. For example, AGS3 interacts with Gαt and blocks rhodopsin-induced dissociation of GDP.[85] The selectivity between GPSM proteins and Gα protein subtypes seems to be influenced by amino acid residues outside of the core domain of the GPSM domain, as well as the all-helical domain of Gα.[86]

The GPSM domain has a higher binding affinity for inactive GDP-bound Gα relative to either nucleotide-free or activated Gα.[86] The association between the GPSM domain and Gα-GDP decreases the rate of nucleotide exchange on Gα, thus this leads to an inhibitory effect on Gα-activation.[87] In addition, binding of the GPSM motif results in a significant conformational change of the switch regions of the Gα subunit and impairs the binding of Gβγ. As a result, Gα–GDP–Gβγ and Gα–GDP–GPSM complexes are mutually exclusive.[42,87,88]

Since the GPSM motif and Gβγ are able to compete for Gα binding,[77] it has been hypothesized that the GPSM–Gα interaction may either promote heterotrimer dissociation or interfere with subunit re-association. A GPSM domain-based peptide derived from GPSM1 has the ability to inhibit Gα binding to Gβγ 10 times more effectively than the Gβγ hot spot-binding peptide (SIGK) (which also interferes with the binding between Gβγ and Gα) with an IC_{50} of 250 nM. In addition, this GPSM peptide was able to cause a rapid dissociation of Gβγ from Gα about 13-fold higher than the intrinsic K_{off} of Gβγ from Gα.[89] However, on the other hand, full-length RGS14 failed to disrupt pre-formed G protein heterotrimer.[90] Still, a GPSM peptide derived from RGS14 seemed to be able to prevent the reformation of the Gαβγ heterotrimer.[91] Overall, the ability of the GoLoco motif to promote G protein subunit dissociation may depend on the experimental or cellular context and the particular proteins in question.

The original discovery of GPSM motif containing proteins was based on a yeast-based screen for receptor-independent activators of heterotrimeric G protein signaling.[77] In this screen, GPSM1 was identified as being able to "release" Gβγ, and thus promote the selective activation of Gβγ-mediated signaling.[77] In addition, silencing GPSM1 in neuron progenitor cells resulted in a shift in spindle orientation and an abnormal differentiation of the cells into neurons, further suggesting a positive link between GPSM1 and Gβγ function.[92]

Ectopic GPSM3 expression was also found to inhibit $PLC\beta_2$ activation by Gβγ dimer in COS-7 cells.[93] This result is somewhat surprising since one would expect GPSM proteins to activate Gβγ signaling. In this study, it appeared that inhibition was due to an interaction between GPSM3 and the Gβ subunit. However, this interaction was observed in a COS-7 cell overexpression system, thus whether this reflects a direct interaction between GPSM3 and Gβ or an involvement of these two proteins in a complex is unclear.

Full-length GPSM2 (LGN) and constructs expressing only the GPSM domains of this protein were found to activate basal K^+ current, but inhibited dopamine receptor-mediated GIRK channel activation. The molecular mechanisms that control these distinct outcomes may reflect different effects of GPSM motifs on Gα and Gβγ signaling. The elevated basal K^+ current is likely due to a GPSM-promoted dissociation of the heterotrimer and subsequent Gβγ-dependent channel activation. On the other hand, the inhibitory effect on receptor signaling may stem from the GDI activity, which decreases Gα activation.[94]

The effects of GPSM proteins on ion channel activity were also demonstrated in Xenopus oocytes expressing Cav2.1 and κ-opioid receptors. Low levels of GPSM4 (Pcp-2) enhanced the inhibitory effect of opioid receptor has on Cav2.1 current. This effect was blocked by adding the C-terminus of β-adrenergic receptor kinase (GRK2), indicating a Gβγ-dependent pathway. Interestingly, at a higher expression level, GPSM4 exhibited the opposite effect.[95]

Although GPSM proteins appear to regulate Gα activation, they also may disrupt receptor–G protein coupling, ligand binding and signaling, etc. It has been reported that cytosolic GPSM1 (AGS3), but not membrane-associated GPSM1, inhibits 5HT receptor coupling to Gα subunits in a dose-dependent manner by interfering with the membrane association of Gαi1 subunits.[96] This inhibitory effect may in turn lead to a reduction in receptor-mediated cell signaling, however, this hasn't been tested. BRET assays using HEK293 cells suggest that receptor activation leads to a moderate decrease in GPSM1–G protein interaction, apparently due to Gα activation. In addition, a small Gα-dependent BRET signal is also detected between GPSM1 and cell-surface receptor, which indicates the possible formation of a receptor–G protein–GPSM complex.[97] A GPSM peptide derived from GPSM1 inhibited GDP dissociation subsequent to rhodopsin-dependent activation of transducin. This inhibitory effect may be due to its GDI activity; however, the possibility that the GPSM domain competes with the receptor for binding to G protein again cannot be excluded.[85]

GPSM3 (a.k.a. G18 or AGS4) is another multiple GPSM protein of great interest. Our lab has shown that beside the GPSM motif, which acts as a GDI on Gαi, the N-terminal proline-rich domain of GPSM3 may also serve as a Gα protein interaction partner; moreover, this domain may have varying biochemical activities toward different G proteins.[98] Not many studies have looked at the effects of GPSM3 on receptor signaling. Similar to GPSM1, a G protein–GPSM3 complex may together serve as a substrate for agonist-induced receptor activation. Upon receptor activation the interaction between GPSM3 and G protein is again reduced.[43] The remaining BRET signal between these two proteins may reflect either a membrane or intracellular localization of the complex. We have shown that the GPSM domains of this protein are able to inhibit the M2 receptor and RGS4-stimulated GTPase activity of Gi in isolated Sf9 membranes. However, whether this effect is due to changes in nucleotide exchange rate *per se* is difficult to know. Alternatively, this inhibitory effect may reflect competition between GPSM3 and the receptor for Gα proteins, or disruption of the G protein heterotrimer. It would be interesting to see if binding between GPSM3 and the G protein is able to alter receptor–G protein-coupling properties in the presence of agonists. In another study, ectopic GPSM3 expression in HEK293 cells inhibited signaling of the agonist-activated lysophophatidic acid receptor to endogenous PLCβ activation and the generation of inositol phosphate.[93]

Taken together, the available data on the effects of GPSM proteins on GPCR signaling would appear to indicate that under basal conditions Gβγ signaling may be stimulated while in the presence of a GPCR agonist signaling both Gα- and Gβγ-mediated signals are decreased due to a reduced nucleotide exchange rate and/or a loss of coupling between the receptor and the G protein heterotrimer. The detailed mechanism of how GPSM proteins affect GPCR signaling still remains largely unknown. It seems like a GPSM–Gα complex may serve as substrate for Ric-8 and together, play an important role in asymmetric cell division.[99] On the other hand, whether a GPSM–Gα protein complex is able to serve as an alternate Gαβγ and mediate receptor signaling also remains apparently untested.

Functional studies of GPSM motif proteins have implied a wide range of physiological roles. Many investigations have focused on the involvement of GPSM proteins in cell division,[100] which can also result in physiological changes. However, the role of GPCR-mediated signaling may also be significant. GPSM proteins, especially GPSM1, may have critical functions in cocaine, heroin, and alcohol addiction. During late withdrawal from repeated cocaine administration, the expression level of GPSM1 was found to be increased. Such changes in turn appear to lead to a decrease in Dopamine and/or opioid receptor-mediated Gi signaling and the activation Gβγ signaling, resulting in behavioral sensitization to cocaine challenge.[101] On the other hand, GPSM1 knock down is associated with the opposite effect.[102,103] The effect of GPSM4 (Pcp-2) on behavioral function has also been studied, and GPSM4 KO mice show a sex-dependent anxiety-like phenotype, with male mice exhibiting an increased anxiety level, similar to 5-HT1AR KO mice.[104] However, whether this is due to an inhibitory effect of GPSM4 on 5HT signaling remains unknown. Although clearly further study is required, the available data are consistent with the notion that GPSM proteins modulate GPCR signals *in vivo*, albeit in ways that are not yet well defined.

5. NONRECEPTOR GEFs

The vast majority of known GEFs for Gα proteins are GPCRs, but over the years it has become evident that other proteins exist that can promote the dissociation of GDP from Gα. These include the resistance to inhibitors of cholinesterase 8 (Ric-8) proteins (Ric-8A, Ric-8B, Ric-8BΔ9),[105] Gα-interacting vesicle-associated protein (GIV),[106] RASD1/AGS1/Dexras1,[107] GAP-43/neuromodulin/B-50,[108] GPSM3/AGS4/G18,[98] cysteine string protein (CSP),[109] human phosphatidylthanolamine-binding protein (hPBP),[110] β-amyloid precursor protein,[111] presenillin-1[112] and the yeast protein Arr4.[113] In addition, nucleobindin 1 and nucleobindin 2 each contain a region of homology to GIV and have been shown to produce measurable increases in Gαi3 steady-state GTPase activity, implying GEF activity.[114] Apart from similarities among the Ric-8 and GIV-like proteins there is no obvious structural resemblance between the noncanonical GEFs, which variously interact with members of all four Gα protein subfamilies. Several of these proteins have been observed to alter GPCR signals, although as discussed below some may have additional properties that might have an indirect impact on signaling. Non-GPCR GEFs could potentially either decrease agonist singling by interfering with receptor–G protein interactions, or alternatively could augment signaling by maintaining receptor-activated G proteins in an activated state, and evidence consistent with both possibilities has been reported. It is expected as well that effects of noncanonical GEFs on GPCR signaling will have physiological and pharmacological consequences, but at present there is little information about such possibilities.

Apart from those noted above, other proteins have been shown to display attributes consistent with GEF activity, but have not been unequivocally shown to promote GDP dissociation from Gα proteins. For example, the human transcription factor E2F8 in *S. cerevisiae*-based reporter assays was found to amplify a-factor GPCR signaling, and to promote signaling in a yeast strain lacking the receptor; positive effects of E2F8 were reduced or lost in strains lacking Gβ or components downstream of the G protein and upstream of the reporter, and reduced in a strain coexpressing RGS4.[115] In addition, E2F8 has been found to co-immunoprecipitate with Gαi (T Baranski, personal communication). These observations suggest that E2F8 can act as a GEF, but this has not yet been demonstrated directly. Another potential nonreceptor GEF, NG-GPA, has been identified in NG108-15 cells; however, even though extracts from these cells have been shown to promote GTPγS binding to purified Gα proteins in a manner that is seemingly distinct from GPCRs, this putative novel GEF has not been successfully isolated.[116] As well, numerous studies have shown that cellular Gα proteins can be turned on in response to activation of receptors that traverse the membrane only once (e.g., growth factor and cytokine receptors), which was postulated to reflect direct effects on nucleotide exchange and/or changes in G protein activity due to tyrosine residue phosphorylation.[117,118] Evidence supporting such mechanisms may be viewed as less than definitive,[119,120] and it remains uncertain which if any "non-G protein-coupled" receptors are capable of directly activating Gα proteins.[118] Other mechanisms may be possible, and in a growing subset of cases, it is clear that Gα can be turned on indirectly via the transactivation of heptahelical receptors by other receptor types.[119,120]

5.1. Ric-8

The Ric-8 proteins are the most thoroughly studied nonreceptor GEFs. Ric-8 was first identified in *Caenorhabditis elegans* as a gene whose deficiency allowed the organism to survive treatment with cholinesterase inhibitors,[121] inferring that the corresponding protein might somehow promote acetylcholine signaling. Single isoforms of Ric-8 have been found in *C. elegans* and *Drosophila*[122], whereas two *Ric-8* genes have been found in mammals. Ric-8A was identified in two independent yeast two-hybrid screens of rat and mouse brain cDNA libraries respectively using GTPase-deficient mutants of Gαo[123] and Gαq[124] as bait. Ric-8B was found using GTPase-deficient Gαs(long)[123] or Gαs(short)[125] against brain cDNA libraries and also using Gαs homologue Gαolf as bait in a yeast two-hybrid screen of an olfactory epithelium library.[126]

Although Ric-8 proteins primarily target free Gα proteins, they can also bind to Gαβγ heterotrimers and/or free Gβγ albeit with relatively low affinity,[105] and overall Gβγ tends to produce negative effects on Ric-8–Gα interactions.[105,123] As noted above (Section 4), Ric-8 proteins can also bind to some GPSM proteins, and thus such proteins fulfill a role in intracellular Ric-8-stimulated Gα functions analogous to that of Gβγ in GPCR signaling at the plasma membrane. Other signaling proteins identified as Ric-8A binding partners include adenylyl cyclase type 5[127] and neural cell adhesion molecule 180.[128]

The two mammalian isoforms of Ric-8 have distinct effects on Gα proteins. Ric-8A acts as a GEF for free Gαi/o, Gαq, and Gα12 proteins, with no apparent effect on Gαs nucleotide exchange.[123] Full length Ric-8B is primarily a GEF for Gαs and its homologue Gαolf, although in vitro Ric-8B also can have appreciable effects on Gαq and Gα13, but not Gαi1.[105] The shorter splice variant Ric-8BΔ9 appears to be a significant GEF for Gαs only,[105] as it shows little or no effect on other G proteins tested including Gαolf.[105,126] Full length Ric-8B and Ric-8A also promote GTPγS dissociation (and presumably GTP dissociation) from Gαs and Gαq, respectively, although this effect is not likely to impede G protein activation at normal cellular GTP concentrations.[105] Ric-8 proteins bind preferentially to their Gα targets in the GDP-bound state, and promote GDP dissociation to form a stable nucleotide-free transition state complex.[105,123] The affinity of Ric-8 for Gα tends to be decreased by Gα activation,[105,123] although in some cases, Ric-8 appears to bind to activated Gα.[105,125,129]

Dynamic cellular functions affected by Ric-8 include G protein-mediated signaling, cell division and differentiation/development. In addition, Ric-8 proteins are now recognized to play a major role in G protein processing. The loss of either Ric-8A or Ric-8B results in embryonic death,[130] pointing to a vital role in early development.[131] In cells derived from Ric-8 null blastocysts, the lack of either isoform was associated with greatly decreased G protein levels, consistent with other studies showing that cellular G protein levels tend to track with levels of their functional Ric-8 partners.[130] *Ric-8B*$^{-/-}$ cells were largely deficient in Gαs while *Ric-8A*$^{-/-}$ cells similarly exhibited major decreases in Gαi1/2, Gαq, and Gα13 as well as a moderate decrease in Gβ.[130] This likely reflects the ability of Ric-8 proteins to bind to nascent Gα proteins and act as chaperones that prevent their degradation prior to membrane insertion.[130] In addition, Ric-8 may impede Gα degradation by inhibiting its ubiquitination,[132] and it appears that Ric-8 can target Gα proteins to an endomembrane prior to trafficking to the plasma membrane,[129,130] which might facilitate their interactions with GPCRs. Overall Ric-8 proteins can have profound effects on G protein levels and localization, and it follows that cellular effects associated with changes Ric-8 could potentially be due to altered G protein availability, activity, or both. As summarized below, the data currently available strongly suggest but do not prove that Ric-8 proteins increase G protein signaling by virtue of their GEF activities.

Ric-8 proteins have been found to enhance G protein-mediated receptor signals in a number of systems. Evidence suggests an important role in the function of sensory GPCRs, where heightened sensitivity to external stimuli might convey a survival advantage. For example, Ric-8B exhibits a similar expression pattern to the olfactory G protein Gαolf in the nasal cavity and brain.[126] The mouse-odorant receptor mOR-EG when transiently expressed by itself or together with either Gαolf and Ric-8B in HEK-293 T cells failed to respond to an activating ligand (eugenol), but was able to stimulate cAMP production when all three proteins were expressed simultaneously.[133] This may be partly due to the increased presence of Gαolf at the cell periphery[133] and/or its upregulation by Ric-8B.[134] However, in other experiments the stimulation of adenylyl cyclase via β2-adrenergic or D1-dopaminergic receptors was measureable in cells transfected with receptor only, and these signals were greatly enhanced by the coexpression of both Gαolf and Ric-8B not by the coexpression of only one of these.[126] The enhancement of G protein activation in each case could only be observed in the presence of a receptor agonist,[126,133] which suggests the possibility that Ric-8B acts as an amplifier for these GPCR signals. Similar functions have been suggested for Ric-8 in other sensory systems, for example in *C. elegans* Ric-8 appears to work in conjunction with AGS-3 to fine-tune responses to aversive by amplifying and/or prolonging GPCR-initiated Gαo signaling via GPCRs coupled to Gαo.[135]

There is indirect evidence that Ric-8A may potentiate Gαq-mediated GPCR signals, as Ric-8A siRNA treatment of HEK293 T cells decreased signals mediated via endogenous Gq-coupled P2Y purinergic and endothelin receptors.[124] While such changes again could reflect a decrease in endogenous G protein levels associated with the loss of Ric-8A, it was also noted in that study that Ric-8A became translocated to the plasma membrane in an agonist-dependent manner.[124] Although the additional expression of exogenous wild-type Ric-8A did not enhance agonist signaling in this study, a myristoylation sequence-tagged form of the protein that spontaneously localized to the plasma membrane was associated with a significant increase in agonist-stimulated signal.[124] These observations reinforce the findings of an earlier study where Ric-8A colocalized at the plasma membrane with Gαq and Gαs subsequent to agonist activation of Gq- and Gs-coupled GPCRs, respectively.[125] It thus appears that Ric-8 may be recruited, through an undefined mechanism, to GPCR-activated Gα proteins in the plasma membrane to strengthen existing signals.

The mechanistic details of how Ric-8 may enhance GPCR signals are not yet known. Ric-8 proteins appear not to be able to interact productively with Gβγ-associated Gα subunits.[105,123] Gα and Gβγ may remain attached to one another[136] and also to the activating GPCR[21] during signal transduction. Therefore it is unclear how Ric-8 might usurp Gα–effector interactions, but it has been proposed that this may occur in conjunction with GPSM proteins, which are known to promote the dissociation of G protein heterotrimers.[135] This idea is supported by observations in *C. elegans* that the ability of Ric-8 to promote Gαo-mediated responses to aversive stimuli is completely dependent upon GPSM 1 (AGS3).[135] Alternatively, it is conceivable that Ric-8 could promote signaling by docking onto the effector protein, and indeed adenylyl cyclase type V (AC5) is able to bind to Ric-8A, an interaction that appears to facilitate the ability of Gαi to inhibit both forskolin-stimulated and receptor-stimulated cAMP production.[127] An anchoring site for Ric-8 might not always be required for post-GPCR stimulation of Gα signals, as G protein heterotrimer dissociation may be more complete in some cases than in others.[136]

5.2. Gα-interacting vesicle-associated protein (GIV)/Girdin

GIV is a large (1870 amino acid residue) multidomain protein that contains a G protein-binding domain that was identified by its ability to interact with inactive, GDP-bound Gαi3.[137] Interestingly, Gαi3 binding can occur through two separate domains on GIV, one insensitive and the other sensitive to the activation state of the Gα protein.[106] A truncated form of GIV made up of the last ~250 amino acid residues containing the activation-sensitive domain but lacking the activation-insensitive one was found to promote guanine nucleotide exchange on Gαi3; a single point mutation (F1685A) in the G protein-binding domain of this construct was associated with profound decreases in both GEF activity and Gαi3 binding.[138] In addition to its Gα-binding regions, GIV also contains a hook domain that interacts with microtubules, a coiled-coil domain that mediates homodimerization, a phosphatidylinositol 4-monophosphate lipid-binding domain, which interacts with plasma membrane and Golgi, and an Akt/actin/epidermal growth factor-binding domain, which modulates growth factor signaling.[139] GIV has been shown to have important but incompletely understood effects in cell migration, wound healing, and cancer metastasis.[139]

Wild-type GIV transiently expressed in cells promotes both Akt (protein kinase B) activation and cell migration, effects that are not observed with a full-length construct bearing the F1685A substitution.[106] GIV-dependent Akt activation and chemotaxis appear to occur through the Gαi-dependent, Gβγ-mediated stimulation of PI3 kinase, which can be initiated via either GPCR (fMLP, LPA) or receptor tyrosine kinase (insulin, EGF) signaling.[106,137] When heterotrimeric GSH–Gαi3–Gβγ was isolated on glutathione beads and exposed to increasing amounts of the C-terminal region of GIV or the corresponding F1685A mutant, a dose-dependent decrease in the amount of heterotrimer was observed with the former but not the latter.[106] Since an activating nucleotide apparently was not included in this experiment, it would appear that the GEF effect of GIV and consequent G protein activation may not be required for Gβγ-dependent PI3 kinase activation, but rather this could reflect competition between GIV and Gβγ for Gα.

Regardless of the activation state of Gα, it is not clear how either GPCR or RTK activation might lead to the promotion of Gβγ signaling by GIV. The direct binding of RTKs to GIV[139] suggests the intriguing possibility that these receptors could induce a conformational change in GIV that affects its ability to bind to and/or promote nucleotide exchange on Gαi; however, evidence for such a mechanism is lacking. For GPCRs, it is conceivable that GIV could maintain free Gβγ levels by re-activating Gα, but again there are no data to either support or refute this notion.

5.3. RASD1/AGS1/Dexras1

RASD1 (also called AGS1 or Dexras1) is an atypical monomeric G protein that along with its homologue RASD2 (Rhes) belongs to a distinct subclass in the Ras superfamily.[140] The architecture of these two proteins closely resembles that of other Ras proteins, although they are larger due to the presence of two cationic insert regions.[107] Both RASD1[107] and RASD2[141] display atypical guanine nucleotide-binding characteristics, with relatively low affinity for GDP and increased GTP binding under basal conditions compared to other Ras-like G proteins.

RASD1 can activate Erk1/2 in a pertussis toxin-sensitive,[107] Gβγ-dependent[142] manner when transfected into cells and it can act as a guanine nucleotide exchange factor for Gαi/o proteins,[107,143] but it does not appear to interact appreciably with either Gαs or Gα16.[78] Mutations in highly conserved amino acid residues in the predicted guanine nucleotide-binding site of RASD1 do not appear to destabilize the protein, but do prevent it from signaling through Gαi in cells,[78] suggesting that the binding of either GDP or GTP to RASD1 itself may be necessary for its activating effects. RASD2 is not known to act as a GEF but it has been reported to interact with Gα proteins.[144] To complicate matters further, both RASD1 and RASD2 can apparently bind to Gβγ dimers.[140,145] In addition to their interactions with heterotrimeric G proteins or subunits thereof, RASD1 and RASD2 can bind to several other signaling proteins. For example, RASD1 binds to the scaffolding protein CAPON (carboxyl-terminal PDZ ligand of neuronal nitric oxide synthase) through which it can form a ternary complex with nNOS, and this juxtaposition facilitates *S*-nitrosylation-mediated increases in the binding of GTP to and the GEF activity of RASD1.[146] Like other Ras proteins, RASD2 can activate p85-dependent PI3 kinase.[141,147]

Multiple effects of RASD1 and RASD2 on GPCR-dependent and GPCR-independent heterotrimeric G protein signaling have been described, but in general the underlying mechanisms are not well understood. In contrast to its ability to independently stimulate Gαi/Gβγ-dependent ERK activation,[107,142] RASD1 has been found to inhibit D2 dopaminergic-[148] or *N*-formyl peptide receptor-activated ERK signaling.[142] RASD1 also decreased the ability of the *N*-formyl peptide receptor agonist f-MLF to promote GTPγS binding to membranes containing this receptor.[142] RASD1 was also found to inhibit the ability of the Gαi-coupled M2 muscarinic receptor to activate Gβγ-dependent inwardly rectifying potassium (GIRK) currents, whereas a mutant RASD1 containing an inactivating point substitution in its nucleotide-binding site failed to inhibit signaling.[149] In another study, RASD1 and RASD2 both were able to attenuate M2 muscarinic receptor and Gβγ-dependent inhibitory effects on N-type calcium channels, but had no effect on the ability of β2-adrenergic receptors to inhibit this channel via Gαs and Gβγ.[150] The strong similarity between homologues in the latter study suggests that either RASD2 may also be able to act as a GEF on Gαi, or alternatively that GEF activity *per se* does not play an important role in the inhibition of this Gαi- and Gβγ-mediated signaling pathway. Several studies have examined the potential effects on adenylyl cyclase regulation of RASD1 and RASD2, some of which have yielded findings consistent with the general idea that these proteins promote inhibitory Gαi–Gβγ signaling in the absence of GPCR activation.[142,144] Overall, it seems that both RASD1 and RASD2 stimulate signaling via Gi heterotrimers under basal conditions but that this in turn can impede the activating effects of GPCRs; however, the details of the interplay among the various signaling proteins involved in these changes it is not yet clear.

5.4. Other nonreceptor GEFs

While the impact of Ric-8, GIV and RASD1 on GPCR signaling is not well understood, even less is known in this regard about the other nonreceptor GEFs that have been identified to date. GAP-43 is a 43-kDa neuronal protein that was found to promote GDP dissociation from both isolated Gαo protein and Go heterotrimer, with comparable effects on Gαi but not Gαs.[108] The addition of purified GAP-43 to reconstituted vesicles containing purified M2 muscarinic receptors plus heterotrimeric Go increased agonist-stimulated GTPase activity in a greater-than-additive manner, and correspondingly the injection of purified GAP-43 into *Xenopus* oocytes expressing either M2 muscarinic or 5HT1C serotonergic receptors exacerbated agonist effects on chloride channel opening.[151] Similarly, the μ-opioid receptor-mediated activation of GIRK channels in oocytes was augmented by the intracellular injection of the novel Gαi1 GEF hPBP,[110] and the nonreceptor yeast GEF Arr4 was found to increase GPCR signaling stimulated by α-factor.[113] In contrast to these effects, the N-terminal GEF function of GPSM3/AGS4/G18 was not found to increase M2 muscarinic receptor-stimulated Gαi1, but rather decreased this activity, suggesting that the receptor and the nonreceptor GEF may be competing for the Gα protein.[98] Taken together, the available data suggest that nonreceptor GEFs can have direct (and possibly also indirect) effects on GPCR-stimulated G protein signaling, and that this can either augment or impede GPCR signaling.

6. INTERACTIONS BETWEEN G PROTEIN MODULATORY DOMAINS

Although we have largely focused on individual effects of nonreceptor GEFs and RGS and GPSM domains, there are two broad areas that merit further attention, namely interactions *between* different Gα modulators, and *within* the various proteins that contain multiple Gα interaction sites. For example, there are now multiple studies demonstrating coordinated interactions between Ric-8 and GPSM proteins,[99,152–156] and in one case this may have an impact on GPCR signaling.[135] It seems likely that GPSM may serve as an anchoring domain in such cases, analogous to the role played by Gβγ in GPCR signaling.

About half of the GPSM proteins contain multiple GPSM domains,[10] but differences between the roles of the individual domains have not been identified nor is it known how they might function in a coordinated fashion. In a few cases G protein-interacting domains with contrasting activities are found in the same protein. Whether these act on G proteins in a coordinated, sequential, or mutually exclusive manner remains to be investigated. GPSM3 in addition to its three GPSM domains contains a proline-rich N-terminal region that acts as a GEF for Gαi,[98] and it is conceivable that these might function in a coordinated manner analogous to Ric-8–GPSM complexes.

The two largest RGS proteins, RGS12 and RGS14, each contain a GPSM domain in addition to an RGS domain. At first blush this pairing might seem odd, as the GDI function of the GPSM domain would presumably leave little for the RGS domain to do. This does not seem to happen, though, as full length RGS14 actually displays a better GAP effect on GPCR-activated G proteins than the isolated RGS domain.[157] The GDI function of the RGS14 GPSM domain, in contrast, appears to predominate under other conditions.[156] Whether RGS12 functions in the same way, and how the switch between RGS14 G protein-regulatory effects is accomplished, remains to be explained.

7. CONCLUSIONS

The studies summarized herein provide compelling evidence that GPCR signaling is influenced by proteins that modify G protein-activation states, and to a large extent the observed changes in signaling are consistent with the known biochemical effects of these proteins on isolated Gα. GPCR signaling is limited by RGS proteins that promote G protein deactivation, and by GPSM proteins that impede activation; nonreceptor GEFs activate Gα and in some cases augment GPCR signaling. Still, it is seldom clear to what degree changes in the activation state of Gα may underlie the alterations in receptor responses caused by RGS proteins, GPSM proteins, and nonreceptor GEFs, as the corresponding nucleotide-binding state and hydrolytic activity of the G protein are generally not monitored. With all three types of Gα modulators, it seems clear that their ability to physically disrupt interactions between a target Gα protein and its affiliated GPCR, Gβγ, or effector protein can also strongly influence receptor signals. These tendencies are illustrated in Fig. 10.2.

Steric or competitive effects on signaling are most widely recognized with GPSM proteins, which were identified as Group II AGS proteins based on their ability to promote Gβγ signaling and in spite of their propensity to impede Gα activation.[10] Combined, these effects tend to promote Gβγ-dependent signaling under basal conditions but interfere with activation by GPCRs due to heterotrimer disruption and/or negative effects on nucleotide exchange. Nonreceptor GEFs may produce similar stimulatory and inhibitory effects but for partly different reasons, and as noted the effects on heterotrimer stability and binding to targets other than Gα can vary considerably from one nonreceptor GEF to the next. Apart from their incompletely understood effects on Gαs-mediated signaling, RGS proteins are generally presumed to inhibit GPCR signals by virtue of their GAP effects; however, a direct demonstration of this in any system is still lacking and

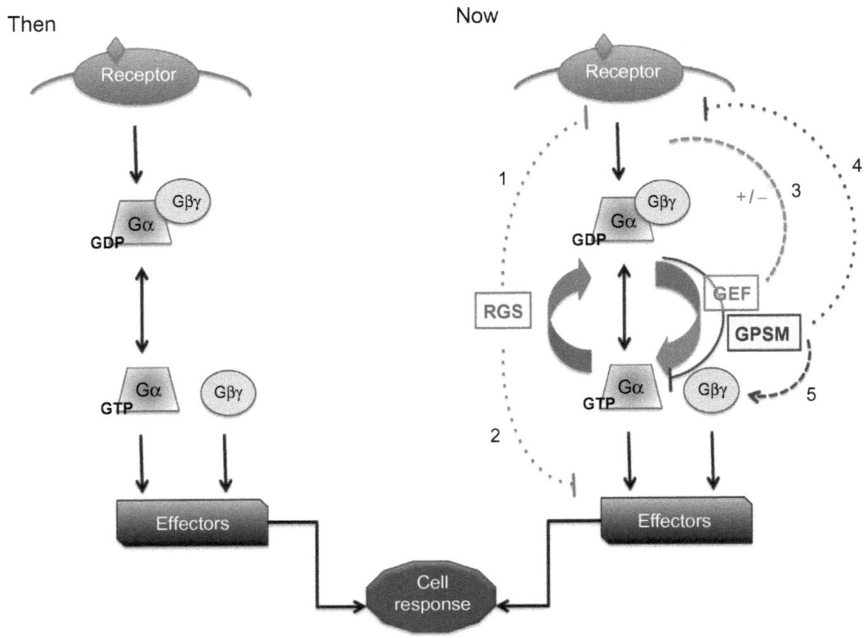

Figure 10.2 Model illustrating the effects of RGS, GPSM and nonreceptor GEF proteins on GPCR signaling. Solid lines indicate effects on G protein activity and dashed lines show potential additional effects on receptor signaling as follows: 1, Direct/indirect interactions between RGS proteins and the receptor may inhibit signaling. 2, RGS proteins can inhibit effector regulation by blocking G protein–effector interactions. 3, Nonreceptor GEFs may either potentiate or decrease receptor signaling respectively by maintaining GPCR-activated Gα in the activated state or by competing with the GPCR for Gα. 4, GPSM proteins may inhibit receptor signaling by interfering with receptor–G protein coupling. 5, GPSM proteins may activate Gβγ signaling. (See Color Insert.)

some findings suggest that non-GAP effects can also play a substantial role. In time hopefully such ambiguities will be resolved, and the impact of these proteins on GPCR signaling will be better understood.

ACKNOWLEDGMENTS

We are grateful to colleagues who shared unpublished data (Thomas Baranski, Washington University), expertise (Lou Luttrell, Medical University of South Carolina), or both (Greg Tall, University of Rochester). P. C. holds a Career Investigator Award from the Heart and Stroke Foundation of Ontario and research in the Chidiac laboratory is funded by the Canadian Institutes of Health Research, the National Sciences and Engineering Research Council of Canada, and the Heart and Stroke Foundation of Ontario.

REFERENCES

1. Bourne HR, Sanders DA, McCormick F. The GTPase superfamily: conserved structure and molecular mechanism. *Nature* 1991;**349**:117–27.
2. Wittinghofer A, Vetter IR. Structure-function relationships of the G domain, a canonical switch motif. *Annu Rev Biochem* 2011;**80**:943–71.
3. Higashijima T, Ferguson KM, Sternweis PC, Smigel MD, Gilman AG. Effects of Mg2+ and the beta gamma-subunit complex on the interactions of guanine nucleotides with G proteins. *J Biol Chem* 1987;**262**:762–6.
4. Clapham DE, Neer EJ. G protein beta gamma subunits. *Annu Rev Pharmacol Toxicol* 1997;**37**:167–203.
5. Chidiac P, Ross EM. Phospholipase C-beta1 directly accelerates GTP hydrolysis by Galphaq and acceleration is inhibited by Gbeta gamma subunits. *J Biol Chem* 1999;**274**:19639–43.
6. Suzuki N, Hajicek N, Kozasa T. Regulation and physiological functions of G12/13-mediated signaling pathways. *Neurosignals* 2009;**17**:55–70.
7. Abramow-Newerly M, Roy AA, Nunn C, Chidiac P. RGS proteins have a signaling complex: interactions between RGS proteins and GPCRs, effectors, and auxiliary proteins. *Cell Signal* 2006;**18**:579–91.
8. Ross EM, Wilkie TM. GTPase-activating proteins for heterotrimeric G proteins: regulators of G protein signaling (RGS) and RGS-like proteins. *Annu Rev Biochem* 2000;**69**:795–827.
9. Carman CV, Parent JL, Day PW, Pronin AN, Sternweis PM, Wedegaertner PB, et al. Selective regulation of Galpha(q/11) by an RGS domain in the G protein-coupled receptor kinase, GRK2. *J Biol Chem* 1999;**274**:34483–92.
10. Blumer JB, Oner SS, Lanier SM. Group II activators of G-protein signaling and proteins containing a G-protein regulatory motif. *Acta Physiol (Oxf)* 2012;**204**:202–18.
11. Werts AD, Goldstein B. How signaling between cells can orient a mitotic spindle. *Semin Cell Dev Biol* 2011;**22**:842–9.
12. Morin X, Bellaiche Y. Mitotic spindle orientation in asymmetric and symmetric cell divisions during animal development. *Dev Cell* 2011;**21**:102–19.
13. Hewavitharana T, Wedegaertner PB. Non-canonical signaling and localizations of heterotrimeric G proteins. *Cell Signal* 2012;**24**:25–34.
14. Chidiac P. Rethinking receptor-G protein-effector interactions. *Biochem Pharmacol* 1998;**55**:549–56.
15. Graeser D, Neubig RR. Compartmentation of receptors and guanine nucleotide-binding proteins in NG108-15 cells: lack of cross-talk in agonist binding among the alpha 2-adrenergic, muscarinic, and opiate receptors. *Mol Pharmacol* 1993;**43**:434–43.
16. Kleuss C, Scherubl H, Hescheler J, Schultz G, Wittig B. Different beta-subunits determine G-protein interaction with transmembrane receptors. *Nature* 1992;**358**:424–6.
17. Kleuss C, Scherubl H, Hescheler J, Schultz G, Wittig B. Selectivity in signal transduction determined by gamma subunits of heterotrimeric G proteins. *Science* 1993;**259**:832–4.
18. Kleuss C, Hescheler J, Ewel C, Rosenthal W, Schultz G, Wittig B. Assignment of G-protein subtypes to specific receptors inducing inhibition of calcium currents. *Nature* 1991;**353**:43–8.
19. Rodbell M. The complex regulation of receptor-coupled G-proteins. *Adv Enzyme Regul* 1997;**37**:427–35.
20. Boivin B, Vaniotis G, Allen BG, Hebert TE. G protein-coupled receptors in and on the cell nucleus: a new signaling paradigm? *J Recept Signal Transduct Res* 2008;**28**:15–28.
21. Hein P, Bunemann M. Coupling mode of receptors and G proteins. *Naunyn Schmiedebergs Arch Pharmacol* 2009;**379**:435–43.

22. Lambert NA. Dissociation of heterotrimeric g proteins in cells. *Sci Signal* 2008;**1**:re5.
23. Scott JD, Pawson T. Cell signaling in space and time: where proteins come together and when they're apart. *Science* 2009;**326**:1220–4.
24. Welch EJ, Jones BW, Scott JD. Networking with AKAPs: context-dependent regulation of anchored enzymes. *Mol Interv* 2010;**10**:86–97.
25. Magalhaes AC, Dunn H, Ferguson SS. Regulation of GPCR activity, trafficking and localization by GPCR-interacting proteins. *Br J Pharmacol* 2012;**165**:1717–36.
26. Suh PG, Hwang JI, Ryu SH, Donowitz M, Kim JH. The roles of PDZ-containing proteins in PLC-beta-mediated signaling. *Biochem Biophys Res Commun* 2001;**288**:1–7.
27. Greenwald M, Peloso PM, Mandel D, Soto O, Mehta A, Frontera N, et al. Further assessment of the clinically effective dose range of etoricoxib: a randomized, double-blinded, placebo-controlled trial in rheumatoid arthritis. *Curr Med Res Opin* 2011;**27**:2033–42.
28. Gazi L, Lopez-Gimenez JF, Strange PG. Formation of oligomers by G protein-coupled receptors. *Curr Opin Drug Discov Devel* 2002;**5**:756–63.
29. Bindslev N. A homotropic two-state model and auto-antagonism. *BMC Pharmacol* 2004;**4**:11.
30. Smith NJ, Milligan G. Allostery at G protein-coupled receptor homo- and heteromers: uncharted pharmacological landscapes. *Pharmacol Rev* 2010;**62**:701–25.
31. Park PS, Filipek S, Wells JW, Palczewski K. Oligomerization of G protein-coupled receptors: past, present, and future. *Biochemistry* 2004;**43**:15643–56.
32. Marshall FH, Jones KA, Kaupmann K, Bettler B. GABAB receptors—the first 7TM heterodimers. *Trends Pharmacol Sci* 1999;**20**:396–9.
33. Lu K, Guidotti G. Identification of the cysteine residues involved in the class I disulfide bonds of the human insulin receptor: properties of insulin receptor monomers. *Mol Biol Cell* 1996;**7**:679–91.
34. Chi SS, Vetiska SM, Gill RS, Hsiung MS, Liu F, Van Tol HH. Transactivation of PDGFRbeta by dopamine D4 receptor does not require PDGFRbeta dimerization. *Mol Brain* 2010;**3**:22.
35. Whorton MR, Bokoch MP, Rasmussen SG, Huang B, Zare RN, Kobilka B, et al. A monomeric G protein-coupled receptor isolated in a high-density lipoprotein particle efficiently activates its G protein. *Proc Natl Acad Sci USA* 2007;**104**:7682–7.
36. Ilkaeva O, Kinch LN, Paulssen RH, Ross EM. Mutations in the carboxyl-terminal domain of phospholipase C-beta 1 delineate the dimer interface and a potential Galphaq interaction site. *J Biol Chem* 2002;**277**:4294–300.
37. Cooper DM, Crossthwaite AJ. Higher-order organization and regulation of adenylyl cyclases. *Trends Pharmacol Sci* 2006;**27**:426–31.
38. Chidiac P, Wells JW. Effects of adenyl nucleotides and carbachol on cooperative interactions among G proteins. *Biochemistry* 1992;**31**:10908–21.
39. Gold MG, Stengel F, Nygren PJ, Weisbrod CR, Bruce JE, Robinson CV, et al. Architecture and dynamics of an A-kinase anchoring protein 79 (AKAP79) signaling complex. *Proc Natl Acad Sci USA* 2011;**108**:6426–31.
40. Hepler JR, Berman DM, Gilman AG, Kozasa T. RGS4 and GAIP are GTPase-activating proteins for Gq alpha and block activation of phospholipase C beta by gamma-thio-GTP-Gq alpha. *Proc Natl Acad Sci USA* 1997;**94**:428–32.
41. Berman DM, Wilkie TM, Gilman AG. GAIP and RGS4 are GTPase-activating proteins for the Gi subfamily of G protein alpha subunits. *Cell* 1996;**86**:445–52.
42. Natochin M, Gasimov KG, Artemyev NO. Inhibition of GDP/GTP exchange on G alpha subunits by proteins containing G-protein regulatory motifs. *Biochemistry* 2001;**40**:5322–8.
43. Oner SS, Maher EM, Breton B, Bouvier M, Blumer JB. Receptor-regulated interaction of activator of G-protein signaling-4 and Galphai. *J Biol Chem* 2010;**285**:20588–94.

44. Roy AA, Nunn C, Ming H, Zou MX, Penninger J, Kirshenbaum LA, et al. Up-regulation of endogenous RGS2 mediates cross-desensitization between Gs and Gq signaling in osteoblasts. *J Biol Chem* 2006;**281**:32684–93.
45. Chatterjee TK, Eapen AK, Fisher RA. A truncated form of RGS3 negatively regulates G protein-coupled receptor stimulation of adenylyl cyclase and phosphoinositide phospholipase C. *J Biol Chem* 1997;**272**:15481–7.
46. Johnson EN, Druey KM. Functional characterization of the G protein regulator RGS13. *J Biol Chem* 2002;**277**:16768–74.
47. Scheschonka A, Dessauer CW, Sinnarajah S, Chidiac P, Shi CS, Kehrl JH. RGS3 is a GTPase-activating protein for g(ialpha) and g(qalpha) and a potent inhibitor of signaling by GTPase-deficient forms of g(qalpha) and g(11alpha). *Mol Pharmacol* 2000;**58**:719–28.
48. Roy AA, Lemberg KE, Chidiac P. Recruitment of RGS2 and RGS4 to the plasma membrane by G proteins and receptors reflects functional interactions. *Mol Pharmacol* 2003;**64**:587–93.
49. Heximer SP, Lim H, Bernard JL, Blumer KJ. Mechanisms governing subcellular localization and function of human RGS2. *J Biol Chem* 2001;**276**:14195–203.
50. Gu S, He J, Ho WT, Ramineni S, Thal DM, Natesh R, et al. Unique hydrophobic extension of the RGS2 amphipathic helix domain imparts increased plasma membrane binding and function relative to other RGS R4/B subfamily members. *J Biol Chem* 2007;**282**:33064–75.
51. Wang J, Tu Y, Woodson J, Song X, Ross EM. A GTPase-activating protein for the G protein Galphaz. Identification, purification, and mechanism of action. *J Biol Chem* 1997;**272**:5732–40.
52. Tang W, Tu Y, Nayak SK, Woodson J, Jehl M, Ross EM. Gbetagamma inhibits Galpha GTPase-activating proteins by inhibition of Galpha-GTP binding during stimulation by receptor. *J Biol Chem* 2006;**281**:4746–53.
53. Mukhopadhyay S, Ross EM. Rapid GTP binding and hydrolysis by G(q) promoted by receptor and GTPase-activating proteins. *Proc Natl Acad Sci USA* 1999;**96**: 9539–44.
54. Cladman W, Chidiac P. Characterization and comparison of RGS2 and RGS4 as GTPase-activating proteins for m2 muscarinic receptor-stimulated G(i). *Mol Pharmacol* 2002;**62**:654–9.
55. Mao H, Zhao Q, Daigle M, Ghahremani MH, Chidiac P, Albert PR. RGS17/RGSZ2, a novel regulator of Gi/o, Gz, and Gq signaling. *J Biol Chem* 2004;**279**:26314–22.
56. Biddlecome GH, Berstein G, Ross EM. Regulation of phospholipase C-beta1 by Gq and m1 muscarinic cholinergic receptor. Steady-state balance of receptor-mediated activation and GTPase-activating protein-promoted deactivation. *J Biol Chem* 1996; **271**:7999–8007.
57. Berstein G, Blank JL, Jhon DY, Exton JH, Rhee SG, Ross EM. Phospholipase C-beta 1 is a GTPase-activating protein for Gq/11, its physiologic regulator. *Cell* 1992;**70**:411–8.
58. Min C, Cheong SY, Cheong SJ, Kim M, Cho DI, Kim KM. RGS4 exerts inhibitory activities on the signaling of dopamine D2 receptor and D3 receptor through the N-terminal region. *Pharmacol Res* 2012;**65**:213–20.
59. Zeng W, Xu X, Popov S, Mukhopadhyay S, Chidiac P, Swistok J, et al. The N-terminal domain of RGS4 confers receptor-selective inhibition of G protein signaling. *J Biol Chem* 1998;**273**:34687–90.
60. Matsuzaki N, Nishiyama M, Song D, Moroi K, Kimura S. Potent and selective inhibition of angiotensin AT1 receptor signaling by RGS2: roles of its N-terminal domain. *Cell Signal* 2011;**23**:1041–9.

61. Sandiford SL, Wang Q, Levay K, Buchwald P, Slepak VZ. Molecular organization of the complex between the muscarinic M3 receptor and the regulator of G protein signaling, Gbeta(5)-RGS7. *Biochemistry* 2010;**49**:4998–5006.
62. Chen Y, Liu Y, Cottingham C, McMahon L, Jiao K, Greengard P, et al. Neurabin scaffolding of adenosine receptor and RGS4 regulates anti-seizure effect of endogenous adenosine. *J Neurosci* 2012;**32**:2683–95.
63. Wang X, Zeng W, Soyombo AA, Tang W, Ross EM, Barnes AP, et al. Spinophilin regulates Ca2+ signaling by binding the N-terminal domain of RGS2 and the third intracellular loop of G-protein-coupled receptors. *Nat Cell Biol* 2005;**7**:405–11.
64. Ruiz de Azua I, Nakajima K, Rossi M, Cui Y, Jou W, Gavrilova O, et al. Spinophilin as a novel regulator of M3 muscarinic receptor-mediated insulin release in vitro and in vivo. *FASEB J* 2012;**26**:4275–86.
65. Xu X, Zeng W, Popov S, Berman DM, Davignon I, Yu K, et al. RGS proteins determine signaling specificity of Gq-coupled receptors. *J Biol Chem* 1999;**274**:3549–56.
66. Wang Q, Liu M, Mullah B, Siderovski DP, Neubig RR. Receptor-selective effects of endogenous RGS3 and RGS5 to regulate mitogen-activated protein kinase activation in rat vascular smooth muscle cells. *J Biol Chem* 2002;**277**:24949–58.
67. Saitoh O, Murata Y, Odagiri M, Itoh M, Itoh H, Misaka T, et al. Alternative splicing of RGS8 gene determines inhibitory function of receptor type-specific Gq signaling. *Proc Natl Acad Sci USA* 2002;**99**:10138–43.
68. Cho H, Harrison K, Schwartz O, Kehrl JH. The aorta and heart differentially express RGS (regulators of G-protein signaling) proteins that selectively regulate sphingosine 1-phosphate, angiotensin II and endothelin-1 signaling. *Biochem J* 2003;**371**:973–80.
69. Benians A, Nobles M, Hosny S, Tinker A. Regulators of G-protein signaling form a quaternary complex with the agonist, receptor, and G-protein. A novel explanation for the acceleration of signaling activation kinetics. *J Biol Chem* 2005;**280**:13383–94.
70. Schneider EH, Seifert R. Histamine H(4) receptor-RGS fusion proteins expressed in Sf9 insect cells: a sensitive and reliable approach for the functional characterization of histamine H(4) receptor ligands. *Biochem Pharmacol* 2009;**78**:607–16.
71. Abramow-Newerly M, Ming H, Chidiac P. Modulation of subfamily B/R4 RGS protein function by 14-3-3 proteins. *Cell Signal* 2006;**18**:2209–22.
72. Anger T, Zhang W, Mende U. Differential contribution of GTPase activation and effector antagonism to the inhibitory effect of RGS proteins on Gq-mediated signaling in vivo. *J Biol Chem* 2004;**279**:3906–15.
73. Bernstein LS, Ramineni S, Hague C, Cladman W, Chidiac P, Levey AI, et al. RGS2 binds directly and selectively to the M1 muscarinic acetylcholine receptor third intracellular loop to modulate Gq/11alpha signaling. *J Biol Chem* 2004;**279**:21248–56.
74. Neer EJ. Heterotrimeric G proteins: organizers of transmembrane signals. *Cell* 1995;**80**:249–57.
75. Kitatsuji C, Kurogochi M, Nishimura S, Ishimori K, Wakasugi K. Molecular basis of guanine nucleotide dissociation inhibitor activity of human neuroglobin by chemical cross-linking and mass spectrometry. *J Mol Biol* 2007;**368**:150–60.
76. Wakasugi K, Nakano T, Morishima I. Oxidized human neuroglobin acts as a heterotrimeric Galpha protein guanine nucleotide dissociation inhibitor. *J Biol Chem* 2003;**278**:36505–12.
77. Takesono A, Cismowski MJ, Ribas C, Bernard M, Chung P, Hazard 3rd S, et al. Receptor-independent activators of heterotrimeric G-protein signaling pathways. *J Biol Chem* 1999;**274**:33202–5.
78. Cismowski MJ, Takesono A, Ma C, Lizano JS, Xie X, Fuernkranz H, et al. Genetic screens in yeast to identify mammalian nonreceptor modulators of G-protein signaling. *Nat Biotechnol* 1999;**17**:878–83.

79. Willard FS, Kimple RJ, Siderovski DP. Return of the GDI: the GoLoco motif in cell division. *Annu Rev Biochem* 2004;**73**:925–51.
80. Granderath S, Stollewerk A, Greig S, Goodman CS, O'Kane CJ, Klambt C. loco encodes an RGS protein required for Drosophila glial differentiation. *Development* 1999;**126**:1781–91.
81. Siderovski DP, Diverse-Pierluissi M, De Vries L. The GoLoco motif: a Galphai/o binding motif and potential guanine-nucleotide exchange factor. *Trends Biochem Sci* 1999;**24**:340–1.
82. Mochizuki N, Cho G, Wen B, Insel PA. Identification and cDNA cloning of a novel human mosaic protein, LGN, based on interaction with G alpha i2. *Gene* 1996;**181**:39–43.
83. Ponting CP. Raf-like Ras/Rap-binding domains in RGS12- and still-life-like signaling proteins. *J Mol Med (Berl)* 1999;**77**:695–8.
84. Blumer JB, Cismowski MJ, Sato M, Lanier SM. AGS proteins: receptor-independent activators of G-protein signaling. *Trends Pharmacol Sci* 2005;**26**:470–6.
85. Natochin M, Lester B, Peterson YK, Bernard ML, Lanier SM, Artemyev NO. AGS3 inhibits GDP dissociation from galpha subunits of the Gi family and rhodopsin-dependent activation of transducin. *J Biol Chem* 2000;**275**:40981–5.
86. Kimple RJ, Kimple ME, Betts L, Sondek J, Siderovski DP. Structural determinants for GoLoco-induced inhibition of nucleotide release by Galpha subunits. *Nature* 2002;**416**:878–81.
87. Siderovski DP, Willard FS. The GAPs, GEFs, and GDIs of heterotrimeric G-protein alpha subunits. *Int J Biol Sci* 2005;**1**:51–66.
88. Bernard ML, Peterson YK, Chung P, Jourdan J, Lanier SM. Selective interaction of AGS3 with G-proteins and the influence of AGS3 on the activation state of G-proteins. *J Biol Chem* 2001;**276**:1585–93.
89. Ghosh M, Peterson YK, Lanier SM, Smrcka AV. Receptor- and nucleotide exchange-independent mechanisms for promoting G protein subunit dissociation. *J Biol Chem* 2003;**278**:34747–50.
90. Mittal V, Linder ME. Biochemical characterization of RGS14: RGS14 activity towards G-protein alpha subunits is independent of its binding to Rap2A. *Biochem J* 2006;**394**:309–15.
91. Webb CK, McCudden CR, Willard FS, Kimple RJ, Siderovski DP, Oxford GS. D2 dopamine receptor activation of potassium channels is selectively decoupled by Galpha-specific GoLoco motif peptides. *J Neurochem* 2005;**92**:1408–18.
92. Sanada K, Tsai LH. G protein betagamma subunits and AGS3 control spindle orientation and asymmetric cell fate of cerebral cortical progenitors. *Cell* 2005;**122**:119–31.
93. Giguere PM, Laroche G, Oestreich EA, Siderovski DP. G-protein signaling modulator-3 regulates heterotrimeric G-protein dynamics through dual association with Gbeta and Galphai protein subunits. *J Biol Chem* 2012;**287**:4863–74.
94. Wiser O, Qian X, Ehlers M, Ja WW, Roberts RW, Reuveny E, et al. Modulation of basal and receptor-induced GIRK potassium channel activity and neuronal excitability by the mammalian PINS homolog LGN. *Neuron* 2006;**50**:561–73.
95. Kinoshita-Kawada M, Oberdick J, Xi Zhu M. A Purkinje cell specific GoLoco domain protein, L7/Pcp-2, modulates receptor-mediated inhibition of Cav2.1 Ca2+ channels in a dose-dependent manner. *Brain Res Mol Brain Res* 2004;**132**:73–86.
96. Ma H, Peterson YK, Bernard ML, Lanier SM, Graber SG. Influence of cytosolic AGS3 on receptor–G protein coupling. *Biochemistry* 2003;**42**:8085–93.
97. Oner SS, An N, Vural A, Breton B, Bouvier M, Blumer JB, et al. Regulation of the AGS3.G{alpha}i signaling complex by a seven-transmembrane span receptor. *J Biol Chem* 2010;**285**:33949–58.

98. Zhao P, Nguyen CH, Chidiac P. The proline-rich N-terminal domain of G18 exhibits a novel G protein regulatory function. *J Biol Chem* 2010;**285**:9008–17.
99. Thomas CJ, Tall GG, Adhikari A, Sprang SR. Ric-8A catalyzes guanine nucleotide exchange on G alphai1 bound to the GPR/GoLoco exchange inhibitor AGS3. *J Biol Chem* 2008;**283**:23150–60.
100. Blumer JB, Smrcka AV, Lanier SM. Mechanistic pathways and biological roles for receptor-independent activators of G-protein signaling. *Pharmacol Ther* 2007;**113**: 488–506.
101. Shippenberg TS, Rea W. Sensitization to the behavioral effects of cocaine: modulation by dynorphin and kappa-opioid receptor agonists. *Pharmacol Biochem Behav* 1997;**57**: 449–55.
102. Bowers MS, McFarland K, Lake RW, Peterson YK, Lapish CC, Gregory ML, et al. Activator of G protein signaling 3: a gatekeeper of cocaine sensitization and drug seeking. *Neuron* 2004;**42**:269–81.
103. Bowers MS. Activators of G-protein signaling 3: a drug addiction molecular gateway. *Behav Pharmacol* 2010;**21**:500–13.
104. Parks CL, Robinson PS, Sibille E, Shenk T, Toth M. Increased anxiety of mice lacking the serotonin1A receptor. *Proc Natl Acad Sci USA* 1998;**95**:10734–9.
105. Chan P, Gabay M, Wright FA, Tall GG. Ric-8B is a GTP-dependent G protein alphas guanine nucleotide exchange factor. *J Biol Chem* 2011;**286**:19932–42.
106. Garcia-Marcos M, Ghosh P, Farquhar MG. GIV is a nonreceptor GEF for G alpha i with a unique motif that regulates Akt signaling. *Proc Natl Acad Sci USA* 2009;**106**: 3178–83.
107. Cismowski MJ, Ma C, Ribas C, Xie X, Spruyt M, Lizano JS, et al. Activation of heterotrimeric G-protein signaling by a ras-related protein. Implications for signal integration. *J Biol Chem* 2000;**275**:23421–4.
108. Strittmatter SM, Valenzuela D, Sudo Y, Linder ME, Fishman MC. An intracellular guanine nucleotide release protein for G0. GAP-43 stimulates isolated alpha subunits by a novel mechanism. *J Biol Chem* 1991;**266**:22465–71.
109. Natochin M, Campbell TN, Barren B, Miller LC, Hameed S, Artemyev NO, et al. Characterization of the G alpha(s) regulator cysteine string protein. *J Biol Chem* 2005;**280**:30236–41.
110. Kroslak T, Koch T, Kahl E, Hollt V. Human phosphatidylethanolamine-binding protein facilitates heterotrimeric G protein-dependent signaling. *J Biol Chem* 2001; **276**:39772–8.
111. Nishimoto I, Okamoto T, Matsuura Y, Takahashi S, Okamoto T, Murayama Y, et al. Alzheimer amyloid protein precursor complexes with brain GTP-binding protein G (o). *Nature* 1993;**362**:75–9.
112. Smine A, Xu X, Nishiyama K, Katada T, Gambetti P, Yadav SP, et al. Regulation of brain G-protein go by Alzheimer's disease gene presenilin-1. *J Biol Chem* 1998;**273**:16281–8.
113. Lee MJ, Dohlman HG. Coactivation of G protein signaling by cell-surface receptors and an intracellular exchange factor. *Curr Biol* 2008;**18**:211–5.
114. Garcia-Marcos M, Kietrsunthorn PS, Wang H, Ghosh P, Farquhar MG. G Protein binding sites on Calnuc (nucleobindin 1) and NUCB2 (nucleobindin 2) define a new class of G(alpha)i-regulatory motifs. *J Biol Chem* 2011;**286**:28138–49.
115. Hagemann IS, Narzinski KD, Baranski TJ. E2F8 is a nonreceptor activator of heterotrimeric G proteins. *J Mol Signal* 2007;**2**:3.
116. Sato M, Ribas C, Hildebrandt JD, Lanier SM. Characterization of a G-protein activator in the neuroblastoma-glioma cell hybrid NG108-15. *J Biol Chem* 1996;**271**:30052–60.

117. Patel TB. Single transmembrane spanning heterotrimeric g protein-coupled receptors and their signaling cascades. *Pharmacol Rev* 2004;**56**:371–85.
118. Marty C, Ye RD. Heterotrimeric G protein signaling outside the realm of seven transmembrane domain receptors. *Mol Pharmacol* 2010;**78**:12–8.
119. Pyne NJ, Tonelli F, Lim KG, Long J, Edwards J, Pyne S. Targeting sphingosine kinase 1 in cancer. *Adv Biol Regul* 2012;**52**:31–8.
120. El-Shewy HM, Luttrell LM. Insulin-like growth factor-2/mannose-6 phosphate receptors. *Vitam Horm* 2009;**80**:667–97.
121. Miller KG, Alfonso A, Nguyen M, Crowell JA, Johnson CD, Rand JB. A genetic selection for Caenorhabditis elegans synaptic transmission mutants. *Proc Natl Acad Sci USA* 1996;**93**:12593–8.
122. Hinrichs MV, Torrejon M, Montecino M, Olate J. Ric-8: different cellular roles for a heterotrimeric G-protein GEF. *J Cell Biochem* 2012;**113**:2797–805.
123. Tall GG, Krumins AM, Gilman AG. Mammalian Ric-8A (synembryn) is a heterotrimeric Galpha protein guanine nucleotide exchange factor. *J Biol Chem* 2003;**278**:8356–62.
124. Nishimura A, Okamoto M, Sugawara Y, Mizuno N, Yamauchi J, Itoh H. Ric-8A potentiates Gq-mediated signal transduction by acting downstream of G protein-coupled receptor in intact cells. *Genes Cells* 2006;**11**:487–98.
125. Klattenhoff C, Montecino M, Soto X, Guzman L, Romo X, Garcia MA, et al. Human brain synembryn interacts with Gsalpha and Gqalpha and is translocated to the plasma membrane in response to isoproterenol and carbachol. *J Cell Physiol* 2003;**195**:151–7.
126. Von Dannecker LE, Mercadante AF, Malnic B. Ric-8B, an olfactory putative GTP exchange factor, amplifies signal transduction through the olfactory-specific G-protein Galphaolf. *J Neurosci* 2005;**25**:3793–800.
127. Wang SC, Lai HL, Chiu YT, Ou R, Huang CL, Chern Y. Regulation of type V adenylate cyclase by Ric8a, a guanine nucleotide exchange factor. *Biochem J* 2007;**406**:383–8.
128. Amoureux MC, Nicolas S, Rougon G. NCAM180 regulates Ric8A membrane localization and potentiates beta-adrenergic response. *PLoS One* 2012;**7**:e32216.
129. Hampoelz B, Hoeller O, Bowman SK, Dunican D, Knoblich JA. Drosophila Ric-8 is essential for plasma-membrane localization of heterotrimeric G proteins. *Nat Cell Biol* 2005;**7**:1099–105.
130. Gabay M, Pinter ME, Wright FA, Chan P, Murphy AJ, Valenzuela DM, et al. Ric-8 proteins are molecular chaperones that direct nascent G protein alpha subunit membrane association. *Sci Signal* 2011;**4**:ra79.
131. Tonissoo T, Lulla S, Meier R, Saare M, Ruisu K, Pooga M, et al. Nucleotide exchange factor RIC-8 is indispensable in mammalian early development. *Dev Dyn* 2010;**239**:3404–15.
132. Nagai Y, Nishimura A, Tago K, Mizuno N, Itoh H. Ric-8B stabilizes the alpha subunit of stimulatory G protein by inhibiting its ubiquitination. *J Biol Chem* 2010;**285**:11114–20.
133. Von Dannecker LE, Mercadante AF, Malnic B. Ric-8B promotes functional expression of odorant receptors. *Proc Natl Acad Sci USA* 2006;**103**:9310–4.
134. Chan P, Gabay M, Wright FA, Kan W, Oner SS, Lanier SM, et al. Purification of heterotrimeric G protein alpha subunits by GST-Ric-8 association: primary characterization of purified G alpha(olf). *J Biol Chem* 2011;**286**:2625–35.
135. Hofler C, Koelle MR. AGS-3 alters Caenorhabditis elegans behavior after food deprivation via RIC-8 activation of the neural G protein G alphao. *J Neurosci* 2011;**31**:11553–62.

136. Digby GJ, Sethi PR, Lambert NA. Differential dissociation of G protein heterotrimers. *J Physiol* 2008;**586**:3325–35.
137. Ghosh P, Garcia-Marcos M, Bornheimer SJ, Farquhar MG. Activation of Galphai3 triggers cell migration via regulation of GIV. *J Cell Biol* 2008;**182**:381–93.
138. Garcia-Marcos M, Kietrsunthorn PS, Pavlova Y, Adia MA, Ghosh P, Farquhar MG. Functional characterization of the guanine nucleotide exchange factor (GEF) motif of GIV protein reveals a threshold effect in signaling. *Proc Natl Acad Sci USA* 2012;**109**:1961–6.
139. Ghosh P, Garcia-Marcos M, Farquhar MG. GIV/Girdin is a rheostat that fine-tunes growth factor signals during tumor progression. *Cell Adh Migr* 2011;**5**:237–48.
140. Hill C, Goddard A, Ladds G, Davey J. The cationic region of Rhes mediates its interactions with specific Gbeta subunits. *Cell Physiol Biochem* 2009;**23**:1–8.
141. Vargiu P, De Abajo R, Garcia-Ranea JA, Valencia A, Santisteban P, Crespo P, et al. The small GTP-binding protein, Rhes, regulates signal transduction from G protein-coupled receptors. *Oncogene* 2004;**23**:559–68.
142. Graham TE, Prossnitz ER, Dorin RI. Dexras1/AGS-1 inhibits signal transduction from the Gi-coupled formyl peptide receptor to Erk-1/2 MAP kinases. *J Biol Chem* 2002;**277**:10876–82.
143. Sato M, Blumer JB, Simon V, Lanier SM. Accessory proteins for G proteins: partners in signaling. *Annu Rev Pharmacol Toxicol* 2006;**46**:151–87.
144. Harrison LM, He Y. Rhes and AGS1/Dexras1 affect signaling by dopamine D1 receptors through adenylyl cyclase. *J Neurosci Res* 2011;**89**:874–82.
145. Hiskens R, Vatish M, Hill C, Davey J, Ladds G. Specific in vivo binding of activator of G protein signaling 1 to the Gbeta1 subunit. *Biochem Biophys Res Commun* 2005;**337**:1038–46.
146. Fang M, Jaffrey SR, Sawa A, Ye K, Luo X, Snyder SH. Dexras1: a G protein specifically coupled to neuronal nitric oxide synthase via CAPON. *Neuron* 2000;**28**:183–93.
147. Bang S, Steenstra C, Kim SF. Striatum specific protein, Rhes regulates AKT pathway. *Neurosci Lett* 2012;**521**:142–7.
148. Nguyen CH, Watts VJ. Dexras1 blocks receptor-mediated heterologous sensitization of adenylyl cyclase 1. *Biochem Biophys Res Commun* 2005;**332**:913–20.
149. Takesono A, Nowak MW, Cismowski M, Duzic E, Lanier SM. Activator of G-protein signaling 1 blocks GIRK channel activation by a G-protein-coupled receptor: apparent disruption of receptor signaling complexes. *J Biol Chem* 2002;**277**:13827–30.
150. Thapliyal A, Bannister RA, Hanks C, Adams BA. The monomeric G proteins AGS1 and Rhes selectively influence Galphai-dependent signaling to modulate N-type (CaV2.2) calcium channels. *Am J Physiol Cell Physiol* 2008;**295**:C1417–C1426.
151. Strittmatter SM, Cannon SC, Ross EM, Higashijima T, Fishman MC. GAP-43 augments G protein-coupled receptor transduction in Xenopus laevis oocytes. *Proc Natl Acad Sci USA* 1993;**90**:5327–31.
152. Afshar K, Willard FS, Colombo K, Johnston CA, McCudden CR, Siderovski DP, et al. RIC-8 is required for GPR-1/2-dependent Galpha function during asymmetric division of C. elegans embryos. *Cell* 2004;**119**:219–30.
153. Couwenbergs C, Spilker AC, Gotta M. Control of embryonic spindle positioning and Galpha activity by C. elegans RIC-8. *Curr Biol* 2004;**14**:1871–6.
154. David NB, Martin CA, Segalen M, Rosenfeld F, Schweisguth F, Bellaiche Y. Drosophila Ric-8 regulates Galphai cortical localization to promote Galphai-dependent planar orientation of the mitotic spindle during asymmetric cell division. *Nat Cell Biol* 2005;**7**:1083–90.

155. Tall GG, Gilman AG. Resistance to inhibitors of cholinesterase 8A catalyzes release of Galphai-GTP and nuclear mitotic apparatus protein (NuMA) from NuMA/LGN/Galphai-GDP complexes. *Proc Natl Acad Sci USA* 2005;**102**:16584–9.
156. Vellano CP, Shu FJ, Ramineni S, Yates CK, Tall GG, Hepler JR. Activation of the regulator of G protein signaling 14-Galphai1-GDP signaling complex is regulated by resistance to inhibitors of cholinesterase-8A. *Biochemistry* 2011;**50**:752–62.
157. Hollinger S, Taylor JB, Goldman EH, Hepler JR. RGS14 is a bifunctional regulator of Galphai/o activity that exists in multiple populations in brain. *J Neurochem* 2001;**79**:941–9.

INDEX

Note: Page numbers followed by "*f*" indicate figures, and "*t*" indicate tables.

B

E

L

M

N

S

T

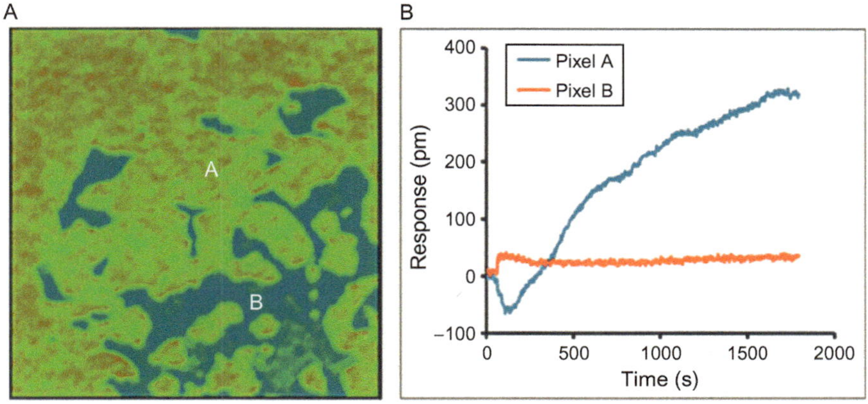

ROCHEVILLE *ET AL.*, FIGURE 3.1

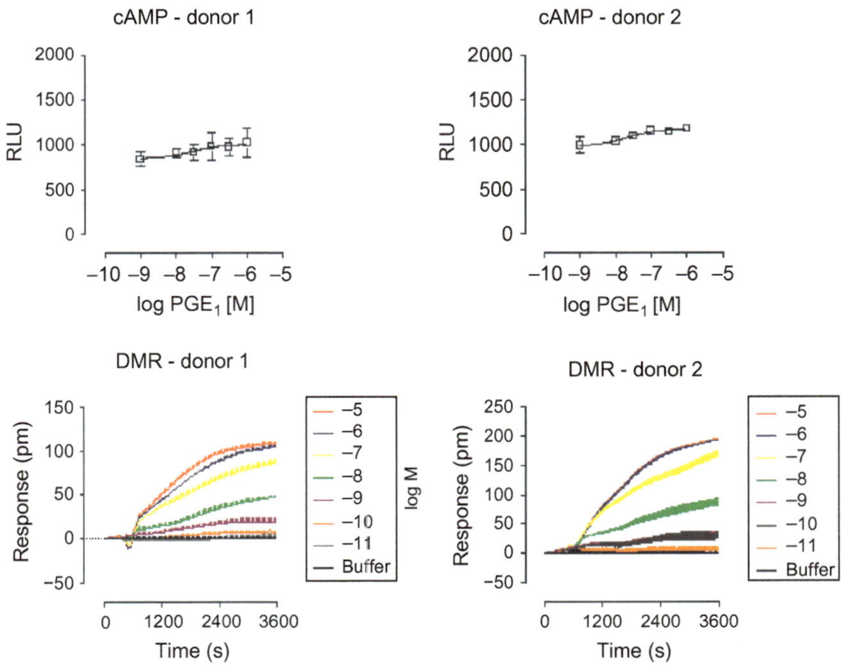

ROCHEVILLE *ET AL.*, FIGURE 3.2

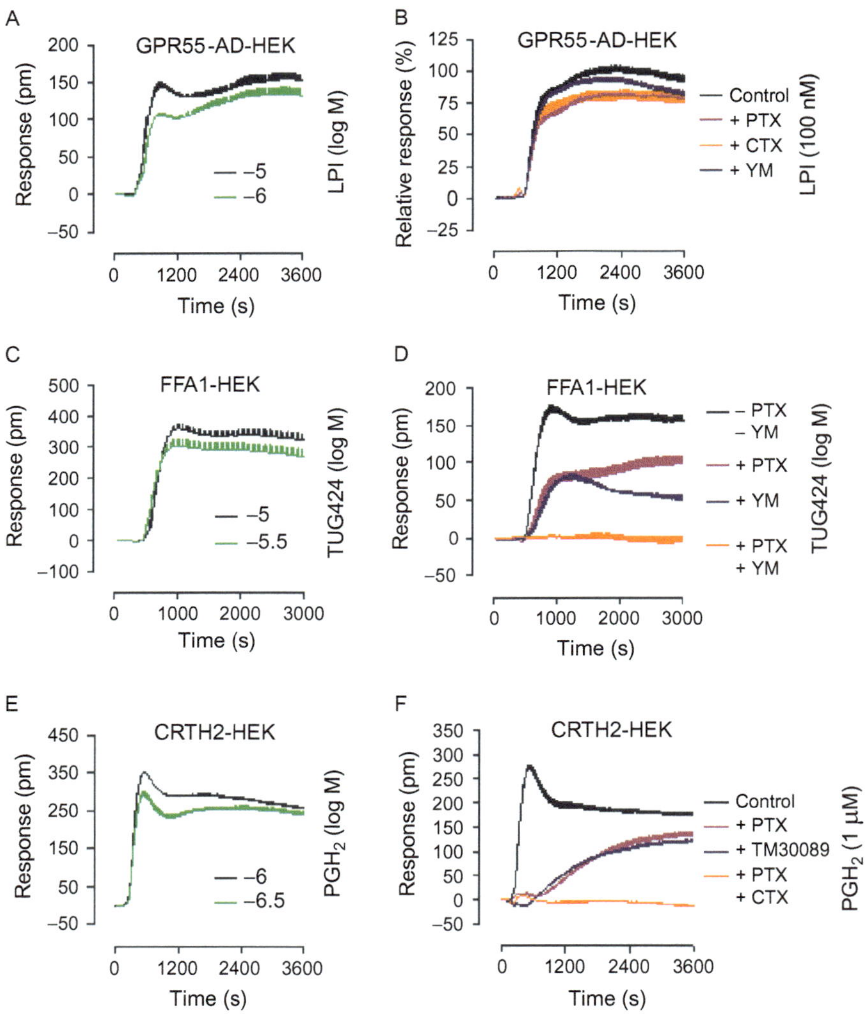

Rocheville *et al.*, Figure 3.3

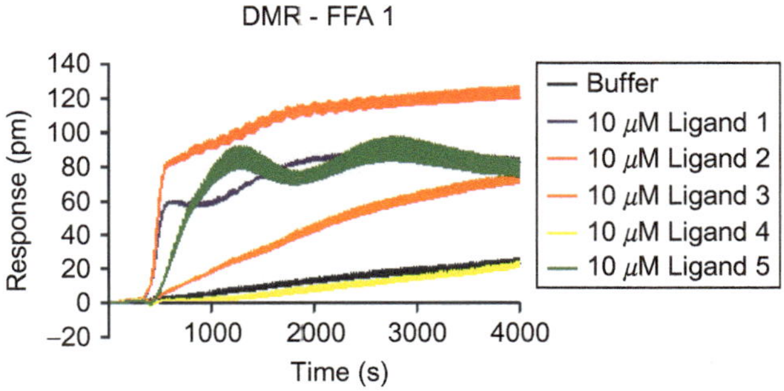

ROCHEVILLE *ET AL.*, FIGURE 3.4

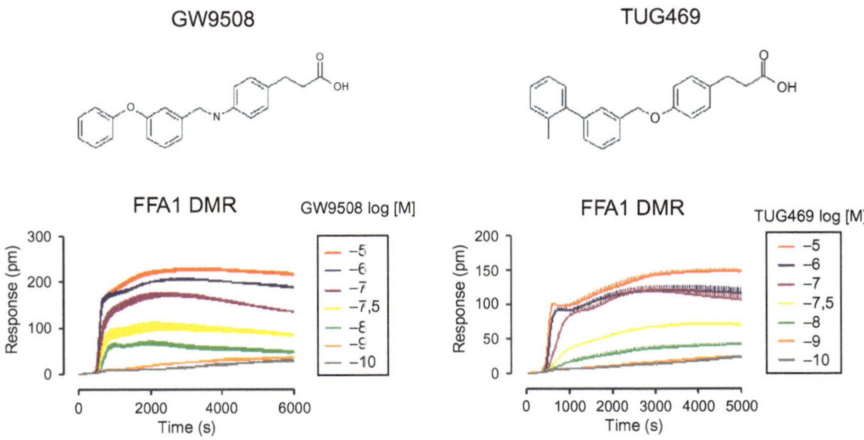

ROCHEVILLE *ET AL.*, FIGURE 3.5

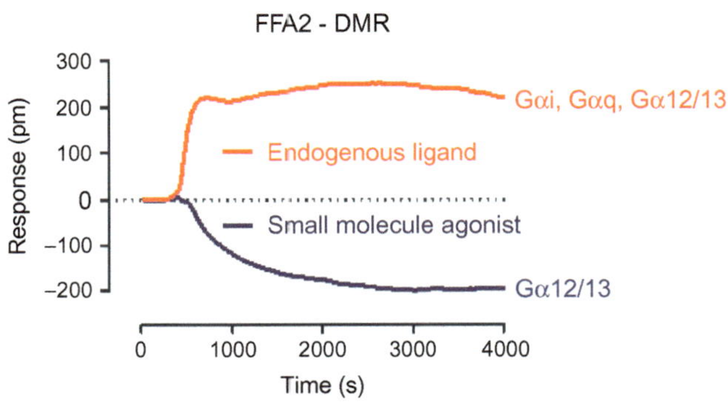

ROCHEVILLE *ET AL.*, FIGURE 3.6

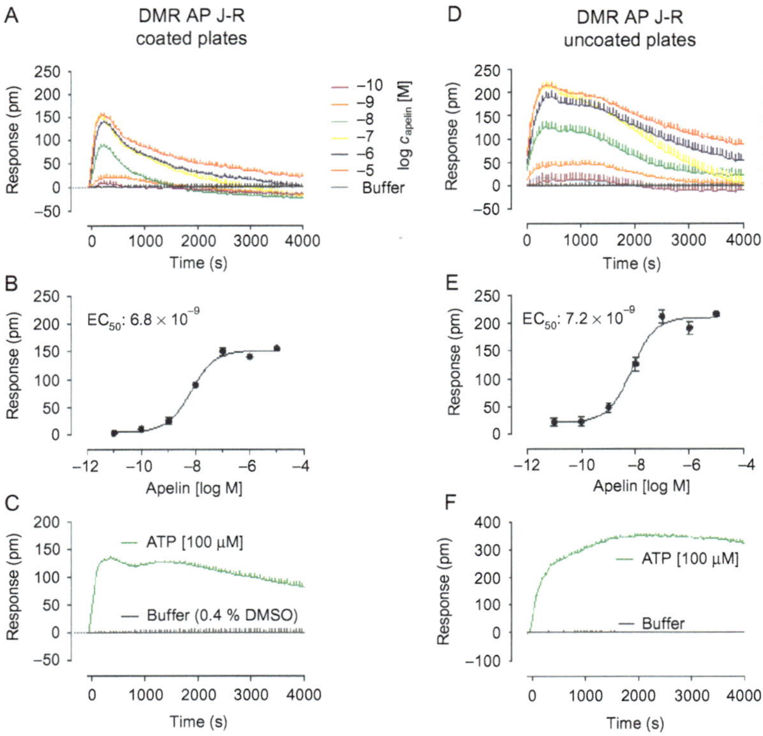

ROCHEVILLE *ET AL.*, FIGURE 3.7

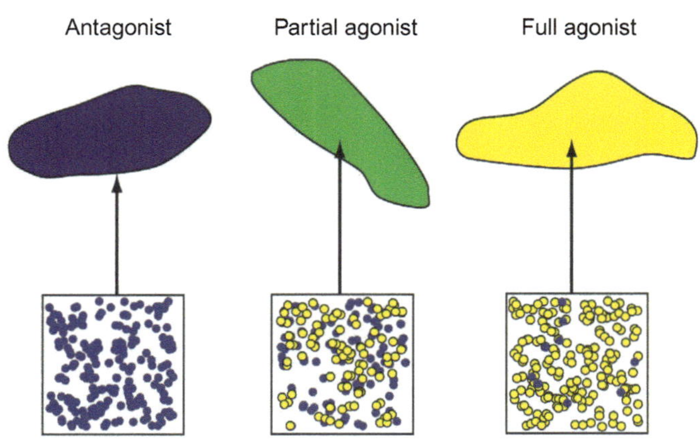

EHLERT, FIGURE 7.2

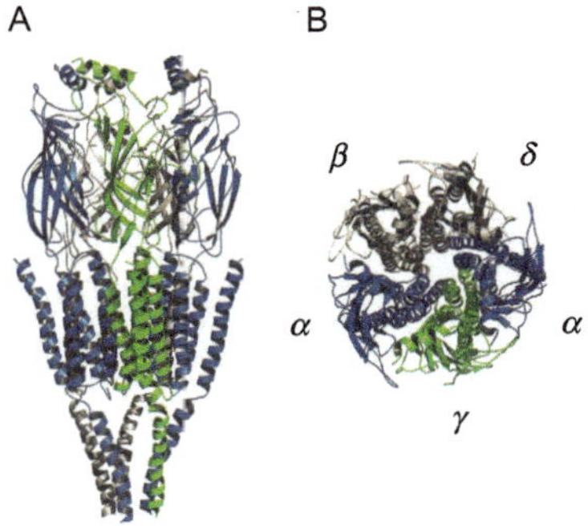

Ehlert, Figure 7.3

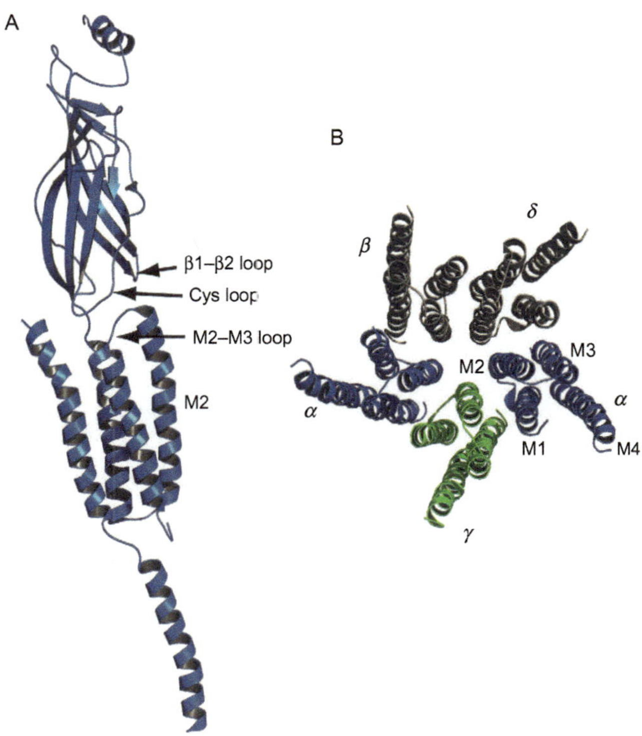

Ehlert, Figure 7.4

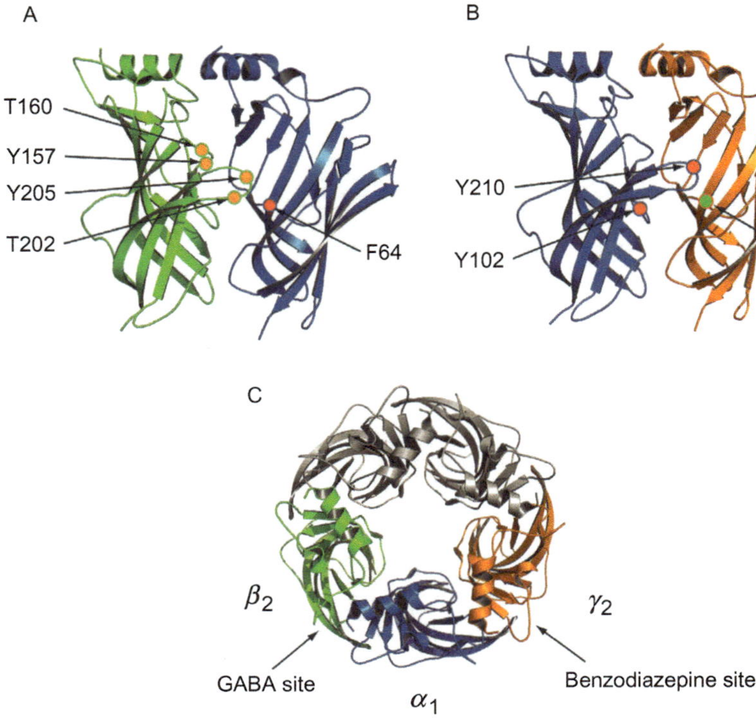

EHLERT, FIGURE 7.5

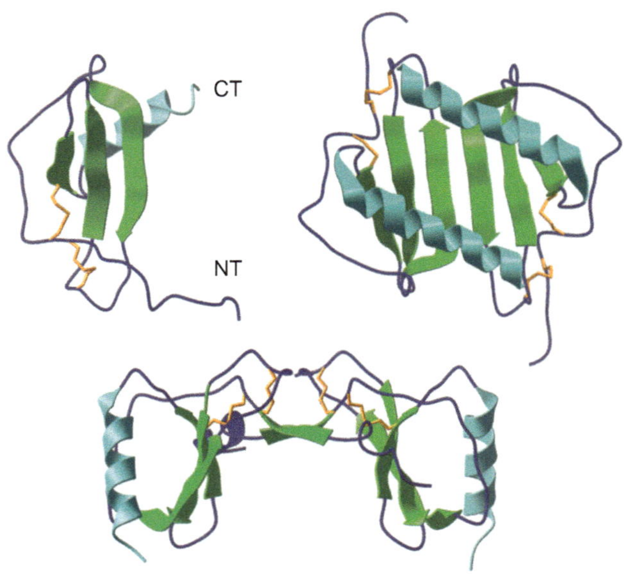

STEPHENS AND HANDEL, FIGURE 9.1

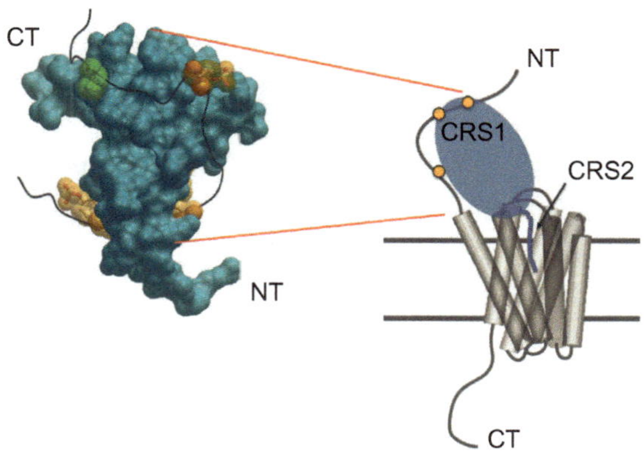

STEPHENS AND HANDEL, FIGURE 9.2

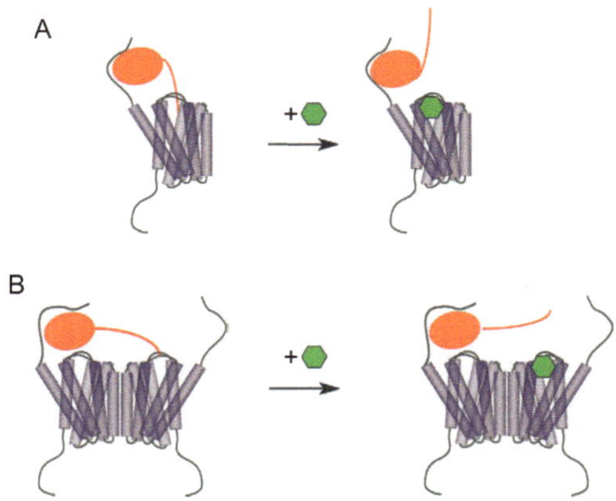

STEPHENS AND HANDEL, FIGURE 9.3

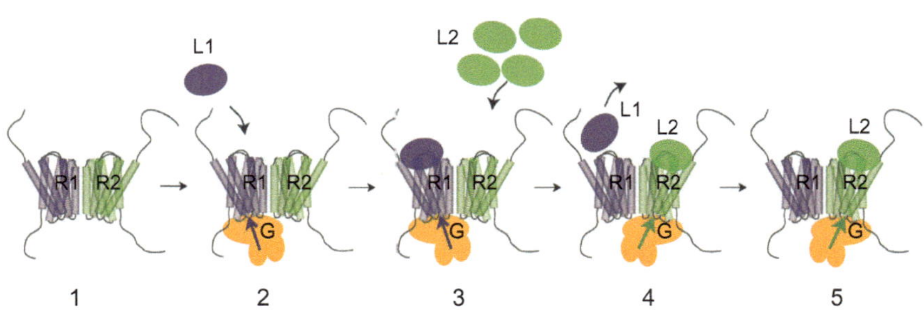

STEPHENS AND HANDEL, FIGURE 9.4

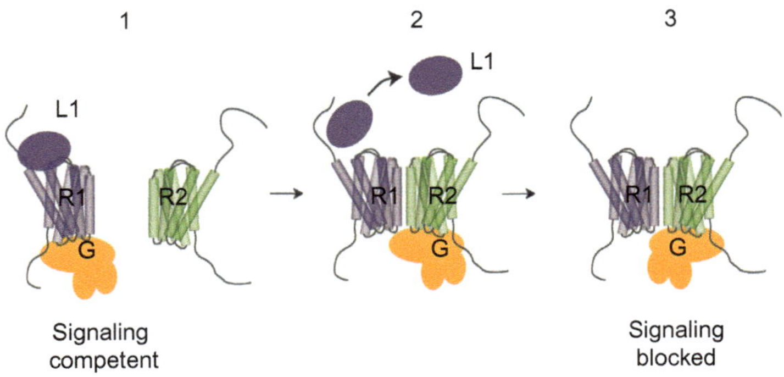

STEPHENS AND HANDEL, FIGURE 9.5

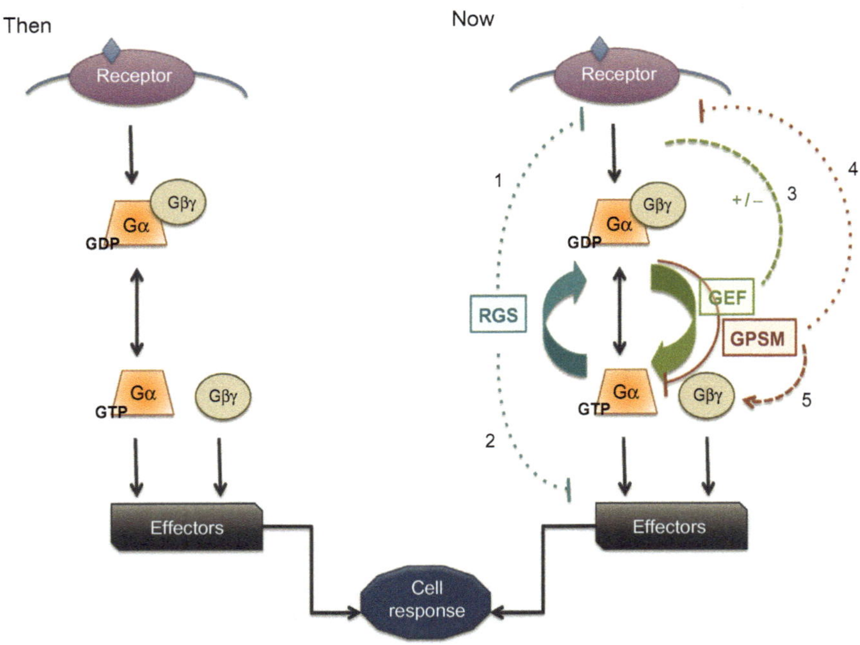

ZHAO ET AL., FIGURE 10.2

FSC
www.fsc.org
MIX
Papier | Fördert gute Waldnutzung
FSC® C083411

Kolibri 360 GmbH
Ferdinand-Jühlke-Straße 7
99095 Erfurt, Deutschland
produktsicherheit@kolibri360.de